g	Geometric gradient: rate of cash flow increase/decrease from period to period.
I	Interest paid on debt. Annual interest payment (uniform over each year).
i	Effective interest rate per interest period.
i_a	Effective interest rate per year (per *annum*).
i_m	Effective interest rate per subperiod.
i_s	Rate of interest "earned" by imaginary sinking fund.
i^*	Internal rate of return.
i_e^*	External rate of return.
i	After tax (internal) rate of return.
IRR	(Internal) rate of return. Sometimes written as *RoR* for "rate of return."
k	Auxiliary interest rate used when computing the external rate of return. The minimum attractive rate of return.
k^*	Cost of capital, reflecting inflation.
L_j	Lease payment at end of period j.
M	Number of compounding subperiods per period (each of which is assumed to be of equal length). Market value of the firm's equity.
MARR	Minimum attractive rate of return.
N	Number of compounding periods (each of which is assumed to be of equal length): the length of the "planning horizon" (study period). Life of investment.
P	Initial investment. Equivalent present value of future cash flow(s). Loan principal.
$\overline{P}$	Amount of money (or equivalent value) flowing continuously and uniformly during the first period of the planning horizon.
P_j	Amount of loan principal unpaid at the start of period j.
Q	Amount of loan.
r	Nominal interest rate per period; usually, the nominal interest rate per year.
S	Net salvage value of capital investment.
SIR	Savings-investment ratio.
SYD	Sum of the years digits.
t	Effective income tax rate.
$\varnothing$	The do-nothing alternative.
λ	Leverage, the ratio of debt to the total value of the firm.

ENGINEERING ECONOMY

CAPITAL
ALLOCATION
THEORY

ENGINEERING ECONOMY

CAPITAL ALLOCATION THEORY

G. A. Fleischer
University of Southern California

PWS Engineering
Boston, Massachusetts

PWS PUBLISHERS

Prindle, Weber & Schmidt • ☙• Duxbury Press •♦• PWS Engineering •⚐• Breton Publishers •⚙
20 Park Plaza • Boston, Massachusetts 02116

Sponsoring Editor: Ray Kingman
Production: Del Mar Associates
Manuscript Editor: Andrea Matyas
Interior Design: John Odam
Cover Design: John Odam
Illustrations: Kim Fraley and Pam Posey
Typesetting: Typothetae
Production Services Coordinator: Bill Murdock

Printed in the United States
of America

10   9   8   7   6   5   4   3

PWS Publishers is a division of Wadsworth, Inc.

Library of Congress Cataloging
in Publication Data

Fleischer, Gerald A.
Engineering economy.

Includes bibliographical references.
I. Engineering economy.   I. Title.
TA177.4.F57 1984 658.1′5 83-24070
ISBN 0-534-02943-4

ISBN 0-534-02943-4

To those of blessed memory
  Louis S. Fleischer, 1900–1940
  Maurice L. Bashkow, 1898–1975
and to my mother
  Rita Bloch Fleischer Bashkow, 1903–

PREFACE

s an area of special concern, the study of optimal investment decisions is of relatively recent interest to both academicians and industrial managers. Although the general problem of allocating limited resources among a variety of competing alternatives must surely have occurred to societies since the beginning of recorded history, it is only within the last century that the evaluation process has been formalized and analytical procedures developed.

Engineering economy, the analysis of economic effects of *engineering* decisions, has its genesis in Arthur M. Wellington's classic text, *The Economic Theory of Railway Location,* published in 1887.* Engineers subsequently expanded and refined the techniques of engineering economy, notably through the work of Professors J. C. L. Fish in the 1920s and E. L. Grant in the 1930s.† In parallel, certain (classical) economists and financial managers were developing scholarly material that, in large measure, both supplemented and complemented the work of the engineering economy community. The resulting literature has been classified under a variety of descriptive terms, usually depending on the professional discipline of the author and/or the target audience: engineering economy (or engineering economics), capital budgeting, economic analysis, life cycle costing, financial decision making, and managerial economics, among others. The common element shared by all titles is concern with the fundamental problem of allocating limited financial resources among competing investment alternatives.

*A. M. Wellington, *The Economic Theory of Railway Location*, (New York; John Wiley & Sons), 1887

†E. L. Grant's first edition of *Principles of Engineering Economy* was published in 1930. He retired from Stanford University in 1962, and at this writing (October 1983) Grant is still active. The 7th edition of *Principles of Engineering Economy* was published in 1982.

This book is a substantially revised version of *Capital Allocation Theory: The Study of Investment Decisions,* originally published by Appleton-Century-Crofts in 1967. The initial title was selected to reflect my view that, although the notation and theoretical structure are most closely related to the classical literature of engineering economy, applications are by no means limited to investment alternatives arising from engineering and technological decisions. Capital allocation problems stem from alternative plans, programs, and projects, irrespective of whether a technology component is present. Nevertheless, since the examples, problems, and exercises are most directly related to engineering decisions, I have elected to adopt the title *Engineering Economy* for this edition. Although the title has changed, the principal emphasis remains a consolidation of current and relevant views of engineering economists, financial managers, and others to present a unified theory of capital allocation appropriate to all levels within the business enterprise or governmental activity.

It should be noted that the theory presented here is as appropriate to individual (personal) investment decisions as to those of private firms, government agencies, or nonprofit organizations. These techniques are suitable whenever decisions must be made concerning the selection from alternative investment opportunities.

This book is intended for upper-division undergraduates or graduate-level students. There are no prerequisites, although prior exposure to accounting and microeconomics would be helpful. Knowledge of integral and differential calculus is not strictly necessary; with few exceptions, development of mathematical models uses only algebra. Some understanding of the elements of probability theory is useful in the discussion of risk and uncertainty (Chapter 8). For students who have not been exposed to this topic previously, an elementary presentation of probability and expectation is provided in the chapter (Section 8.2). Otherwise, this material may be omitted without loss of continuity.

This book may be used as either a primary or a secondary reference in a first course in engineering economy, financial management, capital budgeting, managerial economics, and the like. Most of these applications are found in schools of engineering and business. However, the text may also be used elsewhere when it is desired to explore concepts, principles, and procedures for examining the economic consequences of proposed plans, programs, and projects. Examples include a "systems analysis" course in a school of public administration or short course in "economic analysis" offered by an industrial firm or government agency.

This book begins with the role of capital allocation theory within the larger framework of systems analysis. The introductory chapter presents a qualitative discussion of certain principles from which a unified theory may be developed. The necessary mathematics of compound interest, including discrete and continuous assumptions for cash flows and discounting, are developed in Chapter 2. Chapter 3 presents the principal methods of economic evaluation: annual worth, present worth, (internal) rate of return, and the benefit-cost ratio method. The first three chapters, then, form the theoretical basis for the remainder of the text. Chapter 4 treats the multiple alternative problems, that is, selection of an optimal alternative from a set of alternatives using the methods of Chapter 3. Chapter 5 discusses a number of evaluation techniques that, although widely used in industry and government, are inherently faulty; they are either fundamentally incorrect or they provide only approximations to the true values. Chapter 6 deals with the most common capital allocation application: problems of retirement of assets from service with replacement by new assets.

Chapter 7 is an extensive discussion of economy studies (economic analyses) that considers the effects of income taxes, including investment tax credits and taxes on gains and losses on disposal of depreciable assets. Since cash flows for taxes are related to taxable income, and since taxable income is in part a function of depreciation and depletion expenses, procedures for determining these expenses are presented. (Many engineering economy textbooks include discussion of *personal* income taxes. I have elected to omit this material, however, concentrating instead on the effects of taxes on the *corporate* business enterprise. Once learned, the methodology for after-tax economy studies for corporations is readily transferable to the noncorporate context. If the instructor wishes to include personal income taxes in his or her course, it is suggested that the current edition of *Your Federal Income Tax*, IRS Publication 17, be used as a supplementary reference.)

Chapter 7, "Depreciation, Taxation and After-Tax Economy Studies," is the most lengthy chapter. Other authors frequently separate this material into two chapters: one dealing with depreciation and depletion and the other dealing with cash flows for taxes and after-tax analyses. I have chosen to combine this material into a single chapter, however, because the topics are directly related. Chapters are organized by subject matter—related topics are grouped—but there is no implication that class time should be equal for each chapter. In general, it is expected that about one week (three classroom hours) per chapter will be adequate. But instructors should devote a minimum of two weeks, and perhaps three, to the material in Chapter 7.

The first seven chapters assume that all parameters are known with certainty. This artificiality is akin to assuming a frictionless plane in physics. So Chapter 8 explores a number of techniques for formally considering the noncertain future, including sensitivity analysis, risk analysis, and a variety of principles of choice from decision theory.

The revenue requirement method is presented in Chapter 9. Although this method is equivalent to the more widely used procedures given in Chapter 3—indeed, this is demonstrated in Chapter 9—the revenue requirement method is of special interest to the utility industry, and thus it is free-standing in a separate chapter. Some instructors may choose to omit Chapter 9 because of time constraints and/or their view that the revenue requirement method is limited in application. If so, Chapter 9 may be omitted without loss of continuity with respect to the other topics covered.

Chapter 10 is an extensive treatment of yet another real-world consideration that has become increasingly important in modern society: incorporating price level changes (inflation) into the analysis. Included as an appendix to Chapter 10 is an introduction to index numbers, the statistics used to describe price level changes over time. A variety of methods for computing index numbers are described and contrasted.

Chapter 11 discusses the measurement and use of the ``cost of capital'' concept and describes procedures for measuring the costs of a number of elements of the capital structure. Chapter 11 also includes tax, risk, and inflation elements, which were covered previously. The primary purpose of this chapter is to introduce appropriate procedures for estimating the minimum attractive rate of return, which is the discount rate of critical importance in all economy studies.

Chapter 12, the final chapter, addresses an issue typically not included in traditional engineering economy textbooks: formal consideration in analyses of consequences for which monetary equivalence cannot readily be established. It has been my experience that, all too often, those who complete a course in engineering economy arrive at the misconception that analysis of the economic consequences alone is sufficient to identify the ``optimal'' choice from a set of investment alternatives. This view is naive and should be discouraged. Economic analysis is a decision-*assisting* process, not a decision-*making* process. Although it is not meant to serve as an exhaustive treatment of this subject, Chapter 12 incorporates a variety of procedures that might be employed when ``irreducible'' as well as monetary consequences are to be considered.

Engineering Economy incorporates a number of special features that, I believe, add significantly to the effectiveness of the text. These include

1. *A partial summary of principal notation.* Used consistently throughout the text, this list is shown inside the front cover

for ready reference. An expanded summary of symbols and key abbreviations is included in Appendix A. Moreover, the notation specific to the revenue requirement method (Chapter 9), and the discussion of inflation and index numbers (Chapter 10) are included in the appendixes to those two chapters.

2. *Definitions of key terms and principal notation.* These are generally consistent with those proposed by the American National Standards Institute Committee on Industrial Engineering Terminology (ANSI Committee Z-94) as published in 1983.

3. *A summary of principal mathematical models.* This summary is incorporated immediately inside the back cover to provide ready reference for the user. These models, accompanied by relevant cash flow diagrams, are keyed to the compound interest tables included in Appendix B.

4. *Compound interest tables.* These are included in Appendix B for nineteen separate interest rates, ranging from 1 percent to 50 percent. Each table spans two pages in the text and includes eleven factors: three single-payment factors, six uniform series factors, and two arithmetic gradient series factors. (Three of the eleven factors may be used when finding equivalent values of continuous cash flows under conditions of continuous compounding; these are shaded in the tables for ready identification.) The interest rates are *effective*, not nominal, in all cases. To assist the user in rapid selection of the appropriate table, the margins of the pages in Appendix B are tinted and the various interest rates are clearly marked within the margins.

5. *Problems at the back of each chapter.* There are 343 problems in this book, some of which afford the student an opportunity to work through numerical exercises directly related to the text material. Others are extensions of the text in that they illustrate some new application.

6. *An extensive discussion of depreciation, depletion, amortization, and taxes in Chapter 7.* This discussion incorporates the principal features of the Economic Recovery Tax Act of 1981, including the Accelerated Cost Recovery System (ACRS) required of federal taxpayers. Changes introduced by the Tax Equity and Fiscal Responsibility Act of 1982 are also included where appropriate.

7. *A discussion of the revenue requirement method in Chapter 9.* This discussion is extensive and complex. To assist the reader, a separate glossary and summary of principal equations are included as appendixes to this chapter.

8. *A discussion of relative price change (inflation).* This topic is of considerable importance, and thus this issue is presented extensively in Chapter 10. An appendix to this chapter includes a discussion of *index numbers* used to measure changes over time of prices, quantities, and values. Index numbers are defined and various methods for computing index numbers are summarized. Other appendixes to this chapter include a separate glossary and summary of principal equations.

9. *Computer programs.* These are included at the back of Chapters 2, 3, 4, 6, and 7. The programs for Chapters 2, 3, and 4 are written in BASIC; the program for Chapter 6 is written in Applesoft BASIC; and the program for Chapter 7 is written in IBM BASIC-A. Worked-out examples using the programs are also included. Students will find these programs useful in solving many, but not all, of the problems included at the ends of chapters. (Although computer programs have been included to suggest opportunities for ''automated'' data analysis, the calculations required for problem solving in this book are of modest complexity; computers are helpful but not necessary. Any hand-held slide rule calculator should be adequate in all instances. Indeed, even in cases for which a microcomputer is required or recommended by the instructor, students should solve at least some of the problems ''long hand'' to insure that underlying concepts are understood.)

10. *Answers to problems.* With the exception of Chapters 1 and 12, answers are shown for all odd-numbered problems. The solutions to all problems are provided in a separate Solutions Manual.

I would like to thank the following reviewers for their comments and suggestions: Stanford Baum, University of Utah; Peter Gardiner, University of Southern California; A. K. Mason, California Polytechnic State University, San Luis Obispo; Wayne M. Parker, Mississippi State University; and T. L. Ward, University of Louisville. I am also indebted to my many colleagues, both past and present, without whose encouragement, inspiration, and critical judgment this text would not have been written.

G. A. Fleischer

CONTENTS

CHAPTER EIGHT
RISK AND UNCERTAINTY 263

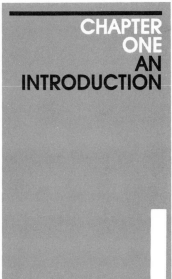

CHAPTER ONE
AN INTRODUCTION

n 1795, the English parliamentarian and man of letters, Edmund Burke, wrote a letter in defense of his own pension, proposed by the prime minister, the Younger Pitt:

It may be new to his Grace, but I beg leave to tell him that mere parsimony is not economy. It is separable in theory from it; and in fact it may or it may not be a part of economy, according to circumstances. Expense, and great expense, may be an essential part in true economy. If parsimony were to be considered as one of the kinds of that virtue, there is, however, another and an higher economy. Economy is a distributive virtue, and consists, not in saving, but in selection. Parsimony requires no providence, no sagacity, no powers of combination, no comparison, no judgement. Mere instinct, and that not an instinct of the noblest kind, may produce this false economy in perfection. The other economy has larger views. It demands a discriminating judgement, and a firm, sagacious mind. [1]

Burke's commentary is a remarkably effective summary of the perspective underlying *Engineering Economy*. This book rests on the proposition that refusing to expend scarce resources is rarely, if ever, the wisest course of action. Rather, the problem is choosing from a variety of investment alternatives in order to best satisfy decision makers' immediate and longer-term objectives. The operative word is *economy,* and the essential ingredient in economy, according to Burke, is *selection.* This book is dedicated to the principles and procedures of the selection process, especially when the economic characteristics of alternatives are of principal or significant concern.

1. From *Edmund Burke: Selected Writings and Speeches,* ed. Peter J. Stanlis (New York: Doubleday Anchor) 1963.

Capital allocation, the theme of this book, is not a decision-*making* process; it is a decision-*assisting* process. Decisions are rarely made on the basis of economic considerations alone. Nor should they be. Consider some common examples: which car to buy; which apartment to rent or which home to purchase; how to spend entertainment dollars; which university to attend, if any; and what portion of society's resources to spend on public education, health and welfare, defense, and public transportation. The choice of options should be based on consideration of all significant consequences that are likely to arise from each alternative.

Occasionally one hears the argument that, since there may be important noneconomic factors relevant to a given problem situation, the economic consequences should be ignored in favor of the ''more important'' factors. This viewpoint is especially prevalent with respect to public-sector investment decisions, although it is frequently encountered in the private sector as well. In any event, it is myopic and potentially wasteful of scarce resources.[2] Unless the economic consequences are clearly insignificant, decision makers should assess and consider them along with any and all other relevant data. Even though a more costly, or economically suboptimal, alternative may ultimately be selected, the analysis will yield the price being paid for the differential noneconomic consequences. This information will be valuable to decision makers as well as to stockholders (private enterprise) and the taxpayers (public sector) to whom decision makers are responsible.

1.2 THE CAPITAL ALLOCATION PROBLEM DEFINED

Many investment opportunities are normally available to individual investors, business firms, and government agencies. For private industry in particular, continued investment and reinvestment is a fact of economic life if a firm is to survive in a competitive economic environment. Supplies and materials must be purchased; new equipment must be acquired and obsolete equipment replaced; physical plant must be maintained and, in many cases, expanded; new products must be researched, developed, and marketed; and so forth. Moreover, investment by one company in another in the form of stock ownership, purchase of government securities, and distribution of cash dividends to a firm's owners represent just a few of the investment possibilities that may be considered as alternatives of capital demand.[3]

Suppose that we lived in a world of unlimited capital re-

---

2. When considering capital allocation decisions within the context of a government agency, the taxpayers or citizens of the government unit in question may be viewed as the owners of the enterprise.
3. Distribution of a firm's earnings in the form of cash dividends is frequently cited as an example of ''disinvestment.'' The semantic distinction between *investment* and *disinvestment* is of little interest to us at this time; the difference will always be clear in context.

sources. In such an ideal environment there would be no need for a book like this. All proposals that require the expenditure of funds would be acceptable, provided that a simple criterion were met: Total cash revenue must exceed total cash expenses. Indeed, even this simple test would be unnecessary in our make-believe world, because having unlimited resources destroys the incentive to increase existing wealth. A person who has plenty to eat—and who is assured of always having plenty to eat—does not devote his or her energies to acquiring more food.

Clearly, in our real world, resources are limited. In all but the most abnormal cases, the available capital supply is insufficient to meet all available demand. Although available capital may be increased by additional borrowing or, in some cases, by taking in additional partners, management is generally prevented from taking advantage of all available investment opportunities by limitations of capital. Thus, the cardinal problem of capital allocation is: *Which of the many available investment alternatives should be selected or rejected in order to maximize the long-term wealth of current owners of the enterpise?*[4] In other words, which alternatives should be budgeted, or funded, in order to produce the optimal capital budget? The following discussion addresses the development and application of a systematic and valid basis for answering this question.

Conceptually, the solution to the capital allocation problem may be viewed as a determination of the point where the *marginal cost* of capital supply is equal to the expected *marginal revenue,* or "payoff," of capital demand. Stated another way, alternatives should be accepted until the point where the last, or marginal, alternative accepted is expected to yield a return that is greater than the cost of the capital necessary to finance it. At this point, no additional capital funds should be obtained and no additional investment proposals should be accepted (see Figure 1.1).

Level Q_1 represents the total cost of all investments and reinvestments necessary to maintain the firm's existence. P_1 is the expected profitability from Q_1, stated as a rate of return on the invested capital, and C_1 is the cost of the capital funds necessary to support these investments. (Methods for measuring the rate of return on invested capital and the cost of capital are discussed in detail later, especially in Section 3.3 and Chapter 11.) Now consider the increment of capital, Q_2-Q_1, necessary to fund the project, resulting in incremental return P_2. Q_2 represents the total number of investment dollars required for level Q_1 plus the next project under consideration. C_2 is the incremental cost of the

---

4. Many authors have discussed the explicit objectives towards which capital management must be directed. Their consensus is that the relevant goal is to maximize the present worth of the current owners of the enterprise. As later chapters will show, *present worth* refers to the discounted present value of long-run net earnings.

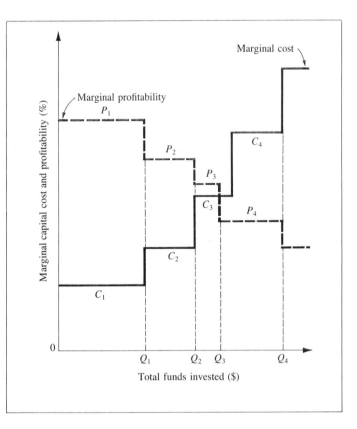

Marginal profitability

P_1

P_2

P_3

C_4

C_3

P_4

C_2

C_1

Marginal cost

Marginal capital cost and profitability (%)

0

Q_1 Q_2 Q_3 Q_4

Total funds invested ($)

Figure 1.1 Marginal Profitability (Rate of Return) and Cost as a Function of Total Funds Invested

incremental funds, $Q_2–Q_1$. Since marginal return P_2 is greater than marginal cost C_2, this incremental project appears economically attractive or acceptable. By similar reasoning, investment of the next increment of funds, $Q_3–Q_2$, is warranted, since marginal return P_3 exceeds marginal cost C_3. The next project, requiring incremental funds $Q_4–Q_3$, should be rejected, because the cost of these funds, C_3 and C_4, exceed anticipated return P_4.

The downward-sloping curve—a step function in Figure 1.1—represents the marginal returns expected from the demand for capital, and the upward-sloping curve represents the cost of the supply of capital. In this simple illustration it is clear that not all projects are affordable; the firm must select certain projects and reject others. Thus, project selection is the central feature of the capital-allocation problem.

Having limited investment funds leads to another type of problem that will be detailed in later chapters: the selection decision when two or more alternatives are available to satisfy a given operational requirement and when each is otherwise affordable. To illustrate, suppose that a firm is considering acquiring production equipment to reduce manufacturing costs. Three alternatives are available: models X, Y, and Z. Each is affordable in that each is expected to produce an economic return exceeding the related investment capital. But because all three satisfy operational requirements, management must select the most desirable alternative.

Note then that there are two major types of problems that, together, are the focus of this book. The first type requires the selection of a subset, or budget, from apparently independent investment proposals. Some alternatives will be accepted and others rejected, because there are insufficient funds to support all the investment possibilities. Given limited funds, the alternatives are not truly independent, because at the margin, the acceptance of one or more alternatives precludes the remainder. The second type of problem requires selection from a set of mutually exclusive alternatives. Examples include which equipment to acquire, what size facility to construct, when to build the facility, whether to lease or purchase equipment, what degree of automation to establish, and where to locate the facility. Both types of problems, of course, are related to the universal condition of limited capital.

Before developing the mathematical models appropriate to evaluating capital proposals, it will be useful to identify the fundamental principles that give rise to the rationale of capital allocation. Moreover, some of these principles lead directly to the quantitative techniques developed in the later chapters.

1.3 FUNDAMENTAL PRINCIPLES OF CAPITAL ALLOCATION

1. *Only feasible alternatives should be considered.* For example, prospective investment in new equipment costing $100,000 is not a feasible alternative if the money is not available and cannot be obtained and if there are no other financing opportunities. Similarly, management need not consider distributing corporate profits as cash dividends if this action has been temporarily prohibited by lenders. The capital budgeting analysis begins with determination of all feasible alternatives, since courses of action that are not feasible, because of certain contractual or technological considerations, are properly excluded.

Feasible alternatives only

2. *Using a common unit of measurement (a common denominator) makes consequences commensurable.* Is a proposal to reduce direct labor by 3,000 man-hours superior to another that is expected to reduce raw-material requirements by 500 tons? Is an investment of $10,000 justified if it promises to shorten production time by three weeks? Clearly, these questions cannot be answered unless the prospective consequences of alternative proposals have been described in common units. All decisions are made in a single dimension, and money units—dollars, francs, pesos,

Common denominator for assessing alternatives

yen, and so forth—seem to be most generally suitable. Thus, after all consequences have been described for each of the feasible alternatives, they should be transcribed into a common denominator to the extent practicable. Of course, not all consequences may be evaluated in money terms. (See principle 9 below.)

Only differences relevant

3. *Only differences are relevant.* The prospective consequences that are common to all contending alternatives need not be considered in an analysis, because including them affects all alternatives equally. This is like adding or subtracting the same number to both sides of an equation; the equation remains undisturbed.

Ignore sunk costs

4. As a direct consequence of principle 3, an important axiom may be stated: *All sunk costs are irrelevant to an economic choice.* A **sunk cost** is an expense or a revenue that has occurred before the decision. When considering whether or not to replace manufacturing equipment, for example, the original cost of the equipment is "sunk" and hence of no direct interest.[5] By the same token, the cost of a major overhaul of equipment that occurred three weeks ago is also irrelevant. All events that take place before a decision are common to all the alternatives, so sunk costs are not differences among alternatives.

5. *All alternatives must be examined over a common planning horizon.* The **planning horizon** is the period of time over which the prospective consequences of various alternatives are assessed. (The planning horizon is often referred to as the **study period** or **period of analysis.**) If a certain alternative impacts on the flow of funds to and/or from an enterprise during a given period of time, the economic impacts of all other alternatives during that same period should be estimated. The context of the decision, the planning horizon, must be uniform for all possible choices.

The time value of money

6. *Criteria for investment decisions should include the time value of money and related problems of capital rationing.* Which of the following is more attractive economically? Alternative A requires an expenditure of $100 today and promises a return of $200 five years from now. Alternative B also requires an initial expenditure of $100, but a payoff of $500 can be expected in ten years. There is no simple, direct answer to this problem, because the returns are expected at different points in time. B cannot be chosen simply because $500 is greater than $200. It should be evident

5. As will be discussed in a later section, the original cost of equipment may be of interest insofar as this information is needed to determine depreciation expenses for tax purposes.

that the appropriate solution must consider the opportunities available to the investor in the interval between year five and year ten. Because resources are limited in our finite world, money can be invested elsewhere, so it has a value, or worth, that increases over time. Therefore, the time value of money must be used as a criterion in selecting from alternative investment opportunities.

7. *Separable decisions should be made separately.* This principle requires the careful evaluation of all capital-allocation problems to determine the number and type of decisions to be made. For example, an analyst may be called on to make recommendations concerning a technological alternative (the equipment to be acquired, for example) as well as a financial alternative (the source of funds to finance the acquisition). If separable decisions are not treated separately, optimal solutions may be obscured in the analysis.

    Separable decisions considered separately

   It is possible, of course, for an interrelationship to exist between certain investment decisions and related financing decisions. For example, the type of equipment selected may affect the range of financing alternatives available. When this occurs, the investment decisions are not entirely separable.

8. *The relative degrees of uncertainty associated with various forecasts should be considered.* All capital-budgeting decisions are based on a series of estimates concerning the future. Because estimates are only predictions of future events, it is probable that the actual outcomes will differ to a greater or lesser degree from the original estimates. The future is uncertain, and any analysis that fails to consider this uncertainty implies false omniscience. On the other hand, one cannot infer that specific solutions are invalid or useless simply because of the uncertainty associated with input data. Formal consideration of the type and degree of uncertainty ensures that the quality of the solution is evident to those responsible for capital-allocation decisions.

    Consider noncertain future

9. *Decisions should give weight to consequences that are not reducible to monetary units.* Selecting candidates for limited capital resources requires that prospective differences among alternatives be clearly specified. Whenever possible, of course, these differences should be reduced to a common unit of measurement, generally a monetary unit, to provide a basis for selection. But not all alternatives can be reduced to monetary units. One proposal may result in an increase in employee comfort and convenience; another may result in a production delay; still another may result in an increased share of the market. Although the analyst

    Consider irreducibles

does not always have the time or resources to reduce such consequences to equivalent money units, decisions may directly be affected by their amount and direction. The irreducible as well as monetary consequences of proposed alternatives should be clearly specified in order to give managers of capital all reasonable data on which to base their decisions.[6]

Procedures integrated throughout organization

10. *The efficacy of capital-budgeting procedures is a function of their implementation at various levels within the organization.* Responsibility for capital-allocation decisions does not rest solely with corporate boards of directors or senior managers. Decisions affecting significant receipts and expenditures are made at many levels within the organization, especially among staff activities such as engineering design, marketing, and new product research and development. Moreover, the technical description of alternatives to be considered by top-level management generally undergoes preselection by lower-level personnel. Thus, capital-allocation procedures must be clearly described and understood at all levels within the organization having responsibility, in whole or in part, for these decisions.

Learn from experience

11. *Postdecision audits improve the quality of decisions.* As indicated above, all capital-allocation analyses require estimates of expected future consequences of various alternatives. Thus the quality of decisions is directly affected by the analyst's ability to forecast the future with reasonable precision. The only way to judge predictive ability, of course, is to audit the results of the decision at a later date. In this way it is possible to determine the extent of an analyst's predictive bias; one may be consistently optimistic, another pessimistic, and still another may exhibit random bias. Auditing can also evaluate the strengths and weaknesses of overall allocation procedures. The capital-allocation function should include postdecision audits in order to provide relevant feedback for improving future decisions.

Note that a postdecision audit is inherently incomplete: If only one of several alternative courses of action is selected, we can never be sure what would have happened if another alternative had been chosen. ''What might have

6. As the title of this book implies, we limit our concern to the *economic* implications of alternative investments. The simultaneous, or aggregate, consideration of monetary and other irreducible consequences is the proper focus of multiattribute analysis. There is substantial literature related to this problem. See, for example, Ralph L. Keeney and Howard Raiffa, *Decisions with Multiple Objectives* (John Wiley and Sons, 1976) and Vira Chankong and Yacov Y. Haimes, *Multiobjective Decision Making* (Elsevier Science Publishing, 1983).

been if . . .'' is conjecture, and all postdecision audits should be made with this reservation in proper perspective.

Capital allocation is the process of selecting investment or disinvestment proposals to optimize the long-term interests of current owners of an enterprise. Although capital allocation is usually thought of as the acquisition of physical assets (capital demand), the sources and amounts of investment funds (capital supply) are also relevant.

The principal problems of owner interest are of two general types. The first type relates to the fact that there is not enough capital to fund all prospective investment opportunities, even though the proposals are independent of one another. There simply isn't enough money to do everything that might otherwise be desirable. The second type of problem occurs when the decision maker must select from two or more mutually exclusive alternatives. Both types of problems concern economy, Edmund Burke's ''distributive virtue [that] consists, not in saving, but in selection.''

Before going on to quantitative techniques and criteria for evaluating alternatives, let's review the principles that form the rational basis for capital-allocation procedures:

1. Consider only feasible alternatives.
2. Use monetary units as a common denominator to make consequences commensurable.
3. Only differences are relevant to the decision-making process.
4. Sunk costs are irrelevant to decisions about the future.
5. All alternatives must be examined over a common planning horizon.
6. Criteria for investment decisions should recognize the time value of money and related problems of capital rationing.
7. Separable decisions should be made separately.
8. The relative degrees of uncertainty associated with various forecasts should be considered.
9. Decisions should give weight to consequences not reduced to money terms.
10. The efficacy of capital-budgeting procedures is a function of their implementation at various levels within the organization.
11. Postdecision audits improve the quality of decisions.

1.1 Decisions are rarely, if ever, based on economic considerations alone. Can you think of any that are? List as many as you can, but remember to identify situations in which *only* economic considerations are relevant to the decision.

1.2 A high school senior is considering application to several universities and colleges. List the various factors that might influence her decision. Assume that she is not infinitely wealthy and that cost is one of the relevant considerations. What are the factors other than cost?

1.3 The director of a university computer center is considering the purchase of 100 terminals for use by undergraduate students. A number of options are available, including manufacturers and different models. List the factors that might influence her decision. Discuss the significance of each factor.

1.4 F. R. Ward has been working for several years for a large manufacturing firm but has decided to go into business as an independent consultant providing engineering services. Ward is considering the purchase of a car that will be used for both business (80 percent) and personal (20 percent) purposes. List the factors that might influence his choice of cars (make, model, new or used). Discuss the significance of each factor.

1.5 Both individuals and businesses have a variety of opportunities to invest their excess capital, the funds not needed for day-to-day expenditures. List ten investment opportunities that might be available to investors in your community. (*Hint:* Consult local newspapers and magazines for advertisements of financial institutions. See especially the financial section of your newspaper.) For each opportunity, identify the rate of return available and discuss the potential risks.

1.6 Choose a city within 500 miles of the community where your school is located. Then prepare an exhibit that lists alternative means of travel from your community to the target city. List only alternatives that are feasible. Also list your constraints—the ways you have determined which alternatives are feasible. (*Example:* "Travel by roller skates is not feasible, because travel time would take twelve days and I must make the trip in one day.") Next, list your criteria for the trip. You might consider cost, time, comfort, and so forth. For each alternative, indicate its performance with respect to each criterion. Is one alternative clearly superior to the others? Explain.

1.7 Consequences that are truly irreducible to monetary units are rare. Although a marketplace may not exist to establish a monetary price for a particular good or service, it is often possible to infer monetary value through indirect measures. Consider the "value of human life." This issue frequently arises when evaluating the benefits of improving a public roadway, for example, as well as in many other contexts. How would you go about placing a monetary value on human life? (*Hint:* There is substantial literature on this subject. See your university library.)

1.8 How would you go about placing a monetary value on the increased opportunity to use recreational resources? Specifically, suppose that the Forest Service could improve an access road into a recreation area so that the number of visitors per day would increase from x to $x + \Delta x$. Can a monetary value be placed on Δx persons per day? Refer to Problem 7. Explain your answer.

1.9 How would you go about placing a monetary value on employee morale? Refer to Problem 7.

1.10 In 1981 the city of Anaheim built a new main scoreboard for Anaheim Stadium, home of the California Angels baseball team and the Los Angeles Rams football team. The cost of installing this massive electronic scoreboard was estimated to be $2.8 million. In a report prepared for the city council, the assistant city manager reported that this cost rose $700,000 from original estimates but that revenues from advertising on the scoreboard were expected to be 245 percent higher than earlier expectations. Eight advertising contracts had been signed and a ninth was in process. "For the ten-year period covered by the eight contracts," he said, "the gross revenues to the city will be $7,850,000, or an average of $785,000 a year." He added that the net profit for Anaheim, after paying a commission for selling the advertising space and scoreboard revenue shares to the Rams and Angels, would be about $4,770,000 for ten years. A ninth advertiser, he claimed, would boost the city's profit to about $5,550,000, nearly twice the cost of the board.

　　Is the profit to the city about 200 percent, 100 percent, or another percentage? What are the uncertainties that might affect the apparent profitability? What can be said about the relationship between current expenses and future revenues? On balance, does this appear to be an attractive investment for the city? Why or why not?

1.11 About four years ago Mr. Anchor purchased 5 ounces of gold for $600 per ounce. His purchase was for investment purposes only; the gold has been stored in his safety-deposit box until some indeterminate date when he can sell it for a "nice profit." Now, Anchor needs some money to buy a new car. He could sell the gold, but the current market value is only $400 per ounce. He is reluctant to sell because he will take a big loss—$1000. Instead, he will borrow money for the car from his credit union at an interest rate of 15 percent annually. What do you think of Anchor's reasoning? What should be the relevant considerations in his decision to keep or sell the gold?

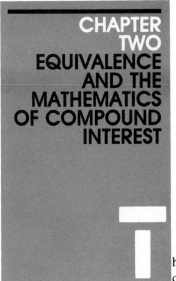

CHAPTER
TWO
EQUIVALENCE
AND THE
MATHEMATICS
OF COMPOUND
INTEREST

The concept of **equivalence** is a critical element in the process of economic analysis. For example, would you prefer $100 or $200, everything else being equal? The obvious answer is to choose the $200. But the question is deceptively simple. The *timing* of the two alternatives is not mentioned, but this bit of information is certainly relevant to the decision.

Suppose, on the other hand, that you were given a choice between $100 today and $200 ten years from now. You might conclude that it is preferable to accept the $100 now so as to enjoy its use, either through consumption or investment, over the next ten years. Indeed, you could accept the $100 today, invest it immediately in a savings account at, say, 8 percent per year, and realize about $216 at the end of ten years. Thus the $100 received today is *equivalent* to $216 ten years from now if funds can be invested at 8 percent per year during the ten-year period.

The principle embedded in the above illustration is clear: In order to choose intelligently among economic alternatives, both the amounts and the timing of expected receipts and disbursements must be estimated, and appropriate adjustments must then be made to account for the differences in timing. These adjustments, effected through the mathematics of compound interest, result in equivalent values that can then be compared directly, as the remainder of this chapter shows.

Before going on, let's define and explain some key words and phrases commonly used in the calculation of compound interest. The first is **cash flow,** a term that refers to receipt or payment of an amount of money. Cash flow is actual income (receipts) or

2.1
CASH FLOWS AND INTEREST

actual expenditures, not merely a financial obligation or accrual. Diagrams and tables are frequently used to show the amount and timing of cash flow in a clear, concise manner.

It is undeniable that we live in a world of limited resources. Except for a few very special cases, companies (as well as individuals, family units, and governments) find it necessary to ration limited wealth. When assets are invested, it is reasonable to expect that a lender will charge a borrower a ''rent'' to reimburse the lender for the lost opportunity to invest elsewhere. **Interest,** the rent charged for the use of borrowed money, would have little importance to a lender who has unlimited capital.

Interest is stated in monetary units and should not be confused with the interest rate, which is a pure number. **Interest rate,** generally written as a percentage rather than a decimal, is the ratio of the interest charged during an **interest period** to the amount of money owed at the beginning of the interest period. The length of the period must be stated or understood. ''Interest rate'' is frequently abbreviated to ''interest.'' For example, one may speak of 10-percent interest on a home mortgage, omitting ''per year.'' (Although strictly improper, this should not be bothersome, because the meaning is usually clear in context.)

Another feature of interest is the risk associated with various lending situations. Since returns from investments may not occur as originally planned, it is reasonable to expect that investors will adjust for uncertainty by adding interest to their loan. Even the most casual observer can see the obvious interdependence of risk and rate of interest. In the same way that investors demand a return to compensate for the loss of alternative uses of their limited capital, they expect some measure of return in anticipation of losses resulting from uncertain investments.

2.2 EQUIVALENT VALUES OF A SINGLE CASH FLOW

Consider an investment of $1,000, to be repaid at the end of five periods with interest computed at the rate of 10 percent per period. *Investment* can be interpreted in a broad sense. One may invest by depositing funds in a savings account, by making a business loan, by purchasing a piece of machinery, by adopting a certain management procedure, and so forth. All involve an original expenditure with possible future returns. Table 2.1 calculates the amount owed at the end of five periods. One thousand dollars today is equivalent to $1,610.51 five periods from now if interest is accumulated at the rate of 10 percent per period.

Table 2.1 reflects **compound interest** calculations. That is, the interest for each period is based on the amount owed at the beginning of the period. But if the interest had been computed only on the amount originally borrowed, the **principal,** the cumu-

Table 2.1
Amount to be Repaid on a $1,000 Loan with Interest Rate of 10 Percent Per Year

Period	Amount Owed at Beginning of Period	Interest During Period	Amount Owed at End of Period
1	$1,000.00	$100.00	$1,100.00
2	1,100.00	110.00	1,210.00
3	1,210.00	121.00	1,331.00
4	1,331.00	133.10	1,464.10
5	1,464.10	146.41	1,610.51

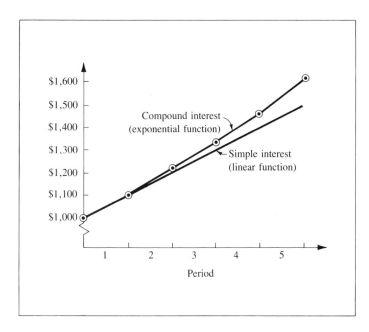

Figure 2.1 Comparison of Compound and Simple Interest

lative debt would have grown more slowly—to only $1,500 after five periods, or $1,000 + ($1,000 × 10% × 5). This type of interest, known as **simple interest,** is of no practical concern, because it does not exist in the real world. No rational investor would lend simple interest. Compound interest is always greater than simple interest because the former computes "interest on interest." See Figure 2.1.

Let's turn now to the development of a general model that will enable us to compute equivalent values of a single cash flow. Let P represent the amount of the original investment, and let F represent the equivalent future amount at the end of N periods if interest is compounded (computed) at the rate i per period. (See Figure 2.2.) The calculations are shown, period by period, in Table 2.2.

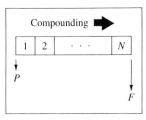

Figure 2.2

Table 2.2

Initial Investment, P, to be Repaid
after N Periods with Interest Rate i

Period	Amount Owed at Beginning of Period	Interest During Period	Amount Owed at End of Period
1	P	iP	$P(1 + i)$
2	$P(1 + i)$	$iP(1 + i)$	$P(1 + i)^2$
3	$P(1 + i)^2$	$iP(1 + i)^2$	$P(1 + i)^3$
$\vdots$	$\vdots$	$\vdots$	$\vdots$
N	$P(1 + i)^{N-1}$	$iP(1 + i)^{N-1}$	$P(1 + i)^N = F$

Figure 2.3

The relationship is clearly shown to be

$$F = P(1 + i)^N \qquad (2.1)$$

Using simple interest, $F = P + NiP = P(1 + Ni)$.

Equation 2.1, with a simple modification, can also be used to determine the equivalent present value, P, of a future cash flow, F, occurring at the end of the Nth period. (See Figure 2.3.) This process, known as **discounting**, results in

$$P = F/(1 + i)^N$$

$$\text{or} \qquad = F\left[\frac{1}{1 + i}\right]^N$$

$$\text{or} \qquad = F(1 + i)^{-N} \qquad (2.2)$$

These fundamental relationships can now be used to solve a wide variety of interesting and practical problems. To illustrate:

1. What is the equivalent future value one year from now of $100 invested at 6 percent per year? (See Figure 2.4.)

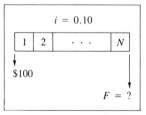

Figure 2.4

$$F = P(1 + i)^N$$
$$= \$100(1.06)^1$$
$$= \$106$$

2. What is the equivalent value five periods from now of $100 invested at 10 percent per period? (See Figure 2.5.)

$$F = P(1 + i)^N$$
$$= \$100(1.10)^5$$
$$= \$161.05$$

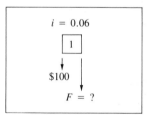

Figure 2.5

3. If interest is compounded at 10 percent per year, how long

will it take to double an amount of money? (See Figure 2.6.)

$$F = P(1 + i)^N$$
$$\$2 = \$1(1.10)^N$$
$$N = 7+ \text{ years}$$

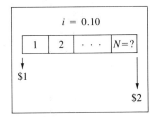

Figure 2.6

4. Using an interest (discount) rate of 6 percent per period, what is the equivalent present value of $1,000 flowing at the end of twenty periods? (See Figure 2.7.)

$$P = F\left[\frac{1}{1 + i}\right]^N$$
$$= \$1,000[1/1.06]^{20}$$
$$\approx \$312$$

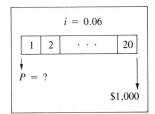

Figure 2.7

5. A current investment of $100 promises to yield $150 after 5 years. (See Figure 2.8.) To what compound interest rate does this investment correspond? (As will be shown elsewhere, this corresponds to the internal rate of return for the proposed investment.)

$$F = P(1 + i)^N$$
$$\$150 = \$100(1 + i)^5$$
$$i \approx 0.085$$

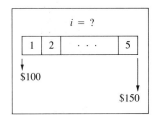

Figure 2.8

In each of the above problems, four parameters are considered: a present cash flow, or equivalent present value, P; a future cash flow, or future equivalent value, F; the number of time periods, N; and the interest rate, i, used for compounding or discounting. In each problem we are given three values and are asked to find the fourth:

Problem	Given	Find
1	P, i, N	F
2	P, i, N	F
3	P, i, F	N
4	F, i, N	P
5	P, F, N	i

All problems, regardless of their complexity, are variations on this theme.

Numerical complexity can be demanding, especially without the use of an adequate calculator. In Problem 4, for example, it was necessary to compute $[1/1.06]^{20}$. To ease this burden, compound-interest tables have been prepared in which values for various factors have been tabulated. A set of compound-interest tables has been included here in Appendix B, with each table

Figure 2.9

| | SINGLE PAYMENT | | |
| | COMPOUND AMOUNT | PRESENT WORTH | |
N	F/P	P/F	$P/\overline{F}$
1	1.010	0.9901	0.9950
2	1.020	.9803	.9852
3	1.030	.9706	.9754
4	1.041	.9610	.9658
5	1.051	.9515	.9562

representing a specific interest rate. The rows within the tables are indexed on the number of periods, N. Each column represents a separate factor; the first three factors are related to the material discussed in this section, the single-payment models. In particular, the first factor tabulates $(1 + i)^N$. (See Figure 2.9.) In the literature of finance, this is known as the **compound amount factor.** It is used to find F given P, written F/P. Similarly, the reciprocal, $(1 + i)^{-N}$, is known as the **present worth factor,** because it is used to find the equivalent present value of a future cash flow—P given F, or P/F. (The third factor, $P/\overline{F}$, will be discussed in Section 2.7.)

If P, F, or N is unknown, the value of the factor can be selected directly from the table for a given value of i. When P, F, and N are given and i is unknown, as in Problem 5 above, the solution can be obtained by searching the tables to find the particular i for which a value of the factor exists. In Problem 5, for example, we determined that $(1 + i)^5 = \$150/\$100 = 1.500$. This is the compound-amount factor, F/P, for $N = 5$. The tables in Appendix B show that the compound-amount factors for $i = 0.14$ and $i = 0.15$ are 1.925 and 2.011, respectively, when $N = 5$. Thus, the solving interest rate, i, is approximately 15 percent. You are invited to solve all of the above examples using the tables in Appendix B.

2.3
EQUIVALENT VALUES OF A UNIFORM SERIES OF CASH FLOWS

Consider problems in which the equivalent present value of a series of end-of-period cash flows, $A_1, A_2, \ldots, A_N$, all of which are identical, must be found. (See Figure 2.10.) Letting $A \equiv A_1 = A_2 = \cdots = A_N$, then

$$P = A_1\left[\frac{1}{1 + i}\right]^1 + A_2\left[\frac{1}{1 + i}\right]^2 + \cdots + A_N\left[\frac{1}{1 + i}\right]^N$$

$$= A\sum_{j=1}^{N}\left[\frac{1}{1 + i}\right]^j$$

$$= A\left[\frac{(1 + i)^N - 1}{i(1 + i)^N}\right] \qquad (2.3)$$

The term in brackets is known as the **series present worth factor,** because it is used to find the equivalent present value of a uniform series of cash flows.[1]

It follows from Equation 2.3 that

$$A = P\left[\frac{i(1 + i)^N}{(1 + i)^N - 1}\right] \tag{2.4}$$

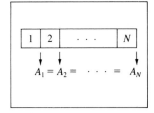

Figure 2.10

This is the **capital recovery factor,** used to convert an initial cash flow to a uniform series. The uniform series "recovers" the initial investment, P, plus an "opportunity cost" reflected by the interest rate, i. This notion will be discussed again in Chapter 3.

For certain problems, it may be useful to compute the equivalent future value, F, given a uniform series of end-of-period cash flows, A. (See Figure 2.11.) This can be accomplished, of course, by first converting from A to P using Equation 2.3. Then by using the compound-amount factor for a single payment, Equation 2.1, we convert from P to F. This can be simplified to one step, however, by using

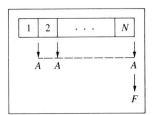

Figure 2.11

$$\begin{aligned} F &= \{P\}(1 + i)^N \\ &= \left\{A\left[\frac{(1 + i)^N - 1}{i(1 + i)^N}\right]\right\}(1 + i)^N \\ &= A\left[\frac{(1 + i)^N - 1}{i}\right] \end{aligned} \tag{2.5}$$

The factor in brackets is known as the **series compound amount factor,** because it finds the equivalent future value, or compound amount, of a uniform series of cash flows.

Finally, it follows from the above that

$$A = F\left[\frac{i}{(1 + i)^N - 1}\right] \tag{2.6}$$

For reasons that will be explained later, the factor in brackets is known as the **sinking fund factor.**

All four of the factors discussed in this section are tabulated in the tables in Appendix B for various values of i and N.

Arithmetic Gradient
A relatively common condition is one in which cash flows increase by a uniform amount from period to period, resulting in an arithmetic progression. Compound interest factors for treating this case are useful. To derive these factors, let G denote the amount, or

**2.4
EQUIVALENT VALUES
OF A GRADIENT
SERIES**

1. In the derivation of Equation 2.3, we use the relationship $\sum_{j=1}^{n} r^j = r(r^n - 1)/(r - 1)$, where $r < 1$.

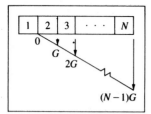

Figure 2.12

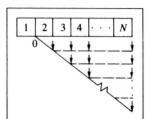

Figure 2.13

arithmetic gradient, by which the cash flows increase uniformly from one period to the next. (See Figure 2.12.) That is, $A_j = (j - 1)G$, for $j = 1, 2, \ldots, N$. This results in a sequence of cash flows, $0, G, 2G, \ldots, (N - 1)G$, at the end of periods 1, 2, 3, $\ldots, N$, respectively. As shown in Figure 2.13, this arithmetic gradient series can be described by $N - 1$ uniform series, each of which has equivalent future value, given by Equation 2.5. Therefore, the equivalent value at the end of period N of the entire arithmetic gradient series is:

$$F = G \sum_{j=1}^{N-1} \left[\frac{(1 + i)^j - 1}{i} \right]$$
$$= \frac{G}{i} \left[\sum_{j=0}^{N-1} (1 + i)^j \right] + \frac{G}{i}(-N)$$

Simplifying[2],

$$F = G\left\{ \frac{1}{i}\left[\frac{(1 + i)^N - 1}{i} \right] - \frac{N}{i} \right\} \tag{2.7}$$

To convert to P, we multiply F by $(1 + i)^{-N}$, with the result

$$P = G\left[\frac{(1 + i)^N - iN - 1}{i^2(1 + i)^N} \right] \tag{2.8}$$

The term in [brackets] is known as the **gradient present worth factor.**

It may also be shown that, to convert from an arithmetic gradient series to an equivalent uniform series, A, one can use the relationship

$$A = G\left[\frac{(1 + i)^N - iN - 1}{i(1 + i)^N - i} \right] \tag{2.9}$$

The term in [brackets] is the **gradient uniform series factor.** Both factors, Equations 2.8 and 2.9, have been tabulated for various values of i and N and are included in the tables in Appendix B.

Geometric Gradient

Another type of cash flow sequence that may be of interest is the **geometric gradient series.** (See Figure 2.14.) Here, the end-of-period cash flows increase at a constant rate, g, from period to period, such that

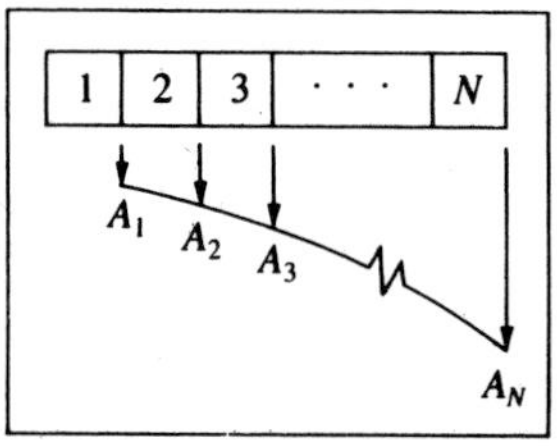

$$A_j = A_{j-1}(1 + g), \text{ for } j = 1, 2, \ldots, N \tag{2.10}$$

Figure 2.14

2. This derivation makes use of the relationship $\sum_{j=0}^{N-1} x^j = (x^N - 1)/(x - 1)$, where $x > 1$.

In terms of the first cash flow

$$A_1 = A_1$$
$$A_2 = A_1(1 + g)$$
$$A_3 = A_2(1 + g) = A_1(1 + g)^2$$
$$\vdots$$
$$A_N = A_1(1 + g)^{N-1}$$

With cash flows discounted at rate i per period, the equivalent present value, P_j, of cash flow A_j is

$$P_j = A_j(1 + i)^{-j}$$
$$= A_1(1 + g)^{j-1}(1 + i)^{-j}$$
$$= A_1(1 + i)^{-1}\left[\frac{1 + g}{1 + i}\right]^{j-1} \tag{2.11}$$

Summing over all j, the present worth, P, of the entire series of cash flows is

$$P = A_1\left[(1 + i)^{-1}\sum_{j=1}^{N}\left(\frac{1 + g}{1 + i}\right)^{j-1}\right] \tag{2.12a}$$

An alternative formula is

$$P = A_1\left[(1 + g)^{-1}\sum_{j=1}^{N}\left(\frac{1 + g}{1 + i}\right)^{j}\right] \tag{2.12b}$$

Closed-form expressions may be useful for evaluating Equations 2.12a and b. We define the ratio $\rho = (1 + g)/(1 + i)$ so that

$$P = A_1\left[(1 + g)^{-1}\sum_{j=1}^{N}\rho^j\right]$$

If $i = g$, then $\rho = 1$ and

$$P = A_1 N(1 + i)^{-1} \tag{2.13}$$

If $i \neq g$, then

$$\sum_{j=1}^{N}\rho^j = \frac{\rho(\rho^N - 1)}{\rho - 1}$$

Substituting and simplifying,

$$P = A_1 \left[\frac{1 - (1 + g)^N (1 + i)^{-N}}{i - g} \right]$$ (2.14)

It may be shown that, as N tends to infinity, the expression in [brackets] in Equation 2.14 is *convergent* if $g < i$; the series is *divergent* if $g \geq i$. The proof, if desired, remains as an exercise for the reader.

The expression in [brackets] in Equation 2.14 is known as the **geometric series present worth factor.** Tables for this factor, based on selected values of i, g, and N, are not included in this book. However, they may readily be generated using Equation 2.14. (See Problems 27 and 28 at the end of this chapter.)

2.5 EFFECTIVE VERSUS NOMINAL INTEREST RATES

An interest rate is meaningful only if it is related to a particular period of time. Nevertheless, the "time tag" is frequently omitted in speech because it is usually understood in context. If someone tells you that he's earning 6 percent on his investments, for example, it is implied that the rate of return is 6 percent *per year*. However, in many cases the interest-rate period is a week, a month, or some other interval of time, rather than the more usual year (per annum). At this point it would be useful to examine the process whereby interest rates and their respective "time tags" are made commensurate.

As before, let i represent the **effective interest rate** per period. As shown in Figure 2.15, let the period be divided into M subperiods of equal length. If interest is compounded at the end of each subperiod at rate i_M per subperiod, then $1 flowing at the start of the first subperiod will have equivalent value of $1(1 + i_M)^M$ at the end of the Mth subperiod. (See Figure 2.16.) If our calculation is based on the period rather than the subperiod, then the end-of-period equivalent value of $1 flowing at the start of the period, as shown in Figure 2.17, is simply $1(1 + i)$.

The relationship between the effective interest rates per period and per subperiod are easily determined. From the above:

$$\$1(1 + i_M) = \$1(1 + i)$$

or

$$i = (1 + i_M)^M - 1$$ (2.15)

The **nominal interest rate** per period, r, is simply the effective interest rate per subperiod times the number of subperiods, or

$$r = M i_M$$ (2.16)

Nominal and effective interest rates are most frequently used when the period is one year and M the number of divisions within

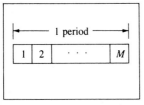

Figure 2.15

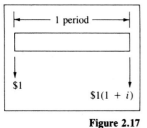

Figure 2.16

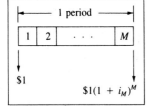

Figure 2.17

the year. Here we let i_a denote the effective interest per year (*per annum*).

To illustrate the above, consider the case where interest is compounded at the rate of 1.5 percent per month. That is, $i_M = 0.015$ and $M = 12$. Thus the nominal annual rate, r, is $12 \times 1.5\% = 18\%$, and the effective annual rate is $i_a = (1.015)^{12} - 1 = 19.56\%$.

Except for the trivial case in which $M = 1$, the annual effective rate will always be larger than the annual nominal rate. Moreover, the difference between the nominal and effective rates increases as r and/or M increase, as illustrated in Table 2.3. Therefore, the appropriate interest rate to use in all equivalence calculations is the effective rate.

Table 2.3

Representative Values for Nominal
and Associated Effective Interest Rates

Nominal interest rate	Number of compounding subperiods	Effective interest rate[1]	Absolute difference	Relative difference
(r)	(M)	(i)	$(i - r)$	$(i - r)/(i)$
0%	any	0%	0%	0.000
10	10	10.46	0.46	0.044
	20	10.49	0.49	0.047
20	10	21.90	1.90	0.087
	20	22.02	2.02	0.092
30	10	34.39	4.39	0.128
	20	34.69	4.69	0.135
40	10	48.02	8.02	0.167
	20	48.59	8.59	0.177

[1]Remember, $i = (1 + r/M)^M - 1$

2.6 THE CONTINUOUS COMPOUNDING CONVENTION

Consider the case in which interest is compounded not once a period, not twice a period, not three times a period, but M times each period, where M is a very large number. Under these circumstances, it may be useful to define the relationship between the nominal interest rate, r, and the effective interest rate, i.

The derivation is straightforward. As before, divide the period into M equal parts. The relationship between i and i_M as given above is

$$i = (1 + i_M)^M - 1, \text{ where } i_M = r/M$$

← effective interest rate.

Rewriting,

$$i = \left[\left(1 + \frac{1}{M/r}\right)^{M/r}\right]^r - 1$$

Letting $k = M/r$, we have

$$i = \left[\left(1 + \frac{1}{k}\right)^k\right]^r - 1$$

For a given r, when M tends to infinity, k also tends to infinity. Thus[3]

$$i = e^r - 1 \qquad (2.17)$$

where e is the base of the Napierian logarithm system, the "exponential," approximately equal to 2.71828. Under the assumption that M is very large, that is, $M \to \infty$, this is known as **continuous compounding** (discounting). The effective interest rate, i, when the nominal rate, r, is 10 percent per period under continuous compounding is

$$i = e^{0.10} - 1 = 0.1052$$

2.7 CONTINUOUS COMPOUNDING AND CONTINUOUS CASH FLOW

There are many instances in which an analyst will evaluate end-of-period cash flows as well as cash flows that occur frequently within a period. For example, suppose that a year is the basic period for an analysis, yet there are other cash flows, say, wages or material costs, that occur weekly or daily. One approach would be simply to accumulate the subperiod cash flows to compute a single value for the period. (See Figure 2.18.) So wages of $400 per week would be roughly equivalent to $20,800 per year, based on a fifty-two-week year. This is only an approximation, however, because there are compounding effects from one subperiod to the next.

An alternative approach is based on the assumption that the number of subperiods, M, within the period is very large. (The question of how large is "very large" will be discussed in the following section.) Let $\overline{A}$ be an amount of money distributed uniformly over M subperiods within the period so that $\overline{A}/M$ is the cash flow occurring at the end of every subperiod. (See Figure 2.19.) From the series compound amount factor, Equation 2.5, the equivalent cash flow, F, at the end of M subperiods is

$$F = \frac{\overline{A}}{M}\left[\frac{(1 + i_M)^M - 1}{i_M}\right]$$

where i_M represents the effective interest rate per *subperiod*. Of course, $i = r/M$, where r is the nominal interest rate per *period*.[4] Substituting in the above equation and rearranging terms gives us

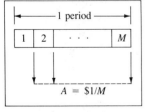

Figure 2.18

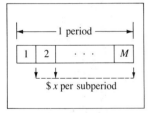

Figure 2.19

---

3. You may recall that $e = \lim\limits_{k \to \infty} \left(1 + \frac{1}{k}\right)^k$.

4. In some textbooks, the term *nominal* rate is used only when the interest period is a *year*. But the nominal rate may always be used for an interest period of any duration, as long as compounding takes place over two or more subperiods within the period.

$$F = \frac{\overline{A}}{M}\left\{\frac{(1 + r/M)^M - 1}{r/M}\right\}$$

$$= \overline{A}\left\{\frac{\left[\left(1 + \dfrac{1}{M/r}\right)^{M/r}\right]^r - 1}{r}\right\}.$$

Taking the limit of this expression as M gets "very large," that is, as $M \to \infty$,

$$F = \overline{A}\left[\frac{e^r - 1}{r}\right] = \overline{A}\left[\frac{i}{\ln(1 + i)}\right] \tag{2.18}$$

where i is the effective interest rate per period.

The factor in [brackets] permits us to adjust all the end-of-period factors to account for continuous cash flows. That is, given a cash flow, $\overline{A}$, flowing uniformly and continuously during the period, the equivalent end-of-period cash flow, A, is given by

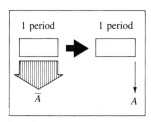

$$A = \overline{A}\left[\frac{i}{\ln(1 + i)}\right] \tag{2.19}$$

Figure 2.20

A and $\overline{A}$ are sketched in Figure 2.20.

To illustrate the use of this adjustment factor, known as the **funds flow conversion factor,** consider the problem represented in Figure 2.21. We want to find the equivalent present value, P, of a sum of money, $\overline{F}$, flowing continuously and uniformly *during* the Nth period. From Equation 2.19, it follows that

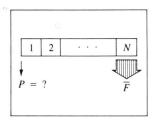

$$F = \overline{F}\left[\frac{i}{\ln(1 + i)}\right]$$

Moreover, since $P = F(1 + i)^{-N}$, we can write

Figure 2.21

$$P = \overline{F}\left\{\left[\frac{i}{\ln(1 + i)}\right]\left[\frac{1}{(1 + i)}\right]^N\right\} \tag{2.20}$$

The expression in {braces} is known as the **continuous single-payment present worth factor.** It is tabulated, for various values of i and N, in Appendix B.

Equivalence models have been developed under two basic cash-flow assumptions: (1) cash flows discretely at the *end* of the interest period and (2) cash flows continuously and uniformly *during* the interest period.

Suppose, for example, that we want to find the equivalent present value of $100 per month, flowing at the end of every

2.8 IMPORTANCE OF END-OF-PERIOD AND CONTINUOUS ASSUMPTIONS

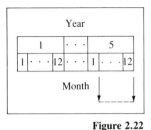

Figure 2.22

month, during the fifth year of operation. (See Figure 2.22.) Given an effective interest rate of 10 percent per year, the exact solution to this problem requires us to first determine the effective interest rate per month, i_M:

$$i_M = \sqrt[M]{1 + i} - 1$$

$$= \sqrt[12]{1.1} - 1$$

$$= 0.007974$$

The equivalent present value at the end of the fifth year is found by applying the compound amount factor, Equation 2.5:

$$F = A\left[\frac{(1 + i)^N - 1}{i}\right]$$

$$= \$100\left[\frac{(1.007974)^{12} - 1}{0.007974}\right]$$

$$= \$1254$$

The equivalent present value, P, at the *beginning* of the five years is found by applying the present worth factor:

$$P = F\left[\frac{1}{1 + i}\right]^N$$

$$= \$1254\left[\frac{1}{1.10}\right]^5$$

$$= \$1254(0.6209)$$

$$= \$778.63$$

A simpler, albeit approximate, solution may be obtained by assuming that the twelve $100 payments in the fifth year all occur at the end of the year. Again, by applying the present worth factor,

$$P = \$100 \times 12 \times 0.6209$$

$$= \$745.08$$

The difference between this approximate solution and the exact solution, $\$778.63 - \$745.08 = \$33.55$, represents an error of approximately 4.3 percent.

Still another approximate solution may be obtained by using the continuous present worth factor, derived in Equation 2.20. Assume that the total of all payments, $1200, is distributed uniformly during the fifth year. The equivalent present worth is

$$P = \overline{F}\left\{\left[\frac{i}{\ln(1+i)}\right]\left[\frac{1}{1+i}\right]^{N}\right\}$$
$$= \$1200\{0.6515\}$$
$$= \$781.80$$

The value in {braces} is taken directly from the 10-percent table in Appendix B, with $N = 5$. Note that the error, $\$781.80 - \$778.63 = \$3.17$, is only 0.4 percent of the true value. Thus, the continuous assumption is much more accurate than the end-of-period assumption in this case.

In general, the assumption of simple aggregation—assuming that all of the cash flows within the period flow at the end of the period—always understates the correct value, because the compounding effects between subperiods are ignored. The continuous assumption always overstates the correct value, because the number of compoundings is assumed to be infinitely large. The relative errors are a function of the interest rate, i, and the number of compounding subperiods. In general, the continuous assumption results in a better approximation when $M \geq 3$. You are invited to test this assertion for values of i, say, over the range $0\% < i \leq 100\%$. Of course, if you desire absolute precision, then use Equation 2.5 as outlined in the previous example.

The compound-interest factors developed in the preceding sections are summarized in Figure 2.23. All of these factors, with the exception of the geometric series present worth factor, shown as the last equation in the figure, are tabulated in Appendix B.

Note that the algebraic form of these factors is, in most cases, rather complex, so many textbook authors and authors of technical articles have chosen to adopt a **mnemonic format.** This is a uniform, simplified set of symbols that assists the reader to readily identify the use of the appropriate factor. The algebraic form is itself unimportant, given that the factor has been tabulated. The user wants to know, for example, that the present worth factor is used to determine the equivalent present value, P, of a future cash flow, A_j. Its algebraic form, $(1 + i)^{-j}$, is relatively unimportant if the numerical value of the factor is readily available in tables.

There are basically two systems of mnemonic formats. One, the **nominative form,** describes the factor with three elements: (1) the initials of the factor name, (2) the interest rate, and (3) the number of interest periods. For example, the mnemonic form of the present worth factor is written

$$(PW, i, N)$$

The principal alternative, the **functional form,** is similar but indicates the factor's function, rather than its name, in the first posi-

2.9
MNEMONIC FORMAT

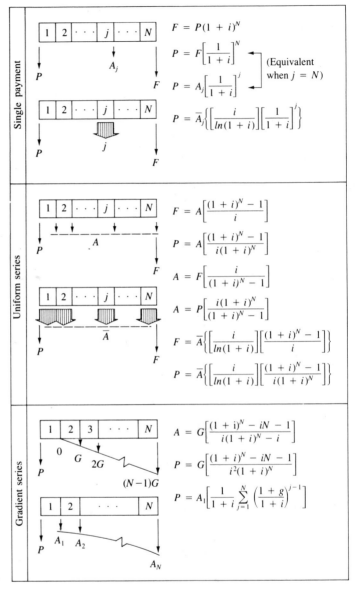

tion. Since the present worth factor is used to find P given F, the format is

$$(P/F, i, N)$$

The symbol P/F is read ''P given F.''

The mnemonic forms of the factors developed in this chapter are presented in Table 2.4 in the order in which they appear in the tables in Appendix B. Both the nominative and functional forms are given to assist those of you who may deal with material

Table 2.4
Compound-Interest Factors and Their Mnemonic Forms

Factor Name	Algebraic Form	Nominative Format	Functional Format
Single-payment			
• Compound amount	$(1 + i)^N$	(CA, i, N)	$(F/P, i, N)$
• Present worth	$\left[\dfrac{1}{1 + i}\right]^N$	(PW, i, N)	$(P/F, i, N)$
• Continuous present worth	$\left\{\left[\dfrac{i}{\ln(1 + i)}\right]\left[\dfrac{1}{1 + i}\right]^N\right\}$	(CPW, i, N)	$(P/\overline{F}, i, N)$
Uniform series			
• Compound amount	$\left[\dfrac{(1 + i)^N - 1}{i}\right]$	(SCA, i, N)	$(F/A, i, N)$
• Continuous compound amount	$\left\{\left[\dfrac{i}{\ln(1 + i)}\right]\left[\dfrac{(1 + i)^N - 1}{i}\right]\right\}$	$(CSCA, i, N)$	$(F/\overline{A}, i, N)$
• Present worth	$\left[\dfrac{(1 + i)^N - 1}{i(1 + i)^N}\right]$	(SPW, i, N)	$(P/A, i, N)$
• Continuous present worth	$\left\{\left[\dfrac{i}{\ln(1 + i)}\right]\left[\dfrac{(1 + i)^N - 1}{i(1 + i)^N}\right]\right\}$	$(CSPW, i, N)$	$(P/\overline{A}, i, N)$
• Sinking fund	$\left[\dfrac{i}{(1 + i)^N - 1}\right]$	(SF, i, N)	$(A/F, i, N)$
• Capital recovery	$\left[\dfrac{i(1 + i)^N}{(1 + i)^N - 1}\right]$	(CR, i, N)	$(A/P, i, N)$
(Arithmetic) gradient series			
• Uniform series	$\left[\dfrac{(1 + i)^N - iN - 1}{i(1 + i)^N - i}\right]$	(GUS, i, N)	$(A/G, i, N)$
• Present worth	$\left[\dfrac{(1 + i)^N - iN - 1}{i^2(1 + i)^N}\right]$	(GPW, i, N)	$(P/G, i, N)$
Geometric gradient series			
• Present worth $i \neq g$	$\left[\dfrac{1 - (1 + g)^N(1 + i)^{-N}}{i - g}\right]$	$(GGPW, i, N)$	$(P/A_1, g, i, N)$

published by other authors. However, throughout the remainder of this book, the functional format will be used.

2.10 LOANS

It may be useful at this point to examine one aspect of compound interest calculations of considerable interest to virtually everyone in our society: loans, the borrowing and lending of money. This examination will afford us the opportunity to use several of the concepts discussed earlier in this chapter as well as to demonstrate the use of the functional format for our compound interest factors.

Consider an amount of money received by a borrower at some point in time. The sum advanced by the lender to the borrower is known as the *principal* of the loan and will be designated by the symbol P in the following discussion. In particular, let P_j represent the amount of principal remaining, or yet to be repaid, at the start of period j, for $j = 1, 2, \ldots, N$. Here, P_1 represents the amount of the original loan.

All loans have at least two critical pieces of information associated with them: (1) the amount and timing of cash flows between the borrower and the lender and (2) the effective interest rate per period, which is a measure of the "rent" paid by the borrower for the temporary use of borrowed funds. In some instances, of course, only the cash flows and their timing are specified; the interest rate is only implied. In any case, the interest rate, whether explicit or implicit, represents the cost to the borrower and the return to the lender.

In general, the interest, I_j, accumulated in period j is given by

$$I_j = iP_j \tag{2.21}$$

where i is the effective interest rate per period.

Consider the case in which the borrower and lender agree on a loan of amount P_1, with repayment of the principal and all accrued interest at the end of N periods. Interest is to be compounded at rate i per period. The amount of the terminal payment, then, is

$$F = P_1(F/P, i, N) \tag{2.22}$$

Suppose that the original loan is $1,000, with payment due at the end of twenty-four months and with interest compounded monthly at the rate of 1 percent per month. Then

$$F = \$1,000(F/P, 1\%, 24)$$
$$= \$1,269.73$$

Another common repayment plan is one in which the borrower promises to repay the loan, P_1, with N equal payments. Some examples are home mortgages, car loans, and in many cases, student loans. If i is the interest rate per compounding period and if the loan payments, A, are to be made at the end of every compounding period, then

$$A = P_1(A/P, i, N) \tag{2.23}$$

The portion of each payment, A, that represents repayment of principal is simply the amount of the payment less the interest incurred in that period, or $A - I_j$. The cash-flow diagram is shown in Figure 2.24. The debt remaining after the periodic payment at the end of period j becomes the principal outstanding at the start of the next period, or

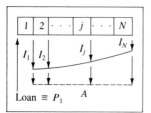

Figure 2.24

$$P_{j+1} = P_j - (A - I_j)$$
$$= P_j - (A - iP_j)$$
$$= P_j(1 + i) - A \tag{2.24}$$

Note a somewhat more useful formulation for determining the unpaid principal: If no payments had been made, the outstanding debt at the start of period $j + 1$ would be $P_1(F/P, i, j)$. However, the uniform payments, A, over the j preceding periods have equivalent value at the start of period $j + 1$ of exactly $A(F/A, i, j)$. The difference represents the amount of principal remaining. So,

$$P_{j+1} = P_1(F/P, i, j) - A(F/A, i, j) \qquad \text{(2.25)}$$

After the last payment has been made, at the end of period N,

$$
\begin{aligned}
P_{N+1} &= P_1(F/P, i, N) - A(F/A, i, N) \\
&= P_1(F/P, i, N) - P_1(A/P, i, N)(F/A, i, N) \\
&= P_1(F/P, i, N) - P_1(F/P, i, N) \\
&= 0 \quad \cdot \quad \cdot \quad \cdot \quad \text{as expected}
\end{aligned}
$$

A numerical example illustrating these calculations is summarized in Table 2.5. Here, a $1,000 loan is repaid over five periods with interest calculated at the rate of 10 percent per period. Using the formula developed above, consider, say, the third payment. The total amount of the payment, from Equation 2.23, is

$$A = \$1,000(A/P, 10\%, 5) = \$263.80$$

Table 2.5
Repayment of $1,000 Loan Assuming Five Uniform Payments at Interest Rate of 10 Percent Per Period

End of Period j	Amount of Payment[1] A	Interest Portion[2] I_j	Principal Portion[3] $A - I_j$	Principal Remaining After Payment[4] P_{j+1}
1	$263.80	$100,00	$163.80	$836.20
2	263.80	83.62	180.18	656.02
3	263.80	65.60	198.20	457.82
4	263.80	45.78	218.02	239.80
5	263.80	23.98	239.82	0

1. $A = \$1,000(A/P, 10\%, 5) = \$263.80$
2. $I_j = iP_{j-1}$, where $P_1 = \$1,000$
3. $A - I_j$
4. $P_{j+1} = P_j - (A - I_j)$

The amount owed, the unpaid principal, at the start of period 3, from Equation 2.25, is

$$
\begin{aligned}
P_3 &= \$1,000(F/P, 10\%, 2) - \$263.80(F/A, 10\%, 2) \\
&= \$1,000(1.21) - \$263.80(2.10) \\
&= \$1,210 - \$553.98 \\
&= \$656.02
\end{aligned}
$$

The amount of the uniform payment at the end of period 3 that is interest, from Equation 2.21, is

$$I_3 = 0.10(\$656.02)$$
$$= \$65.60$$

Several loan problems are included at the end of this chapter.

<h2>2.11 SUMMARY</h2>

Many amounts cannot be fairly compared unless they are evaluated at the same point in time. But since receipts and expenditures, positive and negative cash flows, resulting from alternative capital investments almost always yield dissimilar cash-flow patterns, it is necessary to determine equivalent values through the use of compound interest calculations. Some of the factors most commonly employed for this purpose have been derived and are summarized in Figure 2.23.

The distinction between nominal and effective interest rates should be clearly understood, because the relative difference between these two rates can be significant. In general, these differences will increase as the nominal rate gets larger and as the number of compounding subperiods increases.

Compound interest factors have been developed under two separate sets of assumptions: (1) receipts or disbursements occur only once during the interest period, at the start of the period, and the compounding/discounting takes place at that time, or (2) the cash flows continuously and uniformly throughout the interest period and interest is compounded continuously. The appropriate assumptions for any given analysis should be those that best approximate the specific conditions for any given problem situation.

Principal and interest payments associated with loans provide an especially instructive opportunity to examine the mathematics of compound interest. Several features have been discussed in this chapter.

Unless otherwise indicated, assume discrete cash flows, end-of-period compound/discounting, and effective interest rates.

The Power of Compound Interest

2.1 The headline of an article appearing in the May 16, 1981 edition of the *Detroit Free Press* read, "Widow Gets $3.7 Million for Mate's Death on Job." The article described an out-of-court settlement in which the widow and children of a construction worker killed in an accident agreed to the following settlement:

a. three lump-sum payments to the widow totaling $1,180,000 over the next thirty years
b. another $1,620,000 payable to the widow in 540 consecutive monthly installments of $3,000
c. the sum of $150,000 payable to each of the six children after twenty years

Thus the totals are:

Lump sum payments to widow	$1,180,000
Monthly payments to widow	1,620,000
Lump sum payments to children	900,000
	$3,700,000

$$i_s = \sqrt[12]{1+i} - 1 = 0.00797$$

If the family's "opportunity cost" is, say, 12 percent per year, determine the equivalent present value of this future series of payments. In this calculation, assume that the three lump-sum payments to the widow will occur as follows:

$$P = A(P/A, i_s, 540)$$

paid immediately	$180,000
paid after 15 years	500,000
paid after 30 years	500,000
	$1,180,000

(*Answer:* $695,600)

2.2 In the spring of 1965, Mr. Norton Simon purchased Rembrandt's painting of his son, Titus, for $2,234,400. This famous portrait, painted about 1650, was purchased from Sir Francis and Lady Cook, whose family acquired the portrait in 1915.

It is frequently said that art is a good investment. Ignoring esthetics for the moment and assuming that the painting was purchased for, say, $100,000 in 1915, decide whether or not this was a "good investment." Why or why not? Based on this simple but dramatic example, what may be said about investments in art, real estate, common stock, antique furniture, or rare coins?

2.3 Refer to Problem 1.10 in Chapter 1. Let us examine Anaheim's rate of return on the scoreboard investment in view of what we have learned in Chapter 2. For purposes of the analysis, assume that the ninth advertiser will be signed up, boosting net revenues over the ten-year period to $5,550,000. This figure includes gross revenues less sales commissions and payments to the Rams and the Angels. Assume also that the net *positive* cash flows are $550,000 at the end of each and every year over the ten-year period. There is a *negative* cash flow of $2,800,000 at the start of the planning horizon. Determine the city's rate of return on the investment over a ten-year study period.

(*Answer:* Approximately 14%)

Evaluating Compound-Interest Factors

2.4 The tables in Appendix B are truncated at $N = 100$, or when the single payment present worth factor is equal to or less than 0.0001. Determine the limits of each of the eleven factors in the tables as the number of periods approaches infinity. That is, find:

a. $(F/P, i\%, \infty)$
b. $(P/F, i\%, \infty)$
c. $(P/\overline{F}, i\%, \infty)$
d. $(F/\underline{A}, i\%, \infty)$
e. $(F/\overline{A}, i\%, \infty)$
f. $(P/\underline{A}, i\%, \infty)$
g. $(P/\overline{A}, i\%, \infty)$
h. $(A/F, i\%, \infty)$
i. $(A/P, i\%, \infty)$
j. $(A/G, i\%, \infty)$
k. $(P/G, i\%, \infty)$

2.5 The tables in Appendix B tabulate factors for a variety of typical interest rates. Suitable tables may not always be available, however. Prepare your own tables, if necessary, to compute:

a. $(P/\underline{A}, 9.5\%, 2)$
b. $(P/\overline{A}, 9.5\%, 2)$
c. $(F/P, 9.5\%, 2)$
d. $(A/P, 9.5\%, 2)$
e. $(A/F, 9.5\%, 2)$

(*Answers:* (a) 1.74745, (b) 1.83315 (c) 1.199, (d) 0.5723, (e) 0.4773.)

2.6 It may be shown that $(A/P, 11\%, 10) = 0.1698$. Compute the following to four significant digits:

a. $(P/F, 11\%, 10)$
b. $(F/P, 11\%, 10)$
c. $(P/A, 11\%, 10)$
d. $(A/F, 11\%, 10)$
e. $(F/A, 11\%, 10)$
f. $(P/A, 11\%, \infty)$
g. $(A/P, 11\%, \infty)$
h. Compute the nominal interest rate per year that corresponds to the effective interest rate per year ($i = 0.11$), assuming continuous compounding.

2.7 If the discount rate is 11.4 percent, find the following to four decimal places:
a. $(A/P, 11.4\%, 14)$
b. $(P/F, 11.4\%, 14)$
c. $(F/A, 11.4\%, 14)$
d. Find $(P/\overline{A}, 11.4\%, 14)$, assuming continuous compounding and continuous cash flow, $\overline{A}$.
(*Answers:* (a) 0.1463, (b) 0.2206, (c) 30.9598, (d) 7.2179)

2.8 A discount rate of 9.75 percent was widely used by the State of California during the 1970s. Given $(F/P, 9.75\%, 5) = 1.592$, find the following to at least four significant digits:
a. $(P/F, 9.75\%, 5)$
b. $(P/A, 9.75\%, 5)$
c. $(A/P, 9.75\%, 5)$
d. $(F/A, 9.75\%, 5)$
e. $(A/F, 9.75\%, 5)$
f. $(P/A, 9.75\%, \infty)$
g. $(A/P, 9.75\%, \infty)$
h. Compute the nominal interest rate per year, r, corresponding to the effective interest rate per year, $i = 9.75\%$.
i. Compute $(F/P, r, \infty)$

Equivalence: Discrete Cash Flows

2.9 Assuming an interest rate of 8 percent per period, determine the equivalent present value of each of the following three cash-flow series:

End of period	Series I	Series II	Series III
1	$ 0	$100	$500
2	100	200	400
3	200	300	300
4	300	400	200
5	400	500	100

(*Answers:* (I) $737, (II) $1136, (III) $1259)

2.10 a. Write the equation that will yield the equivalent net present value of the following sequence of cash flows given a 10 percent discount rate. Write your equation using functional notation and the minimum number of factors.

End of period	Cash flow
0–5	100
6	150
7	140
8	130
9	120
10	110

b. Find the solution to the equation.

2.11 Here is a series of cash flows with an interest rate of 8 percent per period:

End of period	Project X	Project Y
1–5	$1,000	$2,000
6–10	2,000	1,000

a. Find the equivalent present values of the two projects.
b. Find the equivalent values of the two projects at the end of 10 periods.
c. Find the equivalent uniform series of the two projects.
(*Answers:* (a) $9428 and $10,703; (b) $20,353 and $23,106; (c) $1405 and $1595)

2.12 Consider the following cash flow sequences:

End of period	A	B	C	D	E
0	100	0	0	0	0
1	100	200	0	200	500
2	100	200	0	300	400
3	100	200	0	400	300
4	100	200	0	500	200
5	0	200	0	600	100
6	0	175	100	0	0
7	0	175	200	0	0
8	0	175	300	0	0
9	0	175	400	0	0
10	0	175	500	0	0

Assuming a discount rate of 10 percent per period, find the equivalent present value of each sequence.

2.13 Consider the cash flows associated with the following alternative projects:

End of period	Alternative X	Alternative Y	Alternative Z
1	$1,000	$ 100	$ 200
2	0	0	200
3	0	0	200
4	0	0	200
5	0	900	200
Total	$1,000	$1,000	$1,000

a. Assuming an interest rate of 5 percent per period, determine the equivalent future value (at the end of five periods) of each of the three projects.
b. Find the interest rate, if there is one, at which the three projects are equivalent.
(*Answers:* (a) $1216, $1022, and $1105; (b) 0 percent)

2.14 Payments on the lease of data-processing equipment are $100, payable at the start of each two-month period. If the effective interest rate is 10 percent per year, what is the equivalent end-of-year payment that corresponds to six bi-monthly payments? Find the interest rate, if there is one, at which the sum of these six payments is less than $600.

2.15 Find the equivalent present value of $100 flowing at the start of every odd-numbered year, beginning in 1985, if the effective interest rate per year is 10 percent. Consider "the present" to be the start of 1985.
(*Answer:* $576.19)

2.16 Determine the following, assuming an interest rate of 6 percent per year:
   a. The initial investment required to yield an income of $1,000 per year five years from now and continuing at the end of each year for the subsequent ten years.
   b. The initial investment required to yield an income of $1,000 per year, at the end of each year, forever.
   c. The amount that must be deposited at the beginning of each year for twenty years in order to provide a fund of $100,000 at the end of twenty years.

2.17 On July 1, 1984, $1,000 is deposited in a fund drawing 4 percent interest. Another $1,000 will be deposited each July 1 up to and including July 1, 1994. The fund must provide a series of uniform annual withdrawals starting July 1, 1999, and the final withdrawal, on July 1, 2003, must exhaust the fund. How much will be withdrawn at the end of each year during the withdrawal sequence?
(*Answer:* $3,544)

2.18 Determine the following, assuming an interest rate of 5 percent per year:
   a. The present investment required to yield an income of $60 per year beginning in seven years and continuing forever.
   b. The value at the end of thirty-five years of a fund created by the year-end investments of $$x$ in year one, $2x$ in year two, $3x$ in year three, and so on, for thirty-five years.
   c. The amount that must be deposited at the beginning of each year for ten years in order to provide a fund of $10,000 at the end of the tenth year.

2.19 Find the annual effective interest rate that should be charged in order to receive $200 at the end of five years for a $100 loan.
(*Answer:* 14.9%)

2.20 With interest at 20 percent per year, how long will it take to recover an initial investment of $10,000 if the investor receives $2,000 annually?

2.21 If interest is compounded at the rate of 10 percent per period, how many periods will it take for a present sum of money to triple?
(*Answer:* approximately twelve periods)

2.22 Consider an infinite series of cash flows of $x flowing at the end of years 5, 10, 15, 20, ... Find the equivalent annual worth of this infinite series if the effective interest rate is 10 percent per year.

2.23 a. Find the sum of money now that will be equivalent to $1,000 four years from now if interest is compounded annually at 8 percent per year.
   b. Find the sum of money now that will be equivalent to $1,000 four years from now if interest is compounded semiannually at 4 percent per six-month period.
   (*Answers:* (a) $735.03, (b) $730.69)

2.24 Determine the present worth of the following cash flow sequence if interest is compounded at 8 percent per year:

End of period	Positive cash flow (in millions of pesos)
1	300
2	400
3	500
4	600
5	600
6	600
7	600

2.25 Using an interest rate of seven percent, determine the present value of the following cash flow sequence:

End of period	Cash flow (in thousands of pesos)
0	− 100
1–5	+ 20
6	+ 18
7	+ 16
8	+ 14
9	+ 12
10	+ 10

(It is possible, of course, to "discount" each of the cash flows one by one. However, using other appropriate factors will greatly simplify the calculations.)
(*Answer:* 23,730 pesos)

2.26 With interest at 10 percent per period, convert the following sequence of cash flows to an equivalent uniform series. Use the series factors where appropriate, and write the simplest equation.

End of period	0	1	2	3	4	5	6	7	8
Cash flow ($)	− 10	20	18	16	14	12	10	8	20

Geometric Series

2.27 Wages in the first year of a five-year project are expected to be $100,000. It is anticipated that wages will increase at the rate of 8 percent per year. If the appropriate discount rate is

10 percent per year, determine the equivalent present value of wage payments over five years. (The present worth factor for a geometric gradient series is not tabulated in Appendix B. These values can be obtained, however, if g, i, and N are given.)

(*Answer:* $438,558)

2.28 Consider a series of cash flows increasing at the rate of 12 percent per period over twenty interest periods. The cash flow at the end of the first period is A_1. With interest compounded/discounted at an effective rate of 5 percent per period, find:

a. the equivalent present value of these twenty cash flows and

b. the uniform annual series equivalent to this series of cash flows.

Effective versus Nominal Interest Rates

2.29 Find the effective interest rate per month that corresponds to an effective rate of 17.5 percent per year.

(*Answer:* 1.35 percent)

2.30 Find the effective interest rate per year that corresponds to an effective interest rate of 0.5 percent per week. Assume a 52-week year.

2.31 An investor's effective "opportunity cost" (discount rate) is 15 percent per year. Find the equivalent effective interest rate per month.

(*Answer:* 1.17 percent)

2.32 A sum of $100 is invested at the end of every quarter (three-month period) for exactly twenty quarters. Invested funds accumulate interest at the rate of 1 percent per month.

a. Find the equivalent effective rate per quarter.

b. Find the equivalent nominal rate per year.

c. Find the equivalent effective rate per year.

d. What effective rate per year corresponds to a rate of 1 percent per month if interest is compounded continuously?

2.33 Find the effective interest rate per year that corresponds to a nominal rate of 12 percent per year

a. compounded semiannually.

b. compounded quarterly.

c. compounded monthly.

d. compounded continuously.

(*Answers:* (a) 12.36 percent, (b) 12.55 percent, (c) 12.68 percent, (d) 12.75 percent)

2.34 Find the effective interest rate per year that corresponds to a nominal rate of 10 percent per year

a. compounded semiannually.

b. compounded continuously.

2.35 You are able to secure a $1,000 loan from the Happy Home Finance Company at an interest of 2 percent quarterly. Interest is compounded every three months. Find the nominal and effective annual rates.
(*Answers:* 8.00 percent and 8.24 percent)

2.36 An investor's effective "opportunity cost" (discount rate) is 18 percent per year. Find the equivalent effective interest rate per month.

2.37 Find the effective rate per year that corresponds to an effective interest rate of 2 percent per month.
(*Answer:* 26.8 percent)

2.38 Find the effective interest rate per period that corresponds to a nominal interest rate of 15 percent per period if interest is compounded continuously.

Continuous Compounding: Continuous Cash Flow

2.39 Consider a cash flow of $10,000 occurring continuously and uniformly during the tenth year of a fifteen-year planning horizon. Find the equivalent present value of this cash flow if the firm's nominal interest rate is 12 percent per year.
(*Answer:* $3,200)

2.40 A firm's discount rate is 22 percent per year.
  a. Find the equivalent effective rate per month.
  b. Find the equivalent present value of $100 flowing during every year forever.
  c. Find the equivalent present value of $100 flowing during every fifth year, beginning in year five and including the fiftieth year. The series is:

During year	5	10	15	. . .	50
Cash flow ($)	100	100	100	. . .	100

2.41 Maintenance expenses of $M are expected to flow continuously and uniformly during every fourth year, beginning with year four and continuing indefinitely. Assuming a nominal interest rate, r, of ten percent per year, find the equivalent present value of this infinite series.
(*Answer:* $2.138M)

2.42 Consider a cash flow of $10,000 occurring continuously and uniformly during the fifth year of a fifteen-year planning horizon. Find the equivalent present value of this cash flow if the firm's *nominal* interest rate is 12 percent per year.

2.43 A sum of $1,000 flows continuously and uniformly during the tenth year of a twenty-year planning horizon. Find the equivalent present value of this cash flow if the discount rate is 8 percent per year.
(*Answer:* $481.48)

2.44 Determine:
   a. $(P/F, 5\%, 5)$ e. $(P/F, 10\%, 5)$
   b. $(P/\overline{F}, 5\%, 5)$ f. $(P/\overline{F}, 10\%, 5)$
   c. $(P/F, 5\%, 10)$ g. $(P/F, 10\%, 10)$
   d. $(P/\overline{F}, 5\%, 10)$ h. $(P/\overline{F}, 10\%, 10)$

   By analyzing the differences between these values, what can be said about the importance of the correct assumption (end-of-period versus continuous) as the interest rate increases? As the number of periods increases?

Loans

2.45 A loan of $5,000 requires monthly payments of $200 over a three-year period. These payments include both principal and interest.
   a. Determine the nominal rate of interest per year.
   b. Determine the effective rate of interest per year.
   c. Find the amount of the unpaid principal on the loan after twelve payments.
   d. Find the total interest paid during the first twelve payments.
   (*Answers:* (a) 25.52 percent, (b) 28.72 percent, (c) $3,729, (d) $1,129)

2.46 An item in the February 25, 1982 edition of the *Wall Street Journal* reported that mortgage lenders are trying to entice homeowners to repay old debts by offering to forgive part of the remaining principal on old, low-interest mortgages if re-paid before the end of the loan period. For example, "In Denville, N.J., Metropolitan Federal Savings offers to for-give 25% of the principal for borrowers who repay mort-gages with old interest rates of 9% or less."
   Consider a homeowner who originally borrowed $30,000 at an effective annual interest rate of 6 percent. The loan is to be repaid in uniform monthly payments over a thirty-year repayment period.
   a. What is the amount of the uniform monthly payments?
   b. How much unpaid principal remains after exactly fifteen years, or after $15 \times 12 = 180$ monthly payments?
   c. Suppose that the homeowner's current opportunity cost is 15 percent per year before income taxes. That is, the homeowner can earn 15 percent on any available in-vestment capital. Moreover, suppose that the 15 percent rate is expected to reamin over the next 15 years. Let p be the percentage of the remaining principal to be for-given by the lender if the homeowner repays the out-standing debt now. What is the minimum value of p such that prepayment would be attractive to the homeowner?

2.47 In order to support his junior year in college, Ivan O. Uni-versal obtains a $5,000 one-time loan from a federal agency. This is a 7 percent loan, which must be repaid in not more than ten years from the date of graduation.

I. O. U. intends to repay the loan in eight uniform annual payments, with the first payment due three years from the date of the loan.

a. Determine the amount of each uniform payment. Assume that the 7-percent cost of the loan begins at the time of graduation, which is two years hence. So Ivan will owe $5,000 in two years. If no payment is made, he will owe $5,000(1.07) = $5,350 in three years, and so on.

b. Determine the true cost of the loan, stated as a percentage, over the ten-year planning horizon. Note that the final, eighth, payment will be made ten years after the date of the original loan. (*Hint:* The rate of return to the lender is the cost to the borrower.)

(*Answers:* (a) $837.50, (b) 4.7 percent)

2.48 In order to finance the purchase of computer equipment, the S. C. Equipment Corporation borrows $100,000, to be repaid in forty equal end-of-month payments at an interest rate of 1.5 percent per month.

a. Determine the amount of the monthly payments.

b. Determine the amount remaining on the loan after two years.

c. Determine the interest portion of the first payment.

d. Determine the principal portion of the twenty-fifth payment.

2.49 A $10,000 car may be purchased with a $2,000 down payment and thirty-six monthly payments. What is the amount of these monthly payments if the interest rate is 17.5 percent per year?

(*Answer:* $281.57)

2.50 A firm borrows $75,000, to be repaid in monthly installments of $720 at the end of every month for thirty years.

a. Determine the nominal interest per year. (*Hint:* Because there are $12 \times 30 = 360$ monthly payments, this uniform series may be approximated by an infinite series. That is, 360 is sufficiently close to infinity to use the perpetual-payment models.)

b. Determine the *effective* interest rate per year.

c. At the end of the five years (sixty payments), the firm wishes to make a lump-sum payment to settle its remaining debt. What should the amount of this payment be?

Computer Programs

The following programs, written in BASIC, may be used to solve the problems in this chapter. They may be used individually, for the simplest problems, or in combination, for some of the more complex problems. The programs are meant to be illustrative only. Students are invited to develop improved versions. In each case, the BASIC program is presented, followed by an example.

NOTE: If Applesoft BASIC is being used, substitute symbol (**) with symbol ($\wedge$). For example, line 85 should read:

$$85 \text{ LET F } = \text{ P}^*(1+i)^{\wedge}N$$

This correction should be made elsewhere in remaining programs where appropriate.

2.1   Single-payment compound amount (given P to find F).

PROGRAM
```
10 PRINT "GIVEN P, TO FIND F"
20 PRINT " "
30 PRINT "                          N"
40 PRINT "THE FORMULA IS F=P(1+i)"
50 PRINT " "
60 INPUT "ENTER PRESENT VALUE P=$";P
70 INPUT "ENTER INTEREST RATE i=";i
80 INPUT "ENTER TOTAL COMPOUNDING PERIOD N=";N
85 LET F=P*(1+i)**N
90 PRINT "THE FUTURE WORTH IS F=$";F
95 END
```

EXAMPLE
```
GIVEN P, TO FIND F
                     N
THE FORMULA IS F=P(1+i)
ENTER PRESENT VALUE P=$ ? 1
ENTER INTEREST RATE i= ? .1
ENTER TOTAL COMPOUNDING PERIOD N= ? 5
THE FUTURE WORTH IS F=$ 1.61051
```

2.2   Single-payment present worth (given F, to find P).

PROGRAM
```
10 PRINT "GIVEN F, TO FIND P"
20 PRINT " "
25 PRINT "                          -N"
30 PRINT "THE FORMULA IS P=F(1+i)"
40 PRINT " "
50 INPUT "ENTER FUTURE VALUE, F=$";F
60 INPUT "ENTER INTEREST RATE i=";i
70 INPUT "ENTER TOTAL COMPOUNDING PERIOD N=";N
80 LET P=F*(1+i)**(-N)
85 PRINT " "
90 PRINT "THE PRESENT WORTH IS P=$";P
95 END
```

EXAMPLE
```
GIVEN F, TO FIND P
                     -N
THE FORMULA IS P=F(1+i)
ENTER FUTURE VALUE, F=$ ? 1
ENTER INTEREST RATE i= ? .1
ENTER TOTAL COMPOUNDING PERIOD N= ? 3
THE PRESENT WORTH IS P=$ 0.7513148
```

2.3 Uniform series present worth (given A, to find P).

PROGRAM
```
10 PRINT "GIVEN A, TO FIND P"
20 PRINT " "
30 PRINT "                              N"
35 PRINT "                         ((1+i)  -1)
40 PRINT "THE FORMULA IS P=A[----------------]"
45 PRINT "                              N"
50 PRINT "                         i(1+i)"
55 PRINT " "
60 INPUT "ENTER UNIFORM ANNUAL VALUE A=$";A
65 INPUT "ENTER INTEREST RATE i=";i
70 INPUT "ENTER COMPOUNDING PERIOD N=";N
75 LET P=A*((1+i)**N-1)*(i*(1+i)**N)**(-1)
80 PRINT " "
85 PRINT "THE PRESENT WORTH IS P=$";P
90 END
```

EXAMPLE
```
GIVEN A, TO FIND P
                              N
                         ((1+i)  -1)
THE FORMULA IS P=A[----------------]
                              N
                         i(1+i)
ENTER UNIFORM ANNUAL VALUE A=$ ? 1
ENTER INTEREST RATE i= ? .1
ENTER COMPOUNDING PERIOD N= ? 10
THE PRESENT WORTH IS P=$ 6.144567
```

2.4 Capital recovery (given P, to find A).

PROGRAM
```
10 PRINT "GIVEN P, TO FIND A"
20 PRINT "                              N"
30 PRINT "                         i(1+i)"
40 PRINT "THE FORMULA IS A=P[----------------]"
50 PRINT "                              N"
60 PRINT "                         (1+i)  -1"
65 PRINT " "
70 INPUT "ENTER PRESENT VALUE P=$";P
75 INPUT "ENTER INTEREST RATE i=";i
80 INPUT "ENTER TOTAL COMPOUNDING PERIOD N=";N
85 LET A=P*(i*(1+i)**N)*((1+i)**N-1)**(-1)
90 PRINT " "
95 PRINT "THE UNIFORM ANNUAL WORTH IS A=$";A
99 END
```

EXAMPLE
```
GIVEN P, TO FIND A
                              N
                         i(1+i)
THE FORMULA IS A=P[----------------]
                              N
                         (1+i)  -1
ENTER PRESENT VALUE P=$ ? 1
ENTER INTEREST RATE i= ? .1
ENTER TOTAL COMPOUNDING PERIOD N= ? 10
THE UNIFORM ANNUAL WORTH IS A=$ 0.1627454
```

2.5 Uniform series compound amount (given A, to find F).

PROGRAM
```
10 PRINT "GIVEN A, TO FIND F"
15 PRINT "                          N"
20 PRINT "                    ((1+i)  -1)"
30 PRINT "THE FORMULA IS F=A[----------------]
40 PRINT "                        i"
50 PRINT " "
60 INPUT "ENTER UNIFORM ANNUAL WORTH A=$";A
65 INPUT "ENTER INTEREST RATE i=";i
70 INPUT "ENTER TOTAL COMPOUNDING PERIOD N=";N
75 LET F=A*((1+i)**N-1)*(i**(-1))
80 PRINT " "
85 PRINT "THE FUTURE WORTH IS F=$";F
90 END
```

EXAMPLE
```
GIVEN A, TO FIND F
                      N
                ((1+i)  -1)
THE FORMULA IS F=A[----------------]
                      i
ENTER UNIFORM ANNUAL WORTH A=$ ? 1
ENTER INTEREST RATE i= ? .1
ENTER TOTAL COMPOUNDING PERIOD N= ? 3
THE FUTURE WORTH IS F=$ 3.31
```

2.6 Sinking fund (given F, to find A).

PROGRAM
```
10 PRINT "GIVEN F, TO FIND A"
20 PRINT "                        i"
30 PRINT "THE FORMULA IS A=F[----------------]"
40 PRINT "                        N"
50 PRINT "                  ((1+i)  -1)"
60 PRINT " "
65 INPUT "ENTER FUTURE VALUE F=$";F
70 INPUT "ENTER INTEREST RATE i=";i
75 INPUT "ENTER TOTAL COMPOUNDING PERIOD N=";N
80 LET A=F*i*((1+i)**N-1)**(-1)
85 PRINT " "
90 PRINT "THE UNIFORM ANNUAL WORTH A=$";A
95 END
```

EXAMPLE
```
GIVEN F, TO FIND A
                      i
THE FORMULA IS A=F[----------------]
                      N
                ((1+i)  -1)
ENTER FUTURE VALUE F=$ ? 1
ENTER INTEREST RATE i= ? .1
ENTER TOTAL COMPOUNDING PERIOD N= ? 10
THE UNIFORM ANNUAL WORTH A=$ 0.0627454
```

2.7 Present worth of arithmetic gradient (given G, to find P).

PROGRAM
```
10 PRINT "GIVEN G, TO FIND P"
20 PRINT "                              N"
30 PRINT "                         {(1+i)   −iN−1}"
40 PRINT "THE FORMULA IS P=G[------------------------]"
50 PRINT "                          2   N"
55 PRINT "                         i (1+i)  "
60 PRINT " "
65 INPUT "ENTER ARITHMETIC GRADIENT G=$";G
70 INPUT "ENTER INTEREST RATE i=";i
75 INPUT "ENTER TOTAL COMPOUNDING PERIOD N=";N
80 LET P=G*((1+i)**N−i*N−1)*(i**2*(1+i)**N)**(−1)
85 PRINT " "
90 PRINT "THE PRESENT WORTH IS P=$";P
95 END
```

EXAMPLE
```
GIVEN G, TO FIND P
                        N
                   {(1+i)  −iN−1}
THE FORMULA IS P=G[--------------------]
                     2   N
                    i (1+i)
ENTER ARITHMETIC GRADIENT G=$ ? 1
ENTER INTEREST RATE i= ? .1
ENTER TOTAL COMPOUNDING PERIOD N= ? 10
THE PRESENT WORTH IS P=$ 22.89133
```

2.8 Arithmetic gradient to uniform series (given G, to find A).

PROGRAM
```
10 PRINT "GIVEN G, TO FIND A"
20 PRINT "                                 N"
30 PRINT "                            {(1+i)   −iN−1}"
40 PRINT "THE FORMULS IS A=G[----------------------------]"
50 PRINT "                             N"
55 PRINT "                            (i(1+i)   −i)"
60 PRINT " "
65 INPUT "ENTER ARITHMETIC GRADIENT G=$";G
70 INPUT "ENTER INTEREST RATE i=";i
75 INPUT "ENTER TOTAL COMPOUNDING PERIOD N=";N
80 LET A=G*((1+i)**N−i*N−1)*(i*(1+i)**N−i)**(−1)
85 PRINT " "
90 PRINT "THE UNIFORM ANNUAL WORTH IS A=$";A
95 END
```

EXAMPLE
```
GIVEN G, TO FIND A
                        N
                   {(1+i)  −iN−1}
THE FORMULA IS A=G[-----------------------]
                        N
                   (i(1+i)  −i)
ENTER ARITHMETIC GRADIENT G=$ ? 1
ENTER INTEREST RATE i= ? .1
ENTER TOTAL COMPOUNDING PERIOD N= ? 10
THE UNIFORM ANNUAL WORTH IS A=$ 3.725453
```

2.9 Nominal to effective interest rate.

PROGRAM
```
10 PRINT "GIVEN NOMINAL INTEREST RATE r, TO FIND EFFECTIVE INTEREST RATE i"
20 PRINT "                          M"
30 PRINT "THIS FORMULA IS i=(1+r/M)   −1"
40 PRINT " "
50 INPUT "ENTER NOMINAL INTEREST RATE r=";r
60 INPUT "ENTER TOTAL COMPOUNDING PERIODS PER YEAR M=";M
70 LET i=(1+r*M**(−1))**M−1
80 PRINT " "
90 PRINT "THE EFFECTIVE INTEREST RATE IS i=";i
95 END
```

EXAMPLE
```
GIVEN NOMINAL INTEREST r, TO FIND EFFECTIVE INTEREST RATE i
                          M
THIS FORMULA IS i=(1+r/M)   −1
ENTER NOMINAL INTEREST RATE r= ? .2
ENTER TOTAL COMPOUNDING PERIODS PER YEAR M= ? 4
THE EFFECTIVE INTEREST RATE IS i= 0.2155062
```

2.10 Effective to nominal interest rate, continuous compounding.

PROGRAM
```
10 PRINT "UNDER CONTINUOUS COMPOUNDING, GIVEN EFFECTIVE INTEREST RATE TO
      FIND NOMINAL INTEREST RATE"
20 PRINT "THE FORMULA IS r=LOG(1+i)"
30 PRINT " "
40 INPUT "ENTER EFFECTIVE INTEREST RATE i=";i
50 LET r=LOG(1+i)
55 PRINT " "
60 PRINT "THE NOMINAL INTEREST RATE IS r=";r
70 END
```

EXAMPLE
```
UNDER CONTINUOUS COMPOUNDING, GIVEN EFFECTIVE INTEREST RATE, TO FIND
NOMINAL INTEREST RATE
THE FORMULA IS r=LOG(1+i)
ENTER EFFECTIVE INTEREST RATE i= ? .1
THE NOMINAL INTEREST RATE IS r= 0.09531017
```

2.11 Nominal to effective rate, continuous compounding.

PROGRAM
```
10 PRINT "UNDER CONTINUOUS COMPOUNDING, GIVEN NOMINAL INTEREST RATE r,
      TO FIND EFFECTIVE INTEREST RATE i"
20 PRINT "                          r"
30 PRINT "THE FORMULA IS i=e  −1"
40 PRINT " "
50 INPUT "ENTER NOMINAL INTEREST RATE r=";r
60 LET i=EXP(r)−1
70 PRINT " "
80 PRINT "THE EFFECTIVE INTEREST RATE i=";i
90 END
```

EXAMPLE
```
UNDER CONTINUOUS COMPOUNDING, GIVEN NOMINAL INTEREST RATE r, TO FIND
   EFFECTIVE INTEREST RATE i
                r
THE FORMULA IS i=e −1
ENTER NOMINAL INTEREST RATE r= ? .1
THE EFFECTIVE INTEREST RATE r= 0.1051709
```

2.12 Present worth of single-period continuous cash flow $(\overline{A}_N)$.

PROGRAM
```
10 PRINT "GIVEN CONTINUOUS CASH FLOW, FIND THE PRESENT WORTH"
20 PRINT "                              i              -N"
30 PRINT "THIS FORMULA IS P=Ā[--------------------](1+i)"
40 PRINT "                         LOG(1+i)"
50 PRINT " "
60 INPUT "ENTER CONTINUOUS CASH FLOW A-BAR  = $";A
70 INPUT "ENTER INTEREST RATE i = ";i
80 INPUT "ENTER TOTAL COMPOUNDING PERIOD N = ";N
85 LET P=A*i*LOG(1+i)**(−1)*(1+i)**(−N)
90 PRINT " "
95 PRINT "THE PRESENT WORTH IS P=$";P
99 END
```

EXAMPLE
```
GIVEN CONTINUOUS CASH FLOW, FIND THE PRESENT WORTH
                      i             −N
THIS FORMULA IS P=Ā[--------------------](1+i)
                   LOG(1+i)
ENTER CONTINUOUS CASH FLOW A-BAR  = $ ? 100
ENTER INTEREST RATE i = ? .1
ENTER TOTAL COMPOUNDING PERIOD N = ? 10
THE PRESENT WORTH IS P=$ 40.45143
```

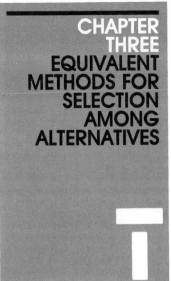

CHAPTER
THREE
EQUIVALENT
METHODS FOR
SELECTION
AMONG
ALTERNATIVES

The concept of equivalence was stressed throughout the preceding chapter: A sum of money at one point in time can be equivalent to another amount at some other point in time, given (1) the number of intervening compounding interest periods and (2) an interest rate that represents alternative investment opportunities during that time interval. (The interest rate is a measure of the minimum return that could be received if the available funds were invested elsewhere. The procedure for estimating this rate, the **minimum attractive rate of return** (*MARR*), is beyond the scope of this chapter; it is discussed in some detail in Chapter 11. Here, the *MARR* is considered as given.)

It is generally incorrect to compare two or more investment alternatives simply by summing their respective cash flows. The *timing,* as well as the *amounts,* of project cash flows must be considered. To illustrate this point, consider the cash flows associated with two alternative investment opportunities:

	Plan I	Plan II
Initial investment	− $100	− $100
Receipts at end of one year	50	100
Receipts at end of two years	100	50
Excess of receipts over disbursements	$50	$50

In this simple example it is clear that the decision maker would not be indifferent about the choice between the two alternatives merely because the cash flow totals are equal. Indeed, assuming no other significant differences between alternatives, Plan II is preferable to Plan I. Of course, "real world" problems are substantially more complicated, and the relative desirability of alternative investments is not, in most cases, evident by inspection. The computation of equivalent values, using the mathematical techniques discussed in Chapter 2, must play a central role in any economic analysis.

This chapter discusses four alternative methods for systematically comparing costs and benefits:

1. the annual worth method
2. the present worth method
3. the rate of return method
4. the benefit-cost ratio method

As you will see, all of these methods, properly used, lead to consistent results. That is, the rank-ordering of competing investment alternatives is perfectly consistent, regardless of which method the analyst uses. The primary reason that all four are presented in this book is simply that they are widely used, in a variety of applications, by economic analysts. (The first three are most common to the private sector; the benefit-cost ratio method, because of its historical antecedents, is most widely used in the public sector, especially by government agencies related to water resources.) All of these methods are commonly referenced in the professional literature, and no one method is used predominantly. Thus it is useful for the analyst to understand thoroughly the proper application of all four procedures.

3.1
THE ANNUAL WORTH METHOD

The procedure discussed here is generally known in the literature of economic analysis as the **annual worth method.** The essence of the annual worth method is the conversion of all cash flows associated with a project to a single figure, namely, the equivalent uniform annual worth (AW). Given any sequence of cash flows, A_j, as indicated in Figure 3.1, an equivalent uniform series, A, is determined using the appropriate compound interest calculations. If all cash flows are end-of-period, then

$$AW = (A/P, i, N) \sum_{j=0}^{N} A_j (1 + i)^{-j} \qquad (3.1)$$

where A_j is the cash flow at the end of period j, and discounting takes place at effective interest rate i per period. There are various other models, of course, to accommodate other assumptions about the timing of cash flows and the type of discounting. For example, if all cash flows are continuous, as indicated in Figure 3.2, and

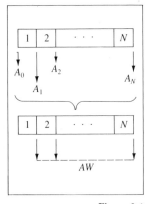

Figure 3.1

discounting takes place continuously at rate i, then

$$AW = (A/P, i, N) \sum_{j=1}^{N} \overline{A}_j \left[\frac{i(1 + i)^{-j}}{\ln(1 + i)} \right] \qquad \textbf{(3.2)}$$

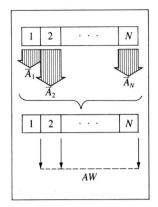

Figure 3.2

where $\overline{A}_j$ is the cash flowing continuously and uniformly during period j.

As used in this context, **worth** is the value of an investment opportunity expressed as an equivalent end-of-period series. The choice of a *uniform* series as a time reference has considerable appeal. Many costs (negative cash flows) and revenues (positive cash flows), such as rental costs of $500 per month, interest income of $200 per quarter, and property taxes of $2,000 per year, occur on a periodic basis. Moreover, even when cash flows do not, or will not, follow a uniform series, it is frequently convenient to refer to them as though they do or will. Budget planning and control, profit determination, and accounting for operations are examples of activities for which forecasts are generally stated in the form of uniform periodic results.

The annual worth method consists of two steps. First, compute the appropriate annual worth, AW, for each alternative. Of course, the same interest rate must be used to evaluate all alternatives. Second, rank the alternatives in decreasing order of AW's. Assuming no other differences among alternatives, our objective is to maximize AW.

The annual worth method is often cited in the literature of engineering economy as the **equivalent uniform annual cost** (*EUAC*) method. *EUAC* is simply the negative of AW, and the objective is to minimize *EUAC*.

Note that, although this is known as the *annual* worth or *annual* cost method, the time interval between equivalent values need not be in years. Days, weeks, months, and so on, do just as well. Thus the term *periodic worth* is more accurate. But we seem to be stuck with the somewhat misleading term *annual worth* because of historical precedent and common usage.

A Numerical Illustration

Consider a proposed equipment purchase requiring an initial investment of $12,000. It is estimated that the equipment will require operating expenses of approximately $6,000 at the end of each and every year for six years and that the residual value of the equipment at the end of six years will be $3,000. The relevant interest rate is 12 percent per year. The resulting cash flow diagram is shown in Figure 3.3.

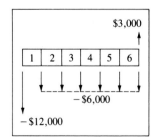

Figure 3.3

Cash flow diagrams identify the amounts and timing of cash flows. By convention, downward arrows indicate negative cash flows, such as the initial cost of an asset and periodic operating costs. Upward arrows indicate positive cash flows, such as sales revenues and positive salvage values.

From Equation 3.1,

$$AC = \$12,000(A/P, 12\%, 6) + \$6,000 - \$3,000(A/F, 12\%, 6)$$
$$= \$12,000(0.24323) + \$6,000 - \$3,000(0.12323)$$
$$= \$8,549$$

Note that, in the absence of any other information, the end-of-period convention is used here, and unless otherwise indicated, it is used throughout this book. Further, annual cost (AC) is an abbreviation for equivalent uniform annual cost $(EUAC)$, since all cash flows have been converted to an equivalent uniform series.

The Significance of Annual Worth

In the previous example, the original set of eight cash flows—an expenditure of $12,000 now and $6,000 at the end of each year for six years, with a receipt of $3,000 at the end of six years—is equivalent to a uniform series of $8,549 occurring at the end of each and every year for six years. Of course, this equivalence is valid only for an interest rate of 12 percent per year. Put somewhat differently, imagine a fund—a bank account, for example—that earns 12 percent per year, using two alternative strategies. The first consists of an investment of $12,000 into the fund now, $6,000 into the fund at the end of each year for six years, and a withdrawal of $3,000 from the fund at the end of the sixth year. The second strategy consists of investments of exactly $8,549 into the fund at the end of each year for six years. Under both strategies, the amount remaining in the fund at the end of six years is identical.[1]

This method results in the substitution of a single figure, AC or AW, equivalent to a larger number of cash flows. Alternatives can now be readily compared using this figure. Suppose, for example, that the proposed equipment purchase outlined previously is intended to replace a manual operation now costing $8,000 per year. How would management know which alternative is economically preferable in the absence of an equivalent uniform annual cost computation? Note that the *average* annual cost per year for the proposed equipment is [$12,000 + 6($6,000) − $3,000]/6 = $7,500. Thus the naive analysis, disregarding the effects of interest, would lead to choosing the new equipment. But the true equivalent uniform annual cost, $8,549, is more ex-

---

1. You may verify that the "terminal value" of the fund is $69,374 in each case.

pensive than the annual cost of the current procedure—$8,000. Thus the equivalent uniform annual cost may be thought of as a *weighted average;* weighted in the sense that the interest cost is included. This equivalent cost of interest is $8,549 − $7,500 = $1,049 annually.

The Cost of Capital Recovery

The cost of capital recovery, generally referred to simply as **capital recovery** (*CR*), is defined as the uniform series equivalent of the original cost of an asset less its salvage value, if any, as indicated in Figure 3.4. Or

$$CR = P(A/P, i, N) - S(A/F, i, N) \tag{3.3}$$

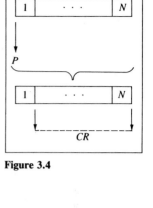

Figure 3.4

where *P* is the initial cost and *S* is the net salvage value after *N* periods. The equivalent uniform annual cost, *AC,* is based on an effective interest rate of *i* per period.

In our previous equipment-purchase example, the cost of capital recovery is

$$
\begin{aligned}
CR &= \$12,000(A/P, 12\%, 6) - \$3,000(A/F, 12\%, 6) \\
&= \$12,000(0.24323) - \$3,000(0.12323) \\
&= \$2,549
\end{aligned}
$$

That is, the weighted average cost per year (capital recovery) is $2,549. But this is not equal to the average cost per year, ($12,000 − $3,000)/6 = $1,500. The difference, $2,549 − $1,500 = $1,049 per year, represents the dollar cost of opportunities forgone. When money has value over time, that is, when $i > 0\%$, it is always true that the cost of capital recovery is greater than the average cost per period. Or

$$CR > \frac{P - S}{N} \quad \text{when} \quad i > 0\%.$$

In Equation 3.3, two factors are used to compute capital recovery, *A/P* and *A/F*. It is possible, however, to use a simplified form by observing that capital recovery consists of two elements: (1) one portion, $P - S$, is "consumed" and must therefore be recovered, and (2) the second portion, *S,* is costly only in the sense that interest on this portion, *iS,* is forgone during the *N* periods when *S* is invested in the asset. (See Figure 3.5.)

$$
\begin{aligned}
CR &= (\text{capital consumed}) + (\text{interest forgone}) \\
&= (P - S)(A/P, i, N) + iS \tag{3.4}
\end{aligned}
$$

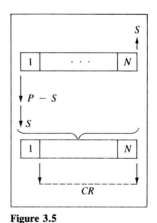

Figure 3.5

Using Equation 3.4 and the data from the previous example,

$$CR = (\$12,000 - \$3,000)(A/P, 12\%, 6) + \$3,000(0.12)$$
$$= \$9,000(0.24323) + \$360$$
$$= \$2,549$$

Equations 3.3 and 3.4 always yield identical results. The latter, however, is somewhat easier to use.

The term $(A/P, i, N)$ is known in the literature of engineering economy as the **capital recovery factor.** The reason, of course, is that it serves an important role in computing capital recovery.

Still another equation used to determine capital recovery may be derived as follows:

$$CR = (P - S)(A/P, i, N) + Si$$

$$= (P - S)\left[\frac{i(1 + i)^N}{(1 + i)^N - 1}\right] + Si$$

$$= \frac{Pi(1 + i)^N - Si(1 + i)^N + Si(1 + i)^N - Si + Pi - Pi}{(1 + i)^N - 1}$$

$$= \frac{Pi - Si}{(1 + i)^N - 1} + \frac{Pi(1 + i)^N - Pi}{(1 + i)^N - 1}$$

$$= (P - S)\left[\frac{i}{(1 + i)^N - 1}\right] + Pi$$

$$= (P - S)(A/F, i, N) + Pi \tag{3.5}$$

This alternative formula is occasionally employed in the public utility industry, although its use is otherwise extremely limited. When capital recovery is computed in this manner, it is generally described as **sinking fund depreciation plus interest on first cost.**

Alternatives with Unequal Lives

One of the most troublesome complications arises when the service lives, or economic lives, of the various alternatives are unequal. It is essential that any specific capital allocation study be made within the context of a uniform planning horizon. Let's examine this issue using a relatively simple numerical example. Consider three mutually exclusive alternatives:

	Alternative X	Alternative Y	Alternative Z
Initial cost	$10,000	$12,000	$20,000
Service life (years)	6	6	12
Salvage value at end of service life	$0	$3,000	$2,000
Annual operating costs	$1,500	$1,600	$900

With the given service lives, it may be shown that the equivalent uniform annual costs for the three alternatives are

$$AC(X) = \$3,932$$
$$AC(Y) = \$4,149$$
$$AC(Z) = \$4,046$$

Now, assuming that only one of these three alternatives can be accepted, what is the appropriate planning horizon, and, given this, how should the analysis be conducted?

These questions may be answered first by assuming an estimated duration of need, which the alternatives are expected to fulfill. For example, it may be nine years, twenty years, a very long time (approximately infinity), and so on. Then, given the duration of the need, it is necessary to make some reasonable assumptions about the consequences of replacement assets.

Suppose that there is reason to believe that the need will continue indefinitely, that is, that the planning horizon approaches infinity. Assume also that replacements have the same first costs, lives, salvage values, and annual operating costs as the current alternatives. These assumptions lead to the series of cash flows outlined in Table 3.1. Under these assumptions it is clear that the total annual costs as previously determined remain applicable and, moreover, the same values are applicable if the planning horizon is any number of periods that is a common multiple of the lives of each of the alternatives.

Although these assumptions allow us to use the original solutions for annual cost without modification, one should resist the temptation to use them, explicitly or implicitly, without carefully considering likely future consequences. If annual costs are compared without considering differences in project lives, the solutions imply identical repetition and a planning horizon equal to a common multiple of project lives. These assumptions should always be questioned in real situations.

Table 3.1
Cash Flows for Alternatives X, Y, and Z,
Assuming Infinite Planning Horizon and Identical Replacements

End of year	Alternative X		Alternative Y		Alternative Z	
0	$-10,000$		$-12,000$		$-20,000$	
1–6		$-1,500$		$-1,600$		-900
6	$-10,000$		$-12,000$	$+3,000$	—	
7–12		$-1,500$		$-1,600$		-900
12	$-10,000$		$-12,000$	$+3,000$	$-20,000$	$+2,000$
13–18		$-1,500$		$-1,600$		-900
18	$-10,000$		$-12,000$	$+3,000$	—	
19–24		$-1,500$		$-1,600$		-900
24				$+3,000$		$+2,000$
		⋮		⋮		⋮

The issue of unequal lives is explored from a somewhat different perspective as the present worth methods is discussed in the following section.

The **present worth method** is probably the most widely used economic evaluation technique. (It is also known in the literature of finance and engineering economy as the *present value* or *discounted cash flow* method.) As illustrated in Figure 3.6, the essential feature of the present-worth method is the *discounting to present value* of all cash flows expected to result from an investment decision. That is, in order to satisfy the basic requirement that alternatives be compared only if money consequences are measured at a common point in time, the "present date" is arbitrarily selected as the point of reference. (In practice, the "present date" is determined relative to the particular problem at hand. It is generally defined as the time at which the project life begins.) The net discounted value of all prospective cash flows is a direct measure of the relative economic attractiveness of the proposed investment.

As with the annual worth method, there are two steps. First, the equivalent present worth, *PW*, of all expected cash flows are computed and aggregated for each alternative. For example, if all cash flows, A_j, are discrete, as shown in Figure 3.6, and the discounting takes place at effective rate i per period, then

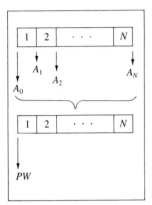

Figure 3.6

$$PW = \sum_{j=0}^{N} A_j(1 + i)^{-j} \tag{3.6}$$

If all cash flows are continuous, as indicated in Figure 3.7, and if discounting takes place continuously at rate i, then

$$PW = \sum_{j=1}^{N} \overline{A}_j\left[\frac{i(1 + i)^{-j}}{\ln(1 + i)}\right] \tag{3.7}$$

where $\overline{A}_j$ is the cash flowing continuously and uniformly during period j. Various other models are used, depending on the assumptions related to the timing of cash flows and the type of discounting.

The second step in the present worth method is to rank-order alternatives on the basis of decreasing *PW*'s. That is, assuming that there are no other differences between alternatives, the objective is to maximize present worth.

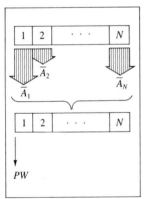

Figure 3.7

Present Worth of the Do-Nothing Alternative
If funds are not invested in a particular proposed project—if the investor chooses to do nothing—it is reasonable to assume that the funds will be invested elsewhere, where they will earn exactly i

per period. (Recall that i is a measure of the minimum return that could be received if the available funds were invested elsewhere. This is the "opportunity cost," stated as a percentage.)

When discounted, the *net* return of funds invested elsewhere is exactly zero. This may be easily demonstrated by observing that an amount P invested at rate i will be worth $P(1 + i)$ at the end of one period, $P(1 + i)^2$ at the end of two periods, . . . , and $P(1 + i)^N$ at the end of N periods. Thus $A_0 = -P$ and $A_N = P(1 + i)^N$. From Equation 3.6,

$$PW = A_0 + A_N(1 + i)^{-N}$$
$$= -P + P(1 + i)^N(1 + i)^{-N}$$
$$= 0$$

Thus, when comparing a single investment proposal with the do-nothing alternative, the investment should be undertaken if its present worth is positive. This indicates that the proposed investment is preferrable to investing the same funds elsewhere at the minimum attractive rate of return, i.

A Numerical Illustration

Consider the equipment purchase example presented in Section 3.1, in which a proposal is under consideration having an initial investment of $12,000, annual expenses of $6,000 occurring at the end of each year over six years, and a residual (salvage) value of $3,000 at the end of six years. (See Figure 3.3.) In this example,

$$A_0 = -\$12,000$$
$$A_1 = A_2 = \cdots = A_5 = -\$6,000$$
$$A_6 = -\$6,000 + \$3,000$$

The discount rate is 12 percent per year.

Note that the six $6,000 annual expenses form a uniform series. Since we have access to a set of present worth factors for the uniform series, the computation can be simplified. Using the functional format, the solution equation can be written

$$PW = -\$12,000 - \$6,000(P/A, 12\%, 6)$$
$$+ \$3,000(P/F, 12\%, 6)$$
$$= -\$12,000 - \$6,000(4.111) + \$3,000(0.5066)$$
$$= -\$35,146$$

The negative net present worth indicates that the cost of this project, stated in equivalent current dollars, is $35,146.

At zero interest rate, that is, if money has *no* value over time, the net present worth is

$$-\$12,000 - 6(\$6,000) + \$3,000 = -\$45,000$$

This is a somewhat larger negative value, $-\$45,000 - (-\$35,146) = -\$9,854$, because the present worth of future consequences is inversely related to the value of the discount rate. In this case, the six future negative cash flows, when discounted at 12 percent, have the value $-\$6,000(4.111) = -\$24,666$, not $-\$36,000$. The discounted value of the salvage value is $\$3,000(0.5066) = \$1,520$. The net difference is

$$-\$36,000 - (-\$24,666) + \$3,000 - \$1,520 = -\$9,854$$

Equivalence of the Annual Worth and Present Worth Methods

It may be interesting to review the equivalent uniform annual cost for this project as determined in Section 3.1. We determined that $AC = \$8,549$. The present worth of each of six end-of-period payments of $-\$8,549$ is:

$$
\begin{aligned}
PW &= -\$8,549(P/A,\ 12\%,\ 6) \\
&= -\$8,549(4.111) \\
&= -\$35,145
\end{aligned}
$$

This is, of course, the same as the net present worth, as determined previously. (The $1 difference is due to rounding.)

The annual worth and present worth methods are directly related. Given a discount rate and a sequence of cash flows occurring over N periods, ranking on the basis of increasing AC (or decreasing AW) is equivalent to ranking on the basis of decreasing PW. This is so because the alternatives' PW's are related to their AW's by the factor $(P/A,\ i,\ N)$, which is constant for given values of i and N. In a given problem situation, the interest rate and the planning horizon are common to all alternatives. So given any pair of alternatives (say, I and II), if $AW(\text{I}) > AW(\text{II})$, then $PW(\text{I}) > PW(\text{II})$, since $PW(\text{I}) = AW(\text{I}) \times (P/A,\ i,\ N)$ and $PW(\text{II}) = AW(\text{II}) \times (P/A,\ i,\ N)$. Thus, maximizing present worth is equivalent to maximizing annual worth or minimizing annual cost.

Capitalized Cost

When the life of a proposed investment is perpetual and/or the planning horizon is extremely long (approaches infinity), it may be necessary to determine the present value of an infinite series of cash flows. In the literature of engineering economy, this is known as the **capitalized cost,** or CC. It represents the amount of money that must be invested today to yield a return of A at the end of each and every period forever, assuming rate of return i per period. (See Figure 3.8.)

Observe the limit of the uniform series present worth factor as N approaches infinity:

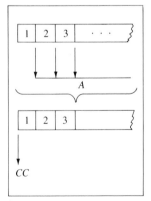

Figure 3.8

$$\lim_{N \to \infty}(P/A, i, N) = \lim_{N \to \infty}\left[\frac{(1 + i)^N - 1}{i(1 + i)^N}\right]$$

$$= \frac{1}{i} \qquad\qquad (3.8)$$

Thus it follows that

$$CC = A(P/A, i, N \to \infty)$$
$$= A(1/i) \qquad\qquad (3.9)$$

To illustrate the use of Equation 3.9, consider a proposed investment of $30,000 now that will require additional costs of $600 at the end of each and every year, perpetually. The discount rate is 12 percent per year. Therefore, the capitalized cost is

$$CC = \$600(1/0.12)$$
$$= \$5,000$$

and the total present worth of costs is

$$\$30,000 + \$5,000 = \$35,000$$

Using Cash Flow Tables to Determine Present Worth

The numerical example on page 51 was solved using Equation 3.6, taking advantage of the fact that the uniform series of cash flows, $6,000 per year for six years, permitted the use of the uniform series factor, P/A. The solution, then, was in the form of an equation. But many analysts prefer to solve for present worth while displaying the data and calculations in a tabular format, a **cash flow table,** especially when cash flow amounts are irregular.

A cash flow table for the example problem is shown in Table 3.2.

Table 3.2
Cash Flow Table for Sample Problem

Period (j)	Cash flows End of period (A_j)	Cash flows During period ($\bar{A}_j$)	Discount factor End of period (P/F, 12%, j)	Discount factor During period ($P/\bar{F}$, 12%, j)	Equivalent present worth (P_j)
0	− $12,000	N.A.	1.0000	N.A.	− $12,000
1	− 6,000		0.8929		− 5,357
2	− 6,000		0.7972		− 4,783
3	− 6,000		0.7118		− 4,271
4	− 6,000		0.6355		− 3,813
5	− 6,000		0.5674		− 3,404
6	− 3,000		0.5066		− 1,520
				Net Present Worth	− $35,148

In general, the table is constructed as follows:

1. Each row in the table represents an interest period, j; $j =$ 0, 1, 2, . . . , N. The period may be a day, a week, a month, a quarter, a year, or any other time period, depending on the characteristics of a particular problem. In most cases, it is convenient to use a year as the interest period. If the table is preprinted in the form of a general worksheet, N should be sufficiently large to accommodate most expected applications.

2. The table contains space for cash flows occurring either at the end of the period or during the period.

3. "End-of-period" and "during period" discount factors are also provided in the table. For a given interest rate, i, these factors may be preprinted.

4. The last column in the table represents either $A_j(P/A, i, j)$ or $\overline{A}_j(P/\overline{A}, i, j)$. This is the equivalent present value of the cash flow occurring in period j.

5. The sum of the products in the last column is the net present worth (PW) for the project.

The table format has several obvious advantages over the equation solution. First, all cash flows are clearly portrayed, period by period. These values may be interesting to managers, especially insofar as cash budgets may be affected. Second, tables may be preprinted as worksheets, thus ensuring uniformity in application from one analysis, or analyst, to another. Third, the internal logic of the analysis is clarified, making it simpler to explain the process to others, such as managers, auditors, and so forth. Finally, this format can be easily accommodated by electronic data processing equipment. With proper programming, the anticipated cash flows can be given as input, the discount factors can either be stored in memory or computed for given values of i and j, and the computer output can be printed in the cash flow table format, as shown in the computer program at the end of this chapter.

Alternatives with Unequal Lives

Failure to consider equal planning horizons for various alternatives leads to incorrect, and frequently absurd, solutions. In the annual worth method, this error may not be obvious, since the resulting solutions *imply* identical repetition of proposals. This is not the case with the present worth method, however. Here the implication is that there are no consequences (that is, no cash flows) after disposal of the shorter-lived asset.

To illustrate this point, consider two alternative investment proposals, L and S, with the following expected cash flows:

End of period	Alternative L	Alternative S
0	− $100	− $120
1	30	50
2	30	50
3	30	38
4	30	
Total cash flow	$ 20	$ 18

Assume that the interest rate is zero. (This is an unrealistic assumption, of course, but it is useful in the context of this discussion. A nonzero interest rate merely complicates the arithmetic; the underlying lesson remains unaffected).

Which alternative, L or S, is preferable? Since the present worth of L exceeds that of S—$PW(L) = \$20$ and $PW(S) = \$18$—it would appear that L is preferable to S. But the equivalent annual worth of L is *less than* that of S—$AW(L) = \$20/4 = \$5$ and $AW(S) = \$18/3 = \$6$. Thus it would appear that S is preferable to L. Which, then, is the correct choice?

The answer lies in the fact that the two alternatives have unequal lives. L generates a net of $20 over four periods, whereas S generates a net of $18 over only three periods. If we are to choose fairly between the alternatives, then *the consequences of all alternatives must be evaluated over a common time interval, or planning horizon*—a fundamental principle of analysis. Thus the relevant question is: What would happen at the end of the third period if alternative S were selected? Would a replacement be purchased? If so, what would the cash flows associated with the replacement be? And are the total time periods associated with S and its replacement(s) now equal to the total time periods associated with L and its replacement(s)?

Let's suppose that there is a continuing need for both L and S over the next twelve periods. Suppose also that the replacements for L and S will have the same lives and cash flows as the projects they replace. *Under these assumptions,* it is clear that the common planning horizon is the least common multiple of 4 and 3, or $4 \times 3 = 12$ periods. Over this twelve-period planning horizon, both the annual worth and present worth methods now yield consistent results:

	Alternative L	Alternative S
PW over 12 periods	$3 \times \$20 = \$60$	$4 \times \$18 = \$72$
AW over 12 periods	$\$60/12 = \$5$	$\$72/12 = \$6$

Note that the ranking of the alternatives, by either the present worth or annual worth method, is exactly the same as the ranking

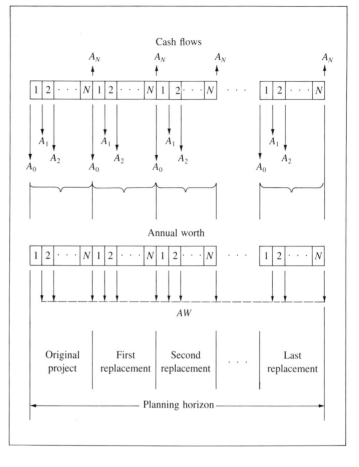

that results from the *AW* method for the original problem. This is
so because of the special character of the replacement assumption.
Given a sequence of cash flows over *N* periods, the equivalent
annual worth remains constant if and only if the identical set of
cash flows is replicated over 2*N* periods, 3*N* periods, . . . , and so
on. This replication is shown in Figure 3.9.

It is tempting to conclude that the problem of unequal project
lives is easily solved if the analyst uses the annual worth method
for all alternatives. But this is true only if the "identical replication" assumption holds. Otherwise, a complete evaluation
should be made over the common planning horizon.

This point can be illustrated using our previous example. But
let's vary the original problem by assuming that replacements for
Project L will increase by $12 every four periods and replacements
for Project S will increase by $11 every three periods. In this case,
the cash flows for the two alternatives over the common planning
horizon are:

End of period	Alternative L		Alternative S	
0	− $100		− $120	
1	30		50	
2	30		50	
3	30		38	− $131
4	30	− $112		50
5		30		50
6		30	− 142	38
7		30	50	
8	− 124	30	50	
9	30		38	− 153
10	30			50
11	30			50
12	30			38
Total cash flow	$24		$6	
AW over 12 periods	$2		$0.5	

Clearly, both the *AW* and *PW* methods indicate that Alternative L is now the preferable one. This is *not* the same solution that results when considering only the annual worth of the original investments:

AW(L) over 4 periods = $5
AW(S) over 3 periods = $6

Again, the analyst should use the "identical replication" assumption with great care. It is a favorite device of textbook authors who need to resolve the problem of unequal lives to show numerical examples and to solve end-of-chapter problems. In the real world, however, analysts should attempt to determine what cash flows, if any, are expected at the end of the life of the shorter-lived alternative.

Now let's take another look at the issue using Alternatives Y and Z from page 54. Recall that

	Alternative Y	Alternative Z
Initial cost	$12,000	$20,000
Service life (years)	6	12
Salvage value at end of service life	$3,000	$2,000
Annual operating costs	$1,600	$900

These data are reflected in Figure 3.10.

Incorrect values for the present worths of the two proposals are

$$PW(Y) = -\$12,000 - \$1,600(P/A, \ 12\%, \ 6)$$
$$+ \$3,000(P/F, \ 12\%, \ 6)$$
$$= -\$12,000 - \$1,600(4.111) + \$3,000(0.5066)$$
$$= -\$17,058$$

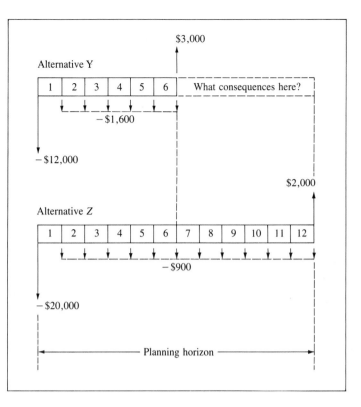

Figure 3.10 Cash Flow Diagrams for Two Alternatives Having Unequal Lives

$$PW(Z) = -\$20,000 - \$900(P/A, 12\% \; 12)$$
$$+ \; \$2,000(P/F, 12\%, 12)$$
$$= -\$20,000 - \$900(6.194) + \$2,000(0.2567)$$
$$= -\$25,061$$

These results indicate that Alternative Y is less costly than Alternative Z, a conclusion opposite to the one obtained when using the annual worth method. The error occurs from failure to account for the difference in lives of the proposals.

Valid solution of this problem requires that all consequences of alternatives be accounted for over the common planning horizon. Assume for example, that Alternative Y may be replaced at the end of six years with an identical replacement. The resulting cash flows for a planning horizon of twelve years are shown in Table 3.3. Based on these assumptions, the appropriate present worth of proposal Y and its successor is

$$PW(Y) = -\$12,000 - \$1,600(P/A, 12\%, 12)$$
$$- \; (\$12,000 - \$3,000)(P/F, 12\%, 6)$$
$$+ \; \$3,000(P/F, 12\%, 12)$$
$$= -\$12,000 - \$1,600(6.194) - \$9,000(0.5066)$$
$$+ \; \$3,000(0.2567)$$
$$= -\$25,700$$

Table 3.3
Cash Flow Table for Alternatives Having Unequal Lives

Assumptions: Requirement for assets at least twelve years			
Identical replacement of shorter-lived asset			

	Alternative Y		Alternative Z
End of year	Original	Replacement	
0	− $12,000		− $20,000
1–6	− 1,600		− 900
6	+ 3,000	− $12,000	
7–12		− 1,600	− 900
12		+ 3,000	+ 2,000

The net present worth of Alternative Z (for a twelve-year planning horizon) is only − $25,061, so it appears that Z is less costly than Y. This conclusion is now consistent with the one obtained ·by the annual worth method when the same assumptions were used.[2] This solution follows directly from our assumption about replacement.

Let's revise this example somewhat. Suppose the operational requirements are such that, if Alternative Y is selected, it will be retained for exactly six years and *not replaced*. Alternative Z, if selected, will be retained for twelve years, sold, and not replaced. The appropriate planning horizon in this case is clearly twelve years, the life of the longer-lived alternative. With no replacement for Alternative Y after six years,

$$PW(Y) \text{ over 12 years} = (PW \text{ of 1st 6 years})$$
$$+ (PW \text{ of 2nd 6 years})$$
$$= -\$17,058 + \$0 = -\$17,058$$
$$PW(Z) \text{ over 12 years} = -\$25,061$$

The replacement assumption has been explicitly considered in this case.

The **rate of return** of a given investment proposal is the interest rate at which the present worth of all cash flows resulting from the investment are equal to zero. Mathematically, it is that value of $i*$ that satisfies the equation

$$PW(i*) = \sum_{j=0}^{N} A_j(1 + i*)^{-j} \equiv 0 \qquad (3.10)$$

3.3
THE RATE OF RETURN METHOD

---

2. *Note:* $AC(Y) = \$25,700(A/P, 12\%, 12) = \$4,149$
   $AC(Z) = \$25,061(A/P, 12\%, 12) = \$4,046$
Compare with the solutions on page 56.

assuming end-of-period cash flows, A_j, and end-of-period discounting/compounding, or

$$PW(i^*) = \sum_{j=1}^{N} \overline{A}_j \left[\frac{i^*(1 + i^*)^{-j}}{\ln(1 + i^*)} \right] \equiv 0 \qquad (3.11)$$

assuming cash flows, $\overline{A}_j$, flowing continuously and uniformly during period j, with continuous discounting/compounding.

An alternative definition of rate of return is that it is the particular interest rate, i^*, for which the present worth of all future cash flows is exactly equal to the initial investment, P. With the assumptions underlying Equation 3.10, for example, with $P = -A_0$, we have

$$PW(i^*) = A_0 + \sum_{j=1}^{N} A_j(1 + i)^{-j} \equiv 0$$

or

$$P = \sum_{j=1}^{N} A_j(1 + i^*)^{-j} \qquad (3.12)$$

Still another definition of rate of return is: that interest rate, i^*, for which the present worth of all receipts (positive cash flows) is exactly equal to the present worth of all disbursements (negative cash flows). Note here that $A_j = R_j - C_j$, where R_j and C_j are the receipts and disbursements, respectively, incurred in period j. Again, using the assumptions underlying Equation 3.10, we have

$$PW(i^*) = \sum_{j=0}^{N} (R_j - C_j)(1 + i^*)^{-j} \equiv 0$$

or

$$\sum_{j=0}^{N} R_j(1 + i^*)^{-j} = \sum_{j=0}^{N} C_j(1 + i^*)^{-j} \qquad (3.13)$$

When rate of return is computed in this manner, the amount as well as the timing of expected costs and revenues are taken into consideration. It is, in a sense, a measure of the profitability of the project. Thus it is an *internal* rate of return. It arises solely from the amounts and timing of the cash flows associated with the investment; there is no relationship to any external factors. It is *not* a rate of return on the *initial* investment.[3] The terms *internal rate of return (IRR)* and *rate of return (RoR)*, are used interchangeably

3. For further discussion of the properties of the (internal) rate of return, see Lynn E. Bussey, *The Economic Analysis of Industrial Projects* (Prentice-Hall, 1978), 213–217. Given a cash flow sequence in which the initial cash flow is negative and all others are positive, Bussey demonstrates that the internal rate of return is the rate of interest earned on the time-varying, unrecovered balances of an investment such that the final balance is zero at the end of the project life.

in the literature of finance and engineering economy. Both terms are used in this book.

The first step in the rate of return method is to determine the internal rate of return for the proposed investment. Next, in order for the investment to be acceptable, the rate of return must be compared to some minimum attractive rate that could be earned if the proposed project were rejected and the funds invested elsewhere. Taken together, determination of the project rate of return and comparison to the minimum attractive rate of return are the essential features of the rate of return method.

A variety of names have been given to the method of calculation described above as the rate of return method. The most prominent are the "interest rate of return," the "return on investment (RoI)," and the "Investor's Method." These methods are equivalent and, when properly applied, produce precisely the same solutions as those that result from the annual worth method and the present worth method. The rate of return method is frequently misused, however. Indeed, surveys of "real world" applications suggest that misuse is more often the rule than the exception. Examples of abuse of the rate of return concept are discussed in some detail in Chapter 4.

A Numerical Illustration

Consider the cash flows given in Table 3.4 for a proposed investment, Project P.

Table 3.4
Rate of Return Solution for Project P

End of period (j)	Cash flow (A_j)	Assuming i = 12%		Assuming i = 15%	
		(P/F, 12%, j)	PW @ 12%	(P/F, 15%, j)	PW @ 15%
0	− $1,000	1.0000	− $1,000	1.0000	− $1,000
1	200	0.8929	179	0.8696	174
2	200	0.7972	159	0.7561	151
3	400	0.7119	285	0.6575	263
4	600	0.6355	381	0.5718	343
Totals	$400		$4		− $69

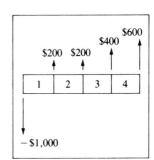

Figure 3.11

The cash flow diagram is shown in Figure 3.11. Assuming end-of-period discounting, the internal rate of return is that value of $i*$ that satisfies the equation

$$\sum_{j=0}^{4} A_j(1 + i*)^{-j} \equiv 0$$

$$= -\$1,000 + \$200(1 + i*)^{-1}$$
$$+ \$200(1 + i*)^{-2}$$
$$+ \$400(1 + i*)^{-3} + \$600(1 + i*)^{-4}$$

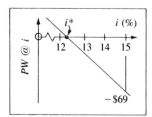

Figure 3.12

There is no simple algebraic solution to this equation, so a trial-and-error approach is necessary. With $i = 12$ percent, the present worth of the cash flows is $4; it is $-$69$ when an interest rate of 15 percent is used. Since the solving interest rate should result in zero present worth, it is clear that we have bracketed the solution. Thus, using straight line interpolation, as indicated in Figure 3.12:

$$i* = 12\% + (15\% - 12\%)\left[\frac{4}{4 - (-69)}\right]$$

$$= 12\% + 3\%\left[\frac{4}{73}\right]$$

$$= 12.2\%$$

The internal rate of return for this proposed investment should now be compared to the minimum attractive rate of return, that is, the rate of return that could be earned by the investor if the funds were invested elsewhere. This is the value i used in the annual worth and present worth methods. Here, the proposed investment of $1,000 is justified if the minimum attractive rate of return is less than 12.2 percent. Put somewhat differently, $AW > 0$ and $PW > 0$ for this sequence of cash flows when $i < 12.2$ percent.

To illustrate, suppose that the *MARR*, i, is 12 percent. Then, from Table 3.3, $PW = \$4$, $AW = \$4(A/P, 12\%, 4) \simeq \$1$, and $IRR(i*) > MARR(i)$. All three methods yield the same conclusion: The investment is preferable to the do-nothing alternative.

Laborious trial-and-error calculations need not be done manually. A variety of computer software programs exist for use with computers, microcomputers, and in some cases, pocket calculators, and may be used to solve for the rate of return given a set of cash flows. An example, written in BASIC, is included at the end of this chapter.

Shortcut Solutions Using One or Two Factors
In certain instances the pattern of cash flows may be such that solutions can be directly determined, avoiding cumbersome trial-and-error approach. For example, suppose that you want to determine the prospective rate of return from an initial investment of $100,000, which results in net revenues of $40,000 per year for four years. The equation that gives the prospective rate of return is

$$\sum_{j=0}^{4} A_j(1 + i*)^{-j} \equiv 0$$

$$= -\$100,000 + \$40,000 \sum_{j=1}^{4}(1 + i*)^{-j}$$

$$= -\$100,000 + \$40,000(P/A, i*, 4)$$

Thus

$$(P/A, i, 4) = \$100,000/\$40,000$$
$$= 2.5$$

From the compound interest tables in Appendix B, $(P/A, 20\%, 4)$ = 2.589 and $(P/A, 25\%, 4)$ = 2.362. Using linear interpolation, the proposed rate of return is found to be approximately 21.9 percent.

The factors for the uniform series and arithmetic gradient series should be used whenever possible to simplify calculations. Consider, for example, the following pattern of cash flows:

End of period (j)	Cash flow (A_j)
0	− \$850
1–20	48
20	1,000

The *solution equation*, the equation that yields the prospective rate of return, is

$$\sum_{j=1}^{20} A_j(1 + i^*)^{-j} \equiv 0$$

$$= -\$850 + \$48(P/A, i^*, 20)$$
$$+ \$1,000(P/F, i^*, 20)$$

Unlike in the previous example, a trial-and-error approach is required here because there is more than one factor in the solution equation. However, using the uniform series present worth factor does simplify the problem; the solution equation is reduced from twenty terms to only two:

$$PW @ 6\% = -\$850 + \$48(11.470) + \$1,000(0.3118)$$
$$= +\$12$$
$$PW @ 7\% = -\$850 + \$48(10.594) + \$1,000(0.2584)$$
$$= -\$84$$

Interpolating,

$$i^* = 6.125\%.$$

Estimating the Rate of Return—First Approximations
When trial-and-error calculations are required, it is obviously helpful if initial trials can result from an "educated guess" rather than from a random choice. There are several ways to improve the first approximation. First, examine the cash flow tables to find out whether the solving rate of return is relatively low or high, or, for that matter, whether there *is* a meaningful solution. Consider the following three examples:

End of period	Project I	Project II	Project III
0	− $100	− $100	− $100
1	20	30	10
2	30	60	30
3	40	40	30
4	20	70	20
Totals	$10	$100	− $10

With an original investment of $100, Project I results in a net gain of $10 after four years, Project II yields a net gain of $100, and Project III results in a net loss. It follows that the rate of return will be low for I and relatively high for II. (There *is* a solution to the present worth equation for Project III, but what does it mean? If the cash flows shown represent those of the lender, then the solution rate of return is that obtained by the borrower. The lender's return is negative in this case.)

Second, an irregular series of cash flows may be approximated by a uniform series. Returning to Alternative P:

End of period	Actual cash flow	Approximate cash flow
0	− $1,000	− $1,000
1	200	350
2	200	350
3	400	350
4	600	350
Totals	$400	$400

The approximate rate of return may now be found.

$$\sum_{j=1}^{4} A_j (1 + i*)^{-j} \equiv 0$$

$$= -\$1,000 + \$350(P/A, i*, 4)$$

From which

$$(P/A, i*, 4) = 2.857$$

From the compound interest tables, $(P/A, 15\%, 4) = 2.855$, so $i*$ is approximately 15 percent.

Third, in conjunction with the result of the uniform series approximation, observe the location of the largest cash flows. The largest cash flows associated with Project P, for example, occur near the end of the project life, during the third and fourth periods. Thus the actual rate of return should be something less than 15 percent, and our first trial might be 12 percent. (As determined earlier, the solution is 12.2 percent.)

Multiple Interest Rates

When properly applied, the internal rate of return method yields the same solutions as those obtained from the annual worth and present worth methods, yet the algebraic structure is such that analysts may easily be led to incorrect solutions. This does not mean that the rate of return method is inherently incorrect—rather, the analyst should be aware of errors that may result from certain situations, such as the "ranking error," discussed in Chapter 4. Our discussion at this point centers on the condition known as "the multiple interest rate problem."

Consider the end-of-period model described by Equation 3.10:

$$\sum_{j=0}^{N} A_j(1 + i^*)^{-j} = 0$$

This expression may also be written as

$$A_0 + A_1x + A_2x^2 + \cdots + A_Nx^N = 0 \qquad (3.14)$$

where $x = (1 + i^*)^{-1}$. Solving Equation 3.14 for x leads to i^*, so we want to find the roots, x, of this Nth-order polynomial expression. Only the real, positive roots are of interest, of course, because any meaningful values of i^* must be real and positive. Descartes's Rule of Signs states that there may be as many real, positive roots, or solutions, as there are changes in sign of the coefficients, the A_j's.[4] Thus, in theory, there are many possible solutions for x and, by extension, for the rate of return, i^*.

In most problems there is only one sign change. That is, the initial investment (a negative cash flow) generally results in a sequence of net revenues, or cost savings (positive cash flows). Such a situation leads to a unique solution. On the other hand, consider the following sequence of cash flows:[5]

End of period (j)	Cash flow (A_j)
0	$1,600
1	− 10,000
2	10,000

Ordinarily, the internal rate of return is that value of i^* that satisfies the equation

$$\$1,600 - \$10,000(1 + i^*)^{-1} + \$10,000(1 + i^*)^{-2} = 0$$

The present worth as a function of the discount rate, i, is graphed in Figure 3.13. You may readily verify that there are two solutions to this equation: $i^* = 0.25$ and $i^* = 4.00$. But what is the significance of these results? If the minimum attractive rate of

4. You may recall Descartes's Rule of Signs from a previous algebra course.
5. This cash flow sequence is based on an example provided by Ezra Solomon, "The Arithmetic of Capital Budgeting Decisions," *Journal of Business* 29 (1956):124.

Figure 3.13 Present Worth as a Function of the Discount Rate

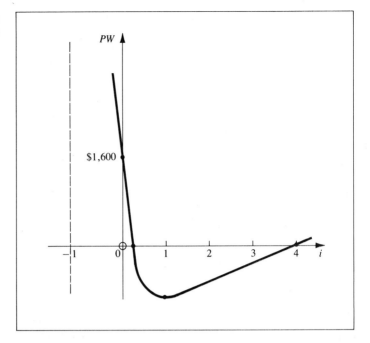

return (*MARR*) is, say, 0.30, should the proposed investment be accepted or rejected?

One possible solution to this problem requires consideration of an **auxiliary interest rate,** the return available from reinvested capital. The positive cash flow at date zero, $1,600, may be reinvested at the minimum attractive rate of return, say, 30 percent. Thus its value at the end of the first period is $2,080 ($1,600 + $480), and the *net* cash flow at the end of the first period is −$7,920 (−$10,000 + $2,080). This manipulation is illustrated in Table 3.5. The revised problem has only one variation in sign:

$$-\$7,920 + \$10,000(1 + i^*)^{-1} = 0$$

which has for its solution $i^* = 26.3$ percent. This result is known as the **external rate of return** (*ERR*). It is "external" in the sense that it is the result of the amounts and timing of cash flows of the

Table 3.5
Using the Auxiliary Interest Rate to Create a Sequence
of Cash Flows with Only One Sign Change

End of period	Original problem (2 sign changes)	Revised problem (1 sign change)
0	$1,600	
1	−10,000	$1,600(1.30) − $10,000 = −$7,920
2	10,000	10,000

original investment as well as of the influence of the auxiliary interest rate.

There are two internal rates of return in this problem: 25 percent and 400 percent; there is only one external rate of return: 26.3 percent. Assuming that the auxiliary interest rate is identical to the minimum attractive rate of return, the investment should be rejected, because *ERR* < *MARR* (26.3 percent < 30 percent).

The preceding example is relatively simple—only three cash flows with two sign changes $(+, -, +)$. The solution procedure can be generalized to any number of sign changes simply by ensuring that *all* positive cash flows occur at the end of the last period, as illustrated in Figure 3.14. As before, let A_j represent the cash flow at the end of period j, for $j = 0, 1, 2, \ldots, N$. Further, let

$$A_j = \begin{cases} R_j & \text{if } A_j > 0 \\ -C_j & \text{if } A_j < 0 \end{cases}$$

The R_j's are the positive cash flows and the C_j's are the negative cash flows. Let k represent the auxiliary interest, assuming that k is also the *MARR*. The equivalent value of all positive cash flows at the end of period N is given by

$$\sum_{j=0}^{N} R_j (1 + k)^{N-j}$$

and the external rate of return, i_e^*, is the solution to the equation

$$\sum_{j=0}^{N} C_j (1 + i_e^*)^{N-j} = \sum_{j=0}^{N} R_j (1 + k)^{N-j} \qquad (3.15)$$

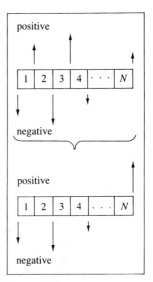

positive

negative

positive

negative

Figure 3.14

In words, the external rate of return is that interest for which the equivalent value of all *negative* cash flows at the end of period N is exactly equal to the equivalent value of all *positive* cash flows at the end of period N, where the equivalent value of the latter is determined by the auxiliary interest rate. The investment is justified if $i_e^* > k$.

To demonstrate this procedure, consider the following sequence of cash flows:

End of period (j)	Cash flow (A_j)	Positive cash flow (R_j)	Negative cash flow (C_j)
0	− $100	$ —	$100
1	50	50	—
2	− 20	—	20
3	100	100	—
4	− 80	—	80
5	100	100	—

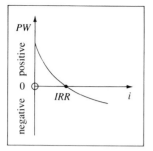

Figure 3.15

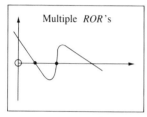

Figure 3.16

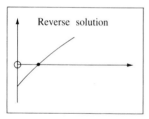

Figure 3.17

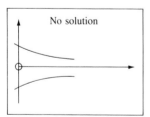

Figure 3.18

Let $k = 0.20$. To find the external rate of return from equation 3.15,

$$\$100(1 + i_e^*)^5 + \$20(1 + i_e^*)^3 + \$80(1 + i_e^*) = \$50(1.2)^4 + \$100(1.2)^2 + \$100$$

There is a unique solution to this problem: $i_e^* = 17.3$ percent. Since this is less than the minimum attractive rate of return, it follows that this investment is not economically justified. (You may find it interesting to determine the AW and PW at $i = 20$ percent for this sample problem. Both the AW and PW, of course, should be negative.)

Additional Comments

In the usual case, where there is an initial investment, $A_0 < 0$), followed by a series of cash revenues or savings ($A_j > 0$ for $j = 1, 2, \ldots, N$), the present worth is a diminishing function of the interest rate. (See Figure 3.15.) The internal rate of return (IRR) is the value of i for which $PW = 0$. The project's internal rate of return should be compared to the minimum attractive rate of return ($MARR$)—the return expected if the funds were invested elsewhere. The $MARR$ should be the discount rate used in the PW equation. Thus, as you saw previously,

$$PW > 0 \quad \text{when} \quad IRR > MARR$$
$$PW < 0 \quad \text{when} \quad IRR < MARR$$

The two methods, therefore, lead to consistent solutions under these conditions.

Some projects have patterns of cash flows that require the most careful interpretation of their internal rate(s) of return. For example, as in the previous problem, solution of the present worth equation may lead to multiple rates of return. (See Figure 3.16.) There are other aberrations as well. As shown in Figure 3.17, there may be "reverse solutions" resulting from positive cash flows followed by a sequence of negative cash flows. Or, as shown in Figure 3.18, the cash flows may be such that there are no solutions (no real roots) to the PW equation.

If there is any question about the interpretation of IRR results, the analyst should prepare a graph of PW as a function of the interest rate. Graphing provides meaningful insight, because the relationship between PW and the rate(s) of return, if any, will be clearly indicated.

**3.4
THE BENEFIT-COST
RATIO METHOD**

The Federal River and Harbor Act of 1902 required the U.S. Army Corps of Engineers to evaluate the costs and benefits of proposed river and harbor projects and to report the desirability of proposals to Congress. The Flood Control Act of 1936 advanced this concept one step further by specifying that "the Federal Government should improve or participate in the improvement of navigable waters or their tributaries, including watersheds thereof, for flood control purposes *if the benefits to whomsoever they may accrue are in excess of the estimated costs* [emphasis added], and if the lives and social security of people are not otherwise adversely affected."[6]

Thus a *criterion* of acceptability for certain federal projects was established by law—namely, that "benefits" must exceed "costs." The formal methodology stemming from this legislation was to become known as the "benefit-cost method," "cost-benefit analysis," and the like. In addition to being employed in flood control and navigation projects, the benefit-cost methodology was employed in many other water resource projects as well as in road and highway projects. Today, benefit-cost methodology is used widely to allocate resources at the federal, state, and local levels of government.[7]

The Acceptance Criterion

The essential element of the **benefit-cost ratio method** is almost trivial, but it can be misleading in its simplicity. An investment is justified only if the incremental benefits, B, resulting from it exceed the resulting incremental costs, C. Of course, all benefits and costs must be stated in equivalent terms, that is, with measurement at the same point(s) in time. Normally, both benefits and costs are stated as "present values" or are "annualized" by using compound interest factors as appropriate.

Clearly, if benefits must exceed costs, then the ratio of benefits to costs must exceed unity. That is, if $B > C$, then $B:C > 1.0$. This statement of the **acceptance criterion** is true only if the incremental costs, C, are positive. It is possible, when evaluating certain alternatives, for the incremental costs to be negative, that is, for the proposed project to result in a reduction of costs. Negative benefits arise when the incremental effect is a reduction in benefits. These possibilities require a somewhat more complex statement of the acceptance criterion, as given in Table 3.6. In summary,

For $C > 0$, if $B:C > 1.0$, accept; otherwise reject.
For $C < 0$, if $B:C > 1.0$, reject; otherwise accept.

In both cases, if the ratio $B:C = 1.0$, the implication is that

6. U.S. Code, 1940 edition (Washington, D.C.: U.S. Government Printing Office), 2964.
7. For additional background reading, see A. R. Prest and R. Turvey, "Cost-Benefit Analysis: A Survey," in *Surveys of Economic Theory, vol. III, Resource Allocation* (New York: St. Martin's Press, 1966). Prepared for the American Economic Association and the Royal Economic Society.

Table 3.6

The Acceptance Criterion

Numerator (B)	Denominator (C)	Ratio (B:C)	Decision
Positive	Positive	>1.0 <1.0	Accept Reject
Positive	Negative	Any	Accept
Negative	Positive	Any	Reject
Negative	Negative	>1.0 <1.0	Reject Accept

benefits equal costs, so we would be indifferent about accepting or rejecting the proposed project.

An Illustration

Consider two alternatives, T and U:

Alternative	Benefits	Costs	B:C
T	$ 7	$2	3.5
U	12	6	2.0

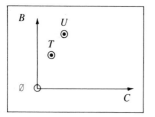

Figure 3.19

The data points are plotted in Figure 3.19. Note that the origin represents the do-nothing alternative, represented by ∅: Invest in neither T nor U but invest the funds elsewhere. The benefits and costs for the two alternatives are measured against the option of doing nothing.

The benefit-cost ratios, 3.5 and 2.0 for T and U, respectively, do *not* imply that T is preferable to U. Because each ratio exceeds unity, we conclude only that both T and U are preferable to the third alternative, do nothing (and invest the funds elsewhere). We cannot conclude from these values and calculations alone that T is preferable to U.

To select between T and U, we must note the consequences if U is selected rather than T. If so, the incremental effects are:

Alternative	Incremental benefits	Incremental costs	Incremental B:C
U rather than T	$12 − $7 = $5	$6 − $2 = $4	$5/$4 = 1.25

Because the incremental ratio exceeds unity, we conclude that U is in fact preferable to T.

This example is graphed in Figure 3.20. Since the scales for benefits (ordinate) and costs (abscissa) are identical, the line drawn at 45 degrees from the origin represents the locus of all points at which benefits are exactly equal to costs. Notice that the *slope* of this line, the ratio of the ordinate to the abscissa, is precisely unity. All alternatives lying above this line are acceptable because their slopes are greater than unity; points below the

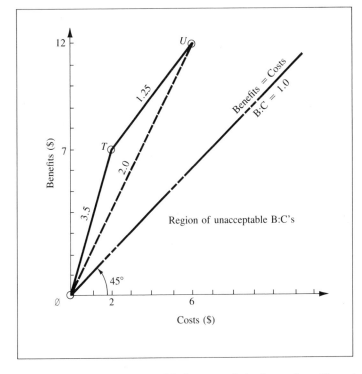

Figure 3.20   Benefit-Cost
Analysis

45-degree line are unacceptable because their slopes (benefit-cost ratios) are less than unity. The point at the origin represents the do-nothing alternative. Note too that the slopes that are of interest are represented by the lines $\emptyset - T$ and $T - U$. In the former instance, T is preferable to doing nothing ($\emptyset$) and U is preferable to T. The slope of the line $\emptyset - U$ is redundant in this case.

Another Example Illustrating the Importance of Incremental Comparisons

Another example serves to illustrate the irrelevance of rank-ordering alternatives solely on the basis of benefit-cost ratios—and the importance of incremental analysis—even when there is a substantial difference between $B:C$ ratios. Consider the data and calculations for two feasible, mutually exclusive alternatives:

Comparison	Benefits	Costs	$B:C$	Decision
$\emptyset \rightarrow V$	$ 800	$ 100	8.0	$V > \emptyset$
$\emptyset \rightarrow W$	2,000	1,000	2.0	$W > \emptyset$
$V \rightarrow W$	1,200	900	1.3	$W > V$

The arrows in the Comparison column indicate that the increments of benefits and costs of Alternative V should be compared with those of Alternative $\emptyset$, the do-nothing alternative. $\emptyset \rightarrow W$ means: Compare the increments of benefits and costs of W and $\emptyset$; and $V \rightarrow W$ means: Compare the increments of benefits and costs of

W and V. In the final column, the symbol ">" means "is preferable to."

In this example, both V and W are preferable to ∅ because their respective benefit-cost ratios exceed unity. Yet V is not preferable to W, even though the $B:C$ for the former is 8.0, compared to only 2.0 for the latter. Examining the *incremental* effects indicates that W is in fact preferable to V. Incremental costs of $900 yield incremental benefits of $1,200, resulting in a $B:C$ of 1.3. Thus the incremental investment in Alternative W is warranted.

The above example notwithstanding, many people find it difficult to understand how an alternative that returns $8 for every $1 invested can possibly be less attractive than another that returns only $2 for every $1 invested. Their confusion arises from a fundamental misunderstanding of the significance of benefit-cost ratios, as well as of the structure of the problem. There are only three alternatives here: Do nothing, choose V, or choose W. We can choose only one V or only one W. We cannot invest in two or more V's; only one is possible. So the question is: Is one V preferable to doing nothing? Since the $B:C$ is greater than unity, the answer is yes. Next, we ask whether one W is preferable to one V. The consequences—an additional cost of $900 leading to additional benefits of $1,200—result in an incremental $B:C$ of 1.3, thereby signifying that the incremental investment is justified. Therefore W is preferable to V.

The Numerator-Denominator Issue in Calculating Benefit-Cost Ratios

Some authors are critical of the benefit-cost ratio method on the grounds that the magnitude of the ratio depends on whether a particular economic consequence is considered in the numerator as a benefit or in the denominator as a "negative cost." (Alternatively, one may choose between including an economic consequence in the denominator as a cost or in the numerator as a "negative benefit.")

This question occurs frequently with respect to facilities designed to result in user savings (benefits) but that require annual costs, such as the cost of maintenance. In particular, consider three major consequences of a proposed improvement: (1) capital costs of construction, (2) benefits accruing to users because of an improved level of service, and (3) costs of facility operation and maintenance. The issue here is whether the operation and maintenance expenses should be deducted from road-user benefits (the numerator), or conversely, added to capital costs (the denominator), because each strategy results in a different benefit-cost ratio.

A simple numerical example serves to illustrate. Let

$B = PW$ of user benefits = $150,000

$C = PW$ of capital costs of the facility = $80,000

$K = PW$ of operation and maintenance expenses = $50,000

Now, in the event that operation and maintenance costs are subtracted first from benefits,

$$B:C = \frac{B-K}{C} = \frac{\$150,000 - \$50,000}{\$80,000} = 1.25$$

If maintenance costs are considered in the denominator,

$$B:C = \frac{B}{C+K} = \frac{\$150,000}{\$80,000 + \$50,000} = 1.15$$

Some critics claim that the differing results make it difficult, if not impossible, to compare two projects by using the benefit-cost ratio method. For example, suppose that we are considering an alternative project, Project II, with a benefit-cost ratio of 1.20. Which, then, is preferable: Project I, with $B:C = 1.25$ (or $B:C = 1.15$), or Project II, with $B:C = 1.20$?

The only characteristic of the benefit-cost ratio that is relevant to the decision-making process is whether the ratio is greater than unity. Otherwise, *the absolute value of the ratio is irrelevant.* Returning to our example, suppose that the benefit-cost ratio of Project II results from the following estimates:

$$B(II) = \$24,000 \qquad C(II) = \$20,000 \qquad K(II) = \$0$$

Now, let's determine which alternative, I or II, is preferable, considering maintenance costs in the numerator (as a "negative benefit") or in the denominator. In the first instance, $B:C = 1.25$, which, because it is greater than unity, leads us to conclude that Project I is preferable to doing nothing, that is, investing elsewhere. But is II preferable to I? To answer this question, determine the benefit-cost ratio for the differences between alternatives:

$$\text{Incremental benefits} = \$24,000 - (\$15,000 - \$5,000)$$
$$= \$14,000$$
$$\text{Incremental costs} = \$20,000 - \$8,000 = \$12,000$$
$$B:C = \$14,000/\$12,000 = 1.17$$

Thus Project II is preferable.

Under the assumption that maintenance costs should be included in the denominator, we have

$$\text{Incremental benefits} = \$24,000 - \$15,000 = \$9,000$$
$$\text{Incremental costs} = \$20,000 - (\$8,000 + \$5,000)$$
$$= \$7,000$$
$$B:C = \$9,000/\$7,000 = 1.29$$

As before, Project II is preferable because the incremental benefit-cost ratio exceeds unity.

To see whether this conclusion holds in all cases, we need only note the decision rule (for positive denominator): *Accept the incremental investment if the resulting incremental benefit-cost ratio exceeds unity; otherwise, reject.* Or:

1. If $B:C = \dfrac{B - K}{C} > 1.0$, accept; otherwise, reject.

The alternative formula is

2. If $B:C = \dfrac{B}{C + K} > 1.0$, accept; otherwise, reject.

Inequalities 1 and 2 lead to identical results. That is,

$$\text{If } \frac{B - K}{C} > 1.0, \text{ then } B > C + K \text{ and } \frac{B}{C + K} > 1.0$$

This result arises from the fact that the direction of an inequality cannot be reversed merely by subtracting a constant from both sides.[8]

3.5 SUMMARY

The annual worth method, also known as the equivalent uniform annual cost method, requires the transcription of all cash flows into a uniform series by using appropriate compound interest factors. Alternatives may be compared directly if annual worth (net annual benefits) or annual costs have been "spread" uniformly over a common time period. This method results in a weighted average worth, or cost, per unit of time.

In the present worth method, all cash flows are discounted to date zero, that is, an equivalent present value is obtained. Again, it is essential that planning horizons for each alternative be of

8. The argument given here holds only if both $C > 0$ and $C + K > 0$. Otherwise, it may be shown that the conditions for *acceptance* are

$$\frac{B - K}{C} > 1 \quad \text{or} \quad \frac{B}{C + K} < 1, \text{ if } C > 0 \text{ and } C + K < 0$$
$$\frac{B - K}{C} < 1 \quad \text{or} \quad \frac{B}{C + K} > 1, \text{ if } C < 0 \text{ and } C + K > 0$$
$$\frac{B - K}{C} < 1 \quad \text{or} \quad \frac{B}{C + K} < 1, \text{ if } C < 0 \text{ and } C + K < 0$$

equal length. Of particular importance to understanding this method is the fact that funds invested elsewhere at rate i result in zero present worth.

The (internal) rate of return method is, in a sense, the inverse of the present worth method in that the internal rate of return for a project is that value of i for which the present worth of a proposed increment of investment is exactly zero. Of course, the rate of return may also be found from an equation that sets the annual worth (or future worth) equal to zero, but the present worth format is the usual case.

The benefit-cost (ratio) method requires an incremental procedure to identify preferable alternatives. For a given pair of alternatives, the incremental benefits and costs, stated as present worths or periodic equivalents, are identified. The incremental investment is economically justified if the resulting ratio of benefits to costs exceeds unity.

Each of the four methods, when properly used, leads to consistent conclusions. To illustrate this point, recall Alternatives X and Y, introduced earlier in this chapter:

	Alternative X	Alternative Y
Initial cost	$10,000	$12,000
Service life (years)	6	6
Salvage value at end of service life	$ 0	$ 3,000
Annual operating costs	$ 1,500	$ 1,600

The minimum attractive rate of return in this problem is $i = 12$ percent per year.

The Annual Worth (Cost) Solution

$$AC(X) = \$10,000(A/P, 12\%, 6) + \$1,500$$
$$= \$3,932$$
$$AC(Y) = (\$12,000 - \$3,000)(A/P, 12\%, 6)$$
$$+ \$3,000(0.12) + \$1,600$$
$$= \$4,149$$

Thus Alternative X is the least-cost solution.

The Present Worth Solution

$$PW(X) = -\$10,000 - \$1,500(P/A, 12\%, 6)$$
$$= -\$16,167$$
$$PW(Y) = -\$12,000 - \$1,600(P/A, 12\%, 6)$$
$$+ \$3,000(P/F, 12\%, 6)$$
$$= -\$17,058$$

Thus Alternative X is the least-cost solution; it has the larger PW.

The (Internal) Rate of Return Solution

Of course, neither X nor Y has a positive rate of return. Neither is preferable to the do-nothing alternative. However, assuming that either X or Y must be selected, note the differences between these alternatives:

Cash flows	X only	Y only	Y instead of X
A_0	− $10,000	− $12,000	− $2,000
A_1–A_6	− 1,500	− 1,600	− 100
A_6		3,000	3,000

The solution equation is

$$-\$2,000 - \$100(P/A,\ i^*,\ 6) + \$3,000(P/F,\ i^*,\ 6) \equiv 0$$

The solution is $i^* = 2.8$ percent. Since $i^* < MARR$, the incremental investment is not justified. Thus Y is not preferable to X.

The Benefit-Cost (Ratio) Solution

The benefit-cost ratio for the incremental investment is

$$\frac{\$3,000(P/F,\ 12\%,\ 6)}{\$2,000 + \$100(P/A,\ 12\%,\ 6)} = \frac{\$1,520}{\$2,411} = 0.63$$

Since the result is less than unity, the incremental investment is not justified and Y is not preferable to X.

All four methods lead to the same result: Alternative X is preferable to Alternative Y. This chapter focuses on the choice between two alternatives. The next chapter explores the appropriate procedures when more than two alternatives are under consideration.

Unless otherwise indicated, assume discrete cash flows, end-of-period compounding/discounting, and effective interest rates.

Digital or analog computers may be used to determine the solutions to most of the following problems. (Indeed, the BASIC program, included at the end of this section, may be especially helpful.) At least some of the problems should be solved manually, however, in order to improve understanding of the underlying procedures.

Annual Worth or Equivalent Uniform Annual Cost

3.1 Consider the following series of cash flows:

End of period	Series A	Series B	Series C
1	$50	$60	$ 0
2	52	58	66
3	54	56	66
4	56	54	66
5	58	52	66
6	60	50	66

Assuming a 10 percent interest rate, find the equivalent annual worth of the three series. The sum of the cash flows for each series is $330.
(*Answers:* $54.45, $55.55, $52.22)

3.2 Consider the following cash flow sequences:

End of period	Sequence D	Sequence E	Sequence F
0	$ 0	$100	− $100
1	20	80	0
2	40	60	0
3	60	40	100
4	80	20	100
5	100	0	100
6	0	− 100	100
Total	$300	$200	$300

Assuming an effective interest rate of 10 percent per period and using the *minimum* number of factors, find the equivalent uniform annual worth of each sequence.

3.3 Consider two mutually exclusive alternatives, G and H, with the following expected cash flows:

End of period	Alternative G	Alternative H
0	− $100	− $120
1	20	50
2	30	50
3	40	50
4	50	50
5	60	80

Assuming that $i = 0.10$, find the equivalent uniform annual cost for each alternative.
(*Answers:* $AW(G) = 11.72 and $AW(H) = 23.26)

3.4 An international funding agency is considering the construction of a road in a certain country. The climate is extremely arid and road use is expected to remain low, so an infinite life is assumed. ("Infinite life" is often assumed when $N \geq$ 50 years.) If the road is constructed, it is expected that net costs to roadway users will be $1,000,000 per year. Maintenance costs of $400,000 will be experienced every five years. Assuming the during-year convention for user costs and the end-of-year convention for maintenance costs, find the equivalent uniform annual cost of this project if the discount rate is 10 percent per year.

3.5 A new warehouse will cost $150,000 and will have an expected life of twelve years with $30,000 net salvage value at the end of that time. Annual disbursements for maintenance, heating, and so forth, are estimated to be $2,500 the first year, $3,000 the second year, and are expected to continue to increase at the rate of $500 per year throughout the life of the warehouse. If the minimum attractive rate of return is 15 percent before income taxes, determine the equivalent uniform annual cost of this proposed investment.
(*Answer:* $31,094)

3.6 The Black Company is considering two alternative plans for erecting a fence around its new central city plant. Galvanized steel "chicken-wire" fencing requires a first cost of $35,000 and estimated annual upkeep costs of $300. The expected life is twenty-five years. A concrete-block wall requires a first cost of only $20,000, but it will need minor repairs every five years at a cost of $1,000 and major repairs every ten years at a cost of $5,000. Assuming an interest rate of 10 percent before taxes and a perpetual (continuing) need, determine the equivalent uniform annual costs for the two plans. Specify all other assumptions.

3.7 The National Park Service (NPS) is developing a certain recreation facility that will require periodic major maintenance at a cost of $10,000 every eighteen months. This federal agency uses an effective interest rate of 10 percent per year. Assume an infinite planning horizon.
a. Find the equivalent uniform cost per month.
b. Find the equivalent uniform annual cost.
(*Answers:* (a) $519; (b) $6,507)

3.8 It is proposed that a facility be constructed in the city of Laketown to garage the Public Works Department's maintenance vehicles. The site may be purchased for $40,000. Construction will take one year and will cost $100,000, distributed uniformly over the year. Over the following ten years, that is, over years 2 through 11, costs to operate and maintain the facility are estimated to be about $1,150 per week, or $60,000 per year. Property taxes are expected to be $1,000 per year, payable at the *end* of each year. Insurance will cost $2,000 per year, payable at the *start* of each year, that is, at the beginning of years 1 through 11. It is expected that the land and the facility will be sold at the end of eleven years for $200,000.

a. Prepare a cash flow diagram for this series of anticipated cash flows.

b. Find the equivalent uniform annual cost (*EUAC*) for this proposed project. Assume an interest rate of 10 percent per year.

3.9 Consider the problem of whether to replace or remodel a small commercial structure. A new building will cost $60,000 and will have an estimated useful life of twenty years. It is expected that the present structure, if repaired, will have an additional useful life of seven years. Remodeling costs are estimated to be $18,000. The existing building may be sold for $8,000 in its present condition, but if remodeled and retained for another seven years, it will have no value on disposal at that time. The terminal salvage value of a replacement structure after twenty years is estimated to be $100,000. All other costs will be unaffected by the replacement decision. If the before-tax minimum attractive rate of return (*MARR*) is 10 percent, determine the equivalent uniform annual costs of replacing and remodeling. (*Answers: EUAC* (replacing) = $5,300; *EUAC* (remodeling) = $5,340)

3.10 A manufacturing plant and its equipment are insured for $700,000. The present annual insurance premium, payable at the beginning of each insured year, is $0.86 per $100 of coverage. A sprinkler system with an estimated life of twenty years and no salvage value at the end of that time can be installed for $18,000. Annual operation and maintenance costs are estimated to be $360. Property taxes, payable at the end of each year, are 1.0 percent of the initial cost of the plant and equipment. If the system is installed and maintained, the premium rate will be reduced to $0.38 per $100 of coverage. If the firm's *MARR* is 10 percent before income taxes, should the sprinkler system be installed? Use the annual worth method.*

3.11 A certain data processing system may be purchased for $100,000. Annual operating costs are expected to be $20,000 the first year, increasing by $2,000 each year. (Assume that operating costs occur uniformly and continuously throughout each year.) Preventive maintenance costs of $5,000 are expected at the end of each and every six-month period throughout the life of the system. It is anticipated that the system may be sold for $25,000 at the end of its five-year service life. The interest rate (opportunity cost) is 10 percent per year.

a. Find the capital recovery cost for the system.

*Adapted from W.J. Fabrycky and G.J. Thuesen, *Economic Decision Analysis*, 2nd Edition (Englewood Cliffs, N.J.: Prentice-Hall, 1980) p. 133.

b. Find the equivalent uniform annual cost of the preventive maintenance.

c. Find the total *EUAC* for this system.

(*Answers:* (a) $22,285; (b) $10,244; (c) $57,311)

3.12 Two possible routes for a power transmission line, one direct and one indirect, are under consideration. Relevant data are:

	Direct route	Indirect route
Length	15 miles	5 miles
Initial cost	$5,000/mile	$25,000/mile
Annual maintenance	$200/mile/year	$400/mile/year
Useful life	15 years	15 years
Salvage value	$3,000/mile	$5,000/mile
Annual cost of power loss	$500/mile	$500/mile
Annual property taxes	2% of first cost	2% of first cost

Assume that all cash flows are end-of-period. Using an 8 percent interest rate, which route is economically preferable?

Present Worth

3.13 Refer to Problem 3.1. Find the present worth of Series A, Series B, and Series C.

(*Answers:* $237.12, $241.93, $227.44)

3.14 Refer to Problem 3.2 Find the present worth of cash flow sequences D, E, and F.

3.15 Refer to Problem 3.3. Find the present worth of cash flows for Alternatives G and H.

(*Answers:* $PW(G) = 44.44 and $PW(H) = 88.17)

3.16 Refer to Problem 3.4. Find the capitalized costs (*CC*) of the proposed project.

3.17 Refer to Problem 3.7. Find the capitalized cost of periodic maintenance.

(*Answer:* $65,066)

3.18 Refer to Problem 3.10. Find the present worth of costs of *both* (a) continuing with the present system, that is, paying annual premiums of ($0.86/$100)($700,000) = $6,020 and (b) purchasing the sprinkler system and taking advantage of the lower insurance rate.

3.19 Consider a prospective investment in a warehouse having a first cost of $300,000, operating and maintenance costs of $35,000 per year, and an estimated net disposal value of $50,000 at the end of sixty years. Assume an 8 percent interest rate.

a. What is the present value of this investment if the planning horizon is sixty years?

b. If replacement structures will have the same first cost, life, salvage value, and operating and maintenance costs as the original building, what is the capitalized cost of perpetual service? Note the difference between a sixty-year life and an infinite life?

(*Answers:* (a) $732,700; (b) $740,027)

3.20 Refer to the data given in Problem 3.8.
    a. Prepare a cash flow table.
    b. Find the net present worth, PW, assuming an interest rate of 10 percent per year.

3.21 The Bigditch Construction Company has a contract to build a major hydroelectric project over a three-year period. Bigditch has most of the requisite equipment, but it will need a mobile van to serve as a field office. If purchased, the van will be sold at the end of three years. Relevant data are:

Initial cost	$54,000
Physical life	6 years
Salvage value at end of 3 years	$20,000
Salvage value at end of physical life	$10,000
Property taxes, payable at the *end* of each year	3% of first cost
Insurance, payable at the *start* of each year	2% of first cost
Operating costs occurring *during* each year	$ 1,000

The firm's *MARR* is 10 percent per year. Find the net present worth of this proposed purchase.
(*Answer:* PWOC = $48,566)

3.22 The payroll for an accounting process is currently $2,000 per week. It is now possible to purchase an electronic data processing system at an initial cost of $100,000 that will have the effect of reducing payroll costs to only $1,200 per week. This system, if purchased, will be maintained by the vendor at an additional cost of $3,000 paid at the beginning of each year. At the end of five years, the equipment may be returned to the vendor for a trade-in value equal to 10 percent of the original cost. Assuming a 10 percent discount rate per year, determine the present worth, PW, of the proposed equipment as compared to the do-nothing alternative. Inasmuch as the payroll costs occur weekly, use the continuous cash flow assumption during each fifty-two-week year. Assume a five-year planning horizon.

3.23 A firm spends $10,000 per year on materials, with the cost spread continuously and uniformly over each year. Annual rental payments for the building and equipment total $30,000 per year, with the payments being made at the beginning of each year. Using an interest rate of 10 percent per year, find the equivalent present worth of ten years of activity.
(*Answer:* $267,240)

3.24 A specialized vending machine may be purchased for $20,000. The expected excess of receipts over disbursements is $4,000 a year for the first four years and $3,000 in the fifth year. It is expected that the machine will be sold at the end of the fifth year for $12,000. Treat the annual receipts and disbursements using the uniform flow, continuous compounding convention. If the effective interest rate is 10 percent per year before taxes, what is the net present worth of this series of cash flows?

3.25 What is the capitalized cost of a series of payments of $200 at the end of each and every period if interest is discounted at the rate of 5 percent per period?
(*Answer:* $4,000)

3.26 A firm's opportunity cost (effective interest rate) is 24 percent per year. Find the capitalized cost of an infinite series of $1,000 payments made at the end of every three-month period. That is, $1,000 payments are made at the end of months 3, 6, 9, . . . , forever.

3.27 Consider a project having a first cost of $1,000, a ten-year life, and no salvage value. Labor expenses during each year of operation are estimated to be $100. A major overhaul will be required five years after the project starts, will last for one year, and will cost a total of $500. (Of course, the project will not be in operation during the overhaul period.) Using the continuous compounding, continuous cash flow convention for labor and overhaul expense, what is the total cost of the project in present dollars if the interest rate is 6 percent per year?
(*Answer:* $2,048)

3.28 A forklift truck may be purchased for $10,000. If purchased, it is expected that the equipment will be retained for ten years and then sold for an expected $1,000. Costs of operation, such as costs for gas, oil, and normal maintenance costs, are expected to be $1,200 per year. Labor costs are expected to be $6,000 per year in the first year, with an increase of $500 per year over the life of the equipment. The company's pretax minimum attractive rate of return is 15 percent. Determine the equivalent present worth of this proposed investment.

3.29 Two years ago, the White Trucking Company leased a terminal site from the Eastern Pacific Railroad and prepaid the rent for a period of five years. The total prepaid rent was $60,000. The terms of the lease permit White Trucking to continue to rent the site for a period of five additional years by paying $10,000 at the beginning of each year of the second five-year period.

The railroad has now decided that it needs funds to finance legal action in connection with a possible merger and it has proposed that White Trucking prepay the rent that was to have been paid year by year in the second five-year period. Assuming interest of 8 percent per year, what is a fair payment to be made now in lieu of the five annual payments?
(*Answer:* $34,232)

3.30 A manufacturing firm is considering the purchase of a heavy-duty punch press that will cost $32,000 and that may be sold at the end of eight years for an expected $16,000. Labor costs are expected to be $20,000 per year to operate the press. Insurance premiums for the press, payable at the start of each year, are expected to be $600 the first year and are expected to decline by $50 each subsequent year. Find the equivalent present value of this proposed investment assuming an interest rate of 15 percent per year. Use the during-period convention for the labor costs.

Unequal Lives

3.31 A certain operation can be performed satisfactorily by both Machine X and Machine Y. Pertinent data for the two alternatives are:

	Machine X	Machine Y
First cost	$6,000	$14,000
Salvage value	$0	20% of first cost
Service life	12 years	18 years
Annual disbursements	$4,000	$2,400

Compare the equivalent total annual costs, assuming a minimum attractive rate of return of 12 percent before taxes. Specify all assumptions.
(*Answers:* $4,969 for Machine X and $4,281 for Machine Y)

3.32 Two gas-powered electric generators are being considered for purchase. Relevant data are:

	Economy generator	Deluxe generator
Initial cost	$12,000	$20,000
Annual operating expense (during year)	$ 1,600	$ 900
Salvage value	$ 3,000	$ 2,000
Operating life	6 years	12 years

Assuming a 10 percent interest rate, determine which alternative is more economical. State all necessary assumptions.

3.33 Two types of conveyor systems, System I and System II, are being considered by the Blue Corporation:

	System I	System II
Initial investment	$5,000	$12,000
Net salvage value	$1,000	$ 0
Annual disbursements	$ 900	$ 600
Estimated life	10 years	15 years

Assuming a pretax interest rate of 20 percent, which system appears to be the better investment, everything else being equal? State all necessary assumptions.
(*Answers:* $2,054 for I and $3,167 for II)

3.34 A firm is considering two investment alternatives, Alpha and Beta. Alpha has an initial cost of $1,000, a three-year life, zero salvage value after three years, and it returns $500 at the end of each and every year over the three-year life. At the end of three years, an identical Alpha will be available for purchase.

Beta has an initial cost of $1,000, a six-year life, and a $3,700 salvage value after six years. There will be an additional cost of $1,000 occurring continuously and uniformly during the third year of ownership.

The firm's pretax minimum attractive rate of return is 10 percent. Use the present worth method to determine which alternative is preferable.

(Internal) Rate of Return

3.35 A firm is considering investment in one of two lathes. The relevant data are:

	Lathe A	Lathe B
Initial cost	$10,000	$15,000
Service life	5 years	10 years
Salvage value	$ 2,000	$ 3,000
Annual receipts	$ 5,000	$ 7,000
Annual disbursements	$ 2,200	$ 4,300

Determine the rate of return on the incremental investment in Lathe B. (That is, since Lathe B is $5,000 more expensive than Lathe A, what is the rate of return on the additional $5,000?) State any necessary assumptions.

(Answer: 8.95 percent)

3.36 A certain data processing operation is now being done manually at a cost of $100,000 annually. The lease of certain automatic equipment that will reduce labor costs to $20,000 annually has been proposed. Assume that labor costs are distributed uniformly over the year. Lease payments are $60,000 per year, payable at the start of each year. It is expected that there will be a need for this operation over the next five years.

a. Determine the rate of return on the proposed lease.

b. If the firm's MARR is 15 percent, should the five-year lease be undertaken?

3.37 The Green Corporation is considering the installation of a semiautomatic freight conveyor requiring an immediate investment of $300,000. It is estimated that, if the conveyor is purchased, it will be kept for ten years and then sold for $12,000. It is also believed that the purchase will result in the following cost savings:

Year	Net savings	Year	Net savings
1	$60,000	6	$55,000
2	60,000	7	50,000
3	60,000	8	45,000
4	60,000	9	40,000
5	60,000	10	35,000

Determine the prospective rate of return of this investment before taxes.

(Answer: 0.132)

3.38 A firm is considering an investment in data processing equipment that has an initial cost of $10,000 and an expected salvage value of zero if retired after ten years. The equipment will result in end-of-year savings of $1,500 each year throughout its ten-year life.
   a. Determine the internal rate of return for this proposed investment.
   b. If the firm's minimum attractive rate of return is 12 percent, is this proposal acceptable?

3.39 Consider the following data for two mutually exclusive alternatives:

	Alternative A	Alternative B
First cost	$10,000	$12,000
Expected life	6 years	6 years
Net salvage value	$ 0	$ 3,000
Annual operating costs	$ 1,500	$ 1,600

Suppose that either A or B must be selected.
   a. What is the expected rate of return on the additional (incremental) $2,000 investment in Alternative B?
   b. If the *MARR* for the company is 12 percent before taxes, is the additional investment justified?
   c. Which alternative should be selected? Is this the same solution as that obtained from the annual worth and present worth methods?
   d. At what interest rate(s), if any, would B be less costly than A?
   (*Answers:* (a) 0.028; (b) no; (c) Alternative A; (d) $i < 2.8$ percent)

3.40 Consider two mutually exclusive alternatives, C and D, with the following end-of-period cash flows:

End of period	Alternative C	Alternative D
0	−$100	−$150
1–5	30	43

   a. Find the internal rates of return for the alternatives.
   b. If the firm's *MARR* is 8 percent, use the rate of return method to determine which alternative should be selected.
   c. For what range of values of *MARR* is Alternative D preferable to Alternative C?

3.41 Consider the following cash flows:

End of year	Alternative E	Alternative F
0	−$10	−$20
1	15	28

Over what range of values of the minimum attractive rate of return (*MARR*) is Alternative F preferable to Alternative E?
(*Answer: MARR* < 30 percent)

3.42 An investor is considering two mutually exclusive alternatives with the following economic data:

	Project G	Project H
Initial capital investment	$58,500	$ 48,500
Net uniform annual benefits	$ 6,648	$ 0
Salvage value 10 years from now	$30,000	$138,000
Internal rate of return	8%	11%

Both projects are based on a ten-year study period.
a. Compute the incremental interest rate(s) from investing in Project G rather than in H. (*Hint:* There are two solving rates of return. One is about 19 percent. Find the other.)
b. Over what range of minimum attractive rate of return (*MARR*) is G preferable to H?

3.43 Refer to Problem 3.3. Find the interest rate at which the decision maker would be indifferent about choosing Alternative P or Alternative Q.
(*Answer:* Approximately 113 percent)

3.44 Refer to Problem 3.2.
a. For each cash flow sequence, plot present worth as a function of the interest rate, i. Display the results over the range $[-1 < i < 0.50]$.
b. What is the rate of return for each cash flow sequence?
c. For what values of i is D preferable to E?
d. For what values of i is F preferable to E?
e. For what values of i is F preferable to D?
Note: Precise answers are not required. You may note the answers by inspecting the graph developed in (a).

Multiple Interest Rates and the External Rate of Return
3.45 Consider an investment proposal requiring a $220 investment at the *end* of the first period and resulting in net revenues of $80 at the *beginning* of the first period and $150 at the *end* of the second period. Or

j	0	1	2
A_j ($)	80	-220	150

a. Find the internal rate(s) of return for this investment. *Hint:* You may use the *quadratic formula:*

$$x = \frac{-b \pm \sqrt{b^2 - 4ac}}{2a} \quad \text{for } ax^2 + bx + c = 0$$

b. If the firm's *MARR* is 0.30, find the external rate(s) of return for this project.
c. If the *MARR* = 0.30, should the project be accepted? Explain your answer.
(*Answers:* (a) 25 percent and 50 percent; (b) 29.6 percent; (c) reject)

3.46 A relatively common form of investment is the so-called "interest only with balloon payment." For such investments the investor assumes a debt whose principal is not repaid until some future date. In the interim, periodic interest payments are made on the outstanding debt.

Mr. Venture has an opportunity to acquire certain commercial property currently available at a price of $100,000. The seller will accept a note payable in full at the end of four years, with interest due and payable each year at the rate of 10 percent per year, or $10,000 annually for four years. The income to Mr. Venture, before interest payments, is estimated to be $15,000 annually. (For simplicity, assume that these receipts will occur at the end of each year.) Mr. Venture expects to keep the property for five years, then sell it for an estimated $100,000. The net cash flows for this proposed investment are:

End of year	1	2	3	4	5
Cash flow ($\times$ $1,000)	5	5	5	-95	115

a. Find the internal rate(s) of return.
b. Assuming a reinvestment rate of 30 percent per year, find the external rate of return.
c. Is this an attractive investment? Explain your answer.

3.47 The cash flows for a certain project are:

End of period	0	1	2	3
Cash flow ($)	-200	300	100	-100

The firm's discount rate (*MARR*) is 10 percent per period.
a. Determine the external rate of return
b. Is the project acceptable? Why or why not?
(*Answers:* (a) 23.1 percent; (b) yes)

The Benefit-Cost Ratio Method

3.48 Suppose that the facility described in Problem 3.8 is intended to replace a contract garage that is currently being rented by the Public Works Department. Rental charges are $50,000, payable semiannually at the end of every six-month period.
a. Find the net present worth, over an eleven-year planning horizon, of this contract. As before, assume an interest rate of 10 percent per year.
b. Using the present worth method, is it preferable to own or lease?
c. What is the benefit-cost ratio if the facility is owned rather than leased?

3.49 A 400-meter tunnel must be constructed as part of a new aqueduct system for a major city. Two alternatives are being considered:

Alternative A—build a full-capacity tunnel now for $500,000 or

Alternative B—build a half-capacity tunnel now for $300,000 and then build a second half-capacity tunnel twenty years from now for $400,000.

The cost of repairing the tunnel lining at the end of every ten years is estimated to be $20,000 for the full-capacity tunnel and $16,000 for each half-capacity tunnel.

Determine whether Alternative A or Alternative B should be constructed now. Use the benefit-cost ratio method with a discount rate of 5 percent and a 50-year planning horizon. (Of course, there will be no tunnel-lining repair at the end of fifty years.)

(*Answer:* Alternative B)

3.50 A municipal agency is considering a proposal from the Jack Janitorial Service to provide service for a number of the city's public facilities. The cost of this service is $100,000 per year, but this contract would save the city $180,000 annually above current costs.

A somewhat more ambitious proposal has been received from the Queen Cleaning Company. This contract would cost the city $150,000 annually, but benefits in the amount of $250,000 per year could be expected.

In summary, the benefits and costs of these proposals are:

Alternative	Benefits	Costs
Jack Janitorial	$180,000	$100,000
Queen Cleaning	250,000	150,000

Use the benefit-cost ratio method to determine which alternative is preferable.

3.51 Consider two mutually exclusive alternatives, R and S. The economic consequences resulting from these two investments are:

	Alternative R	Alternative S
PW of annual benefits to user (B)	$85,000	$110,000
PW of annual operating expenses (K)	0	10,000
PW of initial investment (C)	45,000	50,000

There are two ways of computing benefit-cost ratios:

1. Subtract K in the numerator, or $\left[\dfrac{B-K}{C}\right]$

2. Add K in the denominator, or $\left[\dfrac{B}{C+K}\right]$

Use the benefit-cost ratio method to determine which investment is preferable.

(*Answer:* Alternative S)

3.52 Terry Trojan is an operations analyst for a local municipal agency in which the benefit-cost ratio method is commonly employed for project justification. She has been asked to evaluate a forms processing operation that currently costs the agency $50,000 annually. She is considering two alternatives:

1. System X requires an initial investment of $10,000 for equipment that has a service life of five years. Operating costs will be reduced to $45,000 annually. The equipment will have no salvage value at the end of five years.

2. System Y requires an initial investment of $15,000 for equipment that has a service life of five years. Operating costs will be reduced to $41,000 annually. Annual maintenance costs for the equipment will be $1,000 annually over the five-year service life. There is no salvage value at the end of five years.

The agency uses a 10 percent interest rate as its cost of capital.

a. One formulation of the benefit-cost ratio is to define benefits as the savings in operating costs minus the maintenance costs. That is, the maintenance costs are in the numerator of the $B : C$ ratio. With this formulation, find the $B : C$ ratios for Systems X and Y.

b. An alternative formulation of the $B : C$ ratio is to define benefits solely as the savings in operating costs. That is, maintenance costs are in the denominator of the $B : C$ ratio. With this formulation, find the $B : C$ ratios for Systems X and Y.

c. Consider the three alternatives: do nothing, choose X, or choose Y. Using the benefit-cost ratio method, determine which alternative is preferable. Explain your answer.

d. Suppose that a new alternative, System Z, is available. System Z is a fully automated system that costs $160,000, has a useful life of five years, and has zero salvage value at the end of five years. If acquired, it will entirely eliminate the $50,000 annual operating costs currently being experienced. Is Z preferable to X and Y? Explain your answer.

e. Mr. Bruin, administrative director for the municipal agency, is concerned about the inherent uncertainties associated with the estimates in this analysis. He proposes that raising the cutoff $B : C$ ratio from 1.00 to 1.25 would be prudent, because it would provide a 25 percent "safety margin" for the agency. If Mr. Bruin's proposal is adopted, would System Z be an acceptable alternative? Is Mr. Bruin's proposal reasonable? Explain your answer.

Bond Valuation

3.53 In order to raise capital for acquisition purposes, the Brown Company sells 4 percent bonds to an investment banking firm for $2,800,000. The face value of these bonds is $3,000,000 and they will mature in twenty-five years. Dividend payments will be made quarterly. Initial disbursements connected with the bond issue are $150,000, and each year additional expenses, such as trustee and registrar fees, are expected to be $25,000. Expressed as an interest rate, what is the actual cost of this borrowed money? Note that annual interest is $0.04 \times \$3,000,000 = \$120,000$.
(*Answer:* 5.85 percent)

3.54 Mr. Jones has an opportunity to purchase a bond from the Los Angeles Public Utility Company on February 1, 1984. The terms of the bond are:

$1,000 face value due and payable January 31, 1988

8 percent coupon rate

interest payable quarterly each January 31, April 30, July 31, October 31, up to and including January 31, 1988, for a total of sixteen future interest payments.

a. If Mr. Jones' effective opportunity cost (discount rate) is 3 percent per quarter, determine the equivalent present value of this bond.

b. What is Mr. Jones' effective opportunity cost per year?

3.55 A person wishes to sell a bond that has a face (par) value of $5,000. The bond has a nominal rate of 9 percent with bond premiums paid semiannually. Five years ago, $4,500 was paid for the bond. The most recent premium was paid this morning. Ten years remain to maturity. The owner is considering selling the bond this afternoon. What must the selling price be if the owner is to realize a 10 percent effective annual return over the remaining period of ownership?
(*Answer:* $4,760)

Leases

3.56 A $10,000 car may be leased with a $2,000 down payment and thirty-six monthly payments of $300 at the start of each month. The lessor requires both the first- and last-month's lease payments in advance. The expected residual value for this car at the end of three years is $5,000. Thus the cash flows for the two alternatives, purchase and lease, are:

End of month	Purchase	Lease
0	− $10,000	− $2,600
1–34	0	− 300
36	5,000	0

a. What is the cost of this lease, expressed as an interest rate per month? That is, what is the rate of return on the difference between alternatives? Show your solution to the closest 0.01 percent.

b. If the lessee's opportunity cost is 2 percent per month, is it preferable to lease or purchase? Explain your answer.

3.57 A $10,000 car may be leased with a $2,000 down payment and thirty-six monthly payment of $320 each month. The lessor requires both the first- and last-month's lease payments in advance; all other lease payments are due at the start of each month. The lessor expects that the actual residual value of the car will be substantially less than the book value at the end of three years, so the lessee will pay the lessor three monthly payments, $960, at the end of thirty-six months to make up this difference. The lessee's opportunity cost is 1.5 percent per month. Find the equivalent present value of this lease.

(*Answer: PW* of lease cost = $11,676)

3.58 The L.A. Leasing Company is planning to purchase a vehicle intended for subsequent lease to the firm's clients. The initial cost of this vehicle is $10,000. It will be kept for three years, at the end of which time it will be sold for an expected $4,000. Insurance for the vehicle is $500 per year, with insurance premiums payable at the start of each year. The firm's *MARR* is 20 percent per year.

a. Determine the cost of capital recovery.

b. Determine the equivalent uniform annual cost.

c. If the firm wants to earn a rate of return of 20 percent per year, determine the amount of the monthly lease payments it should charge. Assume that lease payments will be made at the end of each month.

3.59 Certain computer hardware may be purchased outright for $400,000. If purchased, it is expected that the equipment will be retained for five years and then sold for an estimated $100,000. Annual costs of maintenance are expected to be $20,000 during each of the five years of ownership.

Alternatively, the vendor is willing to lease the software at a price of $100,000 per year. Five of these lease payments are to be made at the start of each year. The lease arrangement will include the cost of maintenance, with the vendor (the lessor) providing the maintenance without cost to the customer (the lessee).

What is the cost to the lessee of this lease proposal? That is, if the customer decides to lease the software instead of buying it outright, what price, stated as a percentage, will the customer pay for this lease arrangement? (Note here that you are looking for the rate of return on the difference between alternatives.)

(*Answer:* 9.95 percent)

Computer Program

The following program, written in BASIC, may be used to solve many of the problems in Chapter 3. The user must enter all end-of-period cash flows, A_j, or during-period cash flows, $\overline{A}_j$, occurring in periods j, for $j = 0, 1, 2, \ldots, N$. (Of course, $\overline{A}_0$ has no significance.) The result is a cash flow table consisting of A_j's and $\overline{A}_j$'s.

With the user-specified interest rate, i, the program generates *present worth, annual worth,* and *future worth.* This latter statistic, *FW*, is the equivalent value of the cash flows as measured at the end of period N, the end of the planning horizon. The program also generates the internal rate(s) of return for the given set of cash flows.

NOTE: If Applesoft BASIC is being used, substitute the symbol (**) with the symbol (^) in lines 0510, 0600, and 0700.

PROGRAM

```
0005 DIM CE(100), CFD(100), RT(10), PW(10)
0010 PRINT "THIS PROGRAM IS DEALING WITH"
0020 PRINT "THE CASE OF CASH FLOW FOR 'END OF PERIOD' & 'DURING PERIOD'"
0030 PRINT " "
0040 PRINT " "
0050 PRINT " "
0060 INPUT "ENTER THE LAST COMPOUNDING PERIOD N = ";NP
0080 INPUT "ENTER EFFECTIVE INTEREST RATE i = ";IT
0090 PRINT " "
0100 PRINT "ENTER ALL DATA TO THE FOLLOWING FORMAT. PUT ',' TO SEPARATE
     EACH DATA."
0105 PRINT " "
0107 PRINT " "
0110 PRINT "PERIOD END-OF-PERIOD CASH FLOW DURING-PERIOD CASH FLOW"
0120 PRINT "--------- ------------------------------- -------------------------------"
0130 FOR I = 0 to NP
0140 INPUT I, CE(I), CFD (I)
0150 NEXT I
0170 GOSUB 500
0180 PRINT " "
0190 PRINT " "
0200 PRINT " "
0210 INPUT "IF YOU ARE READY, TYPE YOUR NAME TO GO = "; AA$
0220 PRINT " "
0230 PRINT "THE PRESENT WORTH FOR THESE CASH FLOWS IS, PW = $";PW
0240 GOSUB 600
0250 PRINT " "
0260 PRINT " "
0270 PRINT "THE EQUIVALENT ANNUAL WORTH IS AW = $";AW
0280 GOSUB 700
0290 PRINT " "
0300 PRINT " "
0310 PRINT "THE EQUIVALENT FUTURE WORTH IS FW = $";FW
0320 PRINT " "
0330 PRINT " "
0340 INDEX = 0
0350 GOTO 800
0500 PW = 0
0505 FOR J = 0 TO NP
0510 PW = PW + CE(J)*(1 + IT)**( - J) + CFD(J)*(1 + IT)**( - J)*(IT)*LOG(1 + IT)**( - 1)
0520 NEXT J
0580 RETURN
```

```
0600 AW = PW*(IT*(1 + IT)**NP)*((1 + IT)**NP - 1)**(-1)
0610 RETURN
0700 FW = PW*(1 + IT)**NP
0710 RETURN
0720 PW = 0
0730 FOR K = 0 TO NP
0740 PW = PW + CE(K) + CFD(K)
0750 NEXT K
0760 RETURN
0800 LET IT = 0
0810 GOSUB 720
0820 P1 = PW
0840 IF P1 = 0 THEN 5000
0850 LET IT = 0.1
0860 GOSUB 500
0870 IF PW = 0 THEN 5000
0880 IF SGN(PW) = - SGN(P1) THEN 2000
0890 IF IT> = 20 THEN 6000
0900 IT = IT + 0.1
1000 GOTO 860
2000 LET IT = IT - 0.01
2010 GOSUB 500
2020 IF PW = 0 THEN 5000
2030 IF SGN(PW) = SGN(P1) THEN 3000
2040 IT = IT - 0.01
2050 GOTO 2010
3000 LET IT = IT + 0.001
3100 GOSUB 500
3200 IF PW = 0 THEN 5000
3300 IF SGN(PW) = - SGN(P1) THEN 4000
3400 IT = IT + 0.001
3500 GOTO 3100
4000 LET IT = IT - 0.0001
4100 GOSUB 500
4200 IF PW = 0 THEN 5000
4300 IF SGN(PW) = SGN(P1) THEN 4650
4400 IT = IT - 0.0001
4500 GOTO 4100
4650 IT = IT + 0.00001
4700 GOSUB 500
4750 IF PW = 0 THEN 5000
4800 IF SGN(PW) = - SGN(P1) THEN 5000
4850 IT = IT + 0.00001
4900 GOTO 4700
5000 LET INDEX = INDEX + 1
5100 RT(INDEX) = IT
5200 PW(INDEX) = PW
5300 IT = IT + .01
5400 P1 = - P1
5500 GOTO 860
6000 IF INDEX = 0 THEN 6490
6005 PRINT "THE ANSWER(S) FOR INTERNAL RATE OF RETURN"
6010 PRINT "IS(ARE) AS FOLLOWING:"
6020 PRINT " "
6050 PRINT "INTERNAL RATE(S) OF RETURN IS(ARE): CALCULATED PRESENT  WORTH(S):"
6100 PRINT "------------------------------------------- -------------------------------------"
6200 PRINT " "
6300 FOR L = 1 TO INDEX
6400 PRINT TAB(3);100*RT(L);"%";TAB(28);"$";PW(L)
6450 NEXT L
6470 GOTO 6500
6490 PRINT "THERE IS NO REAL VALUE OF THE INTERNAL RATE OF RETURN"
6500 END
```

Sample Problem

To illustrate the use of this program, consider the problem described by the following cash flows:

$$A_0 = -\$10,000$$
$$A_j = \$2,000 \quad \text{for } j = 1, 2, 3, 4$$
$$A_5 = \$4,000$$
$$A_j = \$1,000 \quad \text{for } j = 1, 2, 3, 4, 5.$$

SOLUTION (USING THE PROGRAM)

THIS PROGRAM IS DEALING WITH
THE CASE OF CASH FLOW FOR 'END OF PERIOD' & 'DURING PERIOD'

ENTER THE LAST COMPOUNDING PERIOD N= ? 5
ENTER EFFECTIVE INTEREST RATE 1= ? .1

ENTER ALL DATA TO THE FOLLOWING FORMAT. PUT ',' TO SEPARATE EACH DATA.

PERIOD	END-OF-PERIOD CASH FLOW	DURING-PERIOD CASH FLOW
? 0,	-10000,	0
? 1,	2000,	1000
? 2,	2000,	1000
? 3,	2000,	1000
? 4,	2000,	1000
? 5,	4000,	1000

IF YOU ARE READY, TYPE YOUR NAME TO GO=LAI

THE PRESENT WORTH FOR THESE CASH FLOWS IS, PW=$ 2800.732

THE EQUIVALENT ANNUAL WORTH IS AW=$ 738.8262

THE EQUIVALENT FUTURE WORTH IS FW=$ 4510.607

THE ANSWER(S) FOR INTERNAL RATE OF RETURN IS(ARE) AS FOLLOWING:

INTERNAL RATE(S) OF RETURN IS(ARE):	CALCULATED PRESENT WORTH(S):
20.296%	$ -0.1306725

CHAPTER FOUR
MULTIPLE ALTERNATIVES

hapter 3 developed several methods for evaluating the economic consequences of various plans, programs, and projects. These methods are equivalent in the sense that, if properly used, they lead to consistent rank-ordering of alternatives. The emphasis previously was on determining the figure of merit:[1]

Present worth

Equivalent uniform annual worth (or cost)

(Internal) rate of return

Benefit-cost ratio

Much of our earlier discussion centered on determining the economic acceptability of a single investment proposal compared to the do-nothing alternative. But in general the capital allocation problem is more concerned with selecting one or more alternatives from a larger number of contenders for limited capital. Thus this chapter now extends the discussion to include multiple alternatives—two or more investment alternatives.

Proper use of the evaluation methods discussed in Chapter 3 is best understood by first outlining the types of problems to which the methods are to be applied. As you will see, although all four methods lead to consistent results *if used properly,* numerical complexity and opportunity for misunderstanding is often a function of the context of the problem. Thus it is useful at this point to explore various contexts, or problem classes.

The simplest, and perhaps most common, class of problems

4.1 CLASSES OF INVESTMENT PROPOSALS

---

1. The *figure of merit,* a term widely used in systems analysis, is simply a statistic, or a number, that represents the quality of a given alternative course of action.

arises when the decision maker must rank-order a set of technologically **mutually exclusive alternatives.** For example:

> How many clerks should staff the billing office? Six, seven, or eight?
>
> Should the new generator be purchased from Vendor A or Vendor B?
>
> Should transmission lines be located over Routes X, Y, or Z?
>
> Should the new "early retirement, no penalty" option be implemented as a personnel policy?

The problem of which one (and only one) proposal to select from a set of competing alternatives is common to all four examples. Because of the form or function of the proposals themselves, selecting one necessarily precludes selecting any of the others. It makes no sense, for example, to staff an office with six clerks *and* seven clerks or to buy a generator from Vendor A *and* Vendor B.

A second type of problem arises when **contingent investment proposals** are being considered. Note, for example, the following alternatives:

> A—do nothing; continue to operate as before.
>
> B—purchase a new photocopy machine.
>
> C—purchase a maintenance contract for the new photocopy machine.

Here, Alternative C is *contingent* upon Alternative B; the maintenance contract cannot be purchased *unless* the photocopy machine is purchased.

But a "contingency" problem can be converted to a "mutually exclusive" problem by combining alternatives. For example, Alternatives A, B, and C can be structured so that three mutually exclusive alternatives result:

> A—do nothing; continue as before.
>
> B'—purchase a new photocopy machine with *no* maintenance contract.
>
> C'—purchase a new photocopy machine *that includes* a maintenance contract.

Thus we must choose one, and only one: *A, B'*, or *C'*.

The third type of problem arises when there are financial constraints. Consider a set of independent alternatives ("independent" in the sense that they are *not* technologically mutually exclusive or contingent). For example, a truck maintenance department may be considering the following purchases:

Alternative	Initial cost
A—Hydraulic hoist	$3,000
B—Paint-spray unit	$4,000
C—Automatic wash unit	$7,000

The proposals are independent; given adequate financial resources, all three could be purchased. (Aside from determining whether each alternative is preferable to doing nothing, sets of independent projects are of no interest in the analysis.) But suppose that funds for new investments are limited to $7,000. Given this financial constraint, it is clear that not all proposals can be included in the final capital budget. Accepting C, for example, precludes accepting A and/or B. Thus they are **financially constrained alternatives.**

As with contingent alternatives, the "financial constraint" problem can be restated in order to develop a set of mutually exclusive alternatives. Simply generate a set of budget "packages," each of which mutually excludes all others. To illustrate, consider the truck maintenance department problem again.

Package	Alternatives	Initial cost
1	do nothing	$ 0
2	A (hydraulic hoist)	3,000
3	B (paint-spray unit)	4,000
4	C (automatic wash unit)	7,000
5	A and B	7,000

There are five feasible, mutually exclusive alternatives to consider. The sixth, A + B + C, requires a total initial investment of $14,000 ($3,000 + $4,000 + $7,000) and thus is not feasible.

Now let's examine a more complex case in which both independent *and* mutually exclusive alternatives coexist, as illustrated by:

Department	Proposal
A	A_1 A_2 A_3 A_4
B	B_1 B_2
C	C_1 C_2 C_3

The proposals in each row (department) are mutually exclusive. Only one proposal can be accepted from any department. However, the selected proposals from each department are independent unless constrained by financial considerations. The maximum number (n) of mutually exclusive budget packages that can be obtained from this group of proposals is

$$n = \prod_{j=1}^{s} (m_j + 1) = (m_1 + 1)(m_2 + 1) \cdots (m_s + 1)$$

where s is the number of sets of proposals (departments in this case) that are independent and m_j is the number of proposals within each set j, where each proposal in the set is mutually exclusive. In the above example,

$$n = (4 + 1)(2 + 1)(3 + 1) = 5 \times 3 \times 4 = 60$$

Even in this relatively simple problem, the number of mutually exclusive alternatives is sizable.

The mutually exclusive alternative is the fundamental unit of interest in economic analysis. Contingent proposals and financial constraints can readily be rearranged in order to obtain a finite set of mutually exclusive alternatives. This set can then be rank-ordered by the techniques discussed in the following sections. (As the previous example shows, rearrangement may produce a very large number of alternatives, making the approach computationally complex. More efficient algorithms exist, but these procedures are beyond the scope of this book.[2])

4.2 RANKING BY INCREASING OR DECREASING FIGURE OF MERIT

Consider four alternatives characterized by the following conditions: (1) they are *mutually exclusive* in that accepting one necessarily precludes accepting any other, (2) they are *exhaustive* in that there are no other candidate alternatives, except, of course, the do-nothing alternative, and (3) they are *feasible* in that there are no considerations, other than economic, that preclude the acceptance of any one or more of them. The cash flows for the four investment opportunities are given in Table 4.1. All four have equal lives: ten periods. Assume that the discount rate, the minimum attractive rate of return, i, is 20 percent per period. Now let's examine the appropriate solution using the four analytical methods discussed in Chapter 3.

Table 4.1
Cash Flows for Four Mutually Exclusive Alternatives

End of period	Alternative I	Alternative II	Alternative III	Alternative IV
0	−$1,000	−$1,000	−$1,100	−$2,000
1–10	0	300	320	550
10	4,000	0	0	0
Net cash flow	$3,000	$2,000	$2,100	$3,500

2. The capital allocation (capital budgeting) problem described in this chapter is occasionally referred to as the Lorie-Savage Problem after an earlier article: J. H. Lorie and L. J. Savage, "Three Problems in Rationing Capital," *Journal of Business*, vol. 28, no. 4 (October 1955): 229–239. Several mathematical programming methods have been proposed for determining the optimal "portfolio" (set of preferred alternatives), among them dynamic programming, linear programming, and zero-one integer programming. References for these programming methods appear in the Bibliography.

The Present Worth Solution

The preferred alternative, based on the economic considerations outlined in Table 4.1, can be determined by computing the equivalent present worths (*PW*'s) of the alternatives, using the appropriate discount rate. These calculations are summarized in Table 4.2.

Table 4.2
Present Worths of Cash Flows from Table 4.1

End of period	Discount factor*	PW's for alternatives			
		I	II	III	IV
0	1.000	− $1,000	− $1,000	− $1,100	− $2,000
1–10	4.192	0	1,258	1,341	2,305
10	0.1615	646	0	0	0
Net present worth		− $ 354	$ 258	$ 241	$ 305

*Discount factors: $(P/A, 20\%, 10) = 4.192$
$(P/F, 20\%, 10) = 0.1615$

Ranking on the basis of decreasing present worth shows that

Alternative IV is preferred to Alternative II because *PW* (IV) = $305 is greater than *PW* (II) = $258

Alternative II is preferred to Alternative III because *PW* (II) = $258 is greater than *PW* (III) = $241

Alternative III is preferred to Alternative Ø (doing nothing) because *PW* (III) = $241 is greater than *PW* (Ø) = $0

Alternative Ø is preferred to Alternative I because *PW* (Ø) = $0 is greater than *PW* (I) = − $354

That is, IV > II > III > Ø > I. (The inequality $x > y$ indicates that Alternative x *is preferred to* Alternative y.)

Ranking on the basis of the sums of cash flows yields quite different results. From Table 4.1, IV > I > III > II > Ø. Ranking is changed by the influence of the timing of the cash flows and the opportunity cost as reflected in the discount rate.

The above example considered only five alternatives, including the do-nothing alternative. The procedure, however, can be extended to any number of alternatives. In each case, simply determine the *PW*'s for all proposals and then rank the alternatives accordingly.

The Equivalent Uniform Annual Worth (Cost) Solution

This method computes the equivalent periodic worth (or cost) for each proposal. Alternatives are then rank-ordered on the basis of decreasing worth, AW's, or increasing costs, AC's. To illustrate, using data from the previous example:

$$
\begin{aligned}
AW\ (I) &= \$4{,}000(A/F,\ 20\%,\ 10) - \$1{,}000(A/P,\ 20\%,\ 10) \\
&= \$4{,}000(0.03852) - \$1{,}000(0.23852) \\
&= \$154 - \$239 \\
&= -\$85
\end{aligned}
$$

$$
\begin{aligned}
AW\ (II) &= \$300 - \$1{,}000(A/P,\ 20\%,\ 10) \\
&= \$61
\end{aligned}
$$

$$
\begin{aligned}
AW\ (III) &= \$320 - \$1{,}100(A/P,\ 20\%\ 10) \\
&= \$320 - \$262 \\
&= \$57
\end{aligned}
$$

$$
\begin{aligned}
AW\ (IV) &= \$550 - \$2{,}000(A/P,\ 20\%,\ 10) \\
&= \$550 - \$477 \\
&= \$73
\end{aligned}
$$

Rank-ordering alternatives on the basis of decreasing annual worth results in IV $>$ II $>$ III $> \emptyset >$ I. This is precisely the same conclusion that results from the present worth procedure. This is so because

$$
AW = PW(A/P,\ i,\ N)
$$

That is, for given values of i and N, the annual worth is simply the equivalent present worth multiplied by a constant, the capital recovery factor. The relative ranking is maintained.

**4.3
RANKING BY
MARGINAL
(INCREMENTAL)
ANALYSIS**

The methods discussed in Section 4.2 permit direct rank-ordering of alternatives on the basis of their respective figures of merit: PW, AW (or AC). This section now turns to two other methods, rate of return and benefit-cost ratio, for which such an approach is not appropriate. As you will see, mutually exclusive alternatives may not, in general, be rank-ordered on the basis of IRR or $B:C$.

The (Internal) Rate of Return Method
The (internal) rate of return for a given investment proposal is that interest rate, i^*, for which the *net* present value of all expected cash flows is precisely zero. When all cash flows are discounted at rate i^*, the equivalent present worth of all benefits exactly equals the equivalent present worth of project costs. As indicated in Chapter 3, one mathematical definition of the internal rate of return is that rate, i^*, that satisfies the equation

$$PW = \sum_{j=0}^{N} A_j (1 + i^*)^{-j} \equiv 0$$

This formula assumes discrete cash flows A_j and end-of-period discounting in periods $j = 1, 2, \ldots, N$.

Recall that the discount rate used in present worth calculations is the opportunity cost—a measure of the return that could be earned on capital if it were invested elsewhere. Thus a given proposed project should be economically attractive if and only if its internal rate of return (*IRR*) exceeds the cost of opportunities forgone as measured by the firm's discount rate. (See Chapter 11 for an additional discussion.)

Returning to our example, the *IRR* for Alternative I is computed as:

$$PW \ (I) \ = \ -\$1{,}000 \ + \ \$4{,}000 \ (P/F, \ i^*, \ 10) \equiv 0$$

$$(P/F, \ i^*, \ 10) \ = \ 0.25$$

$$i^* \ (I) \ \simeq \ 0.15$$

Since the opportunity cost in this example is 0.20, it is clear that Alternative I is *not* attractive; the funds should be invested elsewhere.

The *IRR* for Alternative II is computed as:

$$PW \ (II) \ = \ -\$1{,}000 \ + \ \$300 \ (P/A, \ i^*, \ 10) \equiv 0$$

$$(PW, \ i^*, \ 10) \ = \ 3.3333$$

$$i^* \ (II) \ \simeq \ 0.275$$

Thus, the funds invested in Alternative II produce a higher return (27.5 percent) than if they are invested elsewhere (20 percent).

Similarly, the *IRR*'s for Alternatives III and IV are computed as:

$$PW \ (III) \ = \ -\$1{,}100 \ + \ \$320(P/A, \ i^*, \ 10) \equiv 0$$

$$(P/A, \ i^*, \ 10) \ = \ 3.4375$$

$$i^* \ (III) \ \simeq \ 0.286$$

$$PW \ (IV) \ = \ -\$2{,}000 \ + \ \$550(P/A, \ i^*, \ 10) \equiv 0$$

$$(P/A, \ i^*, \ 10) \ = \ 3.6364$$

$$i^* \ (IV) \ \simeq \ 0.245$$

Thus, Alternatives III and IV are also preferable to doing nothing inasmuch as their respective *IRR*'s exceed 0.20.

At this point it is tempting to rank-order Alternatives II, III, and IV according to their *IRR*'s. But that would be a serious methodological error, because the marginal, or incremental, effects between pairs of projects must be considered.

We know that II > $\emptyset$ and III > $\emptyset$. But is III preferable to II? To resolve this question, note that an *incremental* cost of $100 yields an *incremental* positive cash flow of $20 per period, each and every period, at the end of periods 1 through 10:

	Cash flows		Difference
End of period	Alternative II	Alternative III	III − II
0	− $1,000	− $1,100	− $100
1–10	300	320	20

The rate of return on this incremental investment is

$$PW \ (\text{III} - \text{II}) = -\$100 + \$20(P/A, \ i^*, \ 10)$$

$$(P/A, \ i^*, \ 10) = 5.0$$

$$i^* \ (\text{III} - \text{II}) \simeq 0.15$$

Since this incremental rate of return is less than the minimum required (0.20), the additional investment in Alternative III is not justified; Alternative III is *not* preferable to Alternative II.

Now we must determine whether IV is preferable to II. (The question of whether IV is preferable to III is irrelevant at this point.) The incremental rate of return is determined as:

	Cash flows		Difference
End of period	Alternative II	Alternative IV	IV − II
0	− $1,000	− $2,000	− $1,000
1–10	300	550	250

$$PW \ (\text{IV} - \text{II}) = -\$1,000 + \$250(P/A, \ i^*, \ 10)$$

$$(P/A, \ i^*, \ 10) = 4.0$$

$$i^* \ (\text{IV} - \text{II}) \simeq 0.21$$

Alternative IV is preferable to Alternative II since its incremental rate of return (0.21) exceeds the opportunity cost (0.20).

The analytical procedure outlined above is summarized in Table 4.3. The arrows shown in the first column ($x \rightarrow y$) indicate that we are evaluating the rate of return on the incremental investment of y over x, that is, the incremental rate of return resulting from the cash flows

$$\Delta A_j = A_j(y) - A_j(x) \tag{4.1}$$

Table 4.3

(Internal) Rate of Return Analysis of Alternatives from Table 4.1

Step	Comparison of alternatives	Incremental rate of return	Conclusion (Assuming $MARR$ = 20%)
1	$\emptyset \rightarrow$ I	15%	I $< \emptyset$
2	$\emptyset \rightarrow$ II	27.5	II $> \emptyset$
3	$\emptyset \rightarrow$ III	28.6	III $> \emptyset$
4	$\emptyset \rightarrow$ IV	24.5	IV $> \emptyset$
5	II $\rightarrow$ III	15	III $<$ II
6	II $\rightarrow$ IV	21	IV $>$ II

So

$$PW = \sum_{j=0}^{N} (\Delta A_j)(1 + i^*)^{-j} \equiv 0 \qquad (4.2)$$

Observe that this marginal, or incremental, procedure yields the conclusions IV $>$ II $>$ III $> \emptyset >$ I. This is the identical ranking that resulted from applying both the present worth and annual worth methods. Ranking will *always* be identical if the analyst ensures that only *incremental* rates of return are considered.

Again, it is incorrect to rank-order alternatives solely on the basis of their respective internal rates of return, thereby concluding that alternatives having the higher rate are economically superior. This is known as the *ranking error*. *IRR*'s are *not* indexes of superiority. As the example problem shows, ranking by *IRR*'s yields incorrect results.

Alternative	*IRR* (i^*)	Rank by *IRR*	Correct ranking
$\emptyset$	($MARR$ = 20%)	4	4
I	15%	5	5
II	27.5	2	2
III	28.6	1	3
IV	24.5	3	1

Correctly applying the (internal) rate of return method requires determination of the prospective return on the incremental investment, but irrelevant increments need not be evaluated. Note that Steps 3 and 4 in Table 4.3 are unnecessary; if II $> \emptyset$, then we need not also compare III and IV with the do-nothing alternative. After Step 2 we learn that II $> \emptyset >$ I. All that remains is to compare III and IV with II. In Step 5 we learn that III $<$ II. Since it is now unnecessary to compare IV with III, we compare IV with II in Step 6, concluding that IV $>$ II. Now we learn that IV $>$ II $>$ III and II $> \emptyset >$ I, so we conclude that IV is the preferred alternative. To complete the rank-ordering, if desired, we must compare III with $\emptyset$, as shown in Step 3 of Table 4.3. Since III $> \emptyset$, the completed rank-order is IV $>$ II $>$ III $> \emptyset >$ I.

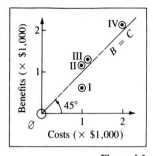

Figure 4.1

Benefits (× $1,000) — vertical axis
Costs (× $1,000) — horizontal axis
45°
IV
III
II
I
B = C

The Benefit-Cost Ratio Method

The necessity for *incremental* analysis when using benefit-cost ratios is first discussed in Chapter 3. Applying this procedure to the current example, the solution is outlined in Table 4.4. The benefits shown in Table 4.4 are the equivalent present values of the positive cash flows given in Table 4.1; the costs are the equivalent present values of the negative cash flows. Benefits and costs are graphed in Figure 4.1. The benefit-cost analysis summarized in Table 4.4 indicates that IV > II > III > Ø > I, the identical ranking obtained by the other three methods.

Table 4.4
The Benefit-Cost Ratio Method in Evaluating Alternatives
from Table 4.1

Step	Comparison of alternatives	Incremental benefits	Incremental costs	Incremental B : C	Conclusion
1	Ø → I	$ 646	$1,000	0.65	I < Ø
2	Ø → II	1,258	1,000	1.29	II > Ø
3	Ø → III	1,341	1,100	1.22	III > Ø
4	Ø → IV	2,305	2,000	1.15	IV > Ø
5	II → III	83	100	0.83	III < II
6	II → IV	1,047	1,000	1.05	IV > II

Again, steps 3 and 4 in Table 4.4 are irrelevant. In Step 2 we determined that II > Ø; thus we need only compare the next proposal, Alternative III, with Alternative II. If III > II, then it must also be true that III > Ø. The same argument holds for alternative IV.

To further emphasize the need for incremental analysis, consider the following example:

Alternative	Benefits	Costs	B : C
T	$300,000	$230,000	1.30
U	360,000	400,000	0.90
V	550,000	500,000	1.10
W	620,000	520,000	1.19

Clearly, of the four alternatives, only U is not preferable to Ø, because its benefit-cost ratio is less than unity. The correct analysis for the remaining three alternatives is shown in Table 4.5. The

Table 4.5
Benefit-Cost Analysis of Example Problem

Step	Comparison of alternatives	Incremental benefits	Incremental costs	Incremental B : C	Conclusion
1	Ø → T	$330,000	$230,000	1.30	T > Ø
2	T → V	250,000	270,000	0.93	V < T
3	T → W	320,000	290,000	1.10	W > T

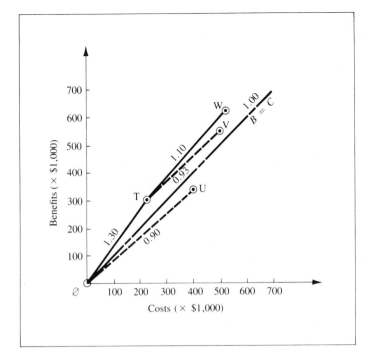

Figure 4.2 Geometric Solution of Example Problem

comparable geometric argument is shown graphically in Figure 4.2, where we find that the preferred choice is *not* the alternative with the highest benefit-cost ratio, $B : C$ (T) $= 1.30$, but rather Alternative W, with $B : C$ (W) $= 1.19$.

That *incremental,* or marginal, analysis must be employed to identify the optimal alternative is demonstrated in the above examples. That incremental analysis *always* results in maximum present worth is illustrated by Figure 4.3. Here, the line $ØMN$ represents the set of all alternatives under consideration. Of course, in real-world application, this line is not continuous, because there is a finite number of alternatives and their incremental benefit-cost ratios are irregular. Nevertheless, this simplification illustrates the point and in no way violates the conclusion.

Benefits and costs are measured by their equivalent present worths. If the ordinate and abscissa are scaled identically, the 45-degree line represents the set of all points at which benefits equal costs; its slope is exactly 1.0.

Assuming that line $ØMN$ is a continuous, differentiable function, we can show that the maximum excess of benefits over costs occurs at the point where the rate of change of benefits relative to costs, the incremental benefit-cost ratio, is exactly equal to unity. That is, we maximize $P = B - C$ by taking the first derivative with respect to costs and setting it equal to zero:

Figure 4.3 The Point of
Maximum Excess of Benefits
Over Costs

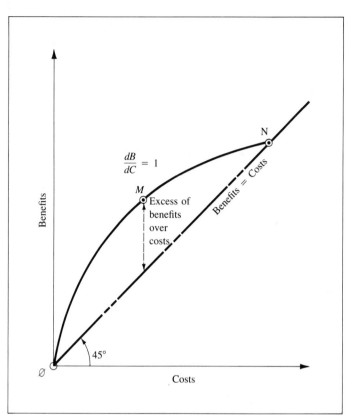

$$\text{Max } P = B - C$$
$$\frac{dP}{dC} = \frac{dB}{dC} - \frac{dC}{dC} = 0$$

Thus

$$\frac{dB}{dC} = 1.0$$

Here, dB/dC is the marginal benefit-cost ratio. Present worth, the excess of benefits over costs, is maximized at the point where dB/dC equals unity.

**4.4
THE PROBLEM
OF PRELIMINARY
SELECTION**

This section explores errors that may occur in preselecting alternatives and shows how improperly using the benefit-cost ratio and present worth methods can lead to incorrect solutions.

Preselection Error—Benefit-Cost Ratio Method
Assume that two alternative designs, P1 and P2, are being considered for a certain facility. The relevant data and analysis are

summarized in the following list. All dollar amounts are equivalent present values, in thousands.

Alternative	Benefits	Costs	B : C	B : C
P1	$180	$100	$ 80	1.80
P2	250	150	100	1.67

If the analyst rank-orders the project using benefit-cost ratios, he will recommend P1 because $B : C$ (P1) $= 1.80$, whereas $B : C$ (P2) $= 1.67$.

Now, assume that three additional projects, Q, R, and S, are being considered in addition to P1 and that only $400,000 is available to invest. The benefits and costs for projects Q, R, and S are shown in Table 4.6 (see Packages 2, 3, and 4). The four projects are *not* technologically mutually exclusive, yet with a budget of only $400,000, not all can be accepted.

Table 4.6
Benefit-Cost Analysis Given a $400,000 Budget Constraint
(dollar values in thousands)

Package	Alternatives	Benefits	Costs*	$PW = B - C$	B : C
1	P1	$ 180	$100	$ 80	1.80
2	Q	400	250	150	1.60
3	R	450	300	150	1.50
4	S	55	50	5	1.10
5	P1 + Q	580	350	230	1.64
6	P1 + R	630	400	230	1.56
7	P1 + S	235	150	85	1.56
8	Q + R	850	550	Not feasible	
9	Q + S	455	300	155	1.51
10	R + S	505	350	155	1.47
11	P1 + Q + R	735	650	Not feasible	
12	P1 + Q + S	635	400	235	1.57
13	P1 + R + S	685	450	Not feasible	
14	Q + R + S	905	600	Not feasible	
15	P1 + R + Q + S	1085	700	Not feasible	

*Assumes that all costs are initial costs

There are $2^4 - 1 = 15$ budget packages, or portfolios, excluding the do-nothing alternative. The relevant data for and analyses of these fifteen packages are given in Table 4.6. Note that Package 1, P1 alone, yields the highest benefit-cost ratio. Of the packages that exactly consume the $400,000 budget, the one with the highest ratio is Package 12 (P1 + Q + S). This, of course, is the best of the feasible packages because it also maximizes the present worth, or $B - C$.

It is important to note that by choosing P1 over P2 we have overlooked what is in fact the optimal package: P2 + Q. With these projects, a cost of $400,000 ($150,000 + $250,000) yields benefits of $650,000 ($250,000 + $400,000). Thus the procedure used at the design level was incorrect. The analyst should *not* have selected the preferred project by rank-ordering on benefit-cost ratios; the ranking error led to the wrong solution.

The benefit-cost method, *properly applied,* would have led to the correct solution.

Comparison of alternatives	Incremental benefits	Incremental costs	Incremental $B:C$	Conclusion
$\emptyset \rightarrow$ P1	$180	$100	1.80	P1 > $\emptyset$
P1 $\rightarrow$ P2	70	50	1.40	P2 > P1

Given this example, it is tempting to conclude that the optimal solution, P2 + Q, was obtained *because of* the proper application of the benefit-cost ratio method. However, as you will see, this conclusion is unwarranted. To illustrate, we will show that the present worth method may also lead to a suboptimal solution owing to **preselection error,** even though the question of rank-ordering on ratios does not apply to this method. Then we will present an additional example wherein the benefit-cost ratio method is applied properly, yet the preselection error persists.

Preselection Error—Present Worth Method

Consider the selection of one project from each of four units. (In general, a "unit" is an operating department, a section, a division, a plan, or a group. It is a budgeting unit wherein a single project must be selected from a set of mutually exclusive alternatives. Proposals are independent *between* units but are mutually exclusive *within* units.) The relevant data and an analysis are summarized in the following table. As in the previous example, all dollar amounts are equivalent present values. Costs, in this case, are assumed to be initial costs.

Unit	Alternative	Benefits	Costs	$PW = B - C$	$B:C$
A	A1	$ 600	$300	$300	2.0
	A2	720	400	320	1.8
B	B1	300	150	150	2.0
	B2	360	200	160	1.8
C	C1	1200	600	600	2.0
	C2	1440	800	640	1.8
D	D1	700	350	350	2.0
	D2	820	450	370	1.8

Clearly, based on maximum *PW* (B − C), Alternatives A2, B2, C2, and D2 are preferable to A1, B1, C1, and D1, respectively. Now suppose that $1,400 is available for the total capital budget. Considering only these four "preferred" alternatives, the optimal feasible package as shown in Table 4.7, is

Table 4.7
Benefit-Cost Analysis Given a $1,400 Budget Constraint

Package	Alternatives	Benefits	Costs*	$PW = B - C$
1	A2	$ 720	$ 400	$ 320
2	B2	360	200	160
3	C2	1,440	800	640
4	D2	820	450	370
5	A2 + B2	1,080	600	480
6	A2 + C2	2,160	1,200	960
7	A2 + D2	1,540	850	690
8	B2 + C2	1,800	1,000	800
9	B2 + D2	1,180	650	530
10	C2 + D2	2,260	1,250	1,010
11	A2 + B2 + C2	2,520	1,400	1,120
12	A2 + B2 + D2	1,900	1,050	850
13	A2 + C2 + D2	2,980	1,650	Not feasible
14	B2 + C2 + D2	2,620	1,450	Not feasible
15	A2 + B2 + C2 + D2	3,340	1,850	Not feasible

*Assumes that all costs are initial costs

A2 + B2 + C2; the PW is $1,120 at a total cost of $1,400. However, reviewing the original data, the same expenditure for A1 + B1 + C1 + D1 produces a PW of $1,400, or an increase of $280.

Alternative	Benefits	Costs	$PW = B - C$
A1	$ 600	$ 300	$ 300
B1	300	150	150
C1	1200	600	600
D1	700	350	350
Total	$2800	$1400	$1400

This is another illustration of the preselection error, proving that preselection *on any basis* can lead to a suboptimal allocation of resources. Here we have shown that preselection using the present worth method results in the incorrect solution.

This point is further illustrated by considering the following example, which shows that through selection on the basis of the maximum benefit-cost ratio in each unit, allocation is suboptimal.

Unit	Alternative	Benefits	Costs	$PW = B - C$	$B : C$
X	X1	$600	$300	$300	2.0
	X2	720	400	320	1.8
Y	Y	110	100	10	1.1
Z	Z	420	400	20	1.05

Assuming that our budget is limited to $800, preselection of X1 (because it has the highest $B : C$) results in seven feasible packages: X1, Y, Z, X1 + Y, X1 + Z, Y + Z, and X1 + Y + Z. Of these, the apparent optimal package is the last one:

Alternative	Benefits	Costs	$PW = B - C$
X1	$ 600	$300	$300
Y	110	100	10
Z	420	400	20
Total	$1,130	$800	$330

The benefit-cost ratio for this package is $1,130/$800 = 1.41.

But if we carefully review this analysis, we see that preselection of X1 leads us to a false optimum solution. In fact, if X2 had been included in our subsequent consideration of budget packages, we would have evaluated X2 + Y and X2 + Z. (X2 + Y + Z is not feasible; total cost = $400 + $100 + $400 = $900.) Of these, the optimal feasible package is X2 + Z.

Alternative	Benefits	Costs	$PW = B - C$
X2	$ 720	$400	$320
Z	420	400	20
Total	$1,140	$800	$340

The benefit-cost ratio for this expenditure of $800 is $1,140/ $800 = 1.45. This example demonstrates two common errors. First, we should not have selected X1 over X2 merely because $B : C$ (X1) $> B : C$ (X2). Second, once X1 was selected, the true optimal combination of independent projects became obscured.

In order to be aware of the preselection problem, it is necessary to restate the appropriate decision rule: (1) When choosing between alternatives, we may rank-order on the basis of present worths, not benefit-cost ratios, assuming that the difference in costs will be invested elsewhere at the minimum attractive rate of return. This is so because, as noted previously, the net discounted present value of funds invested elsewhere is exactly zero. A sum of money, ΔC, invested at rate i results in a future sum, or benefit, of $\$\Delta C(1 + i)^N$ at the end of N periods. If discounting also takes place at rate i, then $\Delta B = [\Delta C(1 + i)^N][(1 + i)^{-N}] = \Delta C$. So $\Delta PW = \Delta B - \Delta C = 0$ and, of course, $\Delta B : C = \Delta B/\Delta C = 1.0$.

To illustrate this point, consider Figure 4.4, the graph of benefits and costs for our previous problem. Recall that $B : C$ (X1) $= 2.0$ and $B : C$ (X2) $= 1.8$. If all *additional* funds yield a benefit-cost ratio of 1.0, then for any given total budget allocation of at least $400, the initial choice of X2 always results in greater net benefits than if we had selected X1. This is so even though $B : C$ (X1) $> B : C$ (X2). Ranking on present worth results in the correct solution, PW (X1) $< PW$ (X2), assuming, of course, that we did not make a preselection error. (2) If either the benefits or the costs of the alternatives are equal, then we may rank-order on either PW or $B : C$. That is, given a pair of alternatives, P and Q,

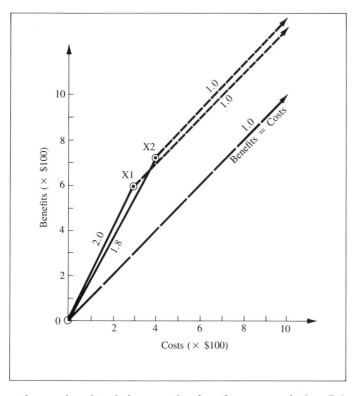

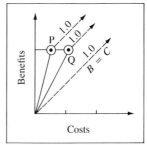

Figure 4.4 Benefits as a Function of Costs for Sample Problem (Alternatives X1 and X2 Only)

and assuming that their respective *benefits are equal*, then P is preferred to Q if $B : C$ (P) $> B : C$ (Q). (See Figure 4.5.) Similarly, assuming that the *costs* of P and Q are *equal*, then P is preferred to Q if $B : C$ (P) $> B : C$ (Q). (See Figure 4.6.) (3) Using either *PW* or $B : C$ may lead to suboptimal decisions if preselection takes place. That is, preselection on the basis of either *PW* or $B : C$ ranking does *not* ensure that the optimal budget package will result. Preliminary selection from a set of mutually exclusive alternatives may yield an incorrect solution because it has been assumed that the excluded alternative(s) should not be considered further. This assumption is valid, however, only when unlimited capital is available for investment. If the analyst does not consider *all* possible alternatives, there is no assurance that the final capital budget represents the optimum employment of limited funds.

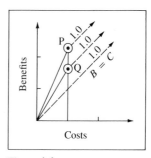

Figure 4.5

Figure 4.6

Central to the problem of economic analysis is the concept of *mutually exclusive alternatives,* which are investment proposals such that the selection of one necessarily precludes all others. Mutually exclusive alternatives arise from two fundamental situations. First, there are those that are inherently mutually exclusive because of their form or function; these are technologically mutually exclusive. Second, some alternatives are mutually exclusive at the margin because limited funds do not permit all proposals to be accepted; these are *financially mutually exclusive.* Selecting some proposals necessarily precludes selecting one or more of the others. Independent alternatives that are financially mutually exclusive may be structured into an exhaustive set of mutually exclusive budget packages from which one and only one optimal package can be selected.

Regardless of whether alternatives are mutually exclusive technologically or financially, valid procedures must be employed for selecting the preferred alternative. These procedures are summarized in Table 4.8. The present-worth method permits the rank-ordering of alternatives on the basis of their respective present worths. By noting that the *PW*'s and *AW*'s of a set of projects are related by the capital recovery factor, $(A/P, i, N)$, which is constant for given i and N, it follows that there is a one-to-one correspondence between the ordered set of *PW*'s and *AW*'s. Thus the annual worth method also permits the direct rank-ordering of alternatives on the basis of their respective annual worths.

The (internal) rate of return and benefit-cost ratio methods both require careful consideration of the incremental consequences between pairs of alternatives. The *IRR* and *B* : *C* for a given project reflect only the differences between that alternative and the do-nothing case. It is necessary to examine the differences between pairs of alternatives to ensure that the incremental *IRR* exceeds the opportunity cost or, alternatively, to ensure that the *B* : *C* exceeds unity. The tests are equivalent. Successive applications of this test ensure that the net present worth is maximized. Analysts are guilty of the ranking error if they rank-order projects solely on the basis of decreasing *IRR*'s or benefit-cost ratios.

Finally, preselection can lead to suboptimal allocation of the capital budget. Analysts must be alert to the possibility that selection of an apparent "best alternative" at some preliminary design or budget level may preclude some future combination that, if examined, may yield a higher present worth. When financially mutually exclusive alternatives are being considered, correct analysis may be based on the definition of an exhaustive set of alternative capital budget packages, each of which is mutually exclusive.[3] Once all the packages have been completely described, the optimum package can be selected by any of the several methods commonly used in economic analysis that take into account the amounts and timing of cash flows and the time value of money.

3. The "budget package" approach described here is relatively inefficient because it is an exhaustive search procedure. Other mathematical programming procedures for this capital budgeting problem are more efficient but are beyond the scope of this book.

Table 4.8
Summary of Investment Criteria

Assumptions: Discrete cash flows
              End-of-period discounting/compounding

Notation: i = effective interest rate per period (opportunity cost); may be written as k (cost of capital)

$i*$ = internal rate of return

B_j = cash flow at end of period j for consequences considered to be "benefits" $(j = 0, 1, 2, \ldots, N)$

C_j = cash flow at end of period j for consequences considered to be "costs" $(j = 0, 1, 2, \ldots, N)$

Present Worth Method
Given: B_j, C_j, N and i

Figure of merit: $PW = \sum_{j=0}^{N} (B_j - C_j)(1 + i)^{-j}$

Test: Accept if $PW > 0$

Remarks: Projects *can* be rank-ordered by decreasing PW's.
        Annual worths (AW's) are directly related
$$AW = PW\,(A/P, i, N)$$
        As in all methods, N must be constant for all alternatives.

(Internal) Rate of Return Method
Given: B_j, C_j, N and k

Figure of merit: $i*$, where $= \sum_{J=0}^{N} (B_j - C_j)(1 + i*)^{-j} \equiv 0$

Test: Accept if $i* > k$

Remarks: Projects *cannot* be rank-ordered by decreasing $i*$'s.

Benefit-Cost Ratio Method
Given: B_j, C_j, N and i

Figure of merit: $B : C$ where $B = \sum_{j=0}^{N} B_j(1 + i)^{-j}$

$$C = \sum_{j=0}^{N} C_j(1 + i)^{-j}$$

Test: Accept if $B : C > 1.0$

Remarks: Projects *cannot* be rank-ordered by decreasing $B : C$'s

---

Note: Unless otherwise indicated, assume discrete cash flows, end-of-period compounding/discounting, and effective interest rates.

PROBLEMS

Present Worth or Annual Worth

4.1   The Comma Corporation is considering the relocation of its western processing plant. Management is presently considering six alternatives:

Alternative	Required initial investment	Estimated annual reduction in disbursements
I	$1,300,000	$ 300,000
II	1,600,000	500,000
III	2,400,000	820,000
IV	2,600,000	840,000
V	3,600,000	1,200,000
VI	5,000,000	1,520,000

Assume that the relocated plant will be used for fifty years. Regardless of which alternative is selected, it is expected that the plant will have a net salvage value of approximately zero at the end of the fifty years. The pretax minimum attractive rate of return used by Comma Corporation is 25 percent. Use the annual worth method to determine which alternative, if any, should be selected.
(*Answer:* Plan V)

4.2 Use the present worth method to make a selection from the alternatives described in Problem 4.1.

4.3 The Asterisk Association is considering the purchase of a small computer for its research department. Several mutually exclusive alternatives are being considered. The estimates for each are:

Computer	First cost of computer	Estimated salvage value 10 years hence	Net annual savings resulting from new computer
A	$280,000	$240,000	$46,000
B	340,000	280,000	56,000
C	380,000	310,000	62,000
D	440,000	350,000	72,000

The company plans to keep the computer for ten years and then sell it. If the pretax minimum attractive rate of return is 15 percent, use the present worth method or the annual worth method to determine which alternative, if any, should be selected.
(*Answer:* Alternative B)

4.4 The Current Lighting System is presently installed at the executive headquarters of the Certain Corporation. Two other systems, Lumenite and Wattsnu, are under consideration, both of which are more costly initially but will increase the time between replacements. Estimates for the proposals are:

	Current system	Lumenite system	Wattsnu system
Time between replacements	2 years	3 years	6 years
Initial cost	$20,000	$30,000	$50,000
Annual operating costs	$ 3,000	$ 3,500	$ 2,500
Net salvage value	$ 4,000	$ 5,000	$10,000

Assuming a minimum attractive rate of return of 10 percent before taxes, use the present worth method or the annual worth method to determine whether the Current System should be retained or replaced.

4.5 Consider the following sequences of cash flows:

End of period	Alternative I	Alternative II	Alternative III
0	− $100	− $120	− $300
1	20	10	90
2	20	20	80
3	40	30	70
4	40	40	60
5	40	50	50

Assume a discount rate of 10 percent per period.
a. Find the present worth for each alternative.
b. Find the annual worth for each alternative.
c. Find the future worth for each alternative.
(*Answers:* (a) $16.92, − $13.49, − $27.43;
       (b) $4.46, − $3.56, − $7.23;
       (c) $27.25, − $21.73, − $44.18)

4.6 The management at the Mars Manufacturing Company is considering a proposal to reduce the cost of heating and cooling its warehouse by installing certain insulation materials. The do-nothing alternative is expected to result in continuing energy costs of $50,000 per year over the next twenty years. Six mutually exclusive insulation plans are under consideration. The expected economic consequences over a twenty-year planning horizon are:

Alternative	Insulation thickness	Initial cost	Annual cost of energy
Ø	(no insulation)		$50,000
U	1.0 cm	$100,000	20,000
V	1.2	120,000	22,000
W	1.4	140,000	12,000
X	1.5	150,000	15,000
Y	1.8	180,000	10,000
Z	2.0	200,000	10,000

The company's pretax minimum attractive rate of return is 20 percent per year. Although the cost of energy varies somewhat from month to month during the year, assume that energy costs occur continuously and uniformly during each year. Determine the preferred alternative using the present worth method.

Internal Rate of Return

4.7 Use the rate of return method to select from the alternatives described in Problem 4.1.
(*Answer:* Plan V)

4.8 Use the rate of return method to select from the four alternatives described in Problem 4.3.

4.9 Use the rate of return method to select from the alternatives described in Problem 4.4.
(*Answer:* Current System)

4.10 Refer to the cash flows for the three alternatives in Problem 4.5.
a. Find the internal rate of return for each alternative.
b. Which of the alternatives, if any, are preferable to the do-nothing alternative.
c. What is the rate of return on the incremental investment in Alternative III as compared to Alternative II?
d. At what interest rate, if any, is Alternative III preferable to Alternative II?

4.11 Refer to Problem 4.6. Determine the preferred alternative using the rate of return method.
(*Answer:* Alternative W)

4.12 The Mars Manufacturing Company is required by the Occupational Safety and Health Administration (OSHA) to provide certain safety equipment in its plant operations. The prospective cash flows for the four plans under consideration are given below. The firm's pretax minimum attractive rate of return is 20 percent per year.

End of year	Plan A	Plan B	Plan C	Plan D
0	− $10,000	− $13,000	− $8,000	− $12,000
1–5	− 2,000	− 1,000	− 3,000	− 1,000
5	4,000	5,000	0	0

a. Choose any method—present worth, annual worth, or rate of return—to rank-order the four alternatives.
b. Find the interest rate at which the firm would be indifferent between Plan C and Plan D.

4.13 A firm is considering three mutually exclusive alternatives as part of a production improvement program:

	Plan A	Plan B	Plan C
Initial cost	$10,000	$15,000	$20,000
Uniform annual benefit	$ 1,625	$ 1,625	$ 1,890
Useful life	10 years	20 years	20 years

The salvage value at the end of the useful lives of all three alternatives is zero. At the end of ten years, Alternative A could be replaced with an identical replacement. The firm's minimum attractive rate of return is 10 percent. The planning horizon is twenty years. Use the rate of return method to determine which alternative should be selected.
(*Answer:* Plan A)

Benefit-Cost Ratio Method
4.14 Refer to Problem 4.6. Determine the preferred alternative using the benefit-cost ratio method.

4.15 Consider three mutually exclusive alternatives with the following cash flows:

End of year	Alternative A	Alternative B	Alternative C
0	−$100	−$150	−$110
1–4	47	66	49

The firm's minimum attractive rate of return is 10 percent. Use the benefit-cost ratio method to determine the preferred alternative.
(*Answer:* Alternative B)

4.16 Consider four mutually exclusive alternatives:

	D	E	F	G
Initial cost	$75	$50	$15	$90
Uniform annual benefit	18.8	13.9	4.5	23.8

Each alternative has a five-year useful life and no salvage value. The minimum attractive rate of return is 10 percent. Use the benefit-cost ratio method to determine which alternative should be selected.

4.17 The Municipal Department of Parks and Recreation is considering four proposals for new landscaping equipment. Analyzing the relevant costs and benefits of these mutually exclusive alternatives yields:

Alternative	Equivalent annual benefits	Equivalent annual costs
H	$182,000	$91,500
I	167,000	79,500
J	115,000	88,500
K	95,000	50,000

Benefits and costs have been annualized using a ten-year assumed life and a 8-percent discount rate. Using the benefit-cost ratio method, determine which alternative is preferable.
(*Answer:* Alternative H)

4.18 A public agency is considering five alternatives for the design of a certain flood-control culvert. The discounted present values of expected benefits and costs are:

Alternatives	Benefits	Costs
L	$ 80,000	$100,000
M	148,000	120,000
N	130,000	110,000
P	180,000	150,000
Q	185,000	160,000

Using the benefit-cost ratio method, determine which alternative, if any, is preferable.

4.19 Five mutually exclusive alternatives, each with a twenty-year useful life with zero salvage value, are summarized below. Assuming a 6-percent discount rate, use the benefit-cost ratio method to determine which alternative should be selected.

Alternative	I	II	III	IV	V
PW of costs	$4000	$2000	$6000	$1000	$9000
Uniform annual benefits	639	410	761	117	785

(*Answer:* Alternative I)

4.20 The public works department for the state of Chaos has prepared an economic analysis of seven different designs for a proposed project. These alternatives are mutually exclusive. The design engineers have determined the economic consequences for each alternative as measured over a twenty-five-year planning horizon, have discounted the expected cash flows using a 10-percent discount rate, and have summarized the results in the following table. (All figures in the table are equivalent present values in millions of dollars.)

	Alternatives						
Consequences	T	U	V	W	X	Y	Z
Initial investment	0.8	3.0	0	2.2	1.0	1.0	3.0
Cost of operations and maintenance	0.2	0	0.5	0.3	1.0	0.4	1.0
Benefits to users	0.5	2.7	2.0	4.0	3.0	3.5	5.2

4.21 The public works department of Small City is considering the construction of a bridge over a flood-control channel. Four alternative designs are being considered. There are three categories of economic consequences: (a) construction costs, (b) expenses associated with periodic maintenance, and (c) benefits to the users of the bridge. All future cash flows over a thirty-year planning horizon have been discounted at a 10-percent discount rate. Resulting equivalent present worths are summarized below.

	Present values			Benefit-cost ratios	
Design	Construction costs (C)	Periodic maintenance (M)	User benefits (B)	$\dfrac{B - M}{C}$	$\dfrac{B}{C + M}$
I	$100,000	$50,000	$200,000	1.50	1.67
II	300,000	0	240,000	0.80	0.80
III	400,000	20,000	480,000	1.15	1.14
IV	300,000	60,000	540,000	1.60	1.50

Using the benefit-cost ratio method in the proper manner, rank the five alternatives, including the do-nothing ($\emptyset$) alternative.

(*Answer:* IV $>$ III $>$ I $>$ $\emptyset$ $>$ II)

4.22 Consider a set of mutually exclusive alternatives for which pertinent data are:

Alternative	Initial cost	Annual benefits
R	$ 0	$ 0
S	100,000	30,000
T	120,000	29,000
U	150,000	38,000
V	150,000	35,000
W	200,000	40,000
X	180,000	40,000

Benefits are measured over a twenty-year planning horizon. The appropriate discount rate is 10 percent before taxes. Use *either* the rate of return method or the benefit-cost ratio method to determine which, if any, of the alternatives is preferable.

4.23 A public agency is considering seven mutually exclusive alternatives. The discounted present values of benefits and costs, in thousands of dollars, are:

Alternative	Benefits	Costs
P	$40	$10
Q	50	30
R	30	20
S	30	40
T	75	40
U	90	60
V	95	70

Using the benefit-cost ratio method in the correct manner, determine the preferable alternative.
(*Answer:* Alternative T)

4.24 An agency is evaluating four mutually exclusive design proposals. Three major categories of consequences and their equivalent present values are:

Proposal	Direct costs (d)	Associated costs (a)	Benefits to users (b)
Q	$1030	$ 800	$2400
R	995	1800	3500
S	1280	0	1960
T	1150	850	2750

Mr. Smith evaluated the alternatives using the benefit-cost ratio method as follows:

Proposal	Benefits net of associated costs (b) − (a)	Costs (d)	Net B : C $\frac{(b) - (a)}{(d)}$
Q	$1600	$1030	1.55
R	1700	995	1.71
S	1960	1280	1.53
T	1900	1150	1.65

Mr. Jones, on the other hand, included the associated costs in the denominator, and thus his analysis was:

Proposal	Benefits (b)	Total costs (d) + (a)	Total B : C (b) / (d) + (a)
Q	$2400	$1830	1.31
R	3500	2795	1.26
S	1960	1280	1.53
T	2750	2000	1.35

Mr. Jones argued that Alternative S is economically superior because it has the highest $B : C$, 1.53. Mr. Smith claimed that Alternative R is preferable because its $B : C$ is highest, 1.71. Why are there differences in the relative as well as in the absolute values of the $B : C$ ratios? Which alternative do you think is preferable? Why?

Independent and Mutually Exclusive Alternatives

4.25 The budget committee of a manufacturing firm has approved the expenditure of $80,000 for improving its manufacturing facilities. The plant manager has received the following proposals for cost reduction programs from four department managers:

Department A submitted four mutually exclusive proposals for improving the receiving department:

Proposal	Initial investment	Net annual savings in disbursements
A1	$10,000	$2,000
A2	15,000	5,000
A3	25,000	7,800
A4	30,000	9,500

Department B submitted three mutually exclusive proposals for improving assembly:

B1	$10,000	$ 4,100
B2	20,000	7,800
B3	30,000	10,000

Department C submitted three mutually exclusive proposals for improving inspection:

C1	$10,000	$4,000
C2	15,000	5,000
C3	20,000	7,500

Department D submitted one proposal for a new forklift truck:

| D1 | $15,000 | $4,600 |

Assume that each proposal will have a five-year life and zero salvage value. The budget committee requires that all alternatives must earn at least 15 percent before taxes. Use the annual worth method or the present worth method to determine which alternatives should be recommended.*

(*Answer:* A2, B2, C3, and D1)

4.26 A comptroller is considering the following investment proposals received from production (E), quality control (F), and shipping (G):

Alternative	Initial investment	Annual excess of receipts over disbursements
E1	$2,000	$ 275
E2	4,000	770
F1	4,000	1,075
F2	8,000	1,750
G1	4,000	1,100

Alternatives E1 and E2 are technologically mutually exclusive; F1 and F2 are also mutually exclusive. Each alternative has an expected life of ten years, and zero salvage values are assumed. A minimum attractive rate of return of 10 percent is used by the firm.

a. Which proposals should be recommended if investment funds are essentially unlimited?

b. Which proposals should be recommended if only $14,000 is available for new investments?

4.27 Use the rate of return method to select from the alternatives described in Problem 4.25.

(*Answer:* A2, B2, C3 and D1)

*Based on a problem that first appeared in E. L. Grant and W. G. Ireson, *Principles of Engineering Economy*, 4th ed. (New York: The Ronald Press, 1960).

Computer Program

The following program, written in BASIC, may be used to solve many of the problems in this chapter. In particular, the user may input end-of-period j cash flows A_{jn} for each alternative n, as well as the effective interest rate i per period. The program determines present worths, annual worths, and benefit-cost ratios using the discount rate provided. The program also determines interest rates, internal rates of return ($i*$), for which the present worth is zero.

NOTE: The Applesoft BASIC version of this program is available in the solutions manual.

```
PROGRAM    00005 DIM PW(30), PWR(30), PWC(30), IRR(30,5), AW(30), BCRATIO(30)
           00008 DIM RT(5), INDEX(5), A(30,30)
           00010 PRINT "THIS PROGRAM DETERMINES PRESENT WORTHS, ANNUAL"
           00020 PRINT "WORTHS, BENEFIT-COST RATIOS, AND INTERNAL RATES"
           00030 PRINT "OF RETURN FOR MULTIPLE ALTERNATIVES."
           00050 PRINT " "
           00055 INPUT "ENTER EFFECTIVE INTEREST RATE OR M.A.R.R., k=";TT
           00060 INPUT "ENTER TOTAL NUMBERS OF ALTERNATIVES M=";MM
           00065 INPUT "ENTER THE MAXIMUM COMPOUNDING PERIODS N=";NP
           00070 PRINT " "
           00075 PRINT " ••••••••••••••••••••••••••••••••••••••••••••••••••••••••••••••••••••••••••• "
           00078 PRINT " "
           00080 PRINT "ENTER ALL THE END-FOR-PERIOD CASH FLOWS FOR EACH"
           00090 INPUT "ALTERNATIVE ACCORDING TO THE FOLLOWING INSTRUCTIONS"
           00095 PRINT " "
           00100 FOR I=1 TO MM
           00105 PWR(I)=0
           00107 PWC(I)=0
           00108 PRINT " "
           00109 PRINT " "
           00110 PRINT "CASH FLOWS FOR ALTERNATIVE";I;":"
           00125 PRINT " "
           00130 PRINT "END OF PERIOD   CASH FLOW"
           00140 PRINT "--------------------   ----------------"
           00150 FOR J=0 to NP
           00160 INPUT J,A(J+1,I)
           00170 IF A(J+1,I)>=0 THEN 200
           00180 PWC(I)=PWC(I)+A(J+1,I)*(1+TT)**(-J)
           00190 GOTO 210
           00200 PWR(I)=PWR(I)+A(J+1,I)*(1+TT)**(-J)
           00210 NEXT J
           00310 PW(I)=PWR(I)+PWC(I)
           00315 BCRATIO(I)=ABS(PWR(I)/PWC(I))
           00320 AW(I)=PW(I)*(TT*(1+TT)**NP)*((1+TT)**NP-1)**(-1)
           00330 GOSUB 1340
           00334 FOR LL=1 TO INDEX(I)
           00336 IRR(I,LL)=RT(LL)
           00338 NEXT LL
           00345 NEXT I
           00350 PRINT " "
           00354 PRINT " ••••••••••••••••••••••••••••••••••••••••••••••••••••••••••••••••••••••••••• "
           00356 PRINT " "
           00360 PRINT "NOW, LET US CHECK IF YOUR INPUT DATA IS CORRECT."
           00370 PRINT "PLEASE CHECK YOUR FOLLOWING INPUT:"
           00375 PRINT " "
```

```
00380 PRINT "END OF                                      ALTERNATIVE"
00382 PRINT "PERIOD";
00384 FOR I=1 TO MM
00390 PRINT USING "######## ";I;
00395 NEXT I
00400 PRINT "------------------------------------------------------------------------------------"
00405 FOR LA=0 TO NP
00408 PRINT USING "###### ";LA;
00410 FOR LB=1 TO MM
00412 PRINT USING "########## ",A(LA+1,LB);
00416 NEXT LB
00417 PRINT
00418 NEXT LA
00420 PRINT " "
00425 PRINT " "
00430 PRINT "IS THE ABOVE CASH FLOW MATRIX CORRECT?"
00440 INPUT "PLEASE TYPE 'Y' IF YES, AND 'N' IF NO.";LA$
00450 IF LA$="N" THEN 50
00455 PRINT " "
00460 PRINT "THE FINAL RESULTS FOR EACH ALTERNATIVE ARE GIVEN"
00470 PRINT "AS FOLLOWS:"
00475 PRINT " "
00480 PRINT "ALTERNATIVE   PRESENT WORTH   ANNUAL WORTH   B/C RATIO"
00500 PRINT "-------------   --------------------   ----------------------   ------------- "
00510 FOR KK=1 TO MM
00520 PRINT TAB(5);KK;"   ";
00522 PRINT USING "###,###,###.##",PW(KK),
00523 PRINT "   ";
00524 PRINT USING "###,###,###.##", AW(KK),
00525 PRINT "   ";
00526 PRINT USING "###.##",BCRATIO(KK)
00527 NEXT KK
00528 PRINT " ********************************************************************************** "
00529 PRINT " "
00530 PRINT "THE INTERNAL RATE(S) OF RETURN FOR EACH ALTERNATIVE ARE AS
   FOLLOWS:"
00540 FOR II=1 TO MM
00550 FOR JJ=1 TO INDEX (II)
00555 PRINT " "
00560 PRINT "ALTERNATIVE ";II;" ";" ";"INTERNAL RATE OF RETURN IS ";(I
   NT(100000*IRR(II,JJ)))/1000;"%"
00570 NEXT JJ
00580 NEXT II
00590 GOTO 6900
01300 REM   THE FOLLOWING IS THE SUBROUTINE TO CALCULATE THE INTERNAL
   RATE(S) OF RETURN
01340 INDEX(I)=0
01345 PW=0
01350 GOTO 1800
01500 PW=0
01505 FOR J=0 TO NP
01510 PW=PW+A(J+1,I)*(1+IT)**(-J)
01520 NEXT J
01580 RETURN
01720 PW=0
01730 FOR K=0 TO NP
01740 PW=PW+A(K+1,I)
01750 NEXT K
01760 RETURN
```

COMPUTER PROGRAM   129

```
01800 LET IT = 0
01810 GOSUB 1720
01820 PW0 = PW
01840 IF PW0 = 0 THEN 5000
01850 LET IT = 0.1
01860 GOSUB 1500
01870 IF PW = 0 THEN 5000
01880 IF SGN(PW) = - SGN(PW0) THEN 2000
01890 IF IT > = 20 THEN 6000
01900 IT = IT + 0.1
01950 GOTO 1860
02000 LET IT = IT - 0.01
02010 GOSUB 1500
02020 IF PW = 0 THEN 5000
02030 IF SGN(PW) = SGN(PW0) THEN 3000
02040 IT = IT - 0.01
02050 GOTO 2010
03000 LET IT = IT + 0.001
03100 GOSUB 1500
03200 IF PW = 0 THEN 5000
03300 IF SGN(PW) = - SGN(PW0) THEN 4000
03400 IT = IT + 0.001
03500 GOTO 3100
04000 LET IT = IT - 0.0001
04100 GOSUB 1500
04200 IF PW = 0 THEN 5000
04300 IF SGN(PW) = SGN(PW0) THEN 4650
04400 IT = IT - 0.0001
04500 GOTO 4100
04650 IT = IT + 0.00001
04700 GOSUB 1500
04750 IF PW = 0 THEN 5000
04800 IF SGN(PW) = - SGN(PW0) THEN 5000
04850 IT = IT + 0.00001
04900 GOTO 4700
05000 LET INDEX (I) = INDEX(I) + 1
05100 RT(INDEX(I)) = IT
05300 IT = IT + .01
05400 PW0 = - PW0
05500 GOTO 1860
06000 IF INDEX(I) = 0 THEN 6200
06100 GOTO 6500
06200 PRINT "THERE IS NO REAL ROOT FOR THE INTERNAL RATE OF RETURN"
06300 PRINT "FOR THIS ALTERNATIVE."
06500 RETURN
06900 END
```

Example

In this problem we consider three mutually exclusive alternatives and a planning horizon of two periods. The minimum attractive rate of return is 10 percent per period. Note that Alternative 2 in this example has two internal rates of return, 100 percent and 200 percent. If a single rate of return is desired, the analyst should use the external rate of return method as described in Section 3.3, Chapter 3.

THIS PROGRAM DETERMINES PRESENT WORTHS, ANNUAL WORTHS, BENEFIT-COST
RATIOS, AND INTERNAL RATES OF RETURN FOR MULTIPLE ALTERNATIVES.

ENTER EFFECTIVE INTEREST RATE OR M.A.R.R., k= ? .1
ENTER TOTAL NUMBERS OF ALTERNATIVES M= ? 3
ENTER THE MAXIMUM COMPOUNDING PERIODS N= ? 2

****

ENTER ALL THE END-FOR-PERIOD CASH FLOWS FOR EACH ALTERNATIVE ACCORD-
ING TO THE FOLLOWING INSTRUCTIONS

CASH FLOWS FOR ALTERNATIVE 1 ;

END OF PERIOD	CASH FLOW
? 0,	−100
? 1,	50
? 2,	90

CASH FLOWS FOR ALTERNATIVE 2 ;

END OF PERIOD	CASH FLOW
? 0,	100
? 1,	−500
? 2,	600

CASH FLOWS FOR ALTERNATIVE 3 ;

END OF PERIOD	CASH FLOW
? 0,	−200
? 1,	100
? 2,	90

THERE IS NO REAL ROOT FOR THE INTERNAL RATE OF RETURN FOR THIS
ALTERNATIVE.

****

NOW, LET US CHECK IF YOUR INPUT DATA IS CORRECT.
PLEASE CHECK YOUR FOLLOWING INPUT:

END OF PERIOD	1	2	ALTERNATIVE 3
0	−100.00	100.00	−200.00
1	50.00	−500.00	100.00
2	90.00	600.00	90.00

IS THE ABOVE CASH FLOW MATRIX CORRECT?

PLEASE TYPE 'Y' IF YES, AND 'N' IF NO. ? Y

THE FINAL RESULTS FOR EACH ALTERNATIVE ARE GIVEN AS FOLLOWS:

ALTERNATIVE	PRESENT WORTH	ANNUAL WORTH	B/C RATIO
1	19.83	11.43	1.20
2	141.32	81.43	1.31
3	− 34.71	−20.00	0.83

****

THE INTERNAL RATE(S) OF RETURN FOR EACH ALTERNATIVE ARE AS FOLLOWS:

ALTERNATIVE 1 , INTERNAL RATE OF RETURN IS 23.108 %
ALTERNATIVE 2 , INTERNAL RATE OF RETURN IS 100 %
ALTERNATIVE 3 , INTERNAL RATE OF RETURN IS 199.999 %

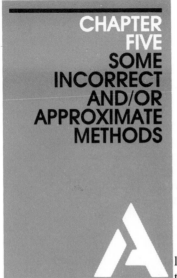

CHAPTER FIVE
SOME INCORRECT AND/OR APPROXIMATE METHODS

lthough the capital allocation techniques discussed in Chapters 3 and 4 are demonstrably valid, they are left generally unused by a large percentage of the business community. Significant evidence exists to indicate that, although valid methodology is being used more frequently, faulty techniques remain widespread in government and industry.

There are a number of reasons for this unhappy state of affairs. First, introduction of these techniques into college curricula is a relatively recent development. (In contrast, academic institutions have long taught the principles of double-entry accounting.) Second, in many cases businessmen call on their ''golden nugget of experience'' to make decisions pertaining to the allocation of limited capital. Differences among alternatives may appear to be sufficiently large that the appropriate course of action is either immediately obvious or follows from a few preliminary calculations. In any event, the decision maker's intuition, based on accumulated experience, frequently leads to the choice of investment alternatives. Finally, certain incorrect and/or approximate methods are often used to select from alternatives in the mistaken view that such methods are valid. Unfortunately, these methods have been institutionalized because of their apparent simplicity. The purpose of the following discussion is to present several of the most common of these methods in order to establish their advantages and disadvantages when employed in the capital allocation process.

The payback method is widely used in American industry to determine the relative attractiveness of investment proposals. The essence of this technique is determination of the *number of periods required to recover an initial investment.* Once this has been done

5.1 THE PAYBACK METHOD

Figure 5.1 Payback (Payout)

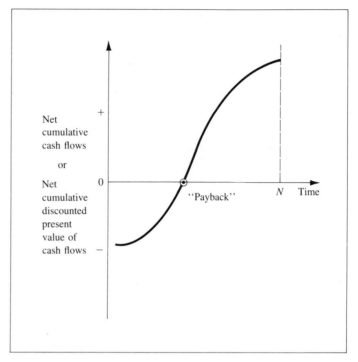

for all alternatives under consideration, a comparison is made on the basis of respective payback periods.

As illustrated in Figure 5.1, **payback,** or **payout,** as it is sometimes known, is the number of periods required for cumulative benefits to exactly equal cumulative costs. Costs and benefits are usually expressed as cash flows, although discounted present values of cash flows may be used. In either case, the payback method is based on the assumption that the relative merit of a proposed investment is measured by this statistic. The smaller the payback (period), the better the proposal.

Undiscounted Payback

The payback method is most often used only in connection with cash flows. In this case, the number of payback periods, or payback, is that value of N^* such that

$$P = \sum_{j=1}^{N^*} A_j \tag{5.1}$$

where P is the initial investment and A_j is the cash flow in period j. Consider, for example, a proposal that has a first cost of $10,000 and net revenues (excess of receipts over disbursements) of $2,000 per year for each and every year over a ten-year period.

FIVE   SOME INCORRECT AND/OR APPROXIMATE METHODS

In this case the number of periods required to "pay back" the original investment is $10,000/$2,000 per year $= 5$ years.

Now consider another project that has a first cost of $9,000 and net revenues of $1,500 per year for seven years. The payback for this proposal is $9,000/$1,500 $= 6$ years. The conclusion, using the payback method, is that the first alternative is superior to the second because its initial cost will be recovered more quickly.

If net cash flows are uniform throughout the planning horizon, as in the above example, then

$$N^* = \frac{P}{A} \qquad (5.2)$$

where A is the net cash flow per period.

Another example of payback appears in Table 5.1. Here, cash flows are not uniform from period to period, and further, N^* is not an integer value.

Table 5.1
Determining Payback for an Initial Investment of $10,000

Period	Net cash flow	Cumulative cash flow
1	$-\$2,000$	$-\$12,000$
2	$-\ 1,000$	$-\ 13,000$
3	0	$-\ 13,000$
4	$2,000$	$-\ 11,000$
5	$4,000$	$-\ 7,000$
6	$6,000$	$-\ 1,000$
7	$4,000$	$3,000$
8	$2,000$	$5,000$

All but $1,000 has been "paid back" after six periods; the initial $10,000 investment has been more than entirely recovered after seven periods. When payback occurs between periods, we can either round N^* to the closest integer (so N^* would be 6) or let N^* assume a fractional value:

$$N^* = 6 + \frac{\$1,000}{\$1,000 + \$3,000} = 6.25 \text{ periods}$$

(As there is little difference between these two conventions, use the second for problems for the sake of uniformity.)

Discounted Payback
An obvious shortcoming of the undiscounted payback method described above is that the time value of money, as reflected in the

discount rate, is ignored. We can adjust for this problem by discounting all cash flows before determining payback, N^*. Assuming that all end-of-period cash flows A_j are discounted at effective rate i per period, N^* is defined by the relationship

$$P = \sum_{j=1}^{N^*} A_j (1 + i)^{-j} \tag{5.3}$$

or

$$\sum_{j=0}^{N^*} A_j (1 + i)^{-j} = 0$$

since $P = -A_0$.

To illustrate, consider the three mutually exclusive alternatives, Q, R, and S, described in Table 5.2. Based on cash flows alone, we see that

$$N^*(Q) = 4, N^*(R) = 2, \text{ and } N^*(S) = 3$$

So it would appear that R is the preferred alternative. Discounted payback, however, presents a very different set of results. Alternative Q results in no definable payback. With a discount rate of 15 percent per period, the initial investment is not entirely recovered over the five-period planning horizon. Furthermore, S now has a shorter discounted payback period than R:

$$N^*(R) = 5 \text{ and } N^*(S) = 4$$

The explanation for this shift in the rank-order is that the major positive cash flow is received earlier by S than by R.

Table 5.2
Determining Discounted Payback for Three Mutually Exclusive Alternatives

End of period	Cash flows			Present value @ i = 15%			Cumulative present value		
	Q	R	S	Q	R	S	Q	R	S
0	−$1000	−$1000	−$1000	−$1000	−$1000	−$1000	−$1000	−$1000	−$1000
1	100	100	900	87	87	783	− 913	− 913	− 217
2	200	900	50	151	680	38	− 762	− 233	− 179
3	300	100	50	197	66	33	− 565	− 167	− 146
4	400	100	255	229	57	146	− 336	− 110	0
5	500	221	100	249	110	50	− 87	0	50
Totals	$ 500	$ 421	$ 355	−$ 87	$ 0	$ 50			
N^*	4	2	3				None	5	4

Theoretical Errors

There are several objections to the payback method. Two are relatively minor; the third is critically important.

The first objection concerns the fact that the usual definition of payback considers cash flows only. But as you saw, this problem can be overcome by using discounted cash flows to determine the payback period.

The second objection concerns the possibility that the payback period, N^*, may appear to be greater than the project life, N. This can occur when Equation 5.2 is used without regard to N. Consider, for example, a project with an initial cost of $10,000 that is expected to yield net returns of $1,000 per year each and every year for six years. The solution, using Equation 5.2 is

$$N^* = \frac{P}{A} = \frac{\$10,000}{\$1,000 \text{ per yr.}} = 10 \text{ years}$$

This equation is meaningless because the expected life of the project is less than the total number of periods required for payback. In general, N^* is meaningful only if $0 < N^* \leq N$, where N is the project life (planning horizon).

The third objection is illustrated by the following example. Let Alternative T have an initial cost of $8,000, annual net revenues of $4,000 each and every year for five years, and no terminal salvage value. Let Alternative U have an initial cost of $9,000, annual net revenues of $3,000 each and every year for five years, and an expected salvage value of $8,000 at the end of the fifth year. Equation 5.2 is appropriate in this case because the annual cash flows are uniform. Therefore

$$N^*(\text{T}) = \frac{\$8,000}{\$4,000 \text{ per yr.}} = 2 \text{ years}$$

$$N^*(\text{U}) = \frac{\$9,000}{\$3,000 \text{ per yr.}} = 3 \text{ years}$$

Both values for N^* are meaningful since $N = 5$.

Because $N^*(\text{T}) < N^*(\text{U})$, it would appear that T is superior to U. But, if the decision maker's minimum attractive rate of return is, say, 10 percent per period, this payback solution may be shown to be incorrect. Calculating the present worths of the two alternatives:

$$PW(\text{T}) = -\$8,000 + \$4,000(P/A, 10\%, 5)$$
$$= -\$8,000 + \$4,000(3.791)$$
$$= \$7,164$$

$$PW(U) = -\$9,000 + \$3,000(P/A,\ 10\%,\ 5)$$
$$+ \$8,000(P/F,\ 10\%,\ 5)$$
$$= -\$9,000 + \$3,000(3.791) + \$8,000(0.6209)$$
$$= \$7,337$$

Alternative U is clearly superior to T by this method. Why the inconsistency? Because the payback method ignores consequences that are beyond the end of the payback periods. In this example the $8,000 salvage value at the end of five years for Alternative T proved to be decisive.

Payback and the Internal Rate of Return

The payback method *may* be useful under the special conditions that the lives of the alternatives are equal and, other than the initial investment, there is a uniform pattern of annual cash flows. However, even under these conditions results may be spurious. To illustrate, consider Alternative V, which has an initial investment of $100 and net annual receipts of $20 each and every year for ten years. The initial cost of Alternative W is $80, and it is expected to produce net receipts of $10 each and every year for ten years. The payback periods are clearly five years for V and eight years for W, thus indicating that V is superior to W.

$$N^*(V) = \frac{\$100}{\$20\ \text{per yr.}} = 5\ \text{years}$$

$$N^*(W) = \frac{\$80}{\$10\ \text{per yr.}} = 8\ \text{years}$$

This solution may be shown to depend on the interest rate, that is, the minimum attractive rate of return, relevant to the problem. Observe that, given a uniform series of cash flows, the number of payback periods is equal to the uniform series present worth factor. From Equation 5.2, $N^* = P/A$. If, at this point, the initial investment (P) exactly equals the discounted present value of a uniform series of cash flows over the range $1 \le j \le N^*$, then it is also true that $P = A(P/A,\ i,\ N^*)$. Therefore it follows that

$$N^* = (P/A,\ i,\ N) \tag{5.4}$$

in the special case of a uniform series of cash flows.
Using the data for Alternatives V and W:

Alternative	P	A	$(P/A,\ i,\ 10)$
V	$100	$20	5.0
W	80	10	8.0

From the compound interest tables in Appendix B, it may be determined that the internal rates of return are 16 percent for Alternative V and 4 percent for Alternative W. That $IRR(V) > IRR(W)$ is not surprising, because $N^*(V) < N^*(W)$. For a given value of N, the interest rate, i, decreases as the factor $(P/A, i, N)$ increases. Put somewhat differently, given two alternatives, I and II, each with uniform cash flows over a common planning horizon, N, then if

$$N^*(I) < N^*(II)$$

it will also be true that

$$IRR(I) > IRR(II)$$

It is particularly important to emphasize, however, that one alternative is not preferable to another simply because it has a higher internal rate of return. This is the ranking error discussed in Chapter 4.

Returning to our example, despite resulting payback periods that are less than the study period ($N = 10$), neither Alternative V nor Alternative W is acceptable if the minimum attractive rate of return, k, is greater than 16 percent. As indicated in Figure 5.2, V should be accepted and W rejected if $4\% < k < 16\%$; if $k < 4\%$, although both alternatives are acceptable, V is preferable to W. Moreover, it may be shown that, although neither alternative is desirable at higher interest rates ($k > 16\%$), W is preferable to V for minimum attractive rates of return in excess of about 49 percent.

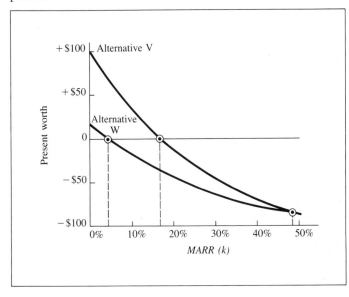

Figure 5.2 Present Worths of Alternatives V and W as a Function of the Minimum Attractive Rate of Return

In summary,

MARR (k)	Correct ranking
$k < 4\%$	V > W > Ø
$4\% < k < 16\%$	V > Ø > W
$16\% < k < 49\%$	Ø > V > W
$49\% < k$	Ø > W > V

Thus rank-ordering by payback is a function of the discount rate appropriate to the problem.

When the periodic cash flows (A) are uniform and the project life (N) is very long, we have the special case where the payback (N^*) is approximately equal to the reciprocal of the internal rate of return. This follows from the fact that

$$\lim_{N \to \infty}(P/A, \, i, \, N) = \frac{1}{i}$$

Possible Usefulness

Since the payback method is a measure of how fast original investments are repaid, it yields information that may be of considerable interest. For example, when future consequences of a prospective investment are highly uncertain, managers may want to alleviate the dangers inherent in uncertainty by recovering the initial cost as quickly as possible. Under these conditions, payback-period computations provide an index of one aspect of the proposed investment.

The payback method may also be useful when a firm's current capital is extremely limited and cash-flow estimates indicate that similar limitations will persist for the next few budgeting periods. In addition, if the firm expects unusually profitable investment opportunities to arise in the near future, then it will want to recover capital as quickly as possible in order to take advantage of these potential reinvestments. Information provided by the payback method should be useful under these conditions, but, in any case, this method should only be used to provide information *in addition to* that given by the valid methods discussed in the preceding chapters.

5.2 PROFITABILITY INDEXES

The term **profitability index** (*PI*) is found in the literature of economic analysis in a variety of contexts, but it has no generally accepted definition. To one analyst, *PI* is simply the internal rate of return on investments; to another it represents the present worth of a set of cash flows; and to a third, it may bear a different meaning. Some meanings are valid; others are not. This section presents three typical forms of the profitability index that are either incorrect or that can easily be abused. Some of the problems one might expect in applying profitability indexes to capital allocation decisions are also discussed.

The Net Benefit-Cost Ratio

The net benefit-cost ratio, $NB : C$, is the ratio of discounted net benefits to the initial cost of the project. Specifically,

$$NB : C = \frac{B - C}{C} \qquad (5.5)$$

where B is the present value of all cash flows over the planning horizon ($j = 1, 2, \ldots, N$) and C is the initial cost of the project.

Note here that $NB : C = B : C - 1$. Thus net benefit-cost ratios should be treated with the same caveats as the benefit-cost ratios discussed in Chapters 3 and 4. In particular, net benefit-cost ratios should *not* be used to rank-order alternative investments.

To illustrate this point, consider two mutually exclusive alternatives, X and Y:

Alternative	B	C	$B - C$	$NB : C$
X	$250	$100	$150	1.50
Y	320	150	170	1.13

Although $NB : C(X) > NB : C(Y)$, this result does not imply that X is preferable to Y. Indeed, since $PW = B - C$, $PW(X) < PW(Y)$, so it follows that Y is preferable to X.

To use net benefit-cost ratios properly, we should examine the differences between Y and X. The incremental benefits and costs are

$$\Delta B = \$320 - \$250 = \$70$$
$$\Delta C = \$150 - \$100 = \$50$$

Thus $\Delta NB : C = (\$70 - \$50)/\$50 = 0.40$. Since the resulting $\Delta NB : C$ is greater than zero, the incremental investment is justified and Y is preferable to X.

Note here that our criterion for acceptance is that the net benefit-cost ratio exceed *zero*, not unity as with the benefit-cost ratio method. This is so because

$$NB : C = \frac{B - C}{C} = \frac{B}{C} - 1$$

Requiring that $B : C > 1$ is the same as requiring that the $NB : C$ be greater than zero.

Premium Worth Percentage

Some years ago an engineering economy textbook introduced the notion of **premium worth percentage,** which, it was claimed, might be helpful in comparative evaluation. Here, the premium worth percentage (*PWP*) is the ratio of the net present worth of all cash flows to the initial investment.

$$PWP = \frac{PW}{P} \tag{5.6}$$

where

$$PW = \sum_{j=0}^{N} A_j(1 + i)^{-j}$$

$$P = \text{initial cost} = -A_0.$$

It was further asserted that the *PWP* is useful in that it is difficult to compare the present worths of the extra profits of various proposals (bonus worths) because they are produced by investments of differing sizes. The terms *extra profit* and *bonus worth* were used in the same sense that this book uses *net present value,* or *present worth.*

The premium worth percentage concept has appeared in various forms over the years, principally because others have shared the concern about direct comparison of present worths when initial costs of alternatives are unequal. This viewpoint warrants examination, so the following is a simple illustration. Consider two mutually exclusive alternatives, I and II. For Alternative I:

Present worth of future cash flows	=	$242,000
Initial investment (*P*)	=	200,000
"Bonus investment" (*PW*)	=	$ 42,000

For Alternative II:

Present worth of future cash flows	=	$130,000
Initial investment (*P*)	=	100,000
"Bonus investment" (*PW*)	=	$ 30,000

With these results, it follows that

$$PWP\,(\text{I}) = \$42{,}000/\$200{,}000 = 0.21$$

and

$$PWP\,(\text{II}) = \$30{,}000/\$100{,}000 = 0.30$$

It does *not* follow that Alternative II is preferable to Alternative I simply because $PWP(\text{II}) > PWP(\text{I})$.

As you saw in Section 3.2, Chapter 3, the present worth of funds invested elsewhere at rate i, when discounted at rate i, is exactly zero. This observation makes it possible to directly compare present worths without adjusting for differences in initial costs. In the present example, we are considering two choices: *either* (1) invest $200,000 in Alternative I, yielding a present worth of $42,000 *or* (2) invest $100,000 in Alternative II, yielding a present worth of $30,000, and invest $100,000 elsewhere at rate of return i, yielding zero present worth.

The premium worth percentage and its variants are irrelevant at best. They add nothing to the analysis and can be misleading to decision makers.

The Savings-Investment Ratio

Although rarely found in the private sector, the **savings-investment ratio** (*SIR*) has enjoyed some prominence in the public sector in recent years, especially in the U.S. Department of Defense.[1] As you will see, the *SIR* is yet another variation of the more familiar benefit-cost ratio ($B:C$). When used solely as a profitability index, the *SIR* can lead to incorrect solutions. If used properly, that is, if used in the same way as the benefit-cost ratio method outlined in Chapters 3 and 4, *incremental SIR*'s may be used to find valid results.

The savings-investment ratio is defined as

$$SIR = \frac{P(E)}{P(I) - P(S)} \qquad (5.7)$$

where $P(E)$ is the discounted present value of all earnings over the life of the project, $P(I)$ is the present value of the initial investment, and $P(S)$ is the discounted present value of the project's terminal salvage value, if any.

Parenthetically, the term *earnings* appears in Equation 5.7, whereas in other formulations, particularly in those used by the Department of Defense, the term *savings* is used instead. The difference in terminology results from the fact that government investment projects are not expected to produce income the way it is produced in the private sector. However, our concern is with the effect of the project on cash flows. Whether positive cash flows are generated by increasing revenues or by reducing costs is irrelevant. Therefore, *earnings* mean net cash flows produced by

1. An early reference is Richard S. Brown, et al., *Economic Analysis Handbook* (Alexandria, Virginia: Naval Facilities Engineering Command, 1975), pp. 23–28.

the project, other than the initial investment and the terminal salvage value.

To illustrate the *SIR,* consider a certain process that currently costs $40,000 per year and is expected to continue at this level over the next fifteen years. An improved process, System Alpha, that costs $60,000 has been proposed. System Alpha will reduce costs to $30,000 per year but is not expected to have any terminal salvage value at the end of fifteen years. The initial investment, $60,000, will occur entirely at the start of the first year, and savings will occur continuously and uniformly during each and every year over the fifteen-year planning horizon. The discount rate is 10 percent per year. Based on Equation 5.7:

$$P(E) = (\$40,000 - \$30,000)(P/\overline{A}, \ 10\%, \ 15)$$
$$= \$10,000(7.980)$$
$$= \$79,800$$
$$P(I) = \$60,000$$
$$P(S) = \$0$$

Therefore

$$SIR \ (\text{Alpha}) = \frac{\$79,800}{\$60,000} = 1.33$$

Since the *SIR* exceeds unity, it appears that investment in System Alpha is warranted. Alpha is preferable to the do-nothing alternative, that is, preferable to continuing with the current process.

Note that the present worth (*PW*) of the proposed investment is given by

$$PW = P(E) - P(I) + P(S) \tag{5.8}$$

If the present worth is greater than zero, it follows that

$$P(E) > P(I) - P(S)$$

and, if the right-hand side of the inequality is positive, then

$$\frac{P(E)}{P(I) - P(S)} > 1.0$$

In other words, if $P(I) - P(S) > 0$, then the investment is warranted if the *SIR* > 1.0.

The problem with using the savings-investment ratio has to do with how *SIR* is applied in judging alternative investments rather than with the statistic, *SIR.* As with the benefit-cost ratio method, care must be taken to ensure that alternatives are not rank-ordered on the basis of their respective *SIR*'s. To illustrate this point, consider a second alternative, System Beta, with $P(E) = $125,000$, $P(I) = $110,000$, and $P(S) = $10,000$. Here

$$SIR \text{ (Beta)} = \frac{$125,000}{$110,000 - $10,000} = 1.25$$

We should not conclude that Alpha is preferable to Beta simply because *SIR* (Alpha) > *SIR* (Beta). Indeed, the correct *incremental* analysis shows the opposite to be true:

	Alpha	Beta	Incremental Alpha → Beta
$P(E)$	$79,800	$125,000	$45,200
$P(I)$	60,000	110,000	50,000
$P(S)$	0	10,000	10,000

If we were to adopt Beta rather than Alpha, the incremental *SIR* would be

$$\Delta SIR = \frac{$45,200}{$50,000 - $10,000} = 1.13$$

Since $\Delta SIR > 1.0$, the incremental investment in Beta is warranted.

Chapter 3 presented the appropriate procedure for determining the (internal) rate of return, based on the analysis of cash flows. There are a number of other approaches that use accounting data (income and expenses) rather than cash flows to determine "rate of return," where income and expense are reflected in the firm's accounting statements. Because the accounting approach to project evaluation is so widely used in industry, this section examines several of the more common procedures.

Consider, for example, two mutually exclusive investment alternatives as shown in Table 5.3.

5.3 ACCOUNTING METHODS FOR COMPUTING RATE OF RETURN

Table 5.3
Cash Flows Before Income Taxes for
Two Mutually Exclusive Alternatives

End of year	Alternative X	Alternative Y
0	− $50,000	− $50,000
1	5,000	14,000
2	10,000	14,000
3	15,000	14,000
4	20,000	14,000
5	25,000	14,000
Total cash flow	$25,000	$20,000

The (internal) rate of return for Alternative X is found from

$$-\$50{,}000 + \$5{,}000(P/A,\ i_X^*,\ 5) + \$5{,}000(P/G,\ i_X^*,\ 5) \equiv 0$$

from which

$$i_X^* = 12.0\%$$

For Alternative Y

$$-\$50{,}000 + \$14{,}000(P/A,\ i_Y^*,\ 5) \equiv 0$$

from which

$$i_Y^* = 12.4\%$$

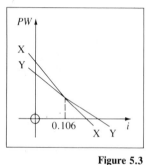

Figure 5.3

Once again, Y is not preferable to X simply because $i_Y^* > i_X^*$. Project preference depends on the minimum attractive rate of return, k. Indeed, it may be shown that the present worth of the differences between alternatives $(Y - X)$ is

$$PW = -\$9{,}000(P/A,\ i,\ 5) + \$5{,}000(P/G,\ i,\ 5)$$

The PW is equal to zero when $i = 10.6\%$. Therefore, as shown in Figure 5.3, X is preferable to Y when $k < 10.6\%$; otherwise Y is preferable to X.

Original Book Method

Using the **original book method,** the project rate of return per year, the RoR, is found by dividing the average annual accounting profit by the original book value of the asset. That is,

$$RoR = \frac{\overline{AP}}{B_0} \tag{5.9}$$

where $\overline{AP}$ = average annual accounting profit and B_0 = original book value of the asset. **Accounting profit** is the excess of income over expenses. In this formulation, assume that all income items are positive cash flows and all expense items, other than depreciation, are negative cash flows. Calculating depreciation expense is an accounting device used to distribute the initial cost of the asset, less expected salvage value, if any, over its estimated useful life. (Depreciation is discussed in detail in Chapter 7.) Therefore, accounting profit is a combination of both cash and noncash elements. **Book value,** another accounting concept, is the original cost basis less accumulated depreciation. For our purposes, assume that the original book value, or cost basis, is simply the initial cost of the asset, P.

Assuming that the company uses *straight line depreciation* in its accounts, the annual depreciation expenses, D, are given by

$$D = \frac{P - S}{N} \qquad\qquad (5.10)$$

where

P = initial cost
S = salvage value after N periods
N = asset life for depreciation purposes

Using data from the previous example:

$$D = \frac{\$50,000}{5} = \$10,000$$

for both Alternative X and Alternative Y.

The contributions to the firm's average annual profits for the two projects are determined in Table 5.4: $\overline{AP}(X) = \$5,000$ and $\overline{AP}(Y) = \$4,000$. Therefore the ratios of return are

$$RoR(X) = \frac{\$5,000}{\$50,000} = 0.10$$

$$RoR(Y) = \frac{\$4,000}{\$50,000} = 0.08$$

Compared to the correct values for X and Y (12 percent and 12.4 percent respectively), this method substantially understates the true rates of return and, incidentally, reverses their relative position.

Table 5.4
Average Annual Accounting Profits
for Alternatives X and Y in Table 5.3

	Alternative X			Alternative Y		
Year	Cash flow	Depreciation	Profit	Cash flow	Depreciation	Profit
1	$ 5,000	$10,000	−$ 5,000	$14,000	$10,000	$ 4,000
2	10,000	10,000	0	14,000	10,000	4,000
3	15,000	10,000	5,000	14,000	10,000	4,000
4	20,000	10,000	10,000	14,000	10,000	4,000
5	25,000	10,000	15,000	14,000	10,000	4,000
Totals	$75,000	$50,000	$25,000	$70,000	$50,000	$20,000
Average			$ 5,000			$ 4,000

Average Book Method

Proponents of the **average book method** argue that the value of the investment is reduced from year to year, thus the rate of return should be based on the *average* book value. In the case of straight line depreciation, the average book value is simply half the sum of the first cost and the expected salvage value. Or

$$RoR = \frac{\overline{AP}}{0.5(P + S)} \tag{5.11}$$

So for Alternatives X and Y:

$$RoR(X) = \frac{\$5,000}{0.5(\$50,000)} = 0.20$$

$$RoR(Y) = \frac{\$4,000}{0.5(\$50,000)} = 0.16$$

By comparing these results to those found by the original book method, we see that the differences arise entirely from the values in the denominators. Moreover, these figures are considerably higher than those obtained by using proper compound interest methods.

Year-by-Year Book Method

Still another method, the **year-by-year book method,** is based on year-by-year ratios of profit after depreciation to book value at the start of the year. Using Alternative X, for example:

Year	Profit after depreciation	Book value at start of year	Rate of return
1	− $ 5,000	$50,000	loss
2	0	40,000	0
3	5,000	30,000	17%
4	10,000	20,000	50%
5	15,000	10,000	150%

A major problem here lies in applying the results in any useful way. What do these values mean? Perhaps the analyst could average the year-by-year rates to compare alternatives, but this strategy leads to some obvious additional problems.

A disadvantage shared by the three preceding methods for computing rates of return is that they all depend on book values, which in turn are functions of the depreciation schedules used by the firm. Although there is a certain inherent appeal in using methods based on book values—basic accounting methods seem to be understood better than compound interest calculations—rate of return solutions should not be influenced by an arbitrary allocation of the investment over accounting periods.

Hoskold's Method for Computing Rate of Return

Hoskold's method, as it is traditionally known in the mining industry, assumes that uniform annual deposits will be made into a sinking fund that will be conservatively invested in order to earn interest at a relatively low rate. The value of the fund at the end of the life of the asset should be just enough for identical replacement. The rate of return is then determined as the ratio of the amount remaining after the sinking fund deposit to the initial investment. (The sinking fund method of depreciation is discussed in Chapter 7.) One formula for determining rate of return by using the Hoskold method is:

$$RoR = \frac{A - D_s}{F} \tag{5.12}$$

where

A = net cash flow before making deposit to the sinking fund

F = amount required for identical replacement after N years

D_s = uniform annual deposit to the sinking fund earning at rate i_s per year

$$= F(A/F, i_s, N) \tag{5.13}$$

(Note in this formulation that A is uniform from year to year. If the annual cash flows A_j are not uniform for all j, the average value may be used.)

To illustrate, let's return to Alternative Y and assume that annual investment into a 5 percent sinking fund. The project rate of return is determined as:

$$RoR = \frac{\$14,000 - \$50,000(A/F, 5\%, 5)}{\$50,000}$$

$$= \frac{\$14,000 - \$9,049}{\$50,000} = 0.099$$

This solution is somewhat less than the correct value, 0.124, because it is an average of two investments, one returning 12.4 percent and the other returning 5 percent. If investment in Alternative Y actually requires annual deposits into a sinking fund earning 5 percent, then determination of the rate of return should consider both investments. Otherwise, the merits of investing in Alternative Y and the sinking fund should be determined independently.

For example, suppose the minimum attractive rate of return for the firm is 11 percent. If a sinking fund is *not* required, Alternative Y is economically desirable because it has an expected return of 12.4 percent; if a sinking fund *is* required along with the investment in Y, then the proposal should be rejected because it returns only 9.9 percent.

The "Truth in Lending" Formula for Rate of Return

At the urging of various consumer groups the federal government and many states have enacted "truth in lending" legislation requiring that interest rates charged by lenders be clearly and uniformly stated. Since these rates represent returns to the lenders as well as costs to the borrowers, they are, in a sense, measures of rate of return. One commonly used definition of the **truth in lending formula** is presented here.

P = amount of original debt (the original cost, or price, less any down payment)

M = number of payments in one year

N = total number of payments necessary

A = dollar amount of each payment, assuming a uniform series of payments

r = nominal interest rate per year

r/M = interest rate per period

I_j = portion of the jth payment that is interest

D = total interest paid

If the original debt is P, the portion of the *first* payment that is interest (I_1) is $P(r/M)$. The *final* payment repays P/N, so the interest on this final payment (I_N) is $(P/N)(r/M)$. Now, let $\bar{I}$ represent the *average interest* paid over the N payments and assume it to be the average of the first and last payments.

$$\bar{I} = 0.5(I_1 + I_N) \tag{5.14}$$

In Figure 5.4, $\bar{I}$ is an approximation of the true average interest. The total interest paid over N periods is approximately $N\bar{I}$, so

$$D = N\bar{I}$$

$$= \frac{N}{2}(I_1 + I_N)$$

$$= \frac{N}{2}\left(\frac{Pr}{M} + \frac{Pr}{NM}\right) \tag{5.15}$$

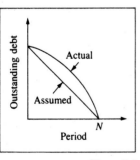

Figure 5.4

FIVE   SOME INCORRECT AND/OR APPROXIMATE METHODS

Solving for r, we have

$$r = \frac{2MD}{P(N + 1)} \qquad (5.16)$$

The "rate of return" is defined in Equation 5.16.

To give an example, assume that a certain item is purchased for which the buyer agrees to repay a debt of $1,000 with twelve equal monthly payments. The buyer is told by the seller or lender that the "interest rate" is 10 percent per year and that finance charges will be added to the purchase price. Total interest charges are $0.10($1,000) = 100. This, added to the $1,000 original debt, requires the buyer, or borrower, to make monthly payments of $(^1/_{12})($1,000 + $100) = 91.67. The cash flow diagram for this exchange is shown in Figure 5.5. In this example:

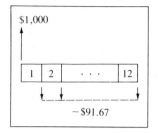

Figure 5.5

$M = 12$

$D = 100

$P = $1,000$

$N = 12$

Therefore, from Equation 5.16,

$$r = \frac{2(12)($100)}{$1,000(12 + 1)} = 0.185$$

In this case, the true annual interest rate per month (i_M) follows from the relationship

$$\$1,000 - \$91.67(P/A, i_M, 12) \equiv 0$$

or

$$(P/A, i_M, 12) = 10.9087$$

From Appendix B, $i_M = 0.015$, so the annual rate is

$$(1.015)^{12} - 1 = 0.186$$

This is very close to the approximate rate given by Equation 5.16. (Of course, both the approximate and true rates are considerably higher than the original 10 percent rate given the borrower by the lender.)

The discussion of the annual worth method in Chapter 3 derived the following expression for computing the cost of capital recovery (CR):

$$CR = (P - S)(A/P, i, N) + Si \qquad (5.17)$$

where P is the first cost, S is the net salvage value, N is the life of the proposal, and i is the minimum attractive rate of return. Moreover, an alternative expression for computing the capital recovery cost was shown to be

$$CR = (P - S)(A/F, i, N) + Pi \qquad (5.18)$$

This equation is generally known as **sinking fund depreciation plus interest on first cost** because of its superficial similarity to the sinking fund method of computing depreciation. (See Chapter 7.) Its proponents claim that it is useful in reconciling the accounts of the firm with the economic analysis when the sinking fund method is used to compute depreciation. This section discusses two other methods that are frequently used with the rationale that they allow reconciliation of capital recovery costs between the analysis and the accounting records of the firm. However, unlike Equation 5.18, these methods are incorrect, as you will see in the following example problem.

A firm is considering investment in a heavy-duty crane. The first cost of this equipment is $120,000, and it is expected to be retained for eight years, at which time it will probably be sold for $20,000. If the company's before-tax minimum attractive rate of return is 10 percent, the capital recovery cost for this proposed investment, using Equation 5.17, is

$$
\begin{aligned}
CR &= (\$120,000 - \$20,000)(A/P, 10\%, 8) \\
&\quad + (\$20,000)(0.10) \\
&= (\$100,000)(0.18744) + \$2,000 \\
&= \$20,744
\end{aligned}
$$

Verifying this solution using Equation 5.18 results in:

$$
\begin{aligned}
CR &= (\$120,000 - \$20,000)(A/F, 10\%, 8) \\
&\quad + (\$120,000)(0.10) \\
&= (\$100,000)(0.08744) + \$12,000 \\
&= \$20,744
\end{aligned}
$$

Straight-Line Depreciation Plus Interest on First Cost
The **straight line depreciation plus interest on first cost method** is frequently recommended for its intuitive appeal as well as for its similarity to accounting results when straight line depreciation is

actually being used for the asset in question. The annual *depreciation* charge, D, is found by using Equation 5.10:

$$D = \frac{P - S}{N}$$

Thus, it is argued, the annual *capital recovery* cost should be the annual (straight line) depreciation charge plus the interest forgone on the original investment. Or

$$\text{"}CR\text{"} = \frac{P - S}{N} + Pi \tag{5.19}$$

Using the data from the crane example,

$$CR = \frac{\$120{,}000 - \$20{,}000}{8} + (\$120{,}000)(0.10)$$

$$= \$12{,}500 + \$12{,}000$$

$$= \$24{,}500$$

The solution is somewhat larger than the correct answer, \$20,744. This will always be the case unless the expected salvage value is equal to or greater than the first cost.

Straight Line Depreciation Plus Average Interest

It may be argued that the straight line depreciation plus interest on first cost method is incorrect in that a portion of the capital recovery cost results from interest on the initial investment. But since the "value" of the asset declines from year to year (as measured by the straight line depreciation), the interest should be averaged over the life of the asset, as in the following table. This is known as the **straight line depreciation plus average cost method.**

Period	Value of asset at start of period		Interest on value of asset at start of period
	Diminishing portion	Nondiminishing portion	
1	$P - S$	S	$i(P - S) + Si$
2	$(P - S)\left(1 - \frac{1}{N}\right)$	S	$i(P - S)\left(1 - \frac{1}{N}\right) + Si$
⋮	⋮	⋮	⋮
N	$(P - S)\left(\frac{1}{N}\right)$	S	$i(P - S)\left(\frac{1}{N}\right) + Si$

Total interest		$i(P - S)\left(\frac{N + 1}{2}\right) + NSi$
Average interest		$i(P - S)\left(\frac{N + 1}{2N}\right) + Si$

The expression for capital recovery now becomes

$$``CR" = \frac{P - S}{N} + i(P - S)\left(\frac{N + 1}{2N}\right) + Si \qquad \textbf{(5.20)}$$

Again, using the data from the crane example:

$$CR = \frac{\$120,000 - \$20,000}{8} + 0.10(\$120,000 - \$20,000)$$

$$\times \left(\frac{9}{16}\right) + \$20,000(0.10)$$

$$= \$12,500 + \$5,625 + \$2,000$$

$$= \$20,125$$

This result, although much closer, is still lower than the correct solution, $20,744. The essential difference is that this method results in an *average* rather than in an *equivalent* annual cost.

5.5 SUMMARY

Approximate or incorrect methods are frequently used in industry to select from alternative investment opportunities. Principal among these is the payback, or payout, method, which measures the number of periods necessary to recover an initial investment. There are two formulations of the payback method: one uses cash flows, the other uses discounted values of the cash flows. In either case, the payback method does provide some valuable information under certain circumstances, but it should never be used as the sole criterion.

So-called profitability indexes are often employed in government and industry to measure the relative desirability of investment proposals. Three typical measures are presented in this chapter. Two of them, the net benefit-cost ratio and the savings-investment ratio, are closely related to the benefit-cost ratio discussed previously. As is the case with the benefit-cost ratio method, these ratios can lead to correct results only if they are used with incremental effects between pairs of alternatives. Ranking on the basis of ratios is incorrect. The premium worth percentage, the third profitability index discussed in this chapter, attempts to adjust for differences in initial costs. This is unnecessary and misleading because the present worth of funds invested "elsewhere" is zero.

A number of methods based on asset book values have been used to determine project rates of return. These methods are fundamentally incorrect in that the rate of return is a function of an accounting device for allocating first cost (less salvage value, if any) over the life of the asset. Thus it is possible to alter apparent profitability by changing the depreciation schedule for the asset in question. (Depreciation schedules *can* affect profitability, but only because of the effect on cash flows for taxes. This matter is discussed in Chapter 7.)

Rate of return computations, such as Hoskold's method, for example, may be erroneous if sinking fund deposits are included when in fact these sinking funds are not essential to the proposal being evaluated. In such cases the resulting project rate of return is an average of the correct rate of return and the earning rate of the sinking fund.

The "truth in lending" formula presented in this chapter is one of several attempts to standardize computation of the cost of consumer loans, but the result is approximate and generally understates the true cost.

There are several methods used for evaluating capital recovery cost, two of which are discussed in this chapter. When interest on first cost is added to straight line depreciation, the resulting capital recovery cost gives too high a result (except when the salvage value is 100 percent or more). On the other hand, when average interest is added to straight line depreciation, the capital recovery cost is too low (except when the salvage value is 100 percent or more). Errors arising from this latter method owe largely to the fact that it is an averaging process rather than an equivalence calculation.

Except where otherwise indicated, assume end-of-period cash flows and discrete discounting at effective interest rate i per period.

PROBLEMS

The Payback Method

5.1 Proposal I has a first cost of $40,000, expected excess of receipts over disbursements of $10,000 each year for seven years, and zero salvage value at the end of that time. Proposal II has a first cost of $30,000, net revenues of $6,000 per year for seven years, and is expected to be sold for 100 percent of the first cost at the end of the seventh year.
   a. Assuming an interest rate of 5 percent, use the present worth method to determine the most economically desirable alternative.
   b. Solve the same problem using the payback method.
   c. Comment on the results of (a) and (b).
   (*Answers:* (a) $PW(\text{I}) = \$17,860$, $PW(\text{II}) = \$26,037$; (b) $N^*(\text{I}) = 4$ years, $N^*(\text{II}) = 5$ years)

5.2 Mister Miser is considering investing $10,000 in one of two alternatives. (Assume that he has other investment opportunities elsewhere that, with similar risk, will return at least 10 percent.) His first alternative requires an initial cost of $10,000 and requires additional annual expenditures of $3,000. However, the investment is expected to yield revenues of $6,000 per year for five years, at the end of which time it will be sold for net salvage value of $100.

The second alternative requires an original investment of $10,000 also, but there are no other cash flow consequences until the twentieth year, at which time it is expected that the asset will be sold for $60,000.

a. Use either the present worth or rate of return method to determine which of the alternatives, if any, is most desirable. Assume that replacements for the first alternative will be made with identical assets.
b. Which alternative is indicated by the payback method?
c. Comment on the results of (a) and (b). In particular, discuss the effects of the replacement assumption in (a).

5.3 Consider the problem presented by the following cash flows for alternatives Alpha and Beta:

End of period	Alpha	Beta
0	− $100	− $220
1	50	100
2	50	100
3	50	100

a. Determine the best alternative using the payback method.
b. Assuming a minimum attractive rate of return of 15 percent, determine the best alternative by either the present worth or the annual cost method.
c. The payback method is independent of the interest rate. However, what must the value of the minimum attractive rate of return be in order to reverse the decision in (b)?
(*Answers*: (a) N^*(Alpha) = 2.0 years, N^*(Beta) = 2.2 years; (b) PW(Alpha) = $14.15, PW(Beta) = $8.30; (c) Approximately 12 percent)

5.4 The anticipated cash flows for two mutually exclusive alternatives are:

End of period	Alternative I	Alternative II
0	− $300	− $300
1	100	200
2	100	100
3	100	100
4	300	100

a. If the appropriate discount rate is 5 percent per year, determine the present worth of both alternatives.
b. Determine the undiscounted payback (periods) for both alternatives.
c. Determine the payback (periods) for both alternatives *after* adjusting for discounting of future benefits.

5.5 The payback method may be modified by subtracting annual straight line depreciation from the annual cash flows before income taxes. For alternatives Alpha and Beta given in Problem 5.3, determine the undiscounted payback period after depreciation. What may be said about the double-counting of depreciation?

(*Answer: N* (Alpha)* = 6.0 years, *N* (Beta)* = 8.25 years)

5.6 A firm is considering two projects, X and Y, each of which will result in a savings of $400 *during* each and every year over a five-year planning horizon. Project X will cost $1,000 and will have no salvage value. Project Y will cost $1,200 initially and will have a 100 percent salvage value ($1,200) at the end of five years. The firm's pretax *MARR* is 20 percent per year.

   a. Find the undiscounted payback for Projects X and Y.
   b. Find the discounted payback for the two projects.
   c. Using the present worth method, which project is preferable?

5.7 Consider three mutually exclusive alternatives with the following cash flows:

End of year	Alternative A	Alternative B	Alternative C
0	−$100	−$150	−$110
1–4	47	66	49

The firm's minimum attractive rate of return is 10 percent.

   a. Apply the (undiscounted) payback period method to determine the preferred alternative.
   b. Use the benefit-cost ratio method to determine the preferred alternative.

(*Answers:* (a) 2.13, 2.27, and 2.24; (b) B > A > C)

Profitability Indexes

5.8 Refer to Problem 4.22, Chapter 4.
   a. Find the net benefit-cost ratios for the seven alternatives.
   b. Using the net benefit-cost ratio method properly, determine the preferred alternative.

5.9 Refer to Problem 4.21, Chapter 4. Using the net benefit-cost ratio method properly, rank the five alternatives (including doing nothing [∅]).

(*Answer:* IV > III > I > ∅ > II)

5.10 Mr. Bearish, the owner of a small business, has an opportunity to purchase a Model A press for $10,000. It is anticipated that the press will result in a savings of $2,500 per year over its ten-year life. There is no salvage value; the press will be worthless after ten years. Mr. Bearish's minimum attractive rate of return is 20 percent per year.

   a. Find the present worth of Model A.
   b. Find the premium worth percentage for Model A.
   c. Find the true rate of return for this proposed investment.

5.11 Refer to Problem 5.10. Mr. Bearish is considering an alternative press, Model B, which has an initial cost of $25,000 and will result in a savings of $6,200 over its ten-year life.

The expected salvage value after ten years is negligible.
a. Find the present worth of Model B.
b. Find the premium worth percentage for Model B.
c. Is Model B preferable to the do-nothing alternative? Why or why not?

(*Answers:* (a) $990; (b) 0.04; (c) yes)

5.12 Refer to Problems 5.10 and 5.11. Which press, A or B, if any, is preferable to the do-nothing alternative? Explain your answer.

5.13 Refer to Problem 4.3, Chapter 4.
a. Find the savings-investment ratios for the four alternative computers.
b. Using incremental savings-investment ratios, find the preferred alternative.

(*Answers:* (a) $SIR(A) = 1.05$, $SIR(B) = 1.04$, $SIR(C) = 1.03$, $SIR(D) = 1.02$; (b) Alternative B)

5.14 Consider a set of five mutually exclusive alternatives. The equivalent present values of earnings, $P(E)$, initial investments, $P(I)$, and salvage values, $P(S)$, are

Alternative	$P(E)$	$P(I)$	$P(S)$
V	$17,000	$ 6,000	$2,000
W	14,000	10,000	3,000
X	9,000	10,000	8,000
Y	24,000	15,000	5,000
Z	26,000	15,000	0

Using the savings-investment ratio method properly, find the preferable alternative.

Accounting Methods

5.15 Two investment proposals, a vacuum still and a product terminal, are competing for limited funds in an oil company.* Each requires an immediate disbursement of $110,000, all of which will be capitalized on the books of account. The after-tax *cash flows* and the after-tax *profits* for the two alternatives are estimated to be those shown in the table at the top of the next page:

a. If the after-tax minimum attractive rate of return is 15 percent per year, determine which alternative is preferable.
b. Find the "rate(s) of return" for both alternatives, using the original book method.
c. Find the "rate(s) of return" for both alternatives, using the average book method.

*Adapted from E. L. Grant and W. G. Ireson, *Principles of Engineering Economy*, 5th edition (New York: The Ronald Press, 1970), p. 181.

End of year	After-tax cash flows			After-tax (accounting) profits	
	Vacuum still	Product terminal		Vacuum still	Product terminal
1	$ 38,000	$ 5,000		$27,000	-$ 6,000
2	34,000	9,000		23,000	- 2,000
3	30,000	13,000		19,000	2,000
4	26,000	17,000		15,000	6,000
5	22,000	21,000		11,000	10,000
6	18,000	25,000		7,000	14,000
7	14,000	29,000		3,000	18,000
8	10,000	33,000		- 1,000	22,000
9	6,000	37,000		- 5,000	26,000
10	2,000	41,000		- 9,000	30,000
	$200,000	$230,000	Total profit	$90,000	$120,000
			Average annual profit	$9,000	$12,000

d. What is the true after-tax internal rate of return for the vacuum still?

(*Answers:* (a) Vacuum still; (b) 8.2 percent and 10.9 percent; (c) 16.4 percent and 21.8 percent; (d) approximately 19 percent)

5.16 A company is considering investment in depreciable property. Estimates for the two alternatives under consideration are:

	Location M	Location N
First cost of property	$1,000,000	$1,400,000
Salvage value	200,000	200,000
Annual cash revenues (net)	120,000	160,000
Useful life	25 years	25 years

a. Determine the correct rates of return for both proposals.
b. If the minimum attractive rate of return for the company is 9 percent before income taxes, are either or both of the alternatives economically acceptable?
c. Using a minimum attractive rate of return of 9 percent, determine whether the extra investment in Location N ($400,000 increment) is justified. That is, find the rate of return on the incremental investment and compare it to the minimum attractive rate of return.
d. Determine the rates of return for Locations M and N using the original book method. Compare your results with a 9 percent minimum attractive rate of return.

5.17 Solve Problem 5.16(d) using the average book method.
(*Answer: RoR*(M) = 14.7 percent, *RoR*(N) = 14.0 percent)
5.18 Solve Problem 5.16(d) by using the average of the year-by-year rates of return as measured by the ratio of profit after depreciation to book value at start of year. Assume straight line depreciation.

Hoskold's Method

5.19 Determine the rates of return for Locations M and N from Problem 5.16 if the firm uses Hoskold's method with a 4 percent sinking fund.
(*Answer: RoR*(M) = 0.126, *RoR*(N) = 0.109)

The Truth in Lending Formula

5.20 The rate of return described by the "truth in lending" formula (Equation 5.16) is not the same as either the true nominal interest rate or the true effective interest rate described in Chapter 2. Show that, in general, the approximate rate of return of Equation 5.16 lies between the true nominal and effective rates.

5.21 A $1,500 item is purchased with a $500 down payment. The balance owed, $1,000, is to be paid in twelve equal end-of-month payments. The seller advertises a "12 percent per month plan," but, in practice, the monthly payments are determined by adding the total finance charge to the original debt and then dividing the total by the number of payment periods.
   a. Determine the amount of the monthly payments.
   b. Determine the annual cost of this debt using the "truth in lending" formula.
   c. Determine the true annual cost of this debt.
(*Answers:* (a) $93.33; (b) 22.2 percent; (c) 23.7 percent)

Approximate Methods for Computing Capital Recovery

5.22 You are considering the use of either of two engine generator sets; one a gas engine driven unit, the other a diesel engine driven unit. The gas engine unit costs $45,000, the diesel engine unit costs $70,000. Assuming the life of both units to be 20 years and zero salvage value, compare the present worth of the cost of 15 years' service, using an interest rate of 4%. Fuel cost for the gas engine is equal to $6,300 per year, for the diesel engine, $4,200 per year; taxes 1% of first cost per year; overhead, maintenance, and operating labor costs, $4,000 for either one. Use straight line depreciation and average interest in solving the problem.*
   a. Solve the problem. Recall that "average interest" is the average of the interest in the first and last periods.
   b. Find the true capital recovery at $i = 4$ percent.
   c. Comment on the differences in your answers to (a) and (b).

5.23 Consider the following alternative materials-handling systems:

	Semiautomatic	Fully automatic
Original cost	$24,000	$40,000
Expected salvage value	0	10,000
Annual operating costs	8,000	6,000

*This problem appeared in an Engineer-in-Training examination given in California.

The selected system will be kept for a period of twelve years.

a. Assuming a minimum attractive rate of return of 10 percent before income taxes, determine the equivalent net annual costs of the two plans using correct methods.

b. Using the straight line depreciation plus interest on first cost method to compute capital recovery, determine the annual costs of the two plans.

(*Answers:* (a) $11,522 and $11,043; (b) $12,400 and $12,500)

5.24 A manufacturing plant is considering the purchase of additional equipment for producing a machine part for the aircraft industry. Machine A with an initial cost of $12,000 and a life of 15 years will produce 20,000 units per year with a monthly maintenance cost of $50. Machine B produced the same part by stamping at the same yearly rate with an initial cost of $15,000 and a life of 30 years. However, the die assembly for stamping must be replaced every 10 years at a cost of $5,000 and a salvage value of $2,000. The monthly maintenance cost for Machine B will be $60. With interest in all cases at 6 percent and using straight line depreciation plus average interest, which machine would you purchase if the salvage value of each machine is considered to be zero at the end of their respective lives?*

a. Using the straight line depreciation plus average interest method, determine the (approximate) annual cost for Machines A and B.

b. Determine the true equivalent uniform annual cost for Machines A and B.

5.25 Use the straight line depreciation plus average interest method to determine the equivalent annual costs of alternatives Alpha and Beta from Problem 5.23.

(*Answers:* $11,300 and $11,125)

*This problem appeared in an Engineer-in-Training examination given in California.

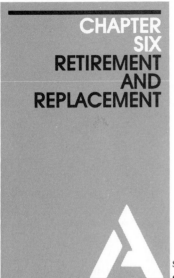

CHAPTER SIX
RETIREMENT AND REPLACEMENT

s noted in preceding chapters, the methodology presented in this book can be used to solve a wide variety of investment problems: which administrative procedure to adopt, which route to select, which equipment to purchase, which lease plan is preferable, whether to lease or purchase, when to implement a new program or initiate construction, and so on. Perhaps the most common of these problems is that of retirement/replacement. Because the retirement/replacement problem occurs so frequently and because of the special analytical techniques appropriate to it, this separate chapter has been developed.

In modern industry, existing assets are not retired merely because they are physically incapable of performing their original function. Rather, retirement generally occurs or is "encouraged" because of changes in economics or the operating environment, as for example when:

1. an existing asset was originally purchased to meet a certain demand, but demand has risen and the equipment can no longer satisfy current or anticipated demand.

2. new, improved equipment is available that is less expensive to operate, say, as a result of lower maintenance costs.

3. new, improved equipment is available that is more efficient than the existing asset.

4. the demand that prompted the purchase of the original equipment no longer exists, so the equipment no longer fills a useful purpose.

5. existing equipment has become a casualty as a result of, say, fire, accident, or major breakdown.

These are only examples. In most practical situations, retirement (and possible replacement) is considered because of a combination of causes.

Every discipline and subdiscipline has its own technical terms. Some are useful; others appear to have little value. The literature of replacement theory is no exception, so before going on, you will find it useful to understand the following terms:

Defender—the existing asset; the one under consideration for retirement[1]

Challenger—the proposed replacement asset

Economic life—that asset life resulting in minimum (equivalent uniform) annual cost. Of course, the economic life must be equal to or less than the asset's maximum physical life.

There is a logical structure to the development of the following sections. Section 6.3 examines **simple retirement,** retirement without replacement. Section 6.4 considers the problem of replacement wherein the currently available challenger is identical to the defender. Section 6.5 develops an evaluation procedure for the situation in which the current challenger is different from the defender but all future challengers are identical to the current challenger. And Section 6.6 briefly discusses the general case wherein all future challengers are different. This development plan is shown in Figure 6.1.

As with all retirement problems, simple retirement concerns only two immediate alternatives: (1) *keep* the existing asset, the defender, or (2) *retire* it. But simple retirement, unlike other retirement problems, concerns only the retirement of the asset, not its replacement. Thus, the analysis is straightforward. If the asset is to be kept in service for N more periods, the present worth of the resulting cash flows would be

$$PW = \sum_{j=1}^{N} A_j(1 + i)^{-j} \qquad \textbf{(6.1}a\textbf{)}$$

where A_j is the cash flow at the end of period j and i is the appropriate discount rate. Note that Equation 6.1 reflects the *end-of-period* assumptions for cash flows and discounting. If cash flows $\overline{A}_j$ are expected to occur continuously and uniformly *during* periods 1 through N, then

1. The terms *defender* and *challenger* originated in George Terborgh, *Dynamic Equipment Policy* (New York: McGraw-Hill, 1949).

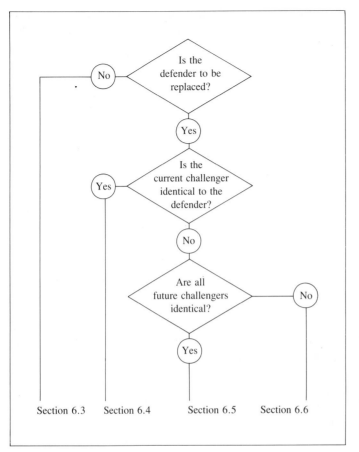

Figure 6.1 Plan of Development for Chapter 6

Section 6.3 Section 6.4 Section 6.5 Section 6.6

$$PW = \left[\frac{i}{\ln(1+i)}\right] \sum_{j=1}^{N} \overline{A}_j (1+i)^{-j} \qquad (6.1b)$$

The alternative is to retire the asset and to invest any funds received from its sale at interest rate i. Letting P represent the current market value of the asset, it follows that retirement is justified if the present worth of the "keep" alternative is less than the present worth of the "retire" alternative. That is, if $PW < P$, retire the asset and invest P elsewhere at rate i; if $PW > P$, keep the asset; and if $PW = P$, we are indifferent between the alternatives.

This rationale may be expressed more succinctly by observing the differences between alternatives. If the asset is kept rather than retired, the net effect is to forgo P now in order to receive cash flows A_j (and/or $\overline{A}_j$) over the remaining N periods. So the decision rule is to keep the asset if

$$[PW] - P > 0 \qquad (6.2)$$

Noting that $P = -A_0$, where A_0 is the cash flow at the start of the first period (end of period 0), and combining Equations 6.1 and 6.2, the rule is to keep the asset if

$$A_0 + \sum_{j=1}^{N} A_j(1 + i)^{-j} + \left[\frac{i}{\ln(1 + i)}\right] \sum_{j=1}^{N} \overline{A_j}(1 + i)^{-j} > 0 \quad (6.3)$$

In other words, the decision is to keep the asset if the present worth of the discrete and continuous cash flows is positive.

To illustrate this procedure, consider an asset that has a current market value of $1,000. If it is retained, it will generate net positive cash flows of $300 per year during each and every year over the next three years. Moreover, at the end of three years it can be sold for $200. The appropriate discount rate is assumed to be 6 percent. To determine whether the asset should be retained, compute the present worth:

Period	Cash flows		Discount factor	Present worth
j	A_j	$\overline{A}_j$		
0	−$1,000	$ 0	1.0000	−$1,000
1–3	0	300	2.752	826
3	200	0	0.8396	168
Totals	−$ 800	$900		−$ 6

Inasmuch as the PW is negative at $i = 6\%$, the asset should be retired. It is economically preferable to dispose of the asset and invest the $1,000 elsewhere at 6 percent.

Note that the **disposal value,** or salvage value, is not necessarily the current market value of the asset. Rather, the value used in the analysis should reflect the value to the firm if the asset is retired from its current activity. If retired, it may be sold. But it may also remain inside the firm to be employed in some other activity, in which case the disposal value is represented by the equivalent present worth if the asset were to be used elsewhere, say, in another operating unit. Put somewhat differently, the disposal value should represent the cost of the foregone opportunity of continuing to employ the asset in its current function—a measure of the monetary cost of the opportunity foregone.

Another example may be helpful in understanding the problem of simple retirement without replacement. Consider the following cost and revenue estimates for an asset that, at the time of the analysis, is three years old and has a remaining physical life of two years:

Age	Revenue	Operating costs	End-of-year salvage vlaue
4	$1,200	$800	$300
5	1,100	900	0

Assume that the asset may currently be sold for $600 and that the pretax minimum attractive rate of return for the firm is 10 percent.

If the asset is kept *one* more year, the relevant consequences are (a) a forgone opportunity to receive $600 "today," (b) net receipts of $400 during the next year, and (c) a salvage value of $300 at the end of the year. Assuming end-of-year cash flows and discounting, the present worth of keeping the asset one more year is

$$PW = -\$600 + (\$400 + \$300)(1.10)^{-1}$$
$$= \$36$$

Thus, as the analysis indicates, the asset should *not* be immediately retired; it should be kept for *at least* one more year.

Although the retirement question has been answered for the moment, it may be of some interest to determine whether, based on current estimates, the asset should be kept for *two* years, that is, through the end of its physical life. The expected consequences are (a) an immediate cost (opportunity forgone) of $600, (b) net receipts from operations of $400 and $200 at the end of years 4 and 5 respectively, and (c) zero net salvage value.

$$PW = -\$600 + \$400(1.10)^{-1} + \$200(1.10)^{-2}$$
$$= -\$71$$

It appears at this point that, unless there are some changes in the estimates for the fifth year, the asset should be retired from service at the end of the fourth year. Thus we have answered the question concerning retirement and we have also determined *how long* the asset should be retained.

Consider an asset that has been in service, say, for N_0 years. Its remaining maximum service life is N_{max} years, so it has a *total* service life of $N_0 + N_{max}$ years. Assume that there currently exists a replacement asset, a challenger, that is identical to the defender in every respect—its total expected service life is the same, the expected cash flows in each and every year of service are identical to those of the defender, and so on.

(It is difficult to conceive of a challenger that is *identical* in every respect to a given defender. However, an approximate situation occurs when the asset under consideration is relatively unchanged by technological advances and prices remain reasonably constant, as is the case with certain hand tools, basic construction equipment, and certain pipes and cables.)

We should focus our attention first on the challenger. Let $AC(C,N)$ represent the (equivalent uniform) *annual cost* of the

6.4
RETIREMENT WITH IDENTICAL REPLACEMENT

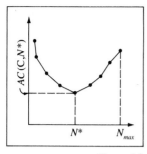

Figure 6.2

challenger if it is purchased and kept in service for exactly N periods. Here, annual cost is the negative of annual worth as defined in Section 3.1. The $AC(C,N)$ is computed for all values of $N = 1, 2, \ldots, N_{max}$. The project life for which the AC is minimized is the economic life, N^*, as shown in Figure 6.2. Once N^* has been determined, the appropriate decision rule for the optimal replacement strategy is straightforward. We need only compare the current age of the defender (DN) with the economic life of the challenger (CN^*). Since the challenger and the defender are identical, the economic life of the challenger is also the economic life of the defender. Has the defender, then, already reached its economic life? If so, it should be replaced. If not, it should be retained. Stated more succinctly:

Keep if $DN < CN^*$

Replace if $DN \geq CN^*$

To illustrate this procedure with a numerical example, consider a challenger that is currently available at a cost of $10,000. The challenger, which is identical to the defender, has a maximum physical life of six years. The expected year-by-year salvage values for years 1 through 6 are given in Column b of Table 6.1. (In general, it is necessary to predict the salvage values for years other than year 6, that is, for each year less than the maximum physical life.) Table 6.1 computes the cost of capital recovery (CR) for $N = 1, 2, \ldots, 6$, based on the equation

$$CR(C,j) = \$10,000(A/P, \, 10\%, \, j) - S_j(A/F, \, 10\%, \, j)$$

$$= (\$10,000 - S_j)(A/P, \, 10\%, \, j) + 0.10S_j$$

$$= (\text{Column c})(\text{Column d}) + (\text{Column e})$$

Note that $CR(C,j) > CR(C, j + 1)$ for all j in this example: the capital recovery decreases as the life of the asset increases.

Table 6.1
Determining Capital Recovery for a $10,000 Asset If It Is Retained
for Varying Periods from 1 through 6 Years, Assuming 10 Percent Interest Rate

Year j (a)	Salvage value at end of year S_j (b)	Initial cost less salvage value $\$10,000 - S_j$ (c)	Capital recovery factor $(A/P, 10\%, j)$ (d)	Interest on salvage value $0.10S_j$ (e)	Capital recovery if kept j years (c)(d) + (e) (f)
1	$7,000	$ 3,000	1.10000	$700	$4,000
2	4,500	5,500	0.57619	450	3,619
3	2,500	7,500	0.40211	250	3,266
4	1,000	9,000	0.31547	100	2,939
5	0	10,000	0.26380	0	2,638
6	0	10,000	0.22961	0	2,296

This will not always be the case, however. The capital recovery factor, $(A/P, i, j)$, always decreases as j increases, but it is not always true that the terms $(C - S_j)$ and iS_j also decrease. Suppose, for example, that S_2 is only $1,000 rather than the $4,500 originally stated. Then the cost of capital recovery if the asset is kept through two years would be

$$CR(C,2) = (\$10,000 - \$1,000)(0.57619) + 0.10(\$1,000)$$

$$= \$5,285$$

This is *larger* than the $AC(C,1)$. If the year-to-year salvage values remain constant, then the cost of capital recovery will steadily decrease. Otherwise, the direction of change depends on the salvage values as well as on the interest rate.

The remaining calculations necessary to the computation of the equivalent uniform annual costs are presented in Table 6.2. The estimated operating costs during years 1 through 6 are given in Column b, and the total annual cost if the asset is kept through N years is calculated. The results are given in Column h and are shown in Figure 6.3.

This pattern is typical of most replacement problems: capital recovery costs tend to decrease over time, whereas operating costs tend to increase. (The cost histories of most private autos are excellent examples of this characteristic.) The minimum total equivalent uniform annual cost occurs at three years, thus three years is the economic life for the challenger.[2] At this life, $AC(C,3)$

2. As shown by both Column h of Table 6.2 and Figure 6.3, the annual costs for lives other than the economic life are only slightly higher than the minimum value. The total cost curve is relatively flat, suggesting that the economic life is rather sensitive to values assumed for input data. This phenomenon appears to occur quite frequently in actual replacement studies.

Table 6.2
Determining Equivalent Uniform Annual Costs
of Asset in Table 6.1

Year j (a)	Operating costs during jth year (b)	Present-worth factor $(P/F, 10\%, j)$ (c)	During jth year (b) × (c) (d)	Through j years $\sum_{k=1}^{j}(d)_k$ (e)	Capital recovery factor $(A/P, 10\%, j)$ (f)	Operating costs (e) × (f) (g)	Capital recovery CR^*	Total annual cost AC (h)
			Present worth of operating costs			Equivalent uniform annual costs if retired after j years		
1	$ 5,000	0.9538	$4,769	$ 4,769	1.10000	$5,246	$4,000	$9,246
2	5,400	0.8671	4,682	9,451	0.57619	5,446	3,619	9,065
3	6,100	0.7883	4,809	14,260	0.40211	5,734	3,266	9,000
4	7,100	0.7166	5,088	19,348	0.31547	6,104	2,939	9,043
5	8,400	0.6515	5,473	24,821	0.26380	6,548	2,638	9,168
6	11,000	0.5922	6,514	31,335	0.22961	7,195	2,296	9,491

*From Column f of Table 6.1.

Figure 6.3 Determining Economic Life (N^*)

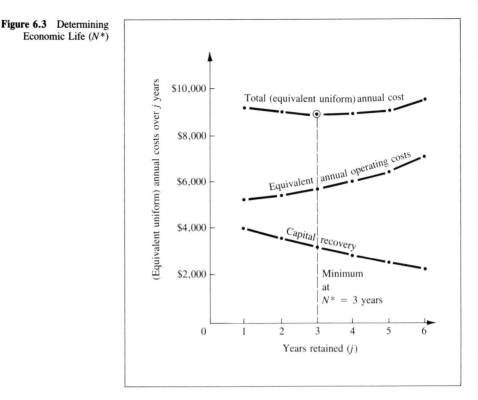

$= \$9,000$. If the defender (identical to the challenger) is at least three years old, it should be replaced; otherwise, it should be retained.

Another way of approaching this problem is to ask: Should the asset be retained for at least one more year (up to its physical life) or should it be replaced with the available challenger? Using the previous example, suppose that the defender is *one* year old at the time of the decision. The cost to keep it one more year (until age 2) is the sum of (a) the loss in market value, (b) the opportunity forgone by not selling the challenger now and investing the proceeds elsewhere at the minimum attractive rate of return, and (c) the operating costs during the next year. Thus the cost to keep a one-year-old asset one more year is

$$(\$7,000 - \$4,500) + \$7,000(0.10) + \$5,400\left(\frac{0.10}{\ln 1.1}\right)$$

$$= \$8,866$$

Since this is less than the AC of the challenger at its economic life, the defender should be kept at least one more year before replacement. (In fact, it should be kept two more years!)

Now suppose that the asset is *four* years old at the time of the decision. The cost if it is kept one more year is

$$(\$1,000 - \$0) + \$1,000(0.10) + \$8,400\left(\frac{0.10}{\ln 1.1}\right) = \$9,913$$

This is larger than $AC(CN^*) = \$9,000$, and replacement is indicated.

In summary, the defender should be replaced with an identical challenger when its economic life has been equaled or exceeded (or, of course, when it has reached its maximum physical life). This rule holds for those cases in which there is an indefinite need for the asset (an infinite planning horizon) or in which the planning horizon is limited but is an exact multiple of the economic life. The appropriate procedure is somewhat more complex in the remaining cases because combinations of assets with different lives must be used.

Consider the case in which an available challenger is under consideration to replace a certain defender. The current challenger is unlike the defender—perhaps it is more or less costly to operate or it is more efficient or it has a different physical life, and so on. Assume that all future challengers are *identical* to the current challenger, so if replacement is forgone at this time, there will be future opportunities to replace the defender with new challengers. Here, however, we are assuming that all future challengers do not differ significantly in any respect from the current challenger.

The appropriate solution procedure is closely related to the one described in Section 6.4:

1. Determine the (equivalent uniform) annual cost for the challenger at its economic life, N^*. (See Figure 6.4.) The procedure for determining the $AC(C,N^*)$ is precisely the same as the one presented in Section 6.4.
2. Determine the annual cost for the defender if it is kept *one* more year. That is, determine $AC(D,1)$.
   a. If $AC(D,1) < AC(C,N^*)$, it is preferable to *keep* the defender for *at least* one more year; thus it should not be replaced at this time. (See Figure 6.5a.)
   b. If $AC(D,1) > AC(C,N^*)$, it is not preferable to keep the defender *exactly* one more year. (See Figure 6.5b.) However, this does not signal replacement. Indeed, the pattern of future cash flows may be such that the annual cost may be reduced if the defender is kept two or more years. Step 3 explores this possibility.
3. Determine $AC(D,2)$.
   a. If $AC(D,2) < AC(C,N^*)$, as in Figure 6.6a, it is preferable to *keep* the defender for at least two more years.

6.5 RETIREMENT WITH AN UNLIKE REPLACEMENT BUT ALL IDENTICAL CHALLENGERS

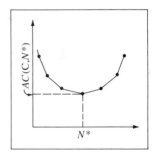

Figure 6.4

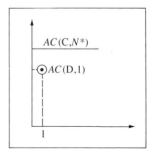

Figure 6.5a

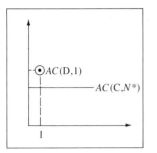

Figure 6.5b

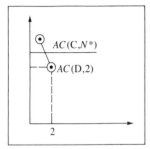

Figure 6.6a

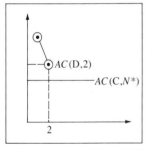

Figure 6.6b

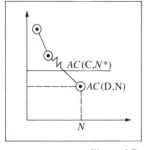

Figure 6.7a

b. If $AC(D,2) > AC(C,N^*)$, using the same reasoning as in Step 2b, it is necessary to determine whether the defender should be kept for at least three additional years. (See Figure 6.6b.)

4. The above procedure should be continued until one of two conditions occurs:

a. Examine the defender if it is kept N more years. If the $AC(D,N) < AC(C,N^*)$, as in Figure 6.7a, the defender should be kept *at least* N more years.

b. Examine the defender throughout its remaining physical life, that is, $N = 1, 2, \ldots, N_{max}$. If there is no life such that $AC(D,N) < AC(C,N^*)$, as in Figure 6.7b, then replacement is indicated.

To illustrate this procedure, consider the challenger described in Section 6.4. Recall that $AC(C,N^*) = \$9,000$. Suppose that the alternative, the defender, has a current market value of \$8,000. If it is retained for one more year, it may be sold for \$5,000 and it will incur operating costs of \$5,100 during the year. Then,

$$AC(C,1) = (\$8,000 - \$5,000)(A/P, 10\%, 1)$$
$$+ \$5,000(0.10) + \$5,100(F/\overline{A}, 10\%, 1)$$
$$= \$3,000(1.10) + \$500 + \$5,100(1.049)$$
$$= \$9,150$$

So the defender should not be kept one more year.

Perhaps the defender should be kept two more years or more. Suppose that the expected salvage value at the end of two years is \$4,000 and operating expenses during the second year are expected to continue at \$5,100. Then,

$$EUAC(C,2) = (\$8,000 - \$4,000)(A/P, 10\%, 2)$$
$$+ \$4,000(0.10) + \$5,100(F/\overline{A}, 10\%, 1)$$
$$= \$4,000(0.57619) + \$400 + \$5,100(1.049)$$
$$= \$8,055$$

Inasmuch as this value is less than \$9,000, it is preferable to keep the defender for at least two more years.

In problems of this type it may be useful to think of the challenger and the defender as sets of mutually exclusive alternatives, as shown by:

C,1 = challenger with life of 1 year

C,2 = challenger with life of 2 years

⋮

C,N_{max} = challenger with life of N_{max} years

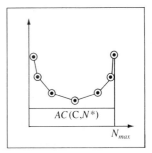

Figure 6.7b

Thus the first step is to determine which alternative represents the optimal challenger. The resulting $AC(C,N^*)$ then serves as the standard against which all defender alternatives are measured. If there are no defender alternatives such that $AC(D,N) < AC(C,N^*)$, then replacement is signaled. Note that the economic life of the defender is not necessarily determined by this procedure. The example showed that replacement should not take place, because the *EUAC* for the two-year defender life is less than the *EUAC* of the challenger. But it is quite possible that the *EUAC* will be still lower if the defender is kept for three more years. This additional information is irrelevant, however, since the question at hand deals only with current replacement, not with the economic life of the defender.

The procedure outlined above is valid when the need for the service provided by the defender and/or its successive challengers is indeterminate, that is, when the planning horizon is infinite. Problems that specify finite planning horizons generally result in complications that are beyond the scope of this discussion.

The preceding sections postulated certain assumptions concerning the character of current and future challengers. Section 6.3 examined retirement with no replacement, Section 6.4 examined the situation wherein the current challenger (and all future challengers) are identical to the defender, and Section 6.5 examined the situation wherein the current challenger is unlike the defender but all future challengers are identical to the current challenger.

6.6 GENERALIZED REPLACEMENT MODEL

Now the assumptions are relaxed in order to establish the most general, and probably the most realistic, case: the case wherein the current challenger is unlike the defender and all future challengers are different from one another.

It is generally expected that future challengers will be superior to those presently available in terms of initial cost, maintainability, useful physical life, and so on. The **generalized replacement model** provides for the broadest possible range of assumptions.

Consider the stream of cash flows generated by the replacement of a defender, the subsequent replacement of the first replacement, the subsequent replacement of the second replacement, and so on, throughout the planning horizon. There are a

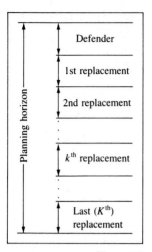

Figure 6.8

total of K such replacements over the entire planning horizon, as shown in Figure 6.8.

Let N_k represent the life of the kth alternative. For the sake of symmetry and completeness, let $k = 0$ specify the defender. The problem, then, becomes one of determining the economic lives (N_k^*) of all $K + 1$ assets $(k = 0, 1, 2, \ldots, K)$. If $N_0^* = 0$ as the result of these calculations, then the defender should be replaced now. That is, the value of N_0^* represents the economic life of the current defender.

In general, the economic lives of the defender and its chain of replacements can be determined by finding those values of N_k such that the present worth of *all* cash flows over the planning horizon is maximized. For $K \geq 1$, the present worth, PW, is given by

$$PW = \sum_{k=0}^{K} [PW(k)](1 + i)^{-(N_0 + N_1 + \cdots + N_{k-1})} \tag{6.4}$$

where $PW(k)$ represents the net present value of the kth replacement in the chain, computed at the time the kth replacement is put into effect. Note that exactly $(N_0 + N_1 + \cdots + N_{k-1})$ periods will have elapsed until the kth replacement is made.

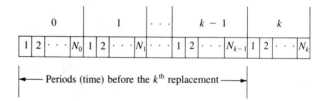

The present worth of the kth replacement is found from

$$PW(k) = -P_k + S_k(1 + i)^{-N_k} + \sum_{j=1}^{N_k} A_{kj}(1 + i)^{-j} \tag{6.5}$$

where

$P_k =$ initial cost of the kth replacement

$S_k =$ expected salvage value of the kth replacement

$A_{kj} =$ expected cash flows from operations in the jth period for the kth replacement

Note that the PW formulation in Equation 6.5 assumes *end-of-period* cash flows, as shown in Figure 6.9. If the cash flows from operations, in particular, are assumed to flow continuously and uniformly *during* the period, then

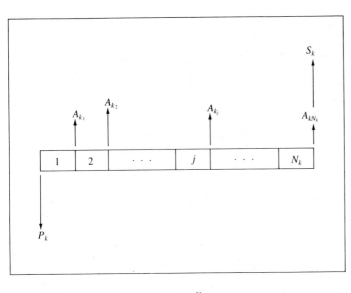

Figure 6.9 Cash Flow Diagram for the kth Replacement of the Generalized Replacement Problem

$$PW(k) = -P_k + S_k(1 + i)^{-N_k}$$

$$+ \left[\frac{i}{\ln(1 + i)}\right] \sum_{j=1}^{N_k} A_{kj}(1 + i)^{-j} \qquad (6.6)$$

Note the adjustment factor, $[i/\ln(1 + i)]$, for the present worth of cash flows from operations.

In summary, the analyst must find the set $(N_0, N_1, \ldots, N_K)$ such that the total present worth is maximized. The optimal values of N_k are indicated by asterisks (*). The values of N_k^* represent the optimal replacement strategy over the planning horizon. In particular, immediate replacement of the current defender is signaled if $N_0^* = 0$.

Unfortunately, this procedure can be tedious if the cost/revenue functions are such that the present worths must be computed for all possible values of k. For example, consider a planning horizon of only three periods. A total of eight separate present worths must be calculated for the eight combinations of lives of the defender and subsequent challengers:

Combination	Defender $(k = 0)$ N_0	Replacements 1 N_1	2 N_2	3 N_3
1	3	0	0	0
2	2	1	0	0
3	1	2	0	0
4	1	1	1	0
5	0	3	0	0
6	0	2	1	0
7	0	1	2	0
8	0	1	1	1

In the first combination, $N_0 = 3$. That is, the defender is retained throughout the three-period planning horizon. In the second combination, the defender is retained for two periods and the first replacement, the challenger available at the end of two periods hence, is retained for one period, and so on. The present worths must be computed for all eight combinations. The maximum PW identifies the optional solution $(N_0^*, N_1^*, N_2^*, N_3^*)$.

In general, given a planning horizon of N periods, there are 2^N possible sets of assumptions, or separate values of PW, that must be calculated. If $N = 10$, for example, 1,024 values of PW must be calculated. Assuming a twenty-period planning horizon, the number of possible values of PW exceed a million! Clearly, the burden of calculation quickly becomes excessive for the general case unless an efficient algorithm can be used. Here, an ''efficient algorithm'' is one that leads to the optimal solution, that is, the set of N_k's such that PW is maximized, without an exhaustive search of all possible combinations. The generalized replacement model may be cast into a convenient solution framework by using **dynamic programming,** although the procedure is beyond the scope of this book.[3]

<table>
<tr><td>6.7
SUMMARY</td><td></td></tr>
</table>

6.7
SUMMARY

Of those problems of interest in the capital allocation context, questions of retirement and replacement probably occur most frequently in the real world. Moreover retirement/replacement problems present special conceptual and computational difficulties.

The simplest retirement case occurs when an existing asset is to be retired but not replaced. In this instance the discounted present value (present worth) of all prospective consequences during the remaining life of the asset should be evaluated. If the present worth of these consequences is greater than the current disposal value, then the asset should be retained; otherwise it should be terminated. This procedure may also be used to determine the economic life of the asset.

An additional complexity is generated when the problem of replacement is added to that of simple retirement. If the existing asset, the defender, is to be replaced by an *identical* asset, the challenger, and if the challenger that will become available in future years is also identical to the defender, the appropriate replacement decision is indicated by determining whether the economic life of the defender has been equaled or exceeded. In this case the defender should be retained if its current age is less than its economic life. (Replacement must occur, of course, if the physical life of the asset has been met.)

---

3. One of the earliest articles that discusses this procedure and includes a numerical example is S. E. Drayfus, ''A Generalized Equipment Replacement Study,'' *Journal of the Society for Industrial and Applied Mathematics*, vol. 8, no. 3 (September 1960): 425–35.

In the event that the current challenger is *different* from the defender but *identical* or superior to all challengers that may become available in the future, the first step in resolving the replacement problem is to determine the economic life of the challenger. The (equivalent uniform) annual cost, AC, of the challenger at this life is then compared to the cost of keeping the defender in service *one* more year. If the AC of the defender is lower, then retention is in order; if it is higher than the AC of the challenger, then the AC of the defender if kept *two* more years must be determined. The comparison is then repeated. If the AC of the defender with a two-year life is less than the AC of the challenger, the defender should not be replaced; if it is greater, then the AC of the challenger if kept *three* more years should be computed. This procedure should be repeated until retention is indicated or the physical life of the defender has been met. If there is no defender life resulting in an AC less than that of the best challenger, the defender should be replaced.

The most difficult problems arise from cases in which the current challenger is different from the defender and succeeding challengers are different from the existing challenger. It is possible to structure a mathematical model that describes the general situation, but optimal solutions for this model are extremely difficult to determine. (For the generalized replacement model, if an improved challenger becomes available each year, there are 2^N possible solutions for a planning horizon of N years.) Certain mathematical programming procedures may be used to search for the optimal solution more efficiently—dynamic programming, for example, can be employed under certain conditions—but these procedures are beyond the scope of this book. References are included in the Bibliography for those who may be interested in additional mathematical programming procedures.

A number of simplified mathematical models have been developed to solve the equipment replacement problem when future challengers are unlike the current challenger. All such models depend on critical assumptions concerning the characteristics of these unlike future challengers. One of the more interesting of these models, the MAPI model, is treated in Chapter 6 Appendix.

The MAPI Model

A principal assumption implicit in Sections 6.4 and 6.5 is the invariability of future challengers, that is, future challengers will in no way differ from the presently available challenger. This assumption is questionable, of course, since in general, it seems likely that improved challengers will become available as time passes. In a competitive economy, where so much productive capacity is directed to developing more efficient and less costly products, future challengers, in most cases, can be expected to be superior to presently available challengers with respect to initial cost, maintainability, useful physical life, and so on. This possibility adds considerable complexity to retirement and replacement analyses, as you saw in Section 6.6. A number of simplified mathematical models have been developed, one of which is discussed here.

The research staff of the Machinery and Allied Products Institute, MAPI, under the direction of George Terborgh, has constructed a number of mathematical models for use in replacement studies. The first appeared in *Dynamic Equipment Policy* (1949), the second in the *MAPI Replacement Manual* (1950), and the third in *Business Investment Policy* (1958). Any treatment of replacement problems would be incomplete without a brief discussion of the principle concepts and procedures presented in these earlier MAPI publications.

Consider a challenger that may be purchased immediately for $1,000, and assume that its salvage value will be zero at all times. Assume also that operating costs for this challenger during the first year will be $2,200 and that these disbursements will increase at the rate of $100 per year thereafter. As a result of technological progress, the initial operating cost of the challenger that will become available next year is expected to be only $2,000. Assume that these patterns will repeat themselves, that is, the first-year costs of future challengers will decrease by $200 each year, annual disbursements of each challenger will increase by $100 each year, and thereafter the salvage values of all challengers will always be zero.

The annual disbursements of the current and next-year challengers, as well as their successors, are shown in Figure 6A.1, in which it is assumed that all challengers will be replaced every four years. Note that the current challenger will be "inferior" to next-year's challenger by $300, of which $200 results from the decrease in initial cost. The other $100 results from the increase in annual disbursements. *Dynamic Equipment Policy* terms the rate at which the current challenger accumulates inferiority to future challengers ($300 in this example) the **inferiority gradient.**

Determination of the economic life of the current challenger, as well as the economic lives of all successive challengers, is illustrated in Table 6A.1. Assuming a pretax minimum attractive

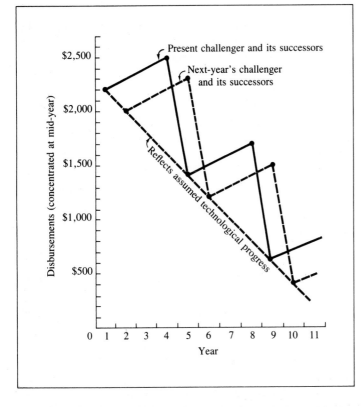

Figure 6A.1 Annual Disbursements for Current and Future Challengers

rate of return of 20 percent, the economic life is shown to be three years. In *Dynamic Equipment Policy* the (equivalent uniform) annual cost at the economic life ($739 in this example) is known as the **challenger's adverse minimum** (*CAM*). The cumulative inferiority in Column b of Table 6A.1 is the difference between the disbursements for the current challenger in the *j*th year and the

Table 6A.1
Determining Economic Life and "Adverse Minimum"

Year j (a)	Cumulative inferiority $300(j-1)$ (b)	Capital recovery cost $1,000(A/P, 20\%, j)$ (c)	Cumulative inferiority annualized $300(A/G, 20\%, j)$ (d)	Equivalent annual cost (c) + (d) (e)
1	$ 0	$1,200	$ 0	$1,200
2	300	655	137	792
3	600	475	264	739
4	900	386	382	768
5	1,200	334	492	826
6	1,500	301	594	895
7	1,800	277	687	964
8	2,100	261	774	1,035

Note: Challenger: Initial cost, $1,000; inferiority gradient, $300; zero salvage value at all times; and assumed interest rate of 20 percent.

disbursements that would occur if a new challenger were to be purchased at the start of the jth year. Note that the economic life occurs at that point where the cumulative inferiority becomes larger than the (equivalent uniform) annual cost.

Dynamic Equipment Policy recommends an approximate formula for determining the challenger's adverse minimum (*CAM*):

$$CAM = \sqrt{2PG} + \frac{iP - G}{2} \qquad (6A.1)$$

where P is the initial cost, G is the inferiority gradient, and i is the interest rate. As before, this formula assumes zero salvage value at all times. It is approximate in that it averages cash flows rather than determining equivalent values using compound interest factors.

The defender in this replacement problem must now be examined. Assume that the defender's next year's disbursements will be $2,500 and that the disbursements of subsequent years will be even higher. Assuming that the defender has no salvage value now and will not have any in the future, it is clear that no year will be as inexpensive as next year. Thus, since there are no capital costs, next-year's operating inferiority to the present challenger is simply $300($2,500 $-$ $2,200), which represents the **defender's adverse minimum.** Since the challenger's adverse minimum, $739, is larger, the defender should be retained. In other words, it would be more costly to replace the defender today with the current challenger than it would be to forgo replacement for one more year and then replace the defender with next year's challenger.

The MAPI methods presented in *Business Investment Policy* (1958) differ considerably from those of the earlier publications. The previous methods compare next-year dollar values, whereas *Business Investment Policy* uses the *MAPI urgency rating*. The **urgency rating** is the next-year after-tax rate of return assuming a 50 percent tax rate and a 3 : 1 equity to debt ratio for capital funds. A 3 percent interest rate is assumed for debt capital; a 10 percent rate of return is required from equity funds. As before, the rate of technological progress is assumed to be constant, but the 1958 techniques also allow the analyst to assume that operating disbursements increase at an increasing or decreasing rate as well as the constant rate used in *Dynamic Equipment Policy*. The analyst may also choose between the straight line or the sum of years digits depreciation methods.

This discussion of the MAPI techniques is necessarily abbreviated. Students interested in in-depth discussions of the MAPI methods should consult the sources directly. Other references to MAPI techniques appear in the Bibliography.

Note: Unless otherwise indicated, assume that all cash flows are end-of-period and that all future challengers will not differ significantly from the current challenger. These problems ignore the effects of federal and state income taxes. (Several problems in Chapter 7 consider the effects of income taxes on retirement and replacement.) Problems 6.18, 6.19, and 6.20 are directly related to the MAPI models discussed in Chapter Six Appendix.

6.1 A firm is contemplating replacing a computer it purchased three years ago for $450,000. Operating and maintenance costs have been $85,000 per year. The computer currently has a trade-in value of $100,000 toward a new computer that costs $650,000 and has a five-year life. The new computer will have annual operating and maintenance costs of $80,000 per year and an expected salvage value of $200,000 at the end of five years.

If the current computer is retained, another small computer will have to be purchased in order to provide the required computing capacity. The smaller computer will cost $300,000, will have a value of $50,000 in five years, and will have annual operating and maintenance costs of $55,000. The current computer will have no value in five years; operating and maintenance costs are expected to remain at $85,000 per year over the next five years.*

Using a before-tax analysis with a minimum attractive rate of return of 20 percent, determine whether the current computer should be replaced.
(*Answer:* Keep the defender)

6.2 The Blue Company owns a machine that cost $26,000 ten years ago. A new machine is available that costs $11,000 and will save $2,000 annually.

If the new machine is purchased, the old machine will be sold for $8,000. If the old machine is retained, it will be scrapped in five years for zero salvage value. Assume that the estimated salvage value of the new machine will be $1,000 at the end of five years. The minimum attractive rate of return before income taxes is 20 percent. Should the existing machine be replaced with the challenger? Why or why not?

6.3 General Hospital is considering the purchase of certain new kitchen equipment to replace its present three-year-old equipment. The total physical life of the present equipment is five years. Cost history for the past three years and estimates for the next two years are:

*Adapted from J. R. Canada and J. A. White, Jr., *Capital Investment Decision Analysis for Management and Engineering* (Englewood Cliffs, N. J.: Prentice-Hall, 1980).

| Year | Past disbursements | | End-of-year |
	Operations	Maintenance	salvage value
1	$ 6,000	$1,000	0
2	8,000	3,000	0
3	10,000	5,000	$14,000
	Estimated future disbursements		
4	12,000	7,000	11,000
5	14,000	9,000	9,000

The kitchen superintendent has suggested that the present equipment be replaced by a new model costing $60,000. The manufacturer claims that the new equipment will require no maintenance and will reduce operating costs by $1,000 per year.

Assuming a five-year physical life for the challenger with a $10,000 terminal salvage value, make a comparison of (equivalent uniform) annual costs using a 10 percent minimum attractive rate of return.

(*Answer:* Replace the defender)

6.4 The manager of the western division of the Brown Manufacturing Company has received a request from the superintendent of the Los Angeles plant for a new heavy-duty generator to replace the existing generator. The manager has forwarded the request to corporate headquarters for approval. The existing generator was purchased five years ago for $5,000. It is believed that it may remain in operation for another eight years before it must be scrapped. The existing generator is specialized equipment but, fortunately, there is a buyer who is willing to pay $600 for it. However this will probably be the last chance to sell it at any price.

The proposed replacement will cost $6,000 and it has an expected useful life of ten years. It is expected that the actual market value of the proposed equipment will be:

End of year	Market value
1	$4,000
2	2,400
3	1,200
4	400
5–10	0

The estimated annual operating costs for the new equipment are $600. Operating costs for the existing equipment are expected to be $1,800 per year. The company plans to sell the plant and all its equipment in five years. The sale price is independent of the present decision concerning the generator.

Assuming a before-tax minimum attractive rate of return of 15 percent, determine whether the new equipment should be purchased. Explain why or why not.

6.5 The Core Corporation is considering replacing or remodeling a small commercial structure. A new building will cost

$60,000 and will have an estimated useful life of twenty years. It is expected that the present structure, if repaired, will have an additional useful life of seven years. Remodeling costs are estimated to be $18,000. The existing building may be sold for $8,000 in its present condition, but if it is remodeled and retained for another seven years, it will have no disposal value at that time. The terminal salvage value of a replacement structure after twenty years is estimated to be $100,000. All other costs are unaffected by the replacement decision.

If the before-tax minimum attractive rate of return is 15 percent, determine the (equivalent uniform) annual costs of the two alternatives.

(*Answer: AC*(remodel) = $6,249, *AC*(build) = $8,610)

6.6 A high-speed, special-purpose, automatic strip-feed punch press costing $15,000 has been proposed to replace three hand-fed presses now in use. The life of this automatic press has been estimated to be five years. Expenditures for labor, maintenance, and so forth, are estimated to be $3,000 per year.

The general-purpose, hand-fed punch presses each cost $2,000 ten years ago and were estimated to have a life of twenty years, with a salvage value of $200 each at the end of that time. Their present net disposal value is $750 each. Operating expenditures for labor, and so forth, will be about $2,850 per machine per year. The net salvage value is expected to decrease by about $100 per press per year. It is expected that the required service will continue for only five more years and that the salvage value of the challenger will be $4,000 at the end of that time.

If the company's before-tax minimum attractive rate of return is 20 percent, determine whether the defenders should be replaced.

6.7 Machines of a given class have the cost characteristics shown below. The cost of a new machine is $100,000.

Machine age (years)	Maintenance and operating costs	Salvage value at end of year
1	$12,000	$60,000
2	17,000	40,000
3	19,000	31,000
4	20,000	24,000
5	22,000	15,000
6	24,000	10,000
7	33,000	2,000
8	37,000	0

If the firm's discount rate is 6 percent, find the economic life of this machine and the (equivalent uniform) annual cost at the economic life.

(*Answer:* $37,538 at six years)

6.8   In this problem assume that $i = 0\%$, an unrealistic assumption, of course, but one that greatly simplifies the arithmetic without sacrificing the underlying principle.

There are two approaches to the treatment of the defender's current disposal value. Approach A treats the current salvage value of the defender as a cost. That is, the defender's current salvage value is an opportunity cost, or a cost of opportunity forgone. With this viewpoint, the cash flows for the defender and the currently available challenger are:

End of year	Defender	Challenger
0	− $50	− $100
1	− 10	+ 50
2	− 20	

(Although the maximum physical life of the challenger is one year, assume that an identical challenger can be purchased at that time.)

Approach B credits the current salvage value to the challenger. Thus the cash flows would be:

End of year	Defender	Challenger
0	$ 0	− $100 + $50 = − $50
1	− 10	50
2	− 20	

a. Determine the optimal policy using Approach A.
b. Determine the optimal policy using Approach B.
c. Which approach is correct? Explain your answer.

6.9   A firm is considering the replacement of a certain asset (the defender) with a certain new asset (the challenger). The relevant economic data are:

| End of year | Defender | | Challenger | |
	Market value	Operating costs	Market value	Operating costs
1980	$8,000	$2,000	$——	$——
1981	6,000	2,200	——	——
1982	4,500	2,400	——	——
1983	3,200	2,600	——	——
1984 (now)	2,500	2,800	6,500	3,000
1985	2,000	3,000	4,000	3,000
1986	1,600	3,200	4,000	3,000
1987	1,300	3,400	4,000	3,000
1988	1,100	3,600	4,000	3,000
1989	1,000	3,800	4,000	3,000
1990	——	——	1,000	3,000
1991	——	——	1,000	3,000
1992	——	——	1,000	3,000
1993	——	——	1,000	3,000
1994	——	——	1,000	3,000

The firm's minimum attractive rate of return is 10 percent. The present time (now) is the end of 1984. Assume that the operating costs at the end of 1984, $2,800, are sunk costs, that is, they were incurred prior to the beginning of the planning horizon. The defender will reach its maximum physical life at the end of 1994.

a. Determine the economic life of the challenger and its (equivalent uniform) annual cost at its economic life.

b. Assume that the need for this type of equipment, whether defender or challenger, will cease at the end of 1989. Given this five-year planning horizon and assuming that all future challengers are identical to the current challenger, determine whether the defender should be replaced? Explain why or why not.

(*Answers:* (a) $3,995 at ten years; (b) Keep the defender)

6.10 Consider a challenger with the following prospective consequences:

End of year	Market value	Operating expenses
0	$10,000	$ 0
1	8,000	2,000
2	7,000	2,500
3	6,500	3,000
4	5,000	3,000

The initial cost of the challenger, as shown above, is $10,000. It has a maximum physical life of four years. Assume that $i = 8\%$ and all future challengers are identical to the current challenger.

a. Find the economic life of the challenger (CN^*). That is, find the CN^* such that the annual cost at that life is minimal.

b. Find the (equivalent uniform) annual cost of the challenger at its economic life.

c. Suppose that the defender is identical to the challenger and that the defender is now exactly one year old. Find the remaining defender life before the defender should be replaced.

6.11 The General Bottling Company is considering the replacement of its twenty-year-old capping machine, which was originally purchased for $30,000. The current net salvage value of the machine is zero; the cost to haul it away is about equal to its scrap value. Operating costs are currently $50,000 per year and are expected to increase at the rate of $5,000 per year. That is, next year's costs will be $50,000, the following year's costs will be $55,000, and so on. The remaining physical life of the machine is five years.

A new capping machine is available for $100,000, but because it is a special-purpose machine, its salvage value will drop sharply:

Age (years)	Salvage value
1	$60,000
2	30,000
3	10,000
More than 3	0

Operating costs for this new machine should hold steady at $40,000 per year over its twenty-year physical life. Assume that all future challengers will be identical to the current challenger. The firm's pretax minimum attractive rate of return is 10 percent.

a. Find the (equivalent uniform) annual cost for the challenger at its economic life.

b. Find the (equivalent uniform) annual cost for the defender if it is kept for five more years.

c. Should the defender be replaced now? Should it be replaced at any time over the next five years? Explain your answers.

(*Answers:* (a) $51,750 at twenty years; (b) $59,050; (c) Keep the defender one more year)

6.12 Consider a finite planning horizon N such that no more than one replacement will be made between now and time N. In other words, either no replacement will be made *or* a challenger will be retained until time N. After time N the requirement will cease. Assume that the operating costs, including maintenance, for both the challenger and the defender do not decrease with time. Show, in general, that replacement should not occur as long as the cost of one additional year of service for the defender is less than the savings that would result from postponing the acquisition of the challenger one year.

6.13 Consider the problem of replacing a certain three-year-old asset (the defender) with an *identical* asset (the challenger). Assume that all future challengers will also be identical to the current challenger and the defender.

The challenger may be purchased for $10,000. Its maximum physical life is five years. End-of-year market values and annual operating costs are:

End of year	Market value	Operating costs
1	$8,000	$10,000
2	4,000	10,000
3	2,000	10,000
4	0	10,000
5	0	10,000

The firm's before-tax minimum attractive rate of return is 10 percent.

a. Find the economic life for the challenger and the (equivalent uniform) annual cost at the economic life.

b. Decide whether the defender should be replaced. Explain why or why not.

(*Answers:* (a) $12,638 at five years; (b) Keep the defender)

6.14 Consider the problem of replacement with an identical asset. An alternative formulation of the decision rule for determining the economic life of the defender is: "The minimum (equivalent uniform) annual cost occurs when the cost of extending service one additional year is greater than the (equivalent uniform) annual cost to date." Why is this true? Demonstrate this principle using the example presented in Tables 6.1 and 6.2. Plot the results.

6.15 Consider the acquisition of a new asset (the challenger) having a five-year physical life. The relevant economic data are:

Year	Estimated market value		Operating costs (at end of year)
	Start of year	End of year	
1	$10,000	$7,000	$1,000
2	7,000	5,000	2,000
3	5,000	4,000	3,000
4	4,000	3,200	4,000
5	3,200	2,500	5,000

The firm's minimum attractive rate of return is 10 percent.

a. What is the (equivalent uniform) annual cost for this proposed asset if it is purchased and kept for the full five years?

b. Determine the economic life for this alternative and the (equivalent uniform) annual cost at the economic life.

(*Answers:* (a) $5,039; (b) $4,750 at three years)

6.16 A prospective acquisition, a capital investment, is currently available at a cost of $13,000. Anticipated salvage values and operating costs are:

End of year	Salvage value	Operating costs
1	$9,000	$2,500
2	8,000	2,700
3	6,000	3,000
4	2,000	3,500
5	0	4,500

Assume a 10 percent discount rate, determine the economic life for this challenger and the (equivalent uniform) annual cost at the economic life.*

*Adapted from L. T. Blank and A. J. Tarquin, *Engineering Economy*, 2nd ed. (New York: McGraw-Hill, 1983), p. 216.

6.17 A new piece of equipment is being considered to replace a certain existing asset. If the defender is replaced, the net effect next year will be that revenue will increase by $1,000 resulting from an improvement in the quality of the product, direct labor costs of operation will decrease by $4,500, maintenance costs will decrease by $500, and insurance costs will increase by $200. The present market value of the defender is $900; the market value will drop to $600 if the defender is kept one more year. The defender is now ten years old. The initial cost of the proposed new equipment is $15,000, and the expected salvage value of the challenger is zero.

    Assuming a pretax minimum attractive rate of return of 20 percent, use the method advocated in *Dynamic Equipment Policy* to determine whether the defender should be immediately replaced. (*Hint:* Recall that the defender's adverse minimum is the sum of next year's operating inferiority and next year's capital costs. The challenger's adverse minimum may be determined by using the approximate zero-salvage-value formula (Equation 6A.1). Since the inferiority gradient is required in the formula, estimate this value by determining the past value and assuming that it will continue into the future. The past gradient is determined by dividing the defender's operating inferiority next year by its age.) What is the challenger's implied economic life?
(*Answer:* 9.3 years)

6.18 Assume that the challenger in Problem 6.16 is being compared to a defender with a current market value of $5,000. Anticipated salvage values and operating costs for the defender are:

End of year	Salvage value	Operating costs
1	$4,000	$3,000
2	2,000	3,500
3	0	4,000

The defender was originally purchased one year ago for $8,000. Its remaining physical life is three years. There is a perpetual need for this equipment, whether challenger or defender. The discount rate is 10 percent.

    Determine the optimal strategy. How many more years, if any, should the defender be retained before it is replaced?

6.19 In the zero salvage value case, the *MAPI Replacement Manual* gives the following formula for the challenger's adverse minimum:

$$CAM = \frac{PNi^2}{iN + (1 + i)^{-N} - 1}$$

where P is the initial cost of the challenger, N is the eco-

nomic life, and i is the interest rate. Assuming a twelve-year economic life for the challenger described in Problem 6.17, use the formula to determine the implied gradient. (The gradient is the adverse minimum divided by the life.)
(*Answer:* $397 per year)

6.20 Show that the formula given in Problem 6.19 may be derived directly from

$$CAM = P(A/P, i, N) + G(A/G, i, N)$$

when the challenger's adverse minimum is equal to NG.

Computer Programs

These computer programs have been written in APPLESOFT BASIC for use with the Apple II personal computer.

The following notation is used in the progams:

$$SP\emptyset = \text{selling price now (the current disposal value)}$$
$$= SA(\emptyset)$$
$$SA(I) = \text{expected salvage value at the end of period I}$$
$$R = \text{expected revenue (end-of-period)}$$
$$C = \text{expected costs (end-of-period)} = OC$$
$$NR = \text{net receipts from operations} = R - C$$
$$MARR = \text{discount rate}$$
$$PNR = \text{present worth of net receipts}$$
$$PSA = \text{present worth of expected salvage value}$$
$$N = \text{number of periods (study period)}$$
$$I = \text{index for period}$$
$$APNR = \text{accumulated } PNR$$
$$PW = \text{net present worth of policy}$$

Simple Retirement

PROGRAM

```
]LOAD SIMPLE RETIRE
]LIST
10  REM
20  REM RETIREMENT WITHOUT
30  REM REPLACEMENT PROGRAM
40  REM
50  REM 2 DECEMBER 1981
60  REM
70  REM LAKSNA THONGTHAI
80  REM
90  DIM R(20),C(20),SA(20),NR(20)
100 DIM PNR(20),APNR(20),PSA(20),PW(20)
110 HOME
120 PRINT "RETIREMENT WITHOUT REPLACEMENT PROGRAM"
130 PRINT
140 PRINT
150 PRINT "SELLING PRICE NOW? (SP0)"
160 INPUT SP0
170 PRINT
180 PRINT
190 PRINT "HOW MANY PERIODS? (N)"
200 INPUT N
210 PRINT
220 PRINT
230 PRINT "MINIMUM ATTRACTIVE RATE OF RETURN? (MARR--%)"
240 INPUT MARR
250 PF = 1 / (1 + MARR / 100)
260 REM--************************************************************************
270 APNR(0) = 0
280 FOR I = 1 TO N
290 PRINT
```

```
300 PRINT
310 PRINT "AT THE END OF PERIOD ";I
320 PRINT "DATA INPUT"
330 INPUT "EXP. REVENUE(R)";R(I)
340 INPUT "EXP. COST(C)";C(I)
350 INPUT "EXP. SALVAGE VALUE(SA)";SA(I)
360 REM NR--NET RECEIPTS FROM OPERATIONS
370 NR(I) = R(I) − C(I)
380 REM PNR--NET RECEIPTS PRESENT WORTH
390 PNR(I) = NR(I) * (PF ∧ I)
400 REM PSA--SALVAGE VALUE PRESENT WORTH
410 PSA(I) = SA(I) * (PF ∧ I)
420 REM APNR ACCUMULATED PNR
430 APNR(I) = APNR(I − 1) + PNR(I)
440 NEXT I
450 LL$ = " "
460 FOR I = 1 TO N
470 REM PW--NET PRESENT WORTH OF THIS POLICY
480 PW(I) = APNR(I) + PSA(I) − SP0
490 IF PW(I) < 0 THEN GOSUB 780
500 NEXT I
510 PRINT
520 PRINT
530 PRINT "END OF"
540 PRINT "PERIOD";: HTAB (11): PRINT "REVENUE";: HTAB (21): PRINT "COST";: HTAB
    (31): PRINT "SALVAGE V"
550 FOR I = 1 TO N
560 PRINT I;: HTAB (11): PRINT R(I);: HTAB (21): PRINT C(I);: HTAB (31): PRINT SA(I)
570 NEXT I
580 PRINT
590 PRINT
600 PRINT "PERIOD(N)","NET PRESENT WORTH"
610 PRINT " ","AFTER (N) PERIODS"
620 PRINT
630 FOR I = 1 TO N
640 PRINT I,PW(I)
650 NEXT I
660 IF LL$ = "Y" THEN 760
670 PRINT : PRINT
680 INPUT "DO YOU WANT A HARD COPY?(Y/N) ";LL$
690 IF LL$ = "N" THEN 770
700 PR# 2
710 PRINT : PRINT : PRINT "RETIREMENT WITHOUT REPLACEMENT"
720 PRINT : PRINT "SELLING PRICE NOW = ";SP0
730 PRINT : PRINT "MINIMUM ATTRACTIVE RATE OF RETURN = ";MARR;"%"
740 PRINT
750 GOTO 460
760 PR# 0
770 END
780 PRINT
790 PRINT
800 PRINT "RETIREMENT SHOULD BE"
810 PRINT "AT THE END OF PERIOD ";(I − 1)
820 PRINT
830 PRINT
840 RETURN
```

Example

Consider a three-year-old asset whose current disposal value is $600. It has a remaining physical life of two years. The relevant data for each of the remaining years are:

End of year	Revenue	Operating costs	Salvage value
1	$1,200	$800	$300
2	1,000	800	0

The minimum attractive rate of return is 10 percent.

EXAMPLE

```
]RUN
RETIREMENT WITHOUT REPLACEMENT PROGRAM

SELLING PRICE NOW? (SP0)
?600

HOW MANY PERIODS (N)
?2

MINIMUM ATTRACTIVE RATE OF RETURN? (MARR--%)
?10

AT THE END OF PERIOD 1
DATA INPUT
EXP. REVENUE(R)1200
EXT. COST(C)800
EXP. SALVAGE VALUE(SA)300

AT THE END OF PERIOD 2
DATA INPUT
EXP. REVENUE(R)1000
EXT. COST(C)800
EXP. SALVAGE VALUE(SA)0

RETIREMENT SHOULD BE
AT THE END OF PERIOD 1

END OF
PERIOD    REVENUE    COST    SALVAGE V
1         1200       800     300
2         1000       800     0

PERIOD(N)    NET PRESENT WORTH
             AFTER (N) PERIODS

1            36.3636365
2            -71.0743802

DO YOU WANT A HARD COPY? (Y/N) Y

RETIREMENT WITHOUT REPLACEMENT

SELLING PRICE NOW = 600

MINIMUM ATTRACTIVE RATE OF RETURN = 10 %
```

RETIREMENT SHOULD BE
AT THE END OF PERIOD 1

END OF PERIOD	REVENUE	COST	SALVAGE V
1	1200	800	300
2	1000	800	0

PERIOD(N)	NET PRESENT WORTH AFTER (N) PERIODS
1	36.3636365
2	-71.0743802

Retirement with Identical Replacement

PROGRAM
```
]LOAD IDENTICAL CHALLENGER
]LIST
10  REM
20  REM RETIREMENT WITH
30  REM IDENTICAL REPLACEMENT
40  REM
50  REM
60  REM LAKSNA THONGTHAI
70  REM
80  REM
90  REM 2 DECEMBER 1981
100 REM
110 REM
120 DIM SA(20),OC(20),AP(20),PF(20)
130 DIM CRC(20),POC(20),SPOC(20),EUAC(20)
140 HOME
150 PRINT "RETIREMENT WITH IDENTICAL REPLACEMENT"
160 PRINT
170 PRINT
180 INPUT "ENTER SELLING PRICE NOW";SA(0)
190 PRINT
200 PRINT
210 INPUT "ENTER MINIMUM ATTRACTIVE RATE OF RETURN (MARR--%)";MARR
220 PRINT
230 PRINT
240 INPUT "ENTER HOW MANY PERIODS (N) ";N
250 PRINT
260 PRINT
270 REM AP--(A/P,I%,N)
280 REM PF--(P/F,I%,N)
290 FOR I = 1 TO N
300 PRINT
310 PRINT
320 PRINT "AT THE END OF PERIOD ";I
330 PRINT
340 INPUT "ENTER SALVAGE VALUE (SA) ";SA(I)
350 PRINT
360 INPUT "OPERATING COST (OC) ";OC(I)
370 AP(I) = ((MARR / 100) * ((1 + MARR / 100) ^ I)) / (((1 + MARR / 100) ^
    I) - 1)
380 PF(I) = 1 / ((1 + MARR / 100) ^ I)
390 NEXT I
```

```
400 REM CRC-CAPITAL RECOVERY COST
410 REM POC-PRESENT WORTH OF OPERATING COST
420 REM SPOC--ACCUMULATED POC
430 SPOC(0) = 0
440 FOR I = 1 TO N
450 CRC(I) = (SA(0) - SA(I)) * AP(I) + SA(I) * (MARR / 100)
460 POC(I) = OC(I) * PF(I)
470 SPOC(I) = SPOC(I - 1) + POC(I)
480 EUAC(I) = SPOC(I) * AP(I) + CRC(I)
490 NEXT I
500 LL$ = " "
510 PRINT
520 PRINT
530 PRINT "PERIOD";: HTAB (11): PRINT "SALVAGE V";: HTAB (25): PRINT "OPER
    COST"
540 PRINT
550 FOR I = 1 TO N
560 PRINT I;: HTAB (11): PRINT SA(I);: HTAB (25): PRINT OC(I)
570 NEXT I
580 PRINT : PRINT
590 IF LL$ = "Y" THEN 620
600 INPUT "PRESS RETURN TO CONTINUE";Z$
610 PRINT : PRINT
620 PRINT "PERIOD";: HTAB (15): PRINT "EUAC AFTER (N) PERIODS"
630 PRINT
640 FOR I = 1 TO N
650 PRINT I;: HTAB (15): PRINT EUAC(I)
660 NEXT I
670 FOR I = 1 TO N
680 IF EUAC(I) < EUAC(I + 1) THEN GOTO 730
690 NEXT I
700 IF LL$ = "Y" THEN 730
710 PRINT : PRINT
720 INPUT "PRESS RETURN TO CONTINUE";Z$
730 PRINT
740 PRINT
750 PRINT "ECONOMIC LIFE OF THIS EQUIPMENT IS ";I;"YEARS"
760 IF LL$ = "Y" THEN 840
770 PRINT : PRINT : INPUT "DO YOU WANT A HARD COPY? (Y/N) ";LL$
780 IF LL$ = "N" THEN 850
790 PR# 2
800 PRINT : PRINT "RETIREMENT WITH IDENTICAL REPLACEMENT"
810 PRINT : PRINT "SELLING PRICE NOW = ";SA(0)
820 PRINT : PRINT "MINIMUM ATTRACTIVE RATE OF RETURN = ";MARR;"%"
830 GOTO 510
840 PR# 0
850 END
```

Example

Consider an asset having an initial cost of $10,000, a six-year physical life, and the following salvage values and operating costs:

End of period	Salvage value	Operating cost
1	$7,000	$5,000
2	4,500	5,600
3	2,500	6,300
4	1,000	7,000
5	0	8,000
6	0	9,000

The discount rate is 10 percent per period.

]RUN
RETIREMENT WITH IDENTICAL
  REPLACEMENT

ENTER SELLING PRICE NOW10000

ENTER MINIMUM ATTRACTIVE RATE OF RETURN (MARR--%)10

ENTER HOW MANY PERIODS (N) 6

AT THE END OF PERIOD 1

ENTER SALVAGE VALUE (SA) 7000

OPERATING COST (OC) 5000

AT THE END OF PERIOD 2

ENTER SALVAGE VALUE (SA) 4500

OPERATING COST (OC) 5600

AT THE END OF PERIOD 3

ENTER SALVAGE VALUE (SA) 2500

OPERATING COST (OC) 6300

AT THE END OF PERIOD 4

ENTER SALVAGE VALUE (SA) 1000

OPERATING COST (OC) 7000

AT THE END OF PERIOD 5

ENTER SALVAGE VALUE (SA) 0

OPERATING COST (OC) 8000

AT THE END OF PERIOD 6

ENTER SALVAGE VALUE (SA) 0

OPERATING COST (OC) 9000

PERIOD	SALVAGE V	OPER COST
1	7000	5000
2	4500	5600
3	2500	6300
4	1000	7000
5	0	8000
6	0	9000

PRESS RETURN TO CONTINUE

```
PERIOD    EUAC AFTER (N) PERIODS

1         8999.99996
2         8904.76186
3         8858.006
4         8834.73385
5         8878.18376
6         8893.97204
```

ECONOMIC LIFE OF THIS
EQUIPMENT IS 4 YEARS

DO YOU WANT A HARD COPY? (Y/N) Y

RETIREMENT WITH IDENTICAL REPLACEMENT

SELLING PRICE NOW = 10000

MINIMUM ATTRACTIVE RATE OF RETURN = 10 %

```
PERIOD    SALVAGE V    OPER COST

1         7000         5000
2         4500         5600
3         2500         6300
4         1000         7000
5         0            8000
6         0            9000
```

```
PERIOD    EUAC AFTER (N) PERIODS
1         8999.99996
2         8904.76186
3         8858.006
4         8834.73385
5         8878.18376
6         8893.97204
```

ECONOMIC LIFE OF THIS
EQUIPMENT IS 4 YEARS

Retirement with an Unlike Replacement but All Identical Challengers

PROGRAM   ]LOAD NEW CHALLENGER

]LIST

```
10    REM
20    REM
30    REM RETIREMENT WITH
40    REM DIFFERENT REPLACEMENT
50    REM
60    REM
70    REM LAKSNA THONGTHAI
80    REM
90    REM 2 DECEMBER 1981
100   REM
110   REM
120   DIM DSA(20),ODC(20),DCRC(20),DPOC(20)
130   DIM SDPOC(20),DEUAC(20),CSA(20),OCC(20)
140   DIM CCRC(20),CPOC(20),SCPOC(20),CEUAC(20)
150   DIM AP(20),PF(20)
```

```
160  REM ************************************************************************
170  HOME
180  PRINT "RETIREMENT WITH DIFFERENT "
190  PRINT
200  PRINT " REPLACEMENT"
210  PRINT
220  INPUT "ENTER MINIMUM ATTRACTIVE RATE OF RETURN (MARR--%) ";MARR
230  REM AP--(A/P,I%,N)
240  REM PF--(P/F,I%,N)
250  FOR I = 1 to 20
260  AP(I) = ((MARR / 100) * ((1 + MARR / 100) ∧ I)) / (((1 + MARR / 100) ∧
     I) − 1)
270  PF(I) = 1 / ((1 + MARR / 100) ∧ I)
280  NEXT I
290  REM ************************************************************************
300  PRINT
310  PRINT "DEFENDER(-D-) INPUT DATA"
320  PRINT
330  INPUT "ENTER DEFENDER SELLING PRICE NOW(DSA(0)) ";DSA(0)
340  PRINT
350  INPUT "ENTER HOW MANY PERIODS(ND) ";ND
360  FOR I = 1 TO ND
370  PRINT
380  PRINT "AT THE END OF PERIOD(-D-) ";I
390  PRINT
400  INPUT "ENTER SALVAGE VALUE ";DSA(I)
410  PRINT
420  INPUT "ENTER OPERATING COST ";ODC(I)
430  PRINT
440  NEXT I
450  REM ************************************************************************
460  REM DCRC--DEFENDER CAPITAL RECOVERY COST
470  REM DPOC--DEFENDER PRESENT WORTH OF OPERATING COSTS
480  REM SDPOC--ACCUMULATED DPOC
490  SDPOC(0) = 0
500  PRINT
510  FOR I = 1 TO ND
520  DCRC(I) = (DSA(0) − DSA(I)) * AP(I) + DSA(I) * (MARR / 100)
530  DPOC(I) = ODC(I) * PF(I)
540  SDPOC(I) = SDPOC(I − 1) + DPOC(I)
550  DEUAC(I) = SDPOC(I) * AP(I) + DCRC(I)
560  NEXT I
570  REM ************************************************************************
580  PRINT
590  PRINT "CHALLENGER(-C-) INPUT DATA"
600  PRINT
610  INPUT "ENTER CHALLENGER SELLING PRICE NOW(CSA(0)) ";CSA(0)
620  INPUT "ENTER HOW MANY PERIODS(NC) ";NC
630  FOR I = 1 TO NC
640  PRINT
650  PRINT "AT THE END OF PERIOD(-C-) ";I
660  PRINT
670  INPUT "ENTER SALVAGE VALUE ";CSA(I)
680  PRINT
690  INPUT "ENTER OPERATING COST ";OCC(I)
700  PRINT
710  NEXT I
720  REM ************************************************************************
```

```
730  REM CCRC--CHALLENGER CAPITAL RECOVERY COST
740  REM CPOC--CHALLENGER PRESENT WORTH OF OPERATING COST
750  REM SCPOC--ACCUMULATED CPOC
760  PRINT
770  SCPOC(0) = 0
780  FOR I = 1 TO NC
790  CCRC(I) = (CSA(0) − CSA(I)) * AP(I) + CSA(I) * (MARR / 100)
800  CPOC(I) = OCC(I) * PF(I)
810  SCPOC(I) =SCPOC(I − 1) + CPOC(I)
820  CEUAC(I) = SCPOC(I) * AP(I) + CCRC(I)
830  NEXT I
840  REM *********************************************************************
850  HOME
860  LL$ = " "
870  PRINT : PRINT
880  PRINT "DEFENDER"
890  PRINT "PERIOD";: HTAB (11): PRINT "SALVAGE V";: HTAB (25): PRINT "OPER
     COST"
900  FOR I = 1 TO ND
910  PRINT I;: HTAB (11): PRINT DSA(I);: HTAB (25): PRINT ODC(I)
920  NEXT I
930  IF LL$ = "Y" THEN 950
940  PRINT : PRINT : INPUT "PRESS RETURN TO CONTINUE";Z$: PRINT : PRINT
950  PRINT : PRINT
960  PRINT "PERIOD";: HTAB (15): PRINT "EUAC AFTER (N) PERIODS": PRINT
970  FOR I = 1 TO ND
980  PRINT I;: HTAB (15): PRINT DEUAC(I)
990  NEXT I
1000 IF LL$ = "Y" THEN 1020
1010 PRINT : PRINT : INPUT "PRESS RETURN TO CONTINUE";Z$: PRINT : PRINT
1020 PRINT : PRINT
1030 REM *********************************************************************
1040 PRINT "CHALLENGER"
1050 PRINT "PERIOD";: HTAB (11): PRINT "SALVAGE V";: HTAB (25): PRINT "OPER
     COST"
1060 FOR I = 1 TO NC
1070 PRINT I;: HTAB (11): PRINT CSA(I);: HTAB (25): PRINT OCC(I)
1080 NEXT I
1090 IF LL$ = "Y" THEN 1110
1100 PRINT : PRINT : INPUT "PRESS RETURN TO CONTINUE";Z$: PRINT : PRINT
1110 PRINT : PRINT
1120 PRINT "PERIOD";: HTAB (15): PRINT "EUAC AFTER (N) PERIODS"
1130 PRINT
1140 FOR I = 1 TO NC
1150 PRINT I;: HTAB (15): PRINT CEUAC(I)
1160 NEXT I
1170 IF LL$ = "Y" THEN 1190
1180 PRINT : PRINT : INPUT "PRESS RETURN TO CONTINUE";Z$: PRINT : PRINT
1190 PRINT : PRINT
1200 REM *********************************************************************
1210 FOR I = 1 TO NC
1220 IF CEUAC(I) < CEUAC(I + 1) THEN GO TO 1240
1230 NEXT I
1240 ZZ = CEUAC(I)
1250 PRINT
1260 PRINT "ECONOMIC LIFE OF CHALLENGER IS ";I;"YRS"
1270 REM *********************************************************************
```

```
1280  FOR I = 1 TO ND
1290  IF ZZ > DEUAC(I) THEN GOTO 1330
1300  NEXT I
1310  PRINT "REPLACE DEFENDER NOW"
1320  GOTO 1350
1330  PRINT "KEEP DEFENDER AT LEAST ONE MORE YEAR"
1340  IF LL$ = "Y" THEN 1430
1350  PRINT : PRINT : INPUT " DO YOU WANT A HARD COPY? (Y/N) ";LL$
1360  IF LL$ = "N" THEN 1440
1370  PR# 2
1380  PRINT : PRINT "RETIREMENT WITH DIFFERENT REPLACEMENT"
1390  PRINT : PRINT "MINIMUM ATTRACTIVE RATE OF RETURN = ";MARR;"%"
1400  PRINT : PRINT "DEFENDER SELLING PRICE NOW = ";DSA(0)
1410  PRINT : PRINT "CHALLENGER SELLING PRICE NOW = ";CSA(0)
1420  GOTO 870
1430  PR# 0
1440  END
```

Example

Consider a certain defender with a remaining life of two years and a current disposal value of \$8,000. Expected operating costs and salvage values are:

End of period	Salvage value	Operating cost
1	\$5,000	\$5,500
2	4,000	6,000

A challenger is available at a cost of \$10,000. If this challenger is acquired, future salvage values and operating costs will be:

End of period	Salvage value	Operating cost
1	\$7,000	\$5,000
2	4,500	5,600
3	2,500	6,300
4	1,000	7,000
5	0	8,000
6	0	9,000

The discount rate is 10 percent per period.

EXAMPLE
```
]RUN
RETIREMENT WITH DIFFERENT
   REPLACEMENT

ENTER MINIMUM ATTRACTIVE RATE OF RETURN (MARR--%) 10

DEFENDER(-D-) INPUT DATA
```

ENTER DEFENDER SELLING PRICE NOW(DSA(0)) 8000

ENTER HOW MANY PERIODS(ND) 2

AT THE END OF PERIOD(-D-) 1

ENTER SALVAGE VALUE 5000

ENTER OPERATING COST 5500

AT THE END OF PERIOD(-D-) 2

ENTER SALVAGE VALUE 4000

ENTER OPERATING COST 6000

CHALLENGER(-C-) INPUT DATA

ENTER CHALLENGER SELLING PRICE NOW(CSA(0)) 10000
ENTER HOW MANY PERIODS(NC) 6

AT THE END OF PERIOD(-C-) 1

ENTER SALVAGE VALUE 7000

ENTER OPERATING COST 5000

AT THE END OF PERIOD(-C-) 2

ENTER SALVAGE VALUE 4500

ENTER OPERATING COST 5600

AT THE END OF PERIOD(-C-) 3

ENTER SALVAGE VALUE 2500

ENTER OPERATING COST 6300

AT THE END OF PERIOD(-C-) 4

ENTER SALVAGE VALUE 1000

ENTER OPERATING COST 7000

AT THE END OF PERIOD(-C-) 5

ENTER SALVAGE VALUE 0

ENTER OPERATING COST 8000

AT THE END OF PERIOD(-C-) 6

ENTER SALVAGE VALUE 0

ENTER OPERATING COST 9000

DEFENDER

PERIOD	SALVAGE V	OPER COST
1	5000	5500
2	4000	6000

PRESS RETURN TO CONTINUE

PERIOD	EUAC AFTER (N) PERIODS
1	9299.99995
2	8442.8571

PRESS RETURN TO CONTINUE

CHALLENGER

PERIOD	SALVAGE V	OPER COST
1	7000	5000
2	4500	5600
3	2500	6300
4	1000	7000
5	0	8000
6	0	9000

PRESS RETURN TO CONTINUE

PERIOD	EUAC AFTER (N) PERIODS
1	8999.99996
2	8904.76186
3	8858.006
4	8834.73385
5	8878.18376
6	8893.97204

PRESS RETURN TO CONTINUE

ECONOMIC LIFE OF CHALLENGER IS 4 YRS
KEEP DEFENDER AT LEAST ONE MORE YEAR

DO YOU WANT A HARD COPY? (Y/N) Y

RETIREMENT WITH DIFFERENT REPLACEMENT

MINIMUM ATTRACTIVE RATE OF RETURN = 10 %

DEFENDER SELLING PRICE NOW = 8000

CHALLENGER SELLING PRICE NOW = 10000

DEFENDER

PERIOD	SALVAGE V	OPER COST
1	5000	5500
2	4000	6000

PERIOD	EUAC AFTER (N) PERIODS
1	9299.99995
2	8442.8571

```
CHALLENGER
PERIOD         SALVAGE V    OPER COST
1              7000         5000
2              4500         5600
3              2500         6300
4              1000         7000
5              0            8000
6              0            9000


PERIOD     EUAC AFTER (N) PERIODS

1          8999.99996
2          8904.76186
3          8858.006
4          8834.73385
5          8878.18376
6          8893.97204


ECONOMIC LIFE OF CHALLENGER IS 4 YRS
KEEP DEFENDER AT LEAST ONE MORE YEAR
```

CHAPTER SEVEN
DEPRECIATION, TAXATION AND AFTER-TAX ECONOMY STUDIES

Most individuals and business firms operating in the private sector are directly influenced by taxation. Cash flows resulting from taxes paid (or avoided) must be included in evaluation models, along with cash flows from investment, maintenance, operations, and so on. Thus private-sector decision makers have a clear interest in cash flows for taxes and related topics.

Decision makers in the public sector may also be directly concerned with cash flows for taxes, and the associated concept, depreciation. Government agencies, including publicly owned utilities, for example, may purchase goods and services from individuals or firms in the private sector. Thus it may be useful to carry out an economy study from the point of view of the private sector in order to determine the after-tax return to the supplier, the vendor, or the customer. Moreover, regulatory rules in many parts of the United States define a utility's revenue requirements in terms of (1) current operating disbursements, other than interest on debt; (2) allowance for depreciation; (3) income taxes, where appropriate; and (4) a ''fair return'' on a rate base that usually approximates depreciated book value. Thus depreciation and taxes are of special concern to regulated utilities.

There is a good deal of misunderstanding about the precise meaning of **depreciation.** In economic analysis, depreciation is *not* a measure of the loss in market value of equipment, land, buildings, and the like. It is *not* a measure of reduced serviceability. Depreciation is strictly an *accounting* concept. Perhaps the best definition is provided by the Committee on Terminology of the American Institute of Certified Public Accountants:

7.1 DEPRECIATION DEFINED

Depreciation accounting is a system of accounting which aims to distribute the cost or other basic value of tangible capital assets, less salvage (if any), over the estimated life of the unit (which may be a group of assets) in a systematic and rational manner. It is a process of allocation, not of valuation. Depreciation for the year is the portion of the total charge under such a system that is allocated to the year.[1]

7.2 REGULATIONS CONCERNING DEPRECIATION ACCOUNTING

Because of its effect on taxes, depreciation accounting is strictly regulated in the United States by the Internal Revenue Service (IRS), as well as by certain other federal, state, and local agencies. The Revenue Act of 1913, enacted after the passage of the Sixteenth Amendment, gave birth to our present tax system. There have been a number of revisions, most notably the radical change in the treatment of depreciation policy resulting from the Internal Revenue Code of 1954. Other substantial changes were effected in 1964, 1969, 1971, 1975, 1978, 1981, and 1982. Clearly, it is beyond the cope of this book to detail all the relevant rules, regulations, and legal interpretations. Only a select number of the most useful concepts are outlined here to provide a framework for the following discussion. For given real-world applications, analysts should seek professional counsel.

Depreciable Assets

Section 167 of the Internal Revenue Code of 1954 gives the following General Rule: "There shall be allowed as a depreciation deduction a reasonable allowance for the exhaustion, wear and tear (including a reasonable allowance for obsolescence) (1) of property used in the trade or business, or (2) of property held for the production of income." Depreciation is allowed for **tangible property,** property that can be seen or touched, but not for inventory, stock in trade, land (apart from improvements), or a depletable natural resource. **Intangible property,** such as patents, copyrights, licenses, franchises, contracts, or similar assets having limited useful lives, are also depreciable.[2] "Good will" is not depreciable, because its useful life cannot clearly be determined.

Depreciable property may be further classified as personal or real. In general, **real property** is anything growing on, attached to, or erected on land. (Land itself is real property, but it is not depreciable, because its life is indeterminable.) **Personal property** is anything that is not real property, such as machinery and equipment.

1. American Institute of Certified Public Accountants, *Accounting Research Bulletin No. 22* (New York: American Institute of Certified Public Accountants, 1944) and American Institute of Certified Public Accountants, *Accounting Terminology Bulletin No. 1* (New York: American Institute of Certified Public Accountants, 1953).

2. Intangible property must be depreciated by the straight-line method. (See Section 7.3.)

Property is depreciable, according to the IRS, if it meets three requirements. First, it must be used in business or held for the production of income. Second, it must have a determinable useful life of more than one year. And third, the property "must be something that wears out, becomes obsolete, or loses value from natural causes."[3]

Cost Basis

The basis for determining depreciation is usually the cost of the property, payable either in cash or other property. If property is materially improved, say, through construction of a special foundation for certain equipment, the additional costs are added to the basis. After the property is acquired, the basis may be increased by all items that may be charged to a capital account (with certain exceptions), including the cost of any improvements having a useful life of more than one year. The basis may be decreased by any items that represent a return of capital. These changes, if any, result in the **adjusted basis.**

Salvage Value

Salvage value is the estimate, at the time of acquisition, of the amount to be realized upon the sale or other disposition of the asset after it is no longer useful in the business or in the production of income and is retired from service. If the asset is expected to be "junk" at the time of disposal, the salvage value may be negligible. On the other hand, if the asset is still in good operating condition, the salvage value may be a large portion of the original cost basis.

Salvage, when reduced by the cost of removal, is called **net salvage.** Either salvage or net salvage may be used to compute depreciation, but the practice must be consistently followed.

Taxpayers may use the **10 percent rule** for computing depreciation for property that has a life of at least three years. That is, the estimate of actual salvage value may be reduced by 10 percent of the basis in the property. For example, consider a depreciable asset having a cost basis of $1,000, an estimated useful life of five years, and an expected value on disposal of $250. For this asset, the taxpayer may elect to lower the estimated salvage value for depreciation purposes by up to $100 (10 percent of $1,000). If the actual estimate is less than 10 percent of the basis, salvage value may be ignored when figuring depreciation.

Depreciable (Useful) Life

As you will see, the profitability of investments can be directly influenced by the **useful life** used for determining the depreciation

---

3. U.S., Department of Treasury, Internal Revenue Service, *Depreciation*, Pub. 534 (Washington, D.C.: Government Printing Office, December 1981), p. 1.

allowance. Thus useful life is of considerable interest to investors as well as to the taxing authority. Prior to July 1962, the determination of useful life was guided by Bulletin F, an item-by-item list of suggested useful lives in each industry. In July 1962, Bulletin F was replaced by the depreciation guidelines of Revenue Procedure 62-21, which were based on categories of assets in broad industry classes. These guidelines included a complex procedure, the reserve ratio test, for determining the reasonableness of a taxpayer's depreciation of a class of assets. But the old Bulletin F has not been used by the IRS for examining depreciation deductions since 1962, and the reserve ratio test was dropped in 1971.

For assets placed in service after 1970 and before 1981, taxpayers have the option of computing depreciation under the general rules using the estimated useful life (see Section 7.3) or electing the **Class Life Asset Depreciation Range** (CLADR) system. Under CLADR, the IRS provides a table that lists the asset guideline classes, descriptions of the assets in each class, and the asset depreciation range (in years) that applies to each class of assets.[4] Taxpayers may select any depreciation period within the asset depreciation range. The CLADR system may no longer be used for recovery property placed in service after December 31, 1980. However, as indicated in Section 7.4, CLADR class lives are still of interest because they serve to define the property classes under the Accelerated Cost Recovery System (ACRS).

It should be emphasized that the useful life of a piece of property is an estimate of how long it can be expected to be used in a trade or business or of how long it can be expected to produce income for the taxpayer. Useful life is not necessarily the same as physical life or even the period of ownership. Rather, the property is no longer useful when it has been removed from service.

7.3 METHODS OF COMPUTING DEPRECIATION FOR ASSETS PLACED IN SERVICE BEFORE 1981

As noted in Section 7.2, federal regulations governing depreciation accounting, and tax policies were significantly affected by the Internal Revenue Code of 1954. Then in 1981, the **Economic Recovery Tax Act (ERTA)** created another set of substantial changes. Thus this chapter presents the 1954–1980 depreciation methods as well as the post-1980 methods. Although the former are termed **"pre-1981" methods,** they are in fact of more than historical interest. There are several important reasons for including them in this book. First, pre-1981 methods are widely used for depreciation accounting in countries other than the United States. Second, they are also specified by tax regulations of a number of states—California, for example, does not (as of mid-1983) recognize the new depreciation accounting methods introduced by

4. See Section 1.167(a)-11 of the Tax Regulations for more information on the CLADR system.

ERTA at the federal level, so the "pre-1981" methods must be used in California when computing taxable income. And third, federal tax law requires that pre-1981 methods be used for property placed in service before 1981 or for property that does not qualify for depreciation by the new methods introduced by ERTA. In summary, then, so-called "pre-1981" methods remain very much in evidence.

The Internal Revenue Code of 1954, as amended, permits any reasonable method, consistently applied, for computing depreciation for depreciable property purchased after December 31, 1953 and before January 1, 1981. Three methods are specifically mentioned: (1) the straight line method, (2) the declining balance method, and (3) the sum of years digits method. In addition, taxpayers were permitted to use "any other consistent method productive of any annual allowance which, when added to all allowances for the period beginning with the taxpayer's use of the property and including the taxable year, does not, during the first two-thirds of the useful life of the property, exceed the total of such allowances which would have been used had such allowances been computed under the (declining balance) method." The first three methods, as well as the sinking fund method, are discussed in the following sections.

The Straight Line Method

For item accounting, that is, for determining the depreciation allowance for a single depreciable asset, the **straight line method** is the simplest method for computing depreciation.[5] The cost or other basis of the asset, less its expected salvage value, is deducted in equal amounts over the period of its estimated useful life.

Let

P = initial cost (basis)

S = salvage value

N = depreciable life

Then the annual depreciation, D, is given by:

$$D = \frac{P - S}{N} \tag{7.1}$$

The **book value** of the asset at any point in time is the initial

---

5. Before 1954, the straight line method was the only method permitted by the IRS for computing depreciation (with two minor exceptions).

cost less the accumulated depreciation. Generally, the book value, B, after j years is given by:

$$B_j = P - jD \qquad (7.2)$$

Note that this expression assumes that the depreciation in the first year includes an entire year's depreciation. But an asset may be purchased at any time during the tax year, so the amount of allowable depreciation may be proportionally reduced.[6] An asset purchased midway through the tax year, for example, is allowed only 50 percent of one year's depreciation. Accounting treatment in these cases is rather complex and beyond the scope of this text, so here, simply assume that all assets are purchased at the start of the tax year and qualify for 100 percent of the depreciation allowance in the first year.

If one were to prepare a graph of book value as a function of time, as in Figure 7.1, the result would be a straight line, because the book value decreases by a constant *amount* from year to year. Hence the name straight line depreciation.

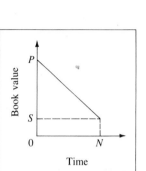

Figure 7.1

The Declining Balance Method

In the **declining balance method,** the amount of depreciation taken each year is subtracted from the book value before the following year's depreciation is computed. A constant depreciation *rate* applies to a smaller, or declining, balance each year.

Let

a = depreciation rate

D_j = depreciation in year j

B_j = book value at the end of year j, that is, *after* the depreciation deduction in year j has been recorded

The annual depreciation expenses are computed as follows (see Figure 7.2):

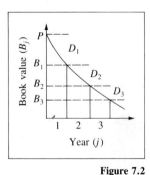

Figure 7.2

$$B_0 = P \qquad\qquad D_1 = aB_0$$

$$B_1 = B_0 - D_1 \qquad D_2 = aB_1$$

$$B_2 = B_1 - D_2 \qquad D_3 = aB_2$$

$$\vdots \qquad\qquad\qquad \vdots$$

$$B_N = B_{N-1} - D_N \qquad D_N = aB_{N-1}$$

---

6. A *tax year* is usually twelve consecutive months. It may either be a *calendar year* (twelve months ending on December 31) or a *fiscal year* (twelve consecutive months ending on the last day of any month other than December).

The depreciation in the jth year is given by

$$D_j = aB_{j-1} = Pa(1 - a)^{j-1} \qquad (7.3)$$

The value selected for the depreciation rate, a, is not entirely arbitrary. Under some circumstances the IRS limits a to twice the rate that would have been used by the straight line method, ignoring salvage value. That is, $a < 2/N$. When $a = 2/N$, the depreciation scheme is known as the **double declining balance method,** or simply DDB. Prior to 1981, federal tax law permitted DDB to be used for personal property acquired new after 1953; for personal property erected, built, or rebuilt after 1953; or for real property that is also new residential rental property. The depreciation rate under the declining balance method is limited to $1.5/N$, or *150% declining balance,* for tangible property that is used personal property or new real estate property that is not residential rental property. Taxpayers may use *125% declining balance,* that is, $a = 1.25/N$, for certain used residential property acquired after July 24, 1969 that has a useful life of at least twenty years.

Property depreciated by the declining balance method must have an estimated useful life of at least three years, that is, $N \geqslant 3$. The reason for this lower bound for N is that, if $N = 2$ under DDB, $a = 100\%$, so all of the depreciation would take place in the first year; there would be no depreciation in the second year, and this result is not permissible.

Salvage value is not deducted from the cost or other basis in determining the annual depreciation allowance, but the asset cannot be depreciated below the expected salvage value. In other words, once book value equals salvage value, no further depreciation may be claimed. As shown in Figure 7.3, book value equals salvage value (S) after n periods, where $n < N$. Thus, no depreciation may be claimed after that time.

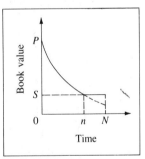

Figure 7.3

In general, it may be shown that

$$B_j = P(1 - a)^j \qquad (7.4)$$

Thus, if taxpayers wish to choose a depreciation rate such that the book value after N periods will be exactly equal to the expected salvage value, as in Figure 7.4, then

$$B_N = S = P(1 - a)^N \qquad (7.5)$$

or

$$a = 1 - \sqrt[N]{S/P} \qquad (7.6)$$

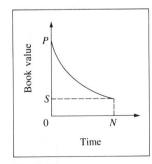

Figure 7.4

The IRS permits taxpayers to change from the declining bal-

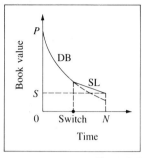

Figure 7.5

ance method to the straight line method on property for which a rate of twice the straight line rate may be used, that is, when DDB may be used. (See Figure 7.5.) This change may be made at any time during the property's useful life without the prior consent of the IRS.[7] When such a change is made, the unrecovered cost less the salvage value is spread over the estimated remaining useful life determined at the time of the change.

The Sum of the Years Digits Method

In the **sum of the years digits method** the annual depreciation allowance is a declining fraction of the cost or other basis of each single-asset account reduced by the estimated salvage value.[8] In particular, the depreciation in year j is given by

$$D_j = \left(\frac{N - j + 1}{SYD}\right)(P - S) \tag{7.7}$$

where SYD is the sum of the years digits, that is

$$SYD = 1 + 2 + \cdots + N = N(N + 1)/2 \tag{7.8}$$

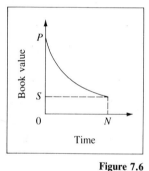

Figure 7.6

For example, consider an asset with an estimated useful life of five years. The sum of the years digits is $5(6)/2 = 15$. The depreciation in the first year is 5/15 of the depreciable amount, the depreciation in the second year is 4/15 of the depreciable amount, and so on. As shown in Figure 7.6, this method results in a declining book value quite similar to that found by using the declining-balance method, except that here we are assured that $B_N = S$; exactly $P - S$ is depreciated over N years.

Both the declining balance and the sum of the years digits methods are called "accelerated" depreciation methods, because they result in greater depreciation in the early years and less depreciation in the later years. The sum of the years digits method may be used only on property that is also subject to double declining balance depreciation.

The Sinking-Fund Method

Whereas the declining balance and the sum of the years digits methods result in relatively greater depreciation in the early years, the **sinking fund method** results in the reverse situation: Depreciation is *lower* in the early years and greater in the later years.

In this method, an *imaginary* sinking fund is established that

7. A change from the declining balance to the straight line method may be made only if such change is not prohibited by a written agreement with the IRS as to useful life and rate of depreciation.

8. A *remaining life* plan is used for group, classified, or composite accounts by applying changing fractions to the unrecovered cost or other basis reduced by expected salvage.

earns *imaginary* interest at a predetermined rate, i_s, on a series of *imaginary* deposits. The sum of these deposits over the life of the asset is exactly equal to the cost or other basis less the expected salvage value. The depreciation charge for the year, D_j, is equal to the imaginary uniform deposit into the imaginary sinking fund, d, plus the accumulated interest on the fund up to that point. That is,

$$D_1 = d = (P - S)(A/F, i_s, N)$$

$$D_2 = d + i_s D_1$$

$$D_3 = d + i_s(D_1 + D_2)$$

$$\vdots$$

$$D_N = d + i_s(D_1 + D_2 + \cdots + D_{N-1})$$

It may be shown that, in general, the depreciation expense in year j is given by

$$D_j = (P - S)(A/F, i_s, N)(1 + i_s)^{j-1} \qquad (7.9)$$

Clearly, for $i_s > 0$, the amount of the depreciation expense increases from year to year. Thus, as shown in Figure 7.7, the resulting book value at any point in time is greater that the book value from using the straight line method.

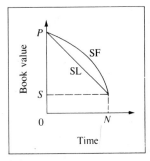

Figure 7.7

Other Depreciation Methods
The straight line, declining balance, and sum of the years digits methods are specifically mentioned in Section 167 of the Internal Revenue Code of 1954. In general, the law also permits "any other consistent methods" for federal income tax purposes if the total depreciation deductions taken during the first two-thirds of the asset's useful life are not more than the total allowable under the declining balance method. One cannot depreciate below salvage value, and "negative depreciation" is not allowed. Figure 7.8 illustrates these constraints. Of course, the sinking fund method is feasible, as previously noted. In addition, two other methods may be used. First, a **"reverse" sum of the years digits method** could be employed, in which

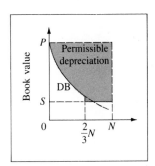

Figure 7.8

$$D_j = \frac{j}{SYD}(P - S) \quad \text{for} \quad j = 1, 2, \ldots, N \qquad (7.10)$$

Another possibility, more frequently used in manufacturing industries, is the **units of production method.** (This is also known as the **service output** method.) Here, depreciation for the period is charged in proportion to the number of units produced during that period. That is,

$$D = \left(\frac{P - S}{U_e}\right)U_j \qquad (7.11)$$

where U_e is the number of units of output expected over the depreciable life of the asset and U_j is the number of units of actual output during period j.

Comparison of Depreciation Methods

A numerical example will help to illustrate the various depreciation methods discussed so far. Consider an asset having a cost basis of $40,000, an expected salvage value of $4,000, and an expected life of eight years. The year-by-year depreciation expenses and book values are displayed in Table 7.1 for the four principal depreciation methods: (1) straight line, (2) declining balance, (3) sum of the years digits, and (4) sinking fund. Book values are shown graphically in Figure 7.9.

Using straight line depreciation,

$$D = \frac{\$40,000 - \$4,000}{8} = \$4,500$$

for each of the eight years.

In the example, the declining balance depreciation is computed using the 200 percent rate: $a = 2/N = 2/8 = 0.25$. Thus the depreciation expenses are:

$$D_1 = 0.25(\$40,000) = \$10,000$$
$$B_1 = \$40,000 - \$10,000 = \$30,000$$
$$D_2 = 0.25(\$30,000) = \$7,500$$
$$B_2 = \$30,000 - \$7,500 = \$22,500$$
$$D_3 = 0.25(\$22,500) = \$5,625$$

and so on.

For the sum of the years digits method, $SYD = (8 \times 9)/2 = 36$, and the amount to be depreciated is $36,000. Thus

$$D_1 = \frac{8}{36}(\$36,000) = \$8,000$$

$$D_2 = \frac{7}{36}(\$36,000) = \$7,000$$

and so on.

Table 7.1

Summary of Depreciation Charges and Book Values Using Four Different Depreciation Methods

	Depreciation expense for year				Book value at end of year			
Year	Straight line	Declining balance	Sum of the years digits	Sinking fund	Straight line	Declining balance	Sum of the years digits	Sinking fund
1	$4,500	$10,000	$8,000	$3,770	35,500	30,000	32,000	36,230
2	4,500	7,500	7,000	3,959	31,000	22,500	25,000	32,272
3	4,500	5,625	6,000	4,156	26,500	16,875	19,000	28,115
4	4,500	4,219	5,000	4,364	22,000	12,656	14,000	23,751
5	4,500	3,164	4,000	4,582	17,500	9,492	10,000	19,169
6	4,500	2,373	3,000	4,812	13,000	7,120	7,000	14,357
7	4,500	1,780	2,000	5,052	8,500	5,340	5,000	9,305
8	4,500	1,335	1,000	5,305	4,000	4,005	4,000	4,000

Note: Assume a first cost of $40,000, expected salvage value of $4,000, and expected life of 8 years

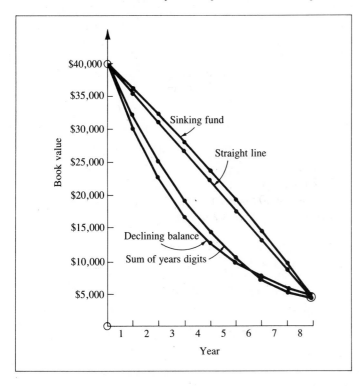

Figure 7.9. Book Values for Example Problem

In general, the declining balance method "writes off" about two-thirds of the original cost of the asset (P) in the first half of the asset's useful life. For the present example, approximately ($40,000 − $12,656)/$40,000 = 0.68 will be written off after the first four years. Similarly, it may be shown that the sum of the years digits method "writes off" about three-fourths of the depreciable amount ($P − S$) during the first half of the expected useful life. In the numerical example in Table 7.1, ($40,000 − $14,000)/$36,000 = 0.72.

Table 7.2

General Forms for Principal Depreciation Methods for Property Placed in Service before 1981

	Rate (a)	Depreciation (D_j)	Book value (B_j)	Book value at half life $(B_{N/2})$
Straight line	$\dfrac{P - S}{NP}$	$\dfrac{P - S}{N}$ for all j	$P - j\left(\dfrac{P - S}{N}\right)$	$\dfrac{P + S}{2}$
Declining balance	$\leq \dfrac{2}{N}$	$aB_{j-1} = aP(1 - a)^{j-1}$	$P(1 - a)^j$	$\approx \dfrac{1}{3}P$
Sum of the years digits	—	$\dfrac{N - j + 1}{SYD}(P - S)$ where $SYD = N(N + 1)/2$	$P - \left(\dfrac{P - S}{SYD}\right)\displaystyle\sum_{k=1}^{j}(N - k + 1)$	approximately $S + \dfrac{1}{4}(P - S)$
Sinking fund	—	$d + i_s \displaystyle\sum_{k=0}^{j-1} D_k$ where $d = (P - S)(A/F, i_s, N)$	$P - d(F/A, i_s, N)$	—

Note: P = initial cost or other basis; S = expected salvage value; N = depreciable life; i_s = interest rate for imaginary sinking fund

For the sinking fund method, the rate used to compute the imaginary sinking fund is assumed to be 5 percent. Thus

$$D_1 = d = (\$40,000 - \$4,000)(A/F, 5\%, 8)$$

$$= \$36,000(0.10472) = \$3,770$$

$$D_2 = \$3,770 + 0.05(\$3,770) = \$3,959$$

$$D_3 = \$3,770 + 0.05(\$3,770 + \$3,959) = \$4,156$$

and so on.

Table 7.2 provides a summary of the general forms of the principal depreciation methods discussed in this section: (1) annual depreciation rate, (2) annual depreciation, (3) end-of-year book value, and (4) book value at half-life.

7.4 COMPUTING DEPRECIATION FOR PROPERTY PLACED IN SERVICE AFTER 1980

The Economic Recovery Tax Act of 1981 (Public Law 97–34) was passed by Congress on August 13, 1981. This legislation has a number of significant features that affect the determination of cash flows for income taxes, and hence after-tax economy studies. Principal among these is the requirement that, with certain exceptions, property placed in service beginning in 1981 must be depreciated by a new method for the purpose of determining federal income taxes. The principal features of this method, the **Accelerated Cost Recovery System** (ACRS), are summarized below.[9]

9. For property placed in service before 1981, taxpayers must continue to use the method of depreciation that was used in the past. (See Section 7.3.) Also, ACRS cannot be used for intangible depreciable property or for certain types of real property. *Regardless of when property is placed in service,* taxpayers may elect to use a method of depreciation that is not based on years of service, such as the units of production method, rather than ACRS.

Classes of Recovery Property

The Economic Recovery Tax Act (ERTA) defines **recovery property** as "tangible property of a character subject to the allowance for depreciation (a) used in the trade or business, or (b) held for the production of income." There are two major types of recovery property based on sections of the Internal Revenue Code of 1954, as amended. The first type, **Section 1245 class property,** includes (1) tangible personal property; (2) special-purpose structures or storage facilities that are also depreciable tangible property, other than buildings or their structural components, used as an integral part of a certain business activity, and a research facility used in connection with this activity; (3) a single-purpose agricultural or horticultural structure; and (4) a storage facility used in connection with the distribution of petroleum or any primary product of petroleum. The second type, **Section 1250 class property,** includes (1) all depreciable real property that is not Section 1245 property and (2) an elevator or an escalator.

For the purpose of applying ACRS, all recovery property is further divided into five classes based on recovery period:

3-year property is Section 1245 class property with a present class life of four years or less or property used in connection with research and experimentation, such as autos, light trucks and other vehicles, machinery and equipment used in research and development, and some special tools.[10]

5-year property is Section 1245 class property that is not 3-year property, 10-year property, or 15-year public utility property, such as machinery and equipment not used in research and development.

10-year property is (a) public utility property, other than Section 1250 class property or 3-year property, with a present class life of more than eighteen years but not more than twenty-five years and (b) Section 1250 class property with a class life of 12.5 years or less, such as certain theme park structures, manufactured homes, and railroad tank cars.

15-year real property is Section 1250 class property that does not have a present class life of 12.5 years or less, such as buildings not designated as either 5-year or 10-year property.

15-year public utility property is public utility property, other than Section 1250 class property or 3-year property, having a present class life of more than twenty-five years.

10. "Present class life" refers to the asset guideline class established by the IRS. For additional information, see IRS Publication 534, *Depreciation.*

Determining Depreciation under ACRS

The annual depreciation expense in a given year under ACRS is the product of the unadjusted basis and the applicable percentage for that year. That is,

$$D_j = B[p_j(k)] \tag{7.12}$$

where

$$D_j = \text{depreciation expense in year } j$$

$$B = \text{unadjusted basis}$$

$$p_j(k) = \text{applicable percentage in year } j \text{ for property class } k$$

The **unadjusted basis** is generally the initial cost of the property; it is unadjusted in the sense that there is no reduction for prior depreciation. The applicable percentages are summarized in Table 7.3 for all classes other than fifteen-year real property.

When depreciating *pre-1981 property* by any of the methods discussed in Section 7.3, it is necessary to prorate the depreciation expense in the years of purchase and disposal, based on the proportion of time during the tax year that the property was owned. For example, if the property is purchased at the start of the fourth month of the tax year, then only 9/12 of the first year's depreciation may be claimed. Similarly, if the property is sold at the start of the fifth month of the tax year, then only 4/12 of that year's depreciation expense, if any, may be claimed. The general rule for pre-1981 property is that depreciation starts when the property is placed in service and stops when the property is retired from service. With respect to ACRS property, however, depreciation in the years of acquisition and retirement are computed somewhat

Table 7.3

Applicable Percentages for Computing
Annual Depreciation Under ACRS

Recovery year	Type of property			
	3-year	5-year	10-year	15-year public utility
1	25%	15%	8%	5%
2	38	22	14	10
3	37	21	12	9
4		21	10	8
5		21	10	7
6			10	7
7			9	6
8			9	6
9			9	6
10			9	6
11–15				6

differently. Under ACRS, the *entire depreciation expense in the year of acquisition may be included* for all property classes except fifteen-year real property. (This exception is discussed below.) *No depreciation expense may be included in the year of disposal* in the event of early disposition of ACRS properties, other than 15-year real property.

To illustrate these rules, consider a 5-year property placed in service in April 1981 and sold ("premature" disposal) in February 1985. The initial cost of the property, or the unadjusted basis, is $20,000. If the taxpayer's tax year is the same as the calendar year, the annual depreciation expenses are

Year	Depreciation expense
1981	$0.15(\$20,000) = \$3,000$
1982	$0.22(\$20,000) = \$4,400$
1983	$0.21(\$20,000) = \$4,200$
1984	$0.21(\$20,000) = \$4,200$
1985	none permitted in year of disposal

The applicable percentages for 15-year real property depend on the month the property is placed in service. These percentages are summarized in Table 7.4 for 15-year real property other than low-income housing.[11] The applicable percentages for each year of recovery are found from the column specified by the month the

11. The applicable percentages for low-income housing are similar to those in Table 7.4, except that the percentages are somewhat higher in the early years and somewhat lower in the later years. The applicable percentages for low-income housing may be found in IRS Publication 534, *Depreciation*.

Table 7.4

Applicable Percentages for Computing Annual Depreciation under ACRS for 15-Year Real Property (other than low-income housing)

Recovery year	Month placed in service											
	1	2	3	4	5	6	7	8	9	10	11	12
1st	12	11	10	9	8	7	6	5	4	3	2	1
2nd	10	10	11	11	11	11	11	11	11	11	11	12
3rd	9	9	9	9	10	10	10	10	10	10	10	10
4th	8	8	8	8	8	8	9	9	9	9	9	9
5th	7	7	7	7	7	7	8	8	8	8	8	8
6th	6	6	6	6	7	7	7	7	7	7	7	7
7th	6	6	6	6	6	6	6	6	6	6	6	6
8th	6	6	6	6	6	6	6	6	6	6	6	6
9th	6	6	6	6	5	6	5	5	5	6	6	6
10th	5	6	5	6	5	5	5	5	5	5	6	5
11th	5	5	5	5	5	5	5	5	5	5	5	5
12th	5	5	5	5	5	5	5	5	5	5	5	5
13th	5	5	5	5	5	5	5	5	5	5	5	5
14th	5	5	5	5	5	5	5	5	5	5	5	5
15th	5	5	5	5	5	5	5	5	5	5	5	5
16th	—	—	1	1	2	2	3	3	4	4	4	5

property is first placed in service. To illustrate, consider a 15-year real property purchased at a cost of $100,000. (Throughout this chapter, it is assumed that a property is placed in service as soon as it is purchased.) Assume that the tax year is the calendar year and that acquisition takes placed in mid-June, the sixth month of the year. Assuming that the property is retained in service for at least fifteen years, the allowable annual depreciation expenses are

Year	Depreciation expense
1st	0.07($100,000) = $ 7,000
2nd	0.11($100,000) = $11,000
⋮	⋮
16th	0.02($100,000) = $2,000

As mentioned earlier, there is generally no ACRS deduction for the year of disposal (or retirement) of recovery property, except for 15-year real property. If there is early disposition of 15-year real property, however, the ACRS deduction for the year of disposition is allocated on the basis of the number of months the property is in use that year.

Returning to the example, suppose that the fifteen-year real property, purchased at a cost of $100,000, is sold at the end of the third month of the fifth year of ownership. Since the property was *purchased* in the sixth month of the tax year, the ACRS deduction for the fifth year would be $7,000 (0.07 × $100,000) if it is retained for the full year. Since it is sold after only three months of ownership in the fifth year, however, the allowable ACRS deduction is prorated: 3/12 × $7,000 = $1,750.

The Alternate ACRS Method

The Economic Recovery Tax Act of 1981 permits taxpayers to elect an alternative recovery percentage determined in a manner similar to that of the straight line method. Using the **alternate ACRS method,** for each property class (k), the applicable percentage for each year, $p(k)$, is

$$p(k) = \frac{1}{n(k)} \tag{7.13}$$

where $n(k)$ is the recovery period selected by the taxpayer for that property class. (Note that *recovery period* under ACRS is comparable to *usable life* under the pre-1981 methods.) The three options available to taxpayers for each property class are

Recovery property	Optional recovery period
3-year	3, 5, or 12 years
5-year	5, 12, or 25 years
10-year	10, 25, or 35 years
15-year real ⎫ 15-year public utility ⎬	15, 35, or 45 years

The alternate ACRS method, like the ACRS method, ignores salvage value when computing annual depreciation expenses.

For *3-*, *5-*, *and 10-year property,* taxpayers must use the *same* method (ACRS or alternate ACRS) and recovery period (in the case of alternate ACRS) for all property in the *same* class placed in service in the *same* tax year. If the alternate ACRS method is adopted for a given property class, the appropriate recovery percentage must apply throughout the recovery period selected. But different methods or recovery periods may be used for property of other classes or for property of the same class placed in service in different tax years. The rule for 15-year real property is somewhat different: Separate acquisitions of 15-year real property in a given tax year may not be grouped together; the alternate method election for real property is made on a property-by-property basis.

Under the alternate ACRS method, the *half-year convention* must be used for the year in which *3-*, *5-*, *and 10-year property* is placed in service. That is, for these classes of property, only half the annual depreciation may be claimed for the initial year of ownership. A full year's deduction may be claimed for subsequent years, and if the property is held for the entire recovery period, $n(k)$, a half-year of depreciation is allowed for the tax year following the end of the recovery period. In general,

Year of ownership	Allowable depreciation
1	one-half
2 through $n(k)$	all
$n(k) + 1$	one-half

To illustrate, consider a $60,000 acquisition classified as a 5-year recovery property, and suppose that the taxpayer elects to use the alternate ACRS method with a recovery period of twelve years. Here, $n(5) = 12$. If this property is retained for at least thirteen years, the annual depreciation expenses for this property would be

Year of ownership	Allowable depreciation
1	$0.5(1/12)(\$60,000) = \$2,500$
2–12	$(1/12)(\$60,000) = \$5,000$
13	$0.5(1/12)(\$60,000) = \$2,500$

The half-year convention does *not* apply to the alternate ACRS method for 15-year real property. Instead, the depreciation in the year of acquisition and year of disposition is prorated on the number of months of ownership during the tax year. Suppose, for example, that a $300,000, 15-year real property is purchased at the beginning of the tenth month of the tax year and that the taxpayer elects to depreciate this property by the alternate ACRS method over a 15-year recovery period. For the second through the fifteenth years of ownership, the annual depreciation expense for this property is $(1/15)(\$300,000) = \$20,000$. Depreciation during the year of acquisition is $(3/12)(\$20,000) = \$5,000$. Depreciation in the sixteenth year would be $(9/12)(\$20,000) = \$15,000$, assuming, of course, that the property is retained in service for at least nine months of the sixteenth year.

The early-disposition rule discussed under ACRS also applies to property for which the alternate ACRS method is elected. That is, no ACRS deduction is allowed in the year of disposal or retirement for 3- , 5- , or 10-year property. If 15-year real property is disposed of or retired before the end of the recovery period, the ACRS deduction is allocated for the number of months the property is in use in the year of disposition.

7.5 PRINCIPAL DIFFERENCES AMONG ACRS, ALTERNATE ACRS, AND PRE-1981 METHODS

As outlined above, there are important differences between the allowable depreciation methods prior to 1981 and the ACRS method introduced by the Economic Recovery Tax Act of 1981. The principal differences, summarized in Table 7.5, relate to (1) depreciable life, (2) cost basis, (3) salvage value, (4) annual depreciation expenses, (5) the depreciation expense in the year of acquisition, and (6) the depreciation expense in the year of disposition if (a) the property is retained for at least as long as the useful life or recovery period or (b) the property is retired or disposed of before the end of the useful life or recovery period. All of the elements summarized in Table 7.5 have been discussed previously in this book. For additional reading see:

> *Tax Guide for Small Business* (Publication 334), Washington, D.C.: U.S. Government, Department of Treasury, Internal Revenue Service (revised annually, no cost).
>
> *Depreciation* (Publication 534), Washington, D.C.: U.S. Government, Department of Treasury, Internal Revenue Service (revised annually, no cost).
>
> *Complete Internal Revenue Code of 1954.* Englewood-Cliffs, N.J.: Prentice-Hall (revised periodically).

7.6 ELECTION TO EXPENSE CERTAIN DEPRECIABLE BUSINESS ASSETS

Before the Economic Recovery Tax Act of 1981, taxpayers were permitted to deduct an additional 20 percent of the cost of certain qualifying property in addition to regular depreciation in the year of acquisition. Qualifying property was defined as new or used tangible personal property with a life of six years or more. The amount of the deduction in any one year was limited to $2,000 for single taxpayers or $4,000 for married taxpayers filing a joint

Table 7.5

Principal Differences among Pre-1981 and ACRS Methods

Features	Pre-1981 methods (straight line, declining balance, SYD, and so forth)	ACRS		Alternate ACRS	
		3-, 5-, and 10 year property	15-year real property	3-, 5-, and 10 year property	15-year real property
Depreciable life	Depreciation taken over useful life of the property	Depreciation taken over specified recovery period		For each property class, taxpayer may select one of three recovery periods	
Basis	Depreciation each year figured on adjusted basis	Deduction each year figured on unadjusted basis			
Salvage value	Accumulated depreciation cannot exceed cost basis less salvage value	Salvage value not taken into account			
Annual depreciation	Function of cost basis, depreciable life, and salvage value (except DB); special rules for each method (see Table 7.2)	Product of recovery percentage and unadjusted basis; recovery percentages specified for each property class (see Tables 7.3 and 7.4)		Product of recovery percentage and unadjusted basis; recovery percentage is reciprocal of recovery period selected by taxpayer for property class	
Depreciation in year of acquisition	Prorated on period of ownership	Full year allowed	Specified percentage (see Table 7.4)	Half-year	Prorated on number of months of ownership
Depreciation in year of disposition (if at least as long as recovery period)	Prorated on period of ownership	None	Specified percentage (see Table 7.4)	Half-year	Prorated on number of months of ownership
Early disposition (if disposition occurs before end of useful life or recovery period)	Prorated on period of ownership	None	Prorated on number of months of ownership	None	Prorated on number of months of ownership

return. However, this **additional first-year depreciation,** as it is known, is no longer permitted for property placed in service after 1980.

In part to replace the advantage offered by the now-repealed provision for addition first-year depreciation, the Economic Recovery Tax Act of 1981 permits taxpayers to deduct as an expense in one tax year a limited amount of the cost of certain depreciable property purchased in that year.[12] This deduction is known as the **Section 179 expense deduction.**

The aggregate cost that may be taken into account in any one tax year is limited:

12. Under Section 179 of the Internal Revenue Code of 1954, as revised by the Economic Recovery Tax Act of 1981, eligible property is ''section 38 property'' purchased for use in the trade or business. In general, this is tangible recovery property that qualifies for the investment tax credit.

Calendar year in which the tax year begins	Maximum amount
1982 and 1983	$ 5,000
1984 and 1985	7,500
1986 and thereafter	10,000

For married taxpayers filing separately, the maximum amount for each is limited to 50 percent of the above figures. For example, consider a certain taxpayer, Ms. Mary Every, who is an industrial engineering consultant. Ms. Every is single; her tax year is the same as her calendar year. In 1984 Ms. Every purchases a personal computer for $3,000 to be used in her business. Assuming that this is her only purchase of qualifying property in the tax year, she may elect to claim a Section 179 expense deduction of $3,000 in 1984.

The amount of the Section 179 expense deduction claimed must be subtracted from the cost basis of the property *prior* to computing the ACRS deduction (see Section 7.4) and the investment tax credit (see Section 7.7). Ms. Every, therefore, would not be entitled to any ACRS deduction in 1984 and subsequent years because, if the Section 179 expense deduction is claimed, the adjusted basis would be zero.

7.7
CHANGING
METHODS OF
FIGURING
DEPRECIATION

The Internal Revenue Service views any change in the method of depreciation (for tax purposes) as a change in the method of accounting, so prior IRS permission is required for certain of these changes. The rules are rather complex, but they are summarized in Table 7.6. In particular, change from the regular ACRS method for recovery property to any other method is not permitted, although it is possible to change from the alternate ACRS method to the regular ACRS method with IRS permission. Taxpayers requesting approval for any changes must file IRS Form 3115, *Application for Change in Accounting Method,* during the first 180

Table 7.6
Changing Methods of Figuring Depreciation

	Change to				
Change from	Straight line	Declining balance	Sum of the years digits	ACRS	Alternate ACRS
Straight line		PR(A)	PR(A)	NP	NP
Declining balance	OK		PR(A)	NP	NP
Sum of years digits	PR(A)	PR(A)		NP	NP
ACRS	NP	NP	NP		NP
Alternate ACRS	NP	NP	NP	PR	

Note: OK = IRS permission is not required
PR(A) = IRS permission must be requested, but change is granted automatically
PR = IRS permission must be requested; approval is not automatic
NP = Not permitted

days of the tax year in which the change is to become effective.

The term **amortization** is frequently confused with the term *depreciation*. Like depreciation, amortization is a procedure for allocating the initial cost or other basis of certain assets over a given period of time, permitting the recovery of such expenditures in a manner similar to that of straight line depreciation. But only certain specified expenditures may be amortized for federal income tax purposes. These include

1. Bond premiums—the amount paid in excess of the face value of the bonds purchased.
2. Organization expenses of a corporation, under certain conditons.
3. Research and experimental expenses. (Alternatively, these may be deducted as current business expenses.)
4. Trademarks and tradename expenditures, under certain conditions.
5. Computer software costs. (Purchased software must be depreciated over the useful life of the hardware if the cost of the software is not separately stated. If separately stated, software costs may be amortized over a five-year period or over such appropriate period as can be established.)
6. Franchises, designs, and drawings and patterns, in some cases.
7. Childcare facilities, under certain conditions.
8. Pollution-control facilities, under certain conditions.

In general, assets that may be amortized fall into three classes: business start-up costs, organizational expenses for a corporation, and organizational expenses for a partnership. Unlike depreciable assets, most amortizable assets have no fixed life.

To determine amortization deductions, the total qualifying investment is divided by the number of months in the amortization period (sixty months or more). The result is the amount that can be deducted each month.

Owners of certain properties—such as mineral deposits; oil, gas, or geothermal wells; standing timber; and other exhaustible natural deposits—may be entitled to an allowance for **depletion.** There are two methods of computing depletion: (1) *cost* depletion and (2) *percentage* depletion. Percentage depletion is not applicable to standing timber or to oil and gas wells (other than certain domestic production).

Cost Depletion

In general, the **cost depletion,** D_{cj}, for minerals for a given time interval is computed as follows:

$$D_{cj} = U_j(P/U_e) \tag{7.14}$$

where

U_j = the number of units sold during the interval[13]

U_e = the total number of recoverable units in the deposit (number of tons, barrels, or thousand board-feet, for example)

P = adjusted basis of the mineral property

The adjusted basis is the original cost of the mineral plus any capitalized costs less all prior depletion allowed or allowable on the property. For timber, the allowance each year is the number of timber units cut that year multiplied by the cost or adjusted basis of the standing timber on hand divided by the total depletable units. Cost does not include any part of the cost of the land.

Percentage Depletion

Percentage depletion is a specified percentage of gross income from the property during the tax year with several limiting conditions:

1. The deduction for percentage depletion, D_{pj}, is limited to 50 percent of taxable income from the property, computed without the deduction for depletion.
2. Rents and royalties paid or incurred on the property must be excluded from the computation of gross income.
3. For purposes of computing the 50 percent limitation, certain deductions may *not* be made: (a) net operating loss and (b) charitable contributions. Moreover, mining expenses must be reduced by any gain reported as ordinary income.
4. The *minimum* depletion allowable under percentage depletion may not be less than it would be under the cost depletion method. That is, if $D_{cj} > D_{pj}$ in any year, then cost depletion must be used for that year.

The percentage depletion rate is controlled by federal law. Currently, the rate varies from 5 percent to 22 percent, depending

---

13. If the cash method of accounting is used, U_j represents the number of units for which payment was received in year j. Otherwise, U_j is the number of units sold in year j.

on the type of resource and whether the deposit is in the United States. As of 1984 the percentage depletion rate is 22 percent for lead, zinc, and nickel deposits in the United States; 15 percent for certain domestic oil and gas property and coal; and 5 percent for gravel, sand, and stone.

A principal (and controversial) provision of percentage depletion is that, even if the cost or other basis of the property has been fully recovered, additional allowances for percentage depletion are permitted. Thus, unlike depreciation, percentage depletion permits recovery of more than the original cost. Cost depletion, on the other hand, does not apply once the basis of the property has been exhausted.

A Numerical Example

Suppose that the owner of a certain mineral deposit estimates, based on engineering studies, that the property contained 100,000 unextracted tons of ore at the start of the current year. During the year, 15,000 tons were extracted. The original cost of the property was $950,000. A total of $350,000 was allowed for depletion in previous years. Thus

$$D_c = U(P/U_e)$$
$$= 15,000 \text{ tons} \left(\frac{\$950,000 - \$350,000}{100,000 \text{ tons}} \right)$$
$$= \$90,000$$

The owner had gross receipts of $500,000 from the sale of the ore during the year. Before deduction of the depletion allowance, the net taxable income is $160,000. Assuming a 15 percent depletion rate for this mineral,

$$D_p = 0.15(\$500,000)$$
$$= \$75,000$$

The limit on percentage depletion this year is 50 percent of net taxable income, or $80,000. Thus D_p as computed above is acceptable.

In this example, cost depletion must be used, since $D_c > D_p$. That is, the taxpayer will declare a depletion expense in the amount of $90,000. The new cost basis for the following year will be $950,000 − $440,000, or $510,000. In the absence of any new engineering data, there are $100,000 − 15,000 = 85,000$ tons of ore remaining in the property at the start of the following year.

The **investment credit** was first instituted by the Revenue Act of 1962, which provided for an initial tax reduction equal to 7 percent of "qualified investment" in certain new and used depreci-

7.10
INVESTMENT CREDIT

able property, including buildings and their structural components installed after 1961.[14] (Initially, the tax credit percentage was only 3 percent for public utilities.) Rules concerning the credit were liberalized somewhat in 1964. Then in October of 1966, the credit was suspended temporarily. It was reinstated the following March, repealed entirely on April 18, 1969, and reinstated once again by the Revenue Act of 1971. There have been several additional revisions, the most recent being the Tax Equity and Fiscal Responsibility Act (TEFRA) of 1982. The following discussion is based on the IRS regulations in force in Fall 1983.

Qualifying Property

In general, the investment credit applies only to ACRS property or other depreciable property having a life of at least three years that is placed in service during the year in a trade or business or for the production of income and that is used predominantly within the United States. There are numerous exceptions, however. Non-qualifying property includes, for example, buildings and their structural components, air conditioning or space-heating units, and boilers fueled by oil or gas.

Amount of Investment Subject to Credit

If the qualifying property is depreciable by ACRS, the amount of the investment subject to the credit depends on the property class. All (100 percent) of the investment is subject to the credit if the recovery property is 5-year, 10-year, or 15-year public utility property, but only 60 percent of the investment in 3-year property is eligible.

The amount of the eligible investment for *other* than ACRS property is a function of the depreciable life (N) in years:[15]

Life	Eligible investment
$N < 3$	0
$3 \leq N < 5$	1/3
$5 \leq N < 7$	2/3
$7 \leq N$	all

The amount of the eligible investment in qualifying *used* property is limited for any one year to $125,000 in tax years beginning in 1981, 1982, 1983, or 1984. After 1984 the limit is increased to $150,000.

---

14. Regulations governing the investment credit are complex. For a more extensive discussion, see the latest revision of IRS Publication 572, *Investment Credit*.

15. IRS publications distinguish between *recovery property* and *nonrecovery property*. The former is property that qualifies for depreciation under ACRS; the latter is property that is depreciated by any method other than ACRS, that is, by the methods outlined in Section 7.3 as "pre-1981 methods."

In summary, consider the following set of properties acquired by a hypothetical investor during the 1984 tax year:

Property	Cost basis*	Portion to be counted	Amount subject to credit
ACRS new 3-year	$600,000	60%	$360,000
ACRS new 5-year	300,000	100%	300,000
Non-ACRS new 6-year life	180,000	2/3	120,000
Non-ACRS used 8-year remaining life	($150,000, but limited to $125,000)		
	125,000	100%	125,000
	Total amount subject to credit		$905,000

*Note here that the *cost basis* reflects the prior adjustment for Section 179 expense deductions, if any.

Determining the Tentative Investment Credit

Beginning in 1962, the investment credit percentage was 3 percent for public utilities and 7 percent for other taxpayers. There have been a number of subsequent rate changes, but effective in 1983 a 10 percent basic rate is to be used by all taxpayers, including public utilities. For example, if the amount of the investment subject to the credit is $905,000, the amount of the investment credit is 10 percent of this amount, or $90,500.

For property placed in service after 1982, the cost basis must be reduced by 50 percent of the regular investment credit (as well as by 50 percent of the *energy* and *certified historic structure* investment credits).

To illustrate, consider a new 5-year ACRS property with an initial cost of $300,000 that fully qualifies for the regular investment credit. This property contributes $30,000 to the investment credit; thus its cost basis must be reduced by half this amount, or $15,000. The cost basis subject to ACRS depreciation, then, is $300,000 − $15,000 = $285,000.

Beginning in 1983, taxpayers may elect to use a basic rate of only 8 percent (rather than 10 percent). If this reduced percentage is used, the 50 percent adjustment to the cost basis of the property is not required. In the above example, the taxpayer has two choices:

	Use 10% rate	Use 8% rate
Regular investment credit	$ 30,000	$ 24,000
Cost basis	285,000 (revised)	300,000
Depreciation—1st year (15%)	42,750	45,000
2nd year (22%)	62,700	66,000
3rd year (21%)	59,850	63,000
4th year (21%)	59,850	63,000
5th year (21%)	59,850	63,000

The taxpayer has the choice of either taking a larger investment credit in the year of acquisition or taking larger depreciation deductions over the recovery period.

The 10 percent and 8 percent rates may be used for ACRS recovery property other than 3-year property. Since only 60 percent of the basis of 3-year recovery property is subject to the regular investment credit, the effective rate is 6 percent (rather than 10 percent). Moreover, if taxpayers elect to take a reduction in the credit instead of the 50 percent reduction to the basis, the reduced percentage allowed is 4 percent (rather than 8 percent). To illustrate, consider a new 3-year recovery property with an initial cost of $600,000 that fully qualifies for the regular investment credit. The taxpayer has two choices:

	Use 6% rate	Use 4% rate
Regular investment credit	$ 36,000	$ 24,000
Cost basis	582,000 (revised)	600,000
Depreciation—1st year (25%)	145,500	150,000
2nd year (38%)	221,160	228,000
3rd year (37%)	215,340	222,000

If the 6 percent rate is used (10 percent of 60 percent), the basis must be revised by half the regular credit; that is, the revised basis becomes $600,000 − 0.5($36,000) = $582,000. If the 4 percent rate is used, the regular investment credit is 4 percent of $600,000, or $24,000, and the basis remains at $600,000. This current year's credit is then added to the investment credit carrybacks and carryovers to this year to determine the total tentative investment credits claimed for the year.

Allowable Credit

The investment credit, if allowable, is taken in the year in which the asset is placed in service, but it is limited to the lesser of (a) the firm's tax liability for the year, or (b) the first $25,000 of the tax liability (as otherwise computed) plus 85 percent of the tax liability in excess of $25,000.[16] To illustrate, suppose that a taxpayer's tax liability (before computing the credit) is $85,000 and that the credit for the year is computed to be $90,500. Assuming no tax credit carried over from prior years, the taxpayer's allowable investment credit (AIC) for the year is

$$AIC = \$25,000 + 0.85(\$85,000 - \$25,000)$$

$$= \$76,000$$

Thus only $76,000 of the $90,500 credit can be used in the current tax year.

16. In 1982, the limit was $25,000 plus 90 percent of the tax liability in excess of $25,000.

Credit Carrybacks and Carryovers

Unused credits are not lost, however, in that a 3-year carryback and a 15-year carryover are provided for credits that may be unused in the current year because of the limitations. In the previous example, the unused credit ($90,500 − $76,000 = $14,500) is carried back to the three prior tax years, and if necessary, any remaining unused credit may be carried over to the fifteen following tax years. The unused credit must be used at the earliest opportunity.

An unused credit *carried back* to the prior tax year may be used to the extent that the limit for the prior year is greater than the sum of (a) the investment credit carryovers to that year, (b) the credit earned for the year, and (c) the investment credit carrybacks from the years prior to the year from which the credit is being carried. An unused credit *carried over* from prior years is used before a credit for the current year to the extent that the unused credit is not more than the limit.

Recapture of Investment Credit

In the event that **nonrecovery property** (other than ACRS) is disposed of, the investment credit must be recomputed, and where appropriate, the investment credit in the year of disposal should be adjusted to reflect the actual period the property was in service. The *new* useful life brackets (3–5 years, 5–7 years, and over 7 years) apply. For example, consider an asset originally placed in service on November 1, 1980. (Since this is prior to 1981, the asset is a nonrecovery property.) The original cost of the asset was $12,000, and at the time it was placed in service it was believed to have a useful service life of ten years. Accordingly, the full $12,000 was considered to be a qualified investment on the 1980 tax return. Assume that the asset will be sold on August 1, 1984. Inasmuch as the actual service life is only three years and nine months, only one-third of the original cost, or $4,000, is applicable. Thus the total qualified investment in 1984 will be reduced by $8,000 as a result of the "premature disposal" of this asset. The firm's income tax liability in 1984 will be increased by $800, assuming no tax credit carryover to the 1984 tax year.

If a firm disposes of **recovery property** (depreciated under ACRS) before the end of the recovery period, a percentage of the credit must be recaptured. The recapture percentage is determined as follows:

Full year within which recovery property is disposed of after being placed in service	Recovery percentage	
	3-year property	5-, 10-, and 15-year property
1st	100%	100%
2nd	66	80
3rd	33	60
4th	——	40
5th	——	20

To illustrate, suppose that a $50,000 qualified investment was made in a certain 5-year recovery property in August 1983, at which time the full regular tax credit was claimed (10% × $50,000 = $5,000). Suppose further that this property is retired "prematurely" in June 1985—that is, within the second full year after it is placed in service. This transaction, then, requires recapture of 80% × $5,000 = $4,000 credit in the 1985 tax year.

If the cost basis was adjusted by half the investment credit at the time the property was placed in service, a similar adjustment may be required if it is removed from service before the end of the recovery period. Using the previous example, suppose that the $5,000 credit in 1983 had the effect of reducing the basis to $50,000 − 0.5($5,000) = $47,500. Thus the depreciation in 1983 for this 5-year recovery property is 0.15($47,500) = $7,125; the depreciation in 1984 is 0.22($47,500) = $10,450; and the book value at the start of 1985 is $47,500 − $7,125 − $10,450 = $29,925. Since there is a credit recovery of $4,000 in 1985, half of this amount, or $2,000, should be added to determine the book value at the time of disposal: $29,925 + $2,000 = $31,925. (Of course, there is no depreciation in 1985, the year of disposal.) In summary:

Initial cost basis (8/83)	$50,000
Less 1/2 investment credit	− 2,500
	$47,500
Less depreciation in 1983	− 7,125
Less depreciation in 1984	− 10,450
	$29,925
Plus 1/2 credit recovery	+ 2,000
Book value (6/85)	$31,925

As you will see, this is an important calculation in that it affects the tax on the gain or loss on disposal of this depreciable asset.

Effect on Investment

Depreciation expenses are deductions from *taxable income;* they affect cash flows indirectly. On the other hand, the primary effect of the *investment credit* is to reduce the taxpayer's income tax liability in the year in which the qualifying property is placed in service. In some cases, the recoverable cost basis is reduced by half the credit, but taxpayers have the option of using a lower rate to compute the credit in return for no reduction in the cost basis. They may choose the most advantageous option. Because of its effect on the amount and timing of cash flows for income taxes, the investment credit makes qualifying investments more profitable. A numerical example demonstrating this effect appears in Section 7.12.

The accounting procedures discussed in Sections 7.3 through 7.9—depreciation, amortization, and depletion—all relate to the determination of taxable income in that they are deductible business expenses and thus have the effect of reducing taxable income. Depreciation, amortization, and depletion are *not* cash flows. Their relevance to economy studies stems from the fact that they directly influence an important cash flow, namely, income taxes. The relationship is as follows:

7.11
INCOME TAXES
FOR CORPORATIONS[17]

$$T = t(R - E) \qquad (7.15)$$

where

T = income tax

t = tax rate

R = revenue

E = expenses $\Big\}$ Taxable income = $R - E$

Of course, *taxable income* is an accounting concept; it is a function of revenue and expense items, some of which are cash flows and others of which are not cash flows. Of particular importance are the accounting expenses that result from depreciation, amortization, and depletion. Having identified procedures for determining these expenses, therefore, let's turn now to the problem of determining the amount and timing of cash flows for income taxes.

Income Tax Rates

As for individual taxpayers, income tax rates for corporations are adjusted from time to time, largely in order to affect the level of economic activity. Currently (1983), federal income tax rates for corporations are related to the firm's taxable income for the tax year:

Range of taxable income (TI)	Corporate income tax
0 < TI ≤ $ 25,000	0.15(TI)
$ 25,000 < TI ≤ $ 50,000	$ 3,750 + 0.18(TI − $ 25,000)
$ 50,000 < TI ≤ $ 75,000	$ 8,250 + 0.30(TI − $ 50,000)
$ 75,000 < TI ≤ $100,000	$15,750 + 0.40(TI − $ 75,000)
$100,000 < TI	$25,750 + 0.46(TI − $100,000)

Many states also impose income taxes. In California, for example, a franchise tax is imposed on all corporations organized (franchised) in California and on out-of-state corporations that do

17. This discussion is limited to income taxes for the incorporated business enterprise. An excellent reference for the single proprietorship or partnership is IRS Publication 334, *Tax Guide for Small Business* (Washington, D.C.: Government Printing Office).

business in California. The tax is based on net income from California sources. There is also a corporation income tax imposed on corporations that derive income from sources within California but that are not subject to the franchise tax. The tax rates for 1981, 1982, and 1983 are determined by a formula based on the level of tax collections. In the early 1980s, tax rates were in the range of 9.3–9.6 percent.

Economy studies are concerned only with the incremental effects of investments. That is, when the effect of a specific investment proposal is being considered, the relevant income tax rate is the one associated with the incremental taxable income. Thus the appropriate rate for federal income taxes is currently 46 percent for firm's that have taxable income in excess of $100,000 during the tax year. Tax rates are summarized in Figure 7.10.

When income is taxed by more than one jurisdiction, the appropriate tax rate for economy studies is a combination of the rates imposed by the jurisdictions. If these rates are independent, they may simply be added. But the combinatorial rule is not quite so simple when there is interdependence. Income taxes paid to local and state governments, for example, are deductible from

Figure 7.10. Corporate Income Tax Rates for Tax Years Beginning in 1983

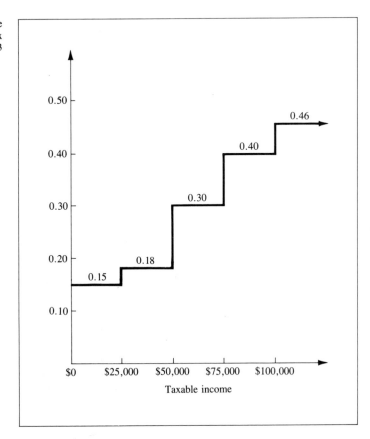

taxable income on federal income tax returns, but the reverse is not true: Federal income taxes are not deductible from local returns. Thus, considering only state (t_s) and federal (t_f) income tax rates, the *combined incremental tax rate* (t) for economy studies is given by

$$t = t_s + t_f(1 - t_s) \qquad (7.16)$$

Consider, for example, a corporation that has an incremental federal rate of 46 percent and an incremental state rate of 10 percent. The combined incremental rate is

$$t = 0.10 + 0.46(1 - 0.10) = 0.514$$

If an investment results in an increase of $1,000 in taxable income, the total cash flow for income taxes is $514, leaving the firm with $486 after income taxes.

Timing of Tax Payments

In some corporations, the tax year (for accounting purposes) coincides with the calendar year. Such firms must file tax returns with the federal government not later than March 15 following the end of the calendar year. If the corporation's tax year is not the same as the calendar year, however, the due date is the fifteenth date of the third month following the close of the tax year. Local and state regulations generally, but not necessarily, parallel those of the federal government.

Because of the time value of money, it is normally advantageous for taxpayers to delay tax payments as long as possible. Conversely, it is to the government's advantage to collect tax payments as quickly as possible, preferably as soon as income is earned. To compromise, the government requires corporations to compute their estimated tax liability in advance of the tax year, and if this estimated tax can reasonably be expected to be $40 or more, periodic installments must be paid. For example, for firms that meet the $40 requirement prior to the fourth month of the tax year, 25 percent of the estimated tax must be paid by the fifteenth day of the fourth, sixth, ninth and twelfth months of the taxable year. The corporation must pay the balance due in full by the due date of the return, that is, before the fifteenth day of the third month following the close of the tax year. This sequence of payments is shown in Figure 7.11.

There may be considerable time lag between the pretax cash flow (for example, the investment) and the resulting effect on cash flow for income taxes. Tax consequences that reduce tax liability (the investment credit, for example) are generally claimed as quickly as possible; consequences that increase taxes are generally

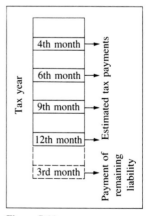

Figure 7.11

delayed as long as possible. The general rule is to accelerate positive cash flows and defer negative cash flows. Authors of economic analysis textbooks usually adopt the following conventions in their illustrations and problems:

Consequence	Timing of tax effect
Initial investment	Beginning of year
Investment credit	Beginning of year
Depreciation and other tax-deductible expenses	End of year
Disposal of asset	End of year

It should be stressed, however, that these are merely simplified textbook assumptions. They are probably appropriate to the vast majority of situations in which the dollar amounts are relatively small and the discount rate is relatively low. In the few cases where these assumptions do not hold, the analyst should identify, insofar as possible, the precise time lag between the pretax events and the cash flows for taxes.

Tax Treatment of Gains and Losses on Disposal

The value of an asset on disposal is rarely equal to its book value at the time of sale or other disposition.[18] When this inequality occurs, a gain or loss on disposal is established and the transaction has certain tax consequences.

Central to understanding the appropriate tax treatment of gains and losses on disposal is the concept of **capital asset.** Everything owned by the taxpayer is a capital asset, with certain exceptions. Principal among these exceptions are real property and depreciable property used in the trade or business (even if fully depreciated). Tax treatments differ for these classes of assets.

All gains and losses on disposal are treated as *ordinary* gains or losses, *capital* gains or losses, or some combination of the two. The rules for determining these amounts are too complex to be discussed adequately here; interested readers should therefore consult a competent expert and/or read the appropriate sections in *Tax Guide for Small Business* (IRS Publication 334) or a similar reference.

As a general rule, with some important exceptions, any gains or losses on the disposition of tangible personal property used in a taxpayer's trade or business are treated as ordinary income, limited to the total amount of depreciation taken up to the time of disposition. That is,

---

18. An exchange of the property is a form of disposition. The value of the item(s) received in exchange represents the salvage value.

Gain	Ordinary income	Capital gain
Less than depreciation	Entire gain	None
Greater than depreciation	Accumulated depreciation	Excess of gain over accumulated depreciation

These relationships are shown graphically in Figure 7.12.

Knowing whether income is treated as ordinary income or capital gain is important, because the tax rates may be quite different. For example, there is special tax treatment for that part

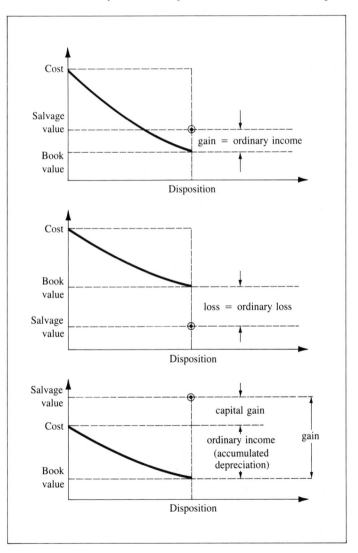

Figure 7.12. Treatment of Gains and Losses on Disposal of Depreciable Assets Used in the Trade or Business

of a total net capital gain during a particular tax year that results from a net long-term capital gain. (A **long-term gain** may result from the disposition of a capital asset held for more than one year.) For corporations, an *alternative tax rate* of only 28 percent may be applied to a corporation's net long-term capital gain. This rate is less than the corporate income tax rate when corporate income is more than $50,000. For *individuals,* only 40 percent of the net long-term capital gain is taxable at the individual's ordinary income tax rate. Because of the significant tax implications, tax law and regulations are quite specific as to the appropriate treatment for the disposition of assets.

7.12 AFTER-TAX ECONOMY STUDIES

The various figures of merit of interest in economy studies (net present worth and internal rate of return, for example) are functions of the amount and timing of prospective cash flows resulting from alternative investment opportunities. Income taxes, therefore, represent an important element in detemining relative profitability.

To illustrate, consider an asset costing $10,000, which, if purchased, will result in a savings of $3,000 each year throughout its life. Assume zero salvage value, a five-year service life, and end-of-year cash flow. The (internal) rate of return *before taxes* is easily determined:

$$\$3,000(P/A, \ i, \ 5) \ - \ \$10,000 \ = \ 0$$

$$(P/A, \ i, \ 5) \ = \ 3.333$$

$$i \ = \ 15.24\%$$

But the effects of cash flows for income taxes on profitability, that is, the (internal) rate of return *after* taxes, must also be determined. Thus the following examples compute the internal rate of return after taxes $(\hat{i})$ for both nonrecovery and ACRS recovery property using the appropriate depreciation and investment credit assumptions. In the first example, it is assumed that the asset is a *nonrecovery property*—it was placed in service before 1981, it does not otherwise qualify for ACRS (on federal tax returns), or ACRS is not permitted (as under certain state tax regulations). In the second example, it is assumed that the asset is an ACRS *recovery property,* and current U.S. federal tax regulations are in effect.

Depreciation of a Nonrecovery Property
Suppose that the example asset may be depreciated by the straight-line method over five years. The annual depreciation expense, therefore, is $10,000/5 = $2,000 and the taxable income is

$3,000 - \$2,000 = \$1,000$ per year. If the corporate income tax rate is 0.50, the effect on cash flow for taxes each year is $0.50(\$1,000) = \$500$ and the effect on net cash flow after taxes is $\$3,000 - \$500 = \$2,500$. Thus the (internal) rate of return *after taxes*, $\hat{i}$, is

$$\$2,500(P/A, \hat{i}, 5) - \$10,000 = 0$$

$$(P/A, \hat{i}, 5) = 4.000$$

$$\hat{i} = 7.93\%$$

Note here that the after-tax rate of return is *not* equal to one minus the tax rate times the before-tax rate of return. That is, $\hat{i} \neq i(1 - t)$. It is unequal because the internal rate of return is not a linear function of the cash flows.

Table 7.7 illustrates the effects of various tax policies on internal rate of return for nonrecovery property (Cases A through E). The numerical values used are summarized below the table. The subscripts d and a distinguish between "for depreciation purposes" and "actual," respectively.

Case A illustrates the effect of straight line depreciation. (See Figure 7.13.) The internal rate of return after taxes is 7.93 percent, as compared to the pretax rate of return of 15.24 percent.

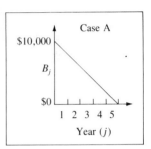

Figure 7.13

Table 7.7

Cash-Flow Tables for Nonrecovery Property Asset under Alternative Tax Treatments

Before-tax *IRR* is 15.24% with basic assumptions	End of year	Effect on cash flow before taxes	Depreciation expense	Effect on taxable income	Effect on cash flow for taxes	Effect on cash flow after taxes	After-tax rate of return
Case A	0	$-\$10,000$ $ ———	$ ———	$ ———	$ ———	$-\$10,000$	
Straight line	1	3,000	2,000	1,000	$-$ 500	2,500	
depreciation	2	3,000	2,000	1,000	$-$ 500	2,500	
	3	3,000	2,000	1,000	$-$ 500	2,500	
$N_d = N_a = 5$	4	3,000	2,000	1,000	$-$ 500	2,500	
$S_d = S_a = 0$	5	3,000	2,000	1,000	$-$ 500	2,500	
	Total	$ 5,000	$10,000	$ 5,000	$-\$2,500$	$ 2,500	$\hat{i} = 7.93\%$
Case B	0	$-\$10,000$ $ ———	$ ———	$ ———	$ ———	$-\$10,000$	
Double declining	1	3,000	4,000	$-$ 1,000	500	3,500	
balance depreciation	2	3,000	2,400	600	$-$ 300	2,700	
	3	3,000	1,440	1,560	$-$ 780	2,200	
$N_d = N_a = 5$	4	3,000	864	2,136	$-$ 1,068	1,932	
$S_d = S_a = 0$	5	3,000	518.40	2,481.60	$-$ 1,240.80	1,759.20	
	5	0	(777.60)*	$-$ 777.60	388.80	388.80	
	Total	$ 5,000	$10,000	$ 5,000	$-\$2,500$	$ 2,500	$\hat{i} = 8.77\%$

Continued on page 238

Note: First cost of asset is $10,000; net operating savings before taxes are $3,000 per year for 5 years; incremental corporate income tax rate is 50%; investment credit is considered only in Case E.

*Book value

Table 7.7 (continued)

Before-tax *IRR* is 15.24% with basic assumptions	End of year	Effect on cash flow before taxes	Depreciation expense	Effect on taxable income	Effect on cash flow for taxes	Effect on cash flow after taxes	After-tax rate of return
Case C	0	− $10,000	$ ——	$ ——	$ ——	− $10,000	
Double declining balance,	1	3,000	4,000	− 1,000	500	3,500	
shift to straight line	2	3,000	2,400	600	− 300	2,700	
	3	3,000	1,440	1,560	− 780	2,220	
$N_d = N_a = 5$	4	3,000	1,080	1,920	− 960	2,040	
$S_d = S_a = 0$	5	3,000	1,080	1,920	− 960	2,040	
	Total	$ 5,000	$10,000	$ 5,000	− $2,500	$ 2,500	$\hat{i}$ = 8.87%
Case D	0	− $10,000	$ ——	$ ——	$ ——	− $10,000	
Sum of years digits	1	3,000	3,333	− 333	167	3,167	
depreciation	2	3,000	2,667	333	− 167	2,833	
	3	3,000	2,000	1,000	− 500	2,500	
$N_d = N_a = 5$	4	3,000	1,333	1,667	− 833	2,167	
$S_d = S_a = 0$	5	3,000	667	2,333	− 1,167	1,833	
	Total	$ 5,000	$10,000	$ 5,000	− $2,500	$ 2,500	$\hat{i}$ = 8.77%
Case E	0	− $10,000	$ ——	$ ——	$ ——	− $10,000	
Straight line depreciation,	1	10% × 2/3 × $10,000 =			667	667	
investment credit	1	3,000	2,000	1,000	− 500	2,500	
	2	3,000	2,000	1,000	− 500	2,500	
	3	3,000	2,000	1,000	− 500	2,500	
$N_d = N_a = 5$	4	3,000	2,000	1,000	− 500	2,500	
$S_d = S_a = 0$	5	3,000	2,000	1,000	− 500	2,500	
	Total	$ 5,000	$10,000	$ 5,000	− $2,500	$ 3,167	$\hat{i}$ = 10.34%
Case F	0	− $10,000	$ ——	$ ——	$ ——	− $10,000	
Straight line depreciation,	1	3,000	3,333	− 333	167	3,167	
shorter "expected life"	2	3,000	3,333	− 333	167	3,167	
	3	3,000	3,333	− 333	167	3,167	
$3 = N_d < N_a = 5$	4	3,000	0	3,000	− 1,500	1,500	
$S_d = S_a = 0$	5	3,000	0	3,000	− 1,500	1,500	
	Total	$ 5,000	$10,000	$ 5,000	− $2,500	$ 2,500	$\hat{i}$ = 9.24%
Case G	0	− $10,000	$ ——	− $10,000	$5,000	− $ 5,000	
Straight line depreciation,	1	3,000	0	3,000	− 1,500	1,500	
asset expensed	2	3,000	0	3,000	− 1,500	1,500	
	3	3,000	0	3,000	− 1,500	1,500	
$0 = N_d < N_a = 5$	4	3,000	0	3,000	− 1,500	1,500	
$S_d = S_a = 0$	5	3,000	0	3,000	− 1,500	1,500	
	Total	$ 5,000	0	$ 5,000	− $2,500	$ 2,500	$\hat{i}$ = 15.24%

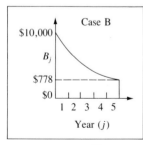

Figure 7.14

Case B illustrates the effect of double declining balance depreciation. (See Figure 7.14.) Note that by using this "accelerated" depreciation method, the after-tax rate of return has increased to 8.77 percent. There are two rows for year 5: The first identifies the effect of depreciation; the second identifies the effect of disposal. That is, a loss on disposal of $777.60 decreases ordinary income by this amount, resulting in tax savings of $0.50($777.60) = $388.80.

Case C illustrates the effect of employing double declining

balance depreciation during the first three years, then shifting to straight line depreciation during the last two years. (See Figure 7.15.) The after-tax rate of return is increased to 8.87 percent.

Case D illustrates the effect of the sum of the years digits depreciation method. (See Figure 7.16.) The after-tax rate of return, 8.77 percent, is about the same as that yielded by the DDB method but is larger than that of the straight line method. Yet in each case, the effect on total net after-tax cash flow is $2,500. The difference in rate of return is caused by the timing, not the amount, of the cash flows for taxes.

Case E reflects the effect of the investment credit. (See Figure 7.17.) Note that there are two rows for year 1: The first reflects the investment credit, and the second identifies the effect of depreciation. (Here, it is assumed that the cash flow from the investment credit is not effected until one year after the investment that led to that credit. It is also assumed that the tax credit has no effect on the cost basis of the asset, as was the case prior to 1983. Finally, straight line depreciation is assumed in this example.) The depreciation schedule is not affected by the investment credit. Reflecting the tax credit, the internal rate of return has increased to 10.34 percent.

Case F illustrates the effect of using a shorter life for depreciation purposes (N_d = 3 years) than is actually experienced (N_a = 5 years). (See Figure 7.18.) Since the tax savings from depreciation are experienced earlier—the *total* is still $2,500—the after-tax rate of return is increased to 9.24%.

Case G represents the extreme: The asset is not depreciated at all. Instead, it is *expensed,* that is, written off entirely in the first year. (See Figure 7.19.) (Here, it is assumed that the tax effect coincides with the intial investment. This assumption may not be realistic in that there may in fact be a time lag between the investment and its tax consequences.) Thus the after-tax rate of return is increased to 15.24 percent, the same as the rate of return before taxes.

Depreciation of an ACRS Recovery Property

After-tax economy studies for recovery property depreciated under ACRS are illustrated in Table 7.8. The basic assumptions here are that (1) the asset is a 5-year recovery property with an unadjusted cost basis of $10,000; (2) net operating savings before income taxes are $3,000 per year; (3) the property will be used for exactly five years, at the end of which time it will be sold; (4) the initial purchase takes place at the start of the first tax year and the annual operating savings occur at the end of each of the first five tax years; (5) income tax consequences occur at the end of each tax year; and (6) the regular investment credit is ignored, except in cases K and L.

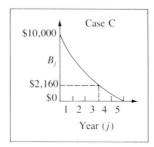

Figure 7.15

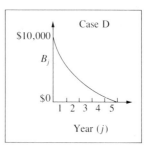

Figure 7.16

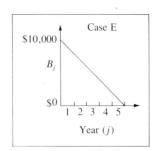

Figure 7.17

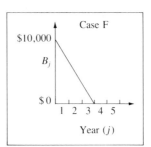

Figure 7.18

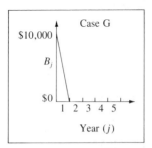

Figure 7.19

Moreover, in the cases described in Table 7.8, it is assumed that the full-year ACRS deduction is allowed in the fifth year. As noted previously, no depreciation expense may be included in the year of disposal for ACRS properties, other than for 15-year real property. Even though the property is retired from service on the last day of the year, no ACRS deduction may be claimed for that year. Therefore, the examples assume that the property is acquired and placed in service on the first day of the first year and sold on the first day of the sixth year; hence the entire deduction may be taken in the fifth year.

Case H assumes ACRS depreciation for a 5-year recovery property. The after-tax (internal) rate of return is 7.8 percent. Note that this figure is somewhat less than the rates of return experienced under the conditions outlined in Table 7.7 for the straight line method (7.93 percent), the double declining balance method with a shift to straight line (8.87 percent), and sum of the years digits method (8.77 percent). But one should *not* conclude that ACRS *always* yields lower rates of return than do the other methods. It is true in this example because a full year's depreciation was allowed in Table 7.7; the asset was assumed to be purchased at the beginning of the tax year. Otherwise, the depreciation in the year of acquistion (for nonrecovery property) must be prorated on the period of ownership during the year. So under other conditions, ACRS could well result in higher after-tax rates of return than would be obtained using the other methods.

Case I assumes that the property is depreciated using the alternate ACRS method with a 5-year recovery period. Since it is assumed that the property is sold on the first day of the sixth tax year, the ACRS deduction in that year is not allowed. (If it were allowed, the half-year convention would result in a $1,000 deduction in the sixth year.) However, the book value of the property at the time of disposal is $1,000, and if the actual salvage value is zero, the loss on disposal is $1,000. This is an ordinary loss and has the effect of reducing taxable income by $1,000. In the present example, it is assumed that the effect of cash flows for income taxes is at the beginning of year 6 or at the end of year 5. This assumption is artificial, of course, because there would normally be some time lag between the disposal of the asset and the time at which the income tax effects are experienced. Under these assumptions, the after-tax rate of return is 7.49 percent.

Case J is similar to Case I except that here the taxpayer elects to use a recovery period of twenty-five years. The after-tax rate of return is only 6.39 percent.

Case K is similar to Case H, except that the 10 percent investment credit has been included. Note that the cost basis is reduced by half the investment credit: $10,000 − 0.5($1,000) = $9,500. No recapture of the credit is necessary here, because the

Table 7.8

Cash-Flow Table for 5-Year ACRS Recovery Property

Before-tax *IRR* is 15.24% with basic assumptions	End of year	Effect on cash flow before taxes	Depreciation expense	Effect on taxable income	Effect on cash flow for taxes	Effect on cash flow after taxes	After-tax rate of return
CASE H	0	− $10,000	$ ——	$ ——	$ ——	− $10,000	
ACRS	1	3,000	1,500	1,500	− 750	2,250	
	2	3,000	2,200	800	− 400	2,600	
with $N_a = 5$	3	3,000	2,100	900	− 450	2,550	
$S_a = 0$	4	3,000	2,100	900	− 450	2,550	
	5	3,000	2,100	900	− 450	2,550	
	Total	$ 5,000	$10,000	$5,000	− $2,500	$ 2,500	i = 7.80%
CASE I	0	− $10,000	$ ——	$ ——	$ ——	− $10,000	
Alternate ACRS	1	3,000	1,000	2,000	− 1,000	2,000	
	2	3,000	2,000	1,000	− 500	2,500	
with $N_d = 5$	3	3,000	2,000	1,000	− 500	2,500	
$N_a = 5$	4	3,000	2,000	1,000	− 500	2,500	
$S_a = 0$	5	3,000	2,000	1,000	− 500	2,500	
	5	0	(1,000)*	− 1,000	500	500	
	Total	$ 5,000	$10,000	$5,000	− $2,500	$ 2,500	i = 7.49%
CASE J	0	− $10,000	$ ——	$ ——	$ ——	− $10,000	
Alternate ACRS	1	3,000	200	2,800	− 1,400	1,600	
	2	3,000	400	2,600	− 1,300	1,700	
with $N_d = 25$	3	3,000	400	2,600	− 1,300	1,700	
$N_a = 5$	4	3,000	400	2,600	− 1,300	1,700	
$S_a = 0$	5	3,000	400	2,600	− 1,300	1,700	
	5		(8,200)*	− 8,200	4,100	4,100	
	Total	$ 5,000	$10,000	$ 5,000	− $2,500	$ 2,500	i = 6.39%
CASE K	0	− $10,000	$ ——	$ ——	$ ——	− $10,000	
ACRS with	1	(Investment credit is 10% of $10,000)			1,000	1,000	
investment credit	1	3,000	1,425	1,575	− 787.5	2,212.5	
	2	3,000	2,090	910	− 455	2,545	
$N_a = 5$	3	3,000	1,995	1,005	− 502.5	2,497.5	
$S_a = 0$	4	3,000	1,995	1,005	− 502.5	2,497.5	
	5	3,000	1,995	1,005	− 502.5	2,497.5	
	Total	$ 5,000	$ 9,500	$ 5,500	− $1,750	$ 3,250	i = 10.64%
CASE L	0	− $10,000	$ ——	$ ——	$ ——	− $10,000	
ACRS with investment	1	(Investment credit is 8% of $10,000)			800	800	
credit and early	1	3,000	1,500	1,500	− 750	2,250	
disposition	2	3,000	2,200	800	− 400	2,600	
	3	3,000	2,100	900	− 450	2,550	
$N_a = 3$	3	4,500	(4,200)	300	− 150	4,350	
$S_a = \$4,500$	3	(Recovery percentage is 40%)			− 320	− 320	
	Total	$ 3,500	$10,000	$ 3,500	− $1,270	$ 2,230	i = 9.33%

Note: Unadjusted cost basis is $10,000; net operating savings before taxes are $3,000 per year for 5 years; incremental corporate income tax rate is 50%; investment credit considered only in Cases K and L.

*Adjusted cost basis = book value at disposal

property is kept in service for the entire recovery period. Assuming that the credit affects income taxes at the end of the first year, the after-tax rate of return is 10.64 percent.

Case L also includes the investment credit for this ACRS property, but now it is assumed that the property will be sold for

$4,500 at the end of the third year (actually, the beginning of the fourth year). In this example, there is the option of using the lower 8 percent rate to compute the investment credit, so it is not necessary to adjust the cost basis. There are two notable effects at the time of disposal. First, the accumulated depreciation at the time of disposal is $5,800, the book value is $4,200, and the $4,500 sale results in a $300 gain on disposal, included as ordinary income. Second, the investment credit ($1,000) was determined initially assuming a five-year recovery period, but the premature disposal after only three years requires recapture of 40 percent of the credit. The after-tax rate of return, under these conditions, is 9.33 percent.

**7.13
SUMMARY**

Depreciation expenses are not cash flows, yet they are significant in economic analysis because of their effect on taxable income, which, in turn, leads to cash flows for income taxes. Before 1982, depreciation accounting for federal income tax purposes was generally limited to the straight line, the declining balance, or the sum of the years digits methods, although certain other methods were also permitted. Beginning in 1982 an entirely new procedure, the ACRS method, was introduced by the federal government, resulting in separate treatments for recovery and nonrecovery properties. The primary difference between depreciation methods is the *timing* of the depreciation expense and, by extension, the effect on taxable incomes and cash flows for taxes.

Depletion may be thought of as a form of depreciation especially applicable to mineral deposits, oil and gas wells, and standing timber. Both cost depletion and percentage depletion should be computed, where permitted, and the method that yields the largest deduction should be used. Only cost depletion may be used for timber deposits.

Amortization is much like straight line depreciation in that it allocates the initial cost or other basis of certain assets uniformly over a given period of time. Generally, assets that may be amortized are certain business start-up costs and organizational expenditures for corporations and partnerships.

The investment credit is a direct deduction from the tax liability. Since it reduces negative cash flows (income taxes), it has the effect of increasing the rate of return for qualifying property. Use of the investment credit is subject to certain limitations. If the 10 percent rate is used to determine the credit, the cost basis of the property must be reduced by half the credit. Taxpayers have the opportunity to use a lower rate, however, in order to avoid the reduction in the cost basis.

Both the absolute and relative economic attractiveness of prospective investment proposals are significantly affected by relevant federal and state tax regulations. It is essential, therefore, that capital allocation analyses consider tax rates, scheduling of tax payments, and other matters that relate to the amounts and timing of cash flows for income taxes. A detailed discussion of tax regulations has been avoided here because they may vary from state to state. In any event, analysts should include all relevant taxes—federal, state, and local—when determining the appropriate tax rate applicable to incremental income to be generated by an investment proposal.

Corporate income is subject to "double taxation" in that after-tax profits, when distributed in the form of dividends, are also taxed as personal income. It is extremely difficult to completely determine incremental tax rates for large, publicly held corporations, but whenever the number of owners is small enough to permit reasonably precise determination of incremental individual income tax rates, both tax structures should be considered.

There is, in general, a substantial lag between the time when the pretax cash flow resulting from the investment occurs and the time when the associated tax effect takes place. Analysts should include these effects when the timing of a specific investment decision appears to be critical; otherwise, a simplifying assumption may reasonably be made.

The prospective rate of return on an investment proposal is clearly affected by tax considerations, but the relative effect is controlled to a large extent by the depreciation schedule used and/or the expected life assumed for depreciation purposes. Table 7.9 summarizes the after-tax rates of return resulting from the various depreciation methods applied to the investment proposal given in Tables 7.7 and 7.8. In each instance the pretax cash flows are identical. In addition to the obvious value of the investment credit, prospective after-tax rates of return are improved (1) by "accelerating" the rate at which the depreciation expense is claimed over the expected life of the asset or (2) by using a depreciable life somewhat less than the expected useful life. The total after-tax cash flows are the same in both instances; only the timing differs.

Table 7.9

Effects of Depreciation on (Internal) Rates of Return

Depreciation method	Depreciable life (nonrecovery property) or Property class (recovery property)			
	Expensed	3 years	5 years	25 years
Straight line	15.24%	9.24%	7.93%	
Declining balance (DDB)			8.77	
DDB, shift to straight line			8.87	
Sum of the years digits			8.77	
Straight line + investment credit			10.34	
ACRS			7.80	
Alternate ACRS			7.49	6.39%
ACRS + investment credit			10.64	

PROBLEMS

Federal regulations governing depreciation accounting, investment credit, and the like are complex, so it is difficult to prepare numerical exercises and problems that are reasonably realistic and yet do not unduly burden the student. The objective of the following problems is to reinforce the concepts outlined in the chapter without requiring expertise in tax accounting. Accordingly, the following assumptions should be used in the problems unless specifically stated otherwise:

1. The taxpayer's tax year is the same as the calendar year: January 1–December 31.

2. The time that depreciable property is placed in service coincides with the time of acquisition, and the property is sold when it is removed from service.

3. When depreciating a specific capital asset, the same depreciation method is used for all levels of government (local, state, and federal) for determining taxable income. Of course, this assumption is invalid for those states (California, for example) where ACRS is not permitted. However, to assume otherwise would require an additional level of complexity that is of little value in learning the principles outlined in the chapter.

4. The investment tax credit is not applicable unless otherwise indicated. When called for, it is used only for federal income taxes.

5. The taxpayer's incremental income tax rate and the minimum attractive rate of return remain constant throughout the study period.

6. Depreciable property is new (not used) when acquired.

7. The taxpayer has no investment credit carryovers from previous years.

8. Other than the proposed investment, investments in plant and equipment are such that the allowable regular investment credit is not exceeded in the year of acquisition.

Depreciation Methods Other than ACRS

7.1 Current federal tax law permits the use of certain depreciation methods (straight line, declining balance, and sum of the years digits) if the depreciable property was placed in service before 1981 or if the property does not qualify for ACRS. In addition, any other method is permitted as long as it is used consistently and, at the end of each tax year, the cumulative depreciation deducted over the years does not exceed the depreciation that would have been allowed had the declining balance method been used.

Consider a depreciable asset with a first cost of $10,000, an estimated useful life of six years, and an expected salvage value of $1,800. The property does not qualify for ACRS, but it could be depreciated using the double declining balance method.

Determine the maximum allowable accumulated depreciation (a) through three years, (b) through four years, (c) through five years, and (d) through six years.
(*Answers:* (a) $7,036; (b) $8,024; (c) $8,200; (d) $8,200)

7.2 Consider a depreciable asset, say an industrial robot, having the following characteristics:

Initial cost	$50,000
Depreciable life	8 years
Estimated salvage value after 8 years	$14,000

Find the allowable depreciation expenses and book values for each of the eight years using (a) the straight line method, (b) the 125% declining balance method, (c) the sum of the years digits method, and (d) the sinking fund method using an 8 percent sinking fund.

7.3 The Comma Corporation is planning the purchase of a high-volume, overhead bulk conveyor. If purchased, it will have an initial cost of $100,000, an estimated useful life of ten years, and an expected salvage value of $20,000. During the ten-year period, it is predicted that the conveyor will handle a total of 400,000 tons of bulk raw materials.

Year	Tons handled	Year	Tons handled
1	20,000	6	60,000
2	30,000	7	50,000
3	40,000	8	40,000
4	50,000	9	30,000
5	60,000	10	20,000

a. With the above projections, determine the expected depreciation charges for each of the ten years if the *units of production* method is used.

b. What is the depreciation charge in the third year if only 35,000 tons are actually handled?
(*Answers:* (a) $4,000, $6,000, . . . , $4,000; (b) $7,101.50)

7.4 Consider a special-purpose punch press that costs $100,000 and has an expected salvage value of $16,000 at the end of six years. This punch press is expected to produce a total of 600,000 units over its six-year useful life, or an average of 100,000 units per year. The actual expected production schedule, however, is

Year	Production
1	80,000
2	100,000
3	120,000
4	120,000
5	100,000
6	80,000

Determine the annual depreciation expenses and end-of-year book values using each of the following methods:
a. Straight line
b. Sum of the years digits
c. 150% declining balance
d. 8% sinking fund

7.5 The XYZ Corporation is considering the purchase of a computer currently owned by the ABC Company. ABC paid $200,000 for this computer and depreciated it by the straight line method using a ten-year estimated useful life and an estimated salvage value of $80,000. The computer was purchased by ABC four years ago. If the sale takes place, XYZ will pay $100,000 for the computer; it will be depreciated by the 150% declining balance method over a five-year remaining useful life, assuming a 20 percent expected salvage value. Assuming that the sale will take place, answer the following questions:
a. What will be the gain (or loss) on disposal to ABC?
b. What will be the depreciation expense in XYZ's *first* year of ownership?
c. What will be the book value of the computer after five years of ownership by XYZ?
d. If the XYZ Company used a rate other than 150%/N for declining balance, it could depreciate the computer down to the expected salvage value after five years. What is this rate?
(*Answers:* (a) $52,000; (b) $30,000; (c) $20,000; (d) 0.27522)

7.6 A certain property was purchased prior to 1981 at a cost of $160,000. The expected service life at the time of acquisition was fifteen years, and the expected salvage value after fifteen years was estimated to be $40,000. Find the allowable depreciation expenses and book values for each of the fifteen years using (a) the straight line method, (b) the double declining balance method, and (c) the sum of the years digits method.

Depreciation of Recovery Property by ACRS
7.7 Consider a depreciable asset purchased for $100,000 at the start of the fourth month of 1984. It is expected that the asset will be retained for exactly four years and then be sold for an estimated $10,000.
a. Determine the allowable depreciation if the asset is depreciated as a 3-year ACRS property.

b. Assume that the tax rate is 40 percent and is constant from 1984 to 1988. The cash flows for taxes for this firm occur at the end of each tax year. If the after-tax cost of capital (discount rate) is 12 percent, determine the present worth of the cash flows from tax savings resulting from ACRS deductions. Assume that the start of the study period is April 1, 1984.

(*Answers:* (a) $25,000, $38,000, $37,000, $0, $0;
(b) $32,488 as of April 1, 1984)

7.8 A firm purchases a 5-year recovery property on September 1, 1984 at a cost of $20,000. The firm expects to keep this property for approximately four years (until September 1, 1988) and then sell it for an anticipated $5,000. The firm's tax year is the same as the calendar year.
a. Determine the ACRS depreciation during each year of ownership.
b. Determine the book value at the time of disposal (September 1, 1988).
c. If the firm uses the *alternate ACRS* method with a twenty-five-year recovery period, determine the depreciation during each year of ownership.

7.9 Consider a certain Section 1250 class property to be depreciated under ACRS as 15-year real property. The initial cost is $10,000, the expected salvage value on disposal is $25,000, and the estimated service life is ten years. The taxpayer's incremental income tax rate is 46 percent; the after-tax minimum attractive rate of return is 12 percent.
a. Determine the year-by-year ACRS deductions over the ten-year service life.
b. Determine the present worth of income tax savings resulting from these ACRS deductions. Assume that present worth is measured at the first day of the tax year.

(*Answers:* (a) $1,200, . . . , $500; (b) $2,121, excluding *PW* of tax on gain on disposal)

7.10 Assume that the 15-year ACRS recovery property in Problem 7.9 is to be depreciated by the *alternate* ACRS method.
a. Determine the present worth of income tax savings resulting from ACRS deductions if a 15-year recovery period is used.
b. Determine the present worth of income tax savings if a 45-year recovery period is used.

7.11 Refer to the 15-year ACRS recovery property described in Problem 7.9. All assumptions are the same as before, with one important exception: The firm is expected to show no taxable income during the first five years of the service life of the property; thus the incremental income tax rate will be zero. The firm will be profitable during the second five years, and the expected incremental income tax rate will be 46 percent. Determine the present worth of income tax savings over the ten-year service life, assuming

a. Regular ACRS deductions
b. Alternate ACRS deductions with a 15-year recovery period.
(*Answers:* (a) — $550; (b) $627, excluding *PW* of tax on gain on disposal

Depletion

7.12 The Gusher Corporation is planning to purchase an oil property for $5,000,000. Initial engineering estimates indicate that the property is capable of producing 1,000,000 barrels. It is estimated that deductible operating expenses, other than depletion, will be $24.50 per barrel. The percentage depletion allowance for this small producer during the first year of operation is 20 percent. If Gusher sells 100,000 barrels during the first year of operation at $30 per barrel, find the maximum allowable depletion expense.

7.13 The Dunn Mining Company owns a small mine that produces a certain metallic mineral. The mine, now ending its third year of operation, was originally purchased for $200,000. A total of $20,000 has been charged as depletion expense for the first two years. At the beginning of the third year it was estimated that 8,000 tons of ore were still available in the mine. During the third year, 2,000 tons were sold at an average price of $155 per ton, and $230,000 in deductible operating expenses (other than depletion) were incurred. Determine:
a. the depletion deduction
b. taxable income for the third year of operation
(*Answers:* (a) $45,000; (b) $35,000)

Incremental Tax Rates

7.14 A corporation is considering the purchase of a certain asset, which, if acquired, will increase the corporation's taxable income during the first year of ownership from $40,000 to $70,000.
a. Determine the increase in federal income taxes attributable to this proposed investment during this tax year.
b. If the asset is acquired, determine the average income tax *rate* paid by the corporation during the year.
c. Suppose that the corporation's incremental *federal* income tax rate is 0.30 and the incremental *state* income tax rate is 0.10. What is the *combined* incremental income tax rate?

7.15 The Fortune Corporation is considering the acquisition of a certain asset. The company pays corporate income taxes both to the state in which it is located and to the federal government. Without this asset, the company's taxable income as reported to the state tax collector would be $40,000 next year; with the asset, taxable income reported to the state would be increased to $60,000. Assume that the state's corporate income tax rate is 11 percent for all taxable income.

a. If the asset is not acquired, determine the amount of next year's combined state and federal income taxes.
b. Determine next year's combined state and federal income taxes if the asset is acquired.
c. What is the total combined incremental income tax rate appropriate to this analysis?
(*Answers:* (a) $10,058; (b) $15,870; (c) 29.1 percent)

Investment Credit

7.16 Consider a 5-year property depreciated under ACRS. The initial cost is $10,000, it has an expected six-year physical life, and the firm expects to sell it after thirty-nine months for $3,000. Assume that the asset is purchased at the midpoint of its tax year, that is, on the first day of the seventh month of the tax year. Assume also that the property fully qualifies for the investment tax credit and that the 10 percent rate will be used.
a. Determine the regular tax credit in the year of acquisition.
b. Determine the depreciation expense in each year of ownership. (*Hint:* The property will be owned during part or all of four tax years.)
c. Determine the tax credit effect, if any, in the year of disposal.

7.17 A small company plans to purchase the following assets during 1986:

Asset	Recovery property	Initial cost
Automobile (used)	3-year	$13,000
Truck (new)	5-year	19,000
Office machine (new)	3-year	12,000
Plant equipment (new)	10-year	25,000

The company expects to claim a Section 179 expense deduction ($10,000) in 1986.
a. Determine the limit on the investment credit if the company's expected tax liability next year is $28,000.
b. If the assets are purchased, determine the regular investment credit in the year of acquisition. Assume that the regular 10 percent credit will be used. (*Hint:* Recall that the cost basis for computing the investment credit must be reduced by the Section 179 expense deduction.)
(*Answers:* (a) $27,550; (b) $4,900)

7.18 In 1980 a contractor purchased an "end-loader" for use in the construction of a certain project. This asset was purchased for $20,000 and was depreciated by the double declining balance (DDB) method over five years. However, the project was completed after two years, at which time the contractor sold the equipment for $10,000. Costs associated with the equipment—property taxes, maintenance, and insurance—were $2,000 per year, payable at the start of each year. The contractor's firm is incorporated. His incremental corporate income tax rates were 40 percent for federal taxes and 10 percent for state taxes. The 10 percent investment credit was relevant.

   a. What is the appropriate combined incremental tax rate for this investment?

   b. What was the dollar amount of the investment credit in the year of purchase?

   c. If the contractor's minimum attractive rate of return is 10 percent after taxes, determine the after-tax equivalent uniform annual cost of this investment over the two years of ownership.

7.19 Section 7.12 evaluated the effect of ACRS depreciation and the 10 percent regular investment credit on 5-year recovery property. (See Case K.) Recall that the initial cost of this property is $10,000, the expected salvage value is zero, the service life is five years, pretax savings are $3,000 per year, and the tax rate is 0.50. Find the after-tax rate of return for this project, assuming that the taxpayer elects to use the optional 8 percent rate for computing the investment credit. Compare your answer with the *IRR* of 10.46 percent obtained using the 10 percent rate for the investment credit. (*Answer:* 10.66%)

7.20 Recall Case J in Section 7.12. The recovery property is depreciated using the alternate ACRS method, and the investment credit is ignored.

   a. Rework Case J assuming the regular 10 percent rate for the investment credit. Find the after-tax (internal) rate of return.

   b. Rework Case J assuming the optional 8 percent rate for the investment credit. Find the after-tax *IRR*.

After-Tax Economy Studies

7.21 A manufacturing firm is considering the purchase of an asset having a first cost of $100,000, an eight-year life, and an estimated salvage value of $20,000 at the end of eight years. If purchased, the asset will result in a pretax savings of $30,000 during each year over the eight-year life. Property taxes must be paid at the end of each year at the rate of 2 percent of the book value of the asset at the start of that year. Insurance premiums of $1,000 must be paid at the start of each year.

    If purchased, the asset will be depreciated by the SYD method using a six-year life for depreciation purposes, assuming $20,000 salvage at the end of six years. (The depreciable life need not, of course, be the same as the expected actual life.) The firm's pretax minimum attractive rate of return is 20 percent; the after-tax *MARR* is 10 percent. The 10 percent investment credit is relevant. The firm's incremental income tax rate is 40 percent.

   a. Find the pretax (before income taxes) rate of return for this proposed investment.

b. Find the after-tax cash flows for each year.
(*Answers:* (a) 23.3%; (b) – $93,933, . . . , $37,760)

7.22 Solve Problem 7.21 assuming that the asset is to be depreciated under ACRS as a 10-year recovery property.

7.23 The White Trucking Company is a large, publicly held common carrier having an incremental tax rate of 40 percent. White is planning to purchase new truck-washing equipment for its Los Angeles maintenance base. This new equipment costs $150,000, has an estimated life of ten years, and is expected to have a terminal salvage value of $40,000.

It is expected that the new equipment will cause a reduction in annual operating expenditures, other than taxes, of $30,000 at the end of each year for the ten-year period. If the equipment is to be depreciated under ACRS as a 10-year recovery property, determine the present worth using an after-tax discount rate of 15 percent.
(*Answer:* – $22,798)

7.24 Missile and Space Electronics, Inc. has just received a four-year contract to manufacture a small, high-frequency pulse generator for the Air Force. Their manufacturing expense estimates for bidding are shown below as Alternative I. Since submitting the bid, the company's engineering division has developed what is possibly a better proposal which uses extensive mechanization, shown below as Alternative II. Which alternative is preferable? The minimum after-tax rate of return is 15 percent.

For this project, assume negligible salvage values. Capitalize the machines and tooling as one group, using straight line depreciation over the four-year project life. Assume that all other items may be expensed in the year in which they are purchased. The incremental income tax rate is 50 percent. Other appropriate input data are

	Alternative I	Alternative II
Initial product development	$ 50,000	$ 50,000
Preproduction engineering	90,000	122,000
Machine tools and tooling	12,000	40,000
Annual labor cost	35,000	15,000
Annual operating expenses for this contract (other than labor)	20,000	20,000
Annual sales revenue from this contract	115,000	115,000

7.25 Solve Problem 7.24 assuming that machines and tooling are depreciated as 3-year recovery property under ACRS.
(*Answer: PW*(I) = $8,137; *PW*(II) = – $46,842)

7.26 The General Corporation is considering the purchase of an industrial robot for a plant operation currently done by human labor. The firm's minimum attractive rate of return is 20 percent before income taxes and 12 percent after income taxes. The firm's effective incremental income tax rate is 60 percent.

The initial cost of the robot is $150,000. If purchased, the robot will be depreciated under ACRS as a 5-year recovery property. However, the operation for which this robot will be used will last for seven years, at the end of which time the firm expects to sell the robot for $25,000. In other words, even though the robot will be depreciated over five years, it will actually be used for seven years.

Property taxes for the robot, computed at 2 percent of the initial cost, are paid at the end of each year. Operating costs, assumed to occur uniformly and continuously throughout each year, are estimated to be $5,000 annually.

a. Prepare a table of after-tax cash flows for the seven-year study period. Include the 10 percent investment tax credit as appropriate.

b. Determine the present worth before income taxes.

c. Determine the equivalent uniform annual cost after income taxes.

d. Let $\bar{x}$ = current cost before income taxes of this operation, expressed as a cash flow occurring continuously and uniformly during each year. (This would be an appropriate assumption in the case of human labor paid, say, biweekly.) At what pretax value of $\bar{x}$ would the proposed robot be economically attractive?

7.27 Three years ago Mr. Master bought a drill press for his small manufacturing plant. He paid $30,000 for the press. For tax purposes, he depreciated the press by the double declining balance method over ten years. The salvage value for depreciation purposes was $1,000. His incremental income tax rate is currently 26 percent for his federal return and 10 percent for his state return. He now has a chance to sell the press for $20,000. If he sells, what will be his after-tax proceeds from the sale?
(*Answer:* $18,450)

7.28 Tammi Trojan, a recent engineering school graduate, is considering the purchase of a minicomputer for her wholly owned consulting firm, TTA. Tammi bought all her office furniture and other necessary equipment previously, thus the minicomputer will be her only business purchase during the tax year. TTA is *not* incorporated; Tammi is not married. Her combined incremental income tax rate is 0.40. If purchased, the minicomputer will be depreciated under ACRS as a 5-year recovery property. The initial cost (cost basis) of the minicomputer is $5,000. If Tammi purchases the minicomputer, she plans to keep it for four years and then sell it for an estimated $2,000. Determine the after-tax equivalent uniform annual cost for this prospective purchase if her after-tax MARR is 8 percent.

7.29 A manufacturing firm is considering the purchase of a certain machine that costs $80,000. It will be depreciated over a five-year life to an estimated $20,000 salvage value using the SYD method. If the machine is purchased, the firm will keep it for six years, however, and will probably be able to receive $30,000 for it if it is sold at the end of six years. That is, although the machine will be depreciated over five years, it will be kept for six years.

Annual maintenance costs are expected to be $5,000 the first year and will increase by $1,000 per year over the machine's six-year life. Annual operating costs are expected to remain constant at $20,000 per year over the six-year life. Assume that both of these classes of expenses occur at end of year.

Property taxes and insurance for this machine are estimated to be $1,000 in the first year and will decrease by $100 per year. These costs are expected to occur at the start of years 1 through 6. The firm's incremental income tax rate is 60 percent.

If the firm's *MARR* is 10 percent after income taxes, find the equivalent uniform annual cost for this proposed investment.

(*Answer:* $19,413)

7.30 Solve Problem 7.29 assuming that the machine will be depreciated as a 5-year recovery property. The alternate ACRS method will be used with a 5-year recovery period. The property fully qualifies for the 10 percent regular investment credit.

7.31 The White Trucking Company is considering the installation of an electronically controlled drag line in its Chicago terminal. The initial investment will be $300,000, but it will cause a reduction in annual operating disbursements of $50,000 a year for twelve years (the expected life of the equipment).

The investment will be depreciated for income tax purposes by the straight line method using a twenty-year life and zero salvage value. However, the equipment will be useless to the company at the end of twelve years and will be disposed of at a zero net salvage value at that time. Determine the prospective after-tax rate of return on this investment if the incremental tax rate is 40 percent.

(*Answer:* 7.4 percent)

7.32 Solve Problem 7.31 assuming that the drag line will be depreciated as fifteen-year recovery property under ACRS.

7.33 The Brown Company is considering the purchase of some special equipment that costs $60,000. If the equipment is purchased, it will cause an expected excess of receipts over disbursements of $16,000 a year for five years; at the end of that time the equipment will be worthless. It is expected that the equipment will be purchased with a down payment of $10,000, and the remainder will be paid off at the rate of $10,000 per year plus 6 percent on the unpaid balance. Determine the after-tax rate of return on the original investment

($10,000) assuming a 50 percent effective tax rate and sum of the years digits depreciation based on a five-year life and zero salvage value.
(*Answer:* 27.5 percent)

7.34 Solve Problem 7.33 assuming that the equipment will be depreciated as 5-year recovery property using the alternate ACRS method with a 5-year recovery period.

7.35 The Black Company is considering a proposal to manufacture a new product. The planning department has been asked to make an economic analysis assuming that production will begin January 3, 198X (zero date) and will terminate after twelve years. An incremental tax rate of 50 percent will be used in the study.

The original cost of machinery and equipment is $90,000. An estimated life of twelve years with a $12,000 salvage value will be used for tax purposes. The sum of the years digits depreciation method will be used. In addition, $30,000 of working capital will be required at zero date. It is assumed that the working capital will be fully recovered at the end of twelve years. Estimated receipts from sale of the product are $60,000 in year 1, $90,000 in year 2, and $120,000 a year from years 3 through 12. Estimated operating disbursements are $48,000 in year 1, $65,000 in year 2, and $80,000 a year from years 3 through 12.

Determine the prospective rate of return after taxes that is indicated by the preceding estimates.
(*Answer:* 15.06 percent)

7.36 Solve Problem 7.35 assuming that the machinery and equipment will be depreciated as 10-year recovery property under ACRS. Assume also that the investment credit is fully applicable but that Black will elect the alternate 8 percent rate for determining the credit.

7.37 A firm is considering the purchase of drilling equipment having an initial cost of $100,000 and an estimated salvage value of $16,000 if sold at the end of six years. Assume that this equipment will actually be kept for a ten-year period and will be sold for $20,000 at the end of that time. However, it will be depreciated over six years by the sum of the years digits method assuming 16 percent salvage value. If the firm's incremental tax rate is 60 percent and the after-tax MARR is 10 percent, find the after-tax present worth of costs.
(*Answer:* $53,745)

7.38 Solve Problem 7.37 assuming that the drilling equipment will be depreciated as 5-year recovery property under ACRS. Assume that the 10 percent investment credit is fully applicable.

7.39 An industrial engineer is developing a facilities layout for a chemical processing plant. Under his "labor intensive" plan, ten people will be required to operate the plant. However, he can reduce the number of operators to only four under a "capital intensive" plan requiring the expenditure of $800,000 for additional instrumentation and controls. The plant is expected to be in operation for five years, at the end of which time the additional instrumentation and controls would add $300,000 to the selling price.

The plant manager feels that the additional expenditure is probably warranted. Since the cost per man-year is $25,000, a savings of thirty man-years (6 people per year × 5 years) results in a savings of $750,000, which more than offsets the $500,000 net cost of additional instrumentation and controls ($800,000 − $300,000). In any case, she asks the industrial engineer to complete an after-tax economy study.

The corporation has a taxable income in excess of $10 million annually. Capital investments range from $2 to $3 million per year. It is likely that these figures will continue in the foreseeable future.

If purchased, the new instrumentation will be depreciated by the double declining balance (DDB) method over the first four years, with a switch to straight line depreciation over the next six years. The salvage value for depreciation purposes is $100,000 at the end of ten years. The firm's incremental income tax rate is 0.60.

Determine the after-tax rate of return on the proposed investment.

(*Answer:* 3.5 percent)

7.40 Solve Problem 7.39 assuming that the instrumentation equipment will be depreciated as 5-year recovery property under ACRS. Assume that the 10 percent investment credit is fully applicable.

7.41 A lumber mill produces 1,000 MBM (1 MBM = 1,000 board feet) of finished lumber per year at a net positive cash flow of $30,000 before depreciation charges and income tax. The nearby source of saw logs will be exhausted this year, however, so hauling logs from a more distant source will cost an additional $10 per MBM of logs. An MBM of logs will produce two-thirds of an MBM of finished lumber.

The mill was installed ten years ago for $120,000 on land that cost $10,000. The mill has been depreciated on a twenty-year, sum of the years digits basis with salvage value originally estimated to be $15,000. The mill owner has received an offer of $40,000 for the mill, which includes its removal from the land, and a separate offer of $30,000 for the land if the mill is removed.

The mill owner is in the 40 percent tax bracket. Assume that the original estimates for mill life and salvage value remain unchanged and that the current offer for the land will be used as its estimated value ten years from now. In a sense, the owner is investing anew in his mill if he forgoes

the opportunity represented by the offer. What is his prospective after-tax rate of return if he refuses to sell?
(*Answer:* 14.3 percent)

Retirement/Replacement*

7.42 An automatic lathe was purchased two years ago for $20,000. It has been depreciated using the straight line method over ten years using an estimated salvage value of $4,000 at the end of the tenth year.

   The machine can now be sold for $15,000. If it is kept for another eight years, as originally planned, it appears that the sale price at disposal will be $5,000. The company has an effective income tax rate of 60 percent. Assume that long-term gains (and losses) of this depreciable asset are taxed at the ordinary income rate. The firm's minimum attractive rate of return is 8 percent after taxes.

   Determine the equivalent uniform annual cost after taxes if the machine is kept for eight more years.

7.43 The Blue Company owns a machine that cost $26,000 ten years ago. It is being depreciated by the straight line method over twelve years to an estimated $2,000 salvage value. A new machine is available that costs $11,000 and, if purchased, will save $2,000 annually. This new machine will be depreciated by the sum of the years digits method over four years to an estimated $1,000 salvage value.

   If the new machine is purchased, the old machine will be sold for $8,000. If the old machine is retained, it will be scrapped in five years for zero salvage value. Assume that the estimated salvage value of the new machine will be $1,000 at the end of five years, that the minimum attractive rate of return after taxes is 12 percent, and that the incremental income tax rate is 50 percent. Should the existing machine be replaced with the challenger described above? Why or why not?
(*Answer:* Replace the defender)

7.44 Refer to Problem 7.43. Repeat the analysis assuming that the challenger is to be depreciated as a 3-year recovery property. All other assumptions are the same.

7.45 The manager of the western division of the Brown Manufacturing Company has received a request from the superintendent of the Los Angeles plant for a new heavy-duty generator to replace existing equipment. The manager has forwarded the request to corporate headquarters for approval. The existing generator was purchased five years ago for $5,000. It was depreciated by the straight line method over five years to zero salvage value.

   Although its present book value is zero, it is believed that it may remain in operation for another eight years before

---

*In these problems, the procedure is the same as outlined in Chapter 4, except that *after-tax* cash flows form the basis of the analysis.

it must be scrapped. The existing generator is specialized equipment, but fortunately there is a buyer who is willing to pay $600 for the equipment. This will probably be the last chance to sell it at any price.

The proposed equipment will cost $6,000 and has an expected useful life of ten years. If purchased, the sum of the years digits depreciation method will be used with a five-year life and zero salvage value. It is expected that the actual market value of the equipment at any point in time will be the same as its book value at that time. The estimated annual operating costs for the new equipment are $600; operating costs for the existing equipment are expected to be $1,800 per year.

The company plans to sell the plant and all its equipment in five years. The sale price is independent of the present decision concerning the generator.

Assuming an after-tax minimum attractive rate of return of 10 percent and an incremental tax rate of 50 percent, determine whether the new equipment should be purchased. Explain why or why not.

(*Answer:* Keep the defender)

7.46 Refer to Problem 7.45. Repeat the analysis assuming that the challenger will be depreciated as a 10-year recovery property using the alternate ACRS method with a 10-year recovery period.

Lease/Purchase Decisions

7.47 The robot in Problem 7.26 may be leased from the manufacturer under the following conditions:

Lease payments are $8,000 quarterly, payable at the start of each quarter over the seven-year period of the lease; lease payments include sales tax.

Property taxes are paid by the manufacturer (the lessor); operating costs, as before, are paid by the General Corporation (the lessee).

Find the after-tax equivalent uniform annual cost (per year) of leasing the robot.

(*Answer:* $17,460)

7.48 South Coast, Inc. is a medium-sized manufacturer of avionics systems for the aircraft manufacturing industry. Its after-tax minimum attractive rate of return is 12 percent. The combined federal and state income tax rate paid by South Coast is 0.55. This rate is expected to remain constant in the foreseeable future.

South Coast is considering the acquisition of a robotic arm for the manufacture of one of its products. The robotic arm may be purchased for $80,000 and depreciated by the

SYD method over a depreciable life of ten years to a $25,000 salvage value. The company anticipates that the market value of the robotic arm after ten years of service will be $75,000, which is very close to its original cost.

If purchased, the robotic arm will result in labor savings of $8,000 in the first year of service; these savings will increase by $2,000 per year over each of the subsequent years. Operating costs for the robotic arm will be $10,000 in the first year, increasing by $1,000 per year during each of the subsequent years.

a. Prepare a table of after-tax cash flows over the ten-year study period.

b. Determine the after-tax present worth for this investment.

7.49 A firm is considering the purchase of certain nonrecovery property having an initial cost of $40,000. The property will be kept for five years and then sold for $10,000. The property, if purchased, will be depreciated by the SYD depreciation method over four years, assuming a 25 percent salvage value. Annual costs of operation and maintenance are estimated to be $12,000 in the first year and will increase at the rate of $1,000 per year throughout the five-year life. The firm's incremental income tax rate is 0.60, which accounts for municipal, state, and federal taxes.

Find the present worth of costs over five years of this proposed investment, assuming a 10 percent after-tax discount rate.

(*Answer:* $39,788)

7.50 The equipment described in Problem 7.49 could be purchased under a special plan: a 25 percent ($10,000) down payment with five equal end-of-year payments and interest computed at 12 percent on the unpaid balance. That is, the debt ($30,000) is to be repaid at the end of each of the five years with uniform payments that include 12 percent interest on the unpaid balance. (Interest payments are a tax-deductible business expense.)

Assuming that all other characteristics of this problem (except the purchase plan) are the same as those described in Problem 7.49, determine the after-tax present worth of costs.

7.51 The equipment described in Problem 7.49 could be leased rather than purchased. The lease (rental) charge is $2,000 per month, payable at the start of each month over the five-year period. Of course, the firm cannot depreciate equipment it does not own. The company providing the equipment, the lessor, will assume all costs of operation and maintenance. The only pretax costs to the firm, the *lessee*, will be the rental charges. Assuming a 10 percent after-tax discount rate as before, find the after-tax present worth of this lease proposal. (Find an "exact" solution; do not use the continuous cash flow convention for the monthly cash flows.)

(*Answer:* $41,247)

Computer Program

The following program, written in IBM BASIC-A, is intended for use with the IBM-PC (personal computer). This version has print capability for use with an on-line printer. The IBM BASIC-A language is quite similar to other forms of BASIC (Applesoft BASIC, for example), so relatively few changes are required to use this program with microcomputers other than the IBM-PC.

This program permits the user to select from four methods: (1) straight line, (2) declining balance, (3) sum of the years digits, and (4) sinking fund. As noted in the chapter, these methods may *not* be used when determining depreciation for ACRS recovery property on federal income tax returns, but they are widely used in all other contexts.

The program assumes that the depreciable property is placed in service at the start of the tax year and is removed from service at the end of the tax year. Therefore depreciation in the years of acquisition and disposition need not be prorated over the periods of ownership in those years.

The program does not permit the option of shifting from declining balance to the straight line method at some point during the depreciable life. This is a fairly simple modification, however; it is left for you to do as an exercise.

Programs to compute after-tax cash flows, using either ACRS or non-ACRS depreciation methods, are necessarily complex. One such program, FLEXICON, incorporates over 1,500 lines of programming code. For further information about FLEXICON, please write to Dr. G. A. Fleischer, Department of Industrial and Systems Engineering, University of Southern California, Los Angeles, California 90089-1452.

PROGRAM

```
10   REM   THIS PROGRAM COMPUTES THE DEPRECIATION CHARGES AND
20   REM   BOOK VALUES USING FOUR DIFFERENT DEPRECIATION METHODS.
30   REM   THESE METHODS ARE:
40   REM
50   REM            1. STRAIGHT LINE METHOD
60   REM            2. DECLINING BALANCE METHOD
70   REM            3. SYD (SUM OF YEARS DIGITS) METHOD
80   REM            4. SINKING FUND METHOD
90   REM
100  REM   THE INITIAL COST, EXPECTED SALVAGE VALUE, AND DEPRECIABLE
110  REM   LIFE ARE REQUIRED TO BE INPUT BY THE USER. IN ADDITION, IN
120  REM   METHOD 2 THE DECLINING BALANCE RATE AND IN METHOD 4 THE
130  REM   SINKING FUND RATE ARE NEEDED.
140  REM
150  REM
160  DIM D[30],B[30]
170  DIM A$[3]
187  PRINT "WHAT IS THE INITIAL COST?"
190  INPUT P
200  PRINT "WHAT IS THE EXPECTED SALVAGE VALUE?"
```

```
210   INPUT S
220   PRINT "WHAT IS THE DEPRECIABLE LIFE( YEARS)?"
230   INPUT N
240   PRINT "WOULD YOU LIKE TO SEE THE MENU OF DEPRECIATION METHODS"
250   PRINT "AVAILABLE IN THIS PROGRAM?"
260   INPUT A$
270   IF A$ = "NO" THEN 360
280   IF A$ = "N" THEN 360
290   PRINT " METHODS AVAILABLE:"
300   PRINT
310   PRINT "              1. STRAIGHT LINE"
320   PRINT "              2. DECLINING BALANCE"
330   PRINT "              3. SUM OF THE YEARS DIGITS (SYD)"
340   PRINT "              4. SINKING FUND"
350   PRINT
360   PRINT "WHICH METHOD (1,2, . . 4)?".
370   INPUT K
380   ON K GOTO 420,480,720,830
390   REM
400   REM =========== STRAIGHT LINE ===========
410   REM
420   FOR J = 1 TO N
430   D[J] = (P − S) /N
440   B[J] = P − J*D[J]
450   NEXT J
460   GOTO 960
470   REM
480   REM ========= DECLINING BALANCE ==========
490   REM
500   PRINT "WHAT DECLINING BALANCE RATE DO YOU WANT TO USE?"
510   PRINT "                          (1.25,1.50,2.00, . . . ?)"
520   INPUT B1
530   A = B1/N
540   D[1] = A*P
550   B[1] = P*(1 − A)
560   FOR J = 2 TO N
570   B[J] = P*((1 − A) ∧ J)
580   IF B[J] < = S THEN 650
590   D[J] = A*B[J − 1]
600   NEXT J
610   GOTO 960
620   REM
630   REM CAN NOT DEPRECIATE BELOW SALVAGE VALUE
640   REM
650   J1 = J
652   B(J1) = S
654   D(J1) = B(J1 − 1) − S
660   FOR J2 = J1 + 1 TO N
670   B[J2] = S
680   D[J2] = 0
690   NEXT J2
700   GOTO 960
710   REM
720   REM =========== SYD METHOD ============
```

```
730   REM
740   S1 = N*(N + 1) /2
750   D[1] = N*(P − S) /S1
760   B[1] = P − D[1]
770   FOR J = 2 TO N
780   D[J] = (N − J + 1)*(P − S) /S1
790   B[J] = B[J − 1] − D[J]
800   NEXT J
810   GOTO 960
820   REM
830   REM ========== SINKING FUND METHOD ==========
840   REM
850   PRINT "WHAT SINKING FUND RATE DO YOU WANT TO USE?"
860   INPUT I1
870   I2 = (1 + I1) ∧ N − 1
880   D[1] = (P − S)*I1/I2
890   B[1] = P − D[1]
900   FOR J = 2 TO N
910   I2 = (1 + I1) ∧ N − 1
920   I3 = (1 + I1) ∧ (J − 1)
930   D[J] = (P − S)*I1*I3/I2
940   B[J] = B[J − 1] − D[J]
950   NEXT J
960   ON K GOTO 970,1010,1050,1090
970   LPRINT
980   LPRINT
990   LPRINT "              METHOD 1:STRAIGHT LINE"
1000  GOTO 1120
1010  LPRINT
1020  LPRINT
1030  LPRINT "              METHOD 2:DECLINING BALANCE"
1040  GOTO 1120
1050  LPRINT
1060  LPRINT
1070  LPRINT "              METHOD 3:SYD"
1080  GOTO 1120
1090  LPRINT
1100  LPRINT "              METHOD 4:SINKING FUND"
1120  LPRINT "         YEAR   DEPRECIATION   BOOK VALUE"
1122  LPRINT USING "          0            0        ######";P
1130  FOR J = 1 TO N
1140  LPRINT USING "         ##       ######       ######";J;D(J);B(J)
1150  NEXT J
1160  PRINT "WOULD YOU LIKE TO TRY ANOTHER METHOD?"
1170  INPUT C$
1180  IF C$ = "YES" THEN 240
1182  IF C$ = "yes" THEN 240
1190  IF C$ = "Y" THEN 240
1192  IF C$ = "y" THEN 240
1200  PRINT "DONE!"
1210  END
```

Example

In order to save space, the "dialogue" between the computer and the user is not shown here. Only the printed results are displayed for this sample problem, for which the initial cost is $40,000, the salvage value is $4,000, and the depreciable life is eight years. In method 2, declining balance, the rate used is 2.00. In method 4, sinking fund, the rate used is 0.05.

EXAMPLE

METHOD 1:STRAIGHT LINE

YEAR	DEPRECIATION	BOOK VALUE
0	0	40000
1	4500	35500
2	4500	31000
3	4500	26500
4	4500	22000
5	4500	17500
6	4500	13000
7	4500	8500
8	4500	4000

METHOD 2:DECLINING BALANCE

YEAR	DEPRECIATION	BOOK VALUE
0	0	40000
1	10000	30000
2	7500	22500
3	5625	16875
4	4219	12656
5	3164	9492
6	2373	7119
7	1780	5339
8	1335	4005

METHOD 3:SYD

YEAR	DEPRECIATION	BOOK VALUE
0	0	40000
1	8000	32000
2	7000	25000
3	6000	19000
4	5000	14000
5	4000	10000
6	3000	7000
7	2000	5000
8	1000	4000

METHOD 4:SINKING FUND

YEAR	DEPRECIATION	BOOK VALUE
0	0	40000
1	3770	36230
2	3958	32272
3	4156	28115
4	4364	23751
5	4582	19168
6	4812	14357
7	5052	9305
8	5305	4000

DONE!

CHAPTER EIGHT
RISK AND UNCERTAINTY

s discussed in Chapter 1, a fundamental principle of economic analysis is that only differences between alternatives are relevant. An immediate corollary is the "sunk cost" principle: Sunk costs are irrelevant. Put somewhat differently, consequences that arise from competing alternatives are significant only to the extent that they differ. If a work force of, say, ten people will be unaffected by the selection of a particular piece of equipment, then the labor cost of these individuals is of no interest in the equipment-selection analysis. And because all past events are common to all future events, the money spent or received in the past (sunk costs) is irrelevant to the current choice among alternatives.

With these principles in mind, it is imperative that the analyst recognize the uncertainty inherent in *all* economy studies. The past is irrelevant, except when it helps predict the future. Only the future is relevant, and *the future is inherently uncertain.*

At this point it will be useful to distinguish between *risk* and *uncertainty,* two terms widely used when dealing with the noncertain future. **Risk** refers to situations in which a probability distribution underlies future events and the characteristics of this distribution are known or can be estimated. For example, although the prior outcome of the flip of a fair coin is unknown, it is reasonable to assume that the likelihood of heads occurring is equal to the probability of tails occurring. These relative likelihoods, or **probabilities,** can be written

Prob [heads] = *Prob* [tails] = 0.5

Another problem that involves risk occurs when the probability of a component failure is known or can be assumed. (Decisions under risk are explored in greater detail in Section 8.3.)

Decisions involving **uncertainty** occur when nothing is known or can be assumed about the relative likelihood, or probability, of future events. Uncertainty situations may arise when the relative attractiveness of various alternatives is a function of the outcome of pending labor negotiations or local elections or when permit applications are being considered by a government planning commission.

A wide spectrum of analytical procedures is available for the formal consideration of risk/uncertainty in analyses. Although space does not permit a comprehensive review of all these procedures, a representative sample of the more widely used procedures is presented in the following sections.

Sensitivity analysis is the process whereby one or more system input variables are changed and corresponding changes in the system output, or figure of merit, are observed. If a decision is changed as a certain input is varied over a reasonable range of possible values, the decision is said to be *sensitive* to that input; otherwise it is *insensitive*.

The term **break-even analysis** is often used to express the same concept. Here, the value of the input variable at which the decision is changed is determined. If the **break-even point** lies within the range of expected values, the decision is said to be sensitive to that input. Thus sensitivity and the break-even point are directly related.

A Numerical Example

By spending $25,000 on automatic controls that have a thirty-year life, a firm can effect annual operating savings of $3,000 during each year over thirty years. If the pretax minimum attractive rate of return is 10 percent, the pretax present worth (PW) of this proposed investment may be easily computed:

$$PW = -\$25,000 + \$3,000(P/\overline{A}, 10\%, 30)$$

$$= -\$25,000 + \$3,000(9.891)$$

$$= \$4,673$$

Since all future consequences are known with certainty, this particular investment proposal appears economically attractive.

Sensitivity to One Parameter

Let's suppose that there is some reason to question the validity of the assumption concerning annual operating savings. Additional investigation, for example, may show that the "certainty estimate" of $3,000 per year is questionable; it now appears that this parameter value could occur anywhere over the range of $2,400 to

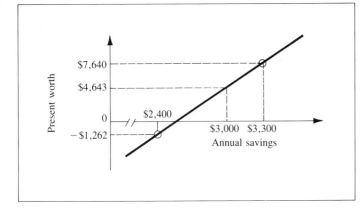

Figure 8.1

$3,000. With this new information we could determine, either algebraically or graphically, as in Figure 8.1, the resulting range of values for the present worth. That is,

$$\text{Min } PW = -\$25,000 + \$2,400(9.891) = -\$1,262$$

$$\text{Max } PW = -\$25,000 + \$3,300(9.891) = \$7,640$$

The break-even point can also be determined. There are two alternatives represented in this problem: (1) accept the proposal or (2) do nothing and invest the funds elsewhere. The break-even point occurs where a change from one alternative to the other is indicated. That is, the break-even point is that value of the parameter in question (here, the annual savings) such that the net present value is zero. Again, this value (x) may be determined either graphically, as in Figure 8.2, or algebraically:

$$-\$25,000 + 9.891x = 0$$

$$x = \$25,000/9.891 = \$2,528$$

In this example, the outcome *is* sensitive to the value assumed for annual savings inasmuch as the break-even point ($2,528) lies within the expected range ($2,400–$3,300). This information, of course, is of considerable interest to the decision maker. If sensitivity does exist, then the decision maker may choose to forgo the investment opportunity, despite the fact that the certainty estimate yields a positive *PW*. A number of other strategies are also available. For example, he or she may choose to invest additional staff resources in an attempt to resolve the uncertainty surrounding the initial estimates or choose to delay the decision (''buy time'') in the expectation that uncertainties will be resolved, at least in part, in the future. If the analysis shows *no* sensitivity, we can conclude that the outcome, or choice among alternatives, is clearly indi-

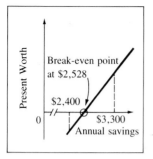

Break-even point
at $2,528

$2,400

0

$3,300
Annual savings

Figure 8.2

cated, regardless of what future values might actually occur for the parameter in question.

The graphic portrayal of sensitivity analysis is usually as illustrated in Figure 8.2: The figure of merit is shown on the vertical axis, the ordinate, and the input variable to be examined is shown on the horizontal axis, the abscissa. An alternative approach is to indicate the relative error of the estimate along the horizontal axis. From the example,

$$PW = -\$25,000 + \$3,000(1 + p)(P/\bar{A}, 10\%, 30)$$
$$= -\$25,000 + \$29,673(1 + p)$$

where p is the percent deviation in our estimate of annual operating savings. Present worth as a function of p is plotted in Figure 8.3. As shown, the break-even point is -15.7 percent, which is, of course, consistent with our previous result:

$$1 - \frac{\$3,000 - \$2,528}{\$3,000} = 0.843$$

Figure 8.3. Present Worth as a Function of Percent Deviation

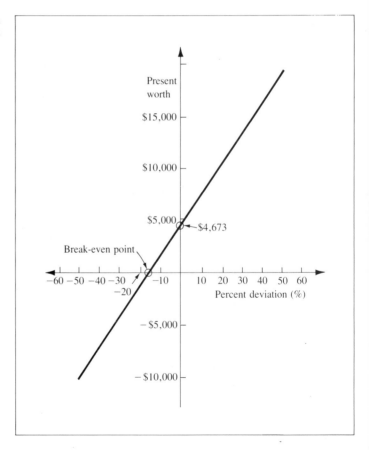

EIGHT   RISK AND UNCERTAINTY

That is, the present worth will be positive if the actual annual operating savings are at least 84.3 percent of the initial $3,000 estimate.

Nonlinear Objective Functions

The equation for present worth (*PW*) for the previous example takes the form

$$PW = -\$25,000 + 9.891x$$

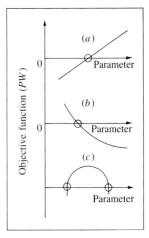

Figure 8.4

where *x* represents the value of annual savings. This equation, known as the **objective function**, is *linear* in *x*. That is, when *PW* is plotted as a function of the parameter *x*, the result is a straight line, as in Figure 8.4*a*. But a straight line is not always the result; the objective function may also be nonlinear. If the *PW* in the previous example had been plotted as a function of the discount rate, the nonlinear graph would appear as in Figure 8.4*b*. In some cases, the nonlinearity of the objective function may result in more than one break-even point between a pair of alternatives, as in Figure 8.4*c*.

More than Two Alternatives

The graphic representation of more than two alternatives may be easily demonstrated by simply depicting a separate objective function for each alternative proposal. In Figure 8.5, for example, the objective functions for three alternatives are displayed. Note that the horizontal axis, the abscissa, represents the present worth of the do-nothing alternative, Alternative ∅. The *PW* of the do-nothing alternative is independent of the parameter values; it is zero at all times.

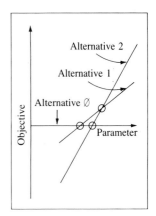

Figure 8.5

The break-even point may be determined either graphically or algebraically. Algebraic solutions are found by equating pairs of objective functions. In Figure 8.5, for example, the break-even point between Alternatives 1 and 2 may be found by solving for that value of the variable in question such that $PW(1) = PW(2)$.

Two Parameters Considered Simultaneously

Now consider the case in which the analyst wishes to test for the sensitivity of the outcome of two parameters, both considered simultaneously. To illustrate using the previous example, suppose that, in addition to the uncertainty about annual savings, there is some question as to which discount rate to use in the analysis. Specifically, the following *range* of values is to be considered:

Parameter	Minimum	Most likely	Maximum
Annual savings (*A*)	$2,400	$3,000	$3,300
Discount rate (*i*)	6%	10%	12%

Figure 8.6

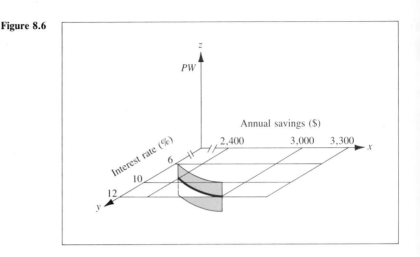

One possible procedure for treating the uncertainty of two parameters simultaneously is to create a three-dimensional graph, or model, as shown in Figure 8.6. *Annual savings* is shown along the *x* axis, the *discount rate* is shown along the *y* axis, and the objective (present worth) is shown along the *z* axis. The set of all possible combinations of outcomes is represented by a surface; the set of break-even points becomes a break-even *line*.

The three-dimensional approach is not particularly useful as an analytical tool because it suffers from two major disadvantages. Preparing a three-dimensional graph is time consuming and complex. But even more important, users of the analysis generally find it difficult, if not impossible, to find meaning in the figure. Thus the principal advantages of the graphic approach, simplicity and ease of understanding, are greatly diminished.

There is, however, an alternative approach, the **family-of-curves approach,** that may be more useful: a family of curves on a two-dimensional graph, as shown in Figure 8.7. One of the variables is held constant at a specific value, say, $i = 10\%$, and the present worth is then plotted as a function of the remaining variable (A). This process is then repeated for other values of i. The values selected for i are usually those that are of greatest interest to the decision maker, namely, the minimum, maximum, and most likely values. The shaded region in Figure 8.7 represents the set of all possible combinations of inputs A and i. Break-even points, as well as the maximum and minimum values for the net present value, can be readily determined from the graph.

Still another approach is to use the **graph of percent deviation,** as shown in Figure 8.3. Since the figure of merit (PW) is plotted as a function of the percent deviation from the most likely value, any number of parameters may be examined simultaneously. The present worth as a function of percent deviation from

Figure 8.7

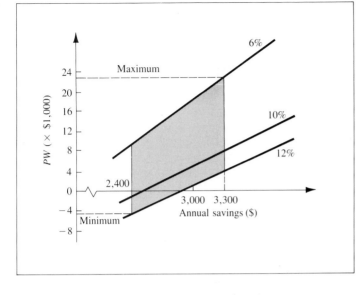

annual savings as well as of interest rate is shown in Figure 8.8. (Ranges of percent deviations are shown as solid lines. Variations outside the ranges are indicated by dashed lines.)

This approach differs in two important ways from the family-of-curves approach. It is advantageous in that it permits the simultaneous plotting of more than two parameters. Its principal disadvantage, however, is that the interactive effects of the parameters are ignored. The family-of-curves approach indicates a region of possible outcomes; the percent deviation approach does not, unless, of course, the minimum and maximum percent deviations are the same for all parameters.

More than Two Parameters

Using the previous example, let's now suppose that, in addition to uncertainty about annual savings and the discount rate, there is uncertainty as to the exact number of periods during which savings can be expected to accrue. Suppose that the following range of values is possible:

Parameter	Minimum	Most likely	Maximum
Annual savings (A)	$2,400	$3,000	$3,300
Discount rate (i)	6%	10%	12%
Project life (N)	25 years	30 years	35 years

Can we develop a four-dimensional graph? Certainly not. How about a three-dimensional graph with a family of *planes*?[1] Hardly

1. If you are a masochist, try this as an exercise. Then try to find someone who can interpret the results!

Figure 8.8. Present Worth as a
Function of Percent Deviation
from Expected Values for Two
Parameters

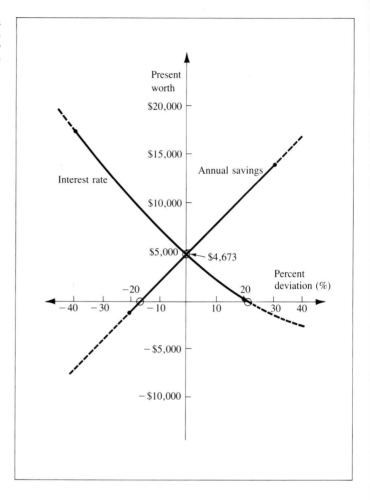

reasonable. Other than the percent deviation graph discussed previously, two approaches remain that, although not entirely satisfactory, may prove useful in certain cases.

The first involves the reduction of the original problem to a *series* of two-dimensional graphs. For this example we might consider three graphs: (1) *PW* as a function of annual savings, assuming $N = 25$ and a family of curves for $i = 6\%$, 10%, and 12%; (2) *PW* as a function of annual savings, assuming $N = 30$ and a family of curves for $i = 6\%$, 10%, and 12%; and (3) *PW* as a function of annual savings, assuming $N = 35$ and a family of curves for $i = 6\%$, 10%, and 12%. This approach suffers from two defects. First, although a series of smaller problems is solved, we are not testing for the sensitivity to *all* parameters *simultaneously*. Second, the number of graphs required grows exponentially as the number of uncertain parameters increases arithmetically:

Number of uncertain parameters	Number of two-dimensional (family-of-curves) graphs required[2]
2	1
3	3
4	9
5	27
$\vdots$	$\vdots$
n	3^{n-2}

A second approach is based on the *a fortiori* ("strength of the argument") **principle.** If it can be shown that a certain course of action is indicated regardless of the input assumptions, then it has been proven, *a fortiori*, that there can be no other possible outcome. To illustrate, the following is a computation of both the minimum and maximum possible values for *PW*, given the ranges for the input assumptions:

$$\text{Min } PW = -\$25,000 + \$2,400(P/\bar{A}, 12\%, 15) \left.\begin{array}{l} \\ \\ \\ \end{array}\right\} \begin{array}{l} A = \$2,400 \\ i = 12\% \\ N = 25 \end{array}$$
$$= -\$25,000 + \$2,400(8.305)$$
$$= -\$5,068$$

$$\text{Max } PW = -\$25,000 + \$3,300(P/\bar{A}, 6\%, 35) \left.\begin{array}{l} \\ \\ \\ \end{array}\right\} \begin{array}{l} A = \$3,300 \\ i = 6\% \\ N = 35 \end{array}$$
$$= -\$25,000 + \$3,300(14.929)$$
$$= \$24,266$$

If both present worths had been negative, we would have proven, *a fortiori*, that the proposal should be rejected on economic grounds. Conversely, if both *PW* values had been positive, an "accept" decision would have been indicated.

Unfortunately, this test of extreme values rarely yields a clear result, and the *a fortiori* argument cannot be used. Nevertheless, analysts would be well advised to try this approach before proceeding further. The calculations can be completed relatively easily, and the few cases for which a clear signal *is* indicated more than justify the time involved.

The Equal Likelihood Assumption
What can be said about instances in which a specific outcome is strictly sensitive, that is, the break-even point does lie within the expected range but is not *very* sensitive? Figure 8.9 illustrates this case. Here, the break-even point lies within the range, albeit very close to the minimum anticipated value. Given the two alternatives implied in Figure 8.9, (1) accept the proposal or (2) do

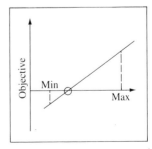

Figure 8.9

2. Assumes that graphs are required for three values: minimum, most likely, and maximum values.

nothing, it would appear that the proposal should be accepted. Why? Because, apparently, the likelihood of obtaining a favorable outcome (for example, positive *PW*) is substantially greater than the likelihood of obtaining an adverse outcome (negative *PW,* for example). This conclusion stems from the **equal likelihood assumption,** the assumption that the relative likelihood that these two outcomes will occur is approximately equal to the relative distances between the break-even point and the minimum and maximum possible values.

Also implicit in this argument is the assumption that specific values for the parameter in question are equally likely through the range of possible values. Let's suppose, for example, that there is supplementary information, as illustrated in Figure 8.10. Note that the future value of the parameter in question (annual saving, in this case) is much more likely to occur at the lower end of the range. If such is the case, then the decision maker may be well advised to reject this particular proposal. It is not slightly sensitive; it is in fact very sensitive.

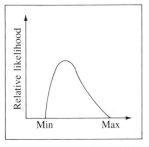

Figure 8.10

In "real world" applications, there is no reason to believe that the relative likelihoods, or probabilities, of future events are distributed symmetrically around some most likely value. Indeed, there are many applications for which highly skewed distributions are quite common, including performance characteristics of certain electrical components, dimensions of manufactured parts after "substandard" units have been rejected, and meteorological data such as rainfall. Thus decision makers and analysts must be cautious about arriving at judgments concerning *degree* of sensitivity.

8.2 FUNDAMENTAL CONCEPTS— PROBABILITY AND EXPECTATION

Sensitivity analysis is a useful procedure when the relative likelihoods (probabilities) of future events are unknown. But if these likelihoods *are* known, an analytical procedure known as **risk analysis** may prove useful.

Two concepts are fundamental to the analysis of decisions under risk: *probability* and *expectation*. If you are unfamiliar with these concepts, this section will be useful. If you are familiar with probability and expectation, move to Section 8.3.

Probability
There are a variety of approaches to the subject of probability, ranging from the most simplistic to the most elegant and esoteric. This discussion is not meant to be exhaustive. Rather, it introduces some fundamental notions of probability so that you can implement risk analysis with reasonable effectiveness. A more

exhaustive treatment can be found in probability textbooks.[3]

PROBABILITY DEFINED. Consider an event, say, rolling a 7 with a pair of dice. The **probability** of the event, E, is the ratio of the number of ways that E can occur divided by the total number of possible outcomes. Or

$$P[E] = \frac{\text{number of ways } E \text{ can occur}}{\text{total number of possible outcomes}} \qquad (8.1)$$

Given a pair of ''fair'' dice, fair in the sense that each face has an equal chance of appearing, there are thirty-six possible outcomes:

1,1	2,1	3,1	4,1	5,1	6,1
1,2	2,2	3,2	4,2	5,2	6,2
1,3	2,3	3,3	4,3	5,3	6,3
1,4	2,4	3,4	4,4	5,4	6,4
1,5	2,5	3,5	4,5	5,5	6,5
1,6	2,6	3,6	4,6	5,6	6,6

Inspecting these outcomes indicates that exactly six of them yield a total of 7 when the two numbers are added. Thus, the probability of rolling a 7, given a pair of fair dice, is exactly 6/36 = 1/6. Note that this is a *relative frequency* definition.

The probabilities of many events can be determined once the underlying physical mechanism is clearly understood. For example, the probability that a 6 will show on the upturned face of a fair die is exactly 1/6; the probability that heads will show, given the flip of a fair coin, is exactly 1/2, or 0.5. (Probabilities are generally written as decimals rather than fractions.) The probability of being dealt a 10 or a facecard from a complete shuffled deck of playing cards is 16/52, or 0.31. Unfortunately, these simple illustrations, treasured for their illustrative value in the classroom, do not reflect the kinds of situations of greatest interest to economic analysts. More relevant questions might be: What is the probability that rainfall in the South Coast Basin will exceed 5 inches next winter? What is the probability that a specific motor will be used for 4,000–4,200 hours three years from now? What is the probability that current negotiations with the union will result in wage increases of 7–8 percent? Clearly, probability estimates concerning future events can be exceedingly difficult to forecast, especially when there is little history from which to extrapolate

---

3. There are a number of excellent textbooks that deal with probability. See William Mendenhall and Richard L. Scheaffer, *Mathematical Statistics with Applications*, (North Scituate, Mass.: Duxbury Press, 1973). For an elementary reference, see Seymour Lipschutz, *Theory and Problems of Probability*, Schaum's Outline Series (New York: McGraw-Hill, 1968).

into the future. Analysts should bear in mind that the quality of risk analysis can be no better than the quality of the underlying probability estimates. The following discussion must be understood in light of this caveat.

PROBABILITY AXIOMS. There are three fundamental probability axioms that analysts will find useful:

> AXIOM 1. *For any event* E, *the probability of that event must be nonnegative.*
>
> $$P[E] \geq 0 \qquad (8.2)$$

> AXIOM 2. *The sum of the probabilities of the set of all possible outcomes (the sample space) is exactly unity:*
>
> $$P[S] = 1 \qquad (8.3)$$
>
> *where* S *is the sum of all possible outcomes in the finite sample space.*

> AXIOM 3. *Given two mutually exclusive events,* E *and* F, *from a finite sample space, the probability that either* E *or* F *will occur is the sum of the probabilities of the two events:*
>
> $$P[E \cup F] = P[E] + P[F] \qquad (8.4)$$

In addition, if two events, E and F, are independent, then the probability of *both* events E *and* F occurring is the product of their respective probabilities. Or

$$P[EF] = P[E]P[F] \qquad (8.5)$$

These axioms can be illustrated by the rolling of a pair of dice. It is evident that

> AXIOM 1 $P[E_i] > 0$ for $E_i = 2, 3, \ldots, 12$
> $\qquad\qquad\quad = 0$ for all other values of E_i
>
> AXIOM 2 $P[S] = P[2] + P[3] + \cdots + P[12] = 1$

To illustrate Axiom 3, consider any subset of outcomes, say rolling a 7 or an 11. These two events are mutually exclusive for any single roll of the dice. Thus the probability of rolling either a 7 or an 11 on any one roll is

$$P[7 \text{ or } 11] = P[7] + P[11] = 6/36 + 2/36 = 2/9$$

Similarly, the probability of rolling a 2, a 3, or a 12 is

$$
\begin{aligned}
P[2, 3, 12] &= P[2] + P[3] + P[12] \\
&= (1 + 2 + 1)/36 = 1/9
\end{aligned}
$$

A variety of probability calculations can be made, of course, using these probability axioms.

PROBABILITY FUNCTIONS. A **random variable** is a function that assigns a value to each outcome in an exhaustive set of all possible outcomes. For example, in a roll of a pair of fair dice, the random variable describing the total number of spots appearing on the upturned faces of the dice can have the values 2, 3, . . . , 12. When a random variable is *discrete,* as in this example, a **probability mass function,** $p(x)$, is used to describe the probability that a random variable will be equal to a particular value. Graphically, the probability mass function appears as in Figure 8.11.

Certain random variables are *continuous* rather than discrete, as are temperature, weight, and distance. For continuous random variables, a **probability density function,** $f(x)$, is used to relate the probability of an event to a value (more properly, a *range* of values) for the random variable. Three well-known probability density functions are illustrated in Figure 8.12.

Figure 8.11

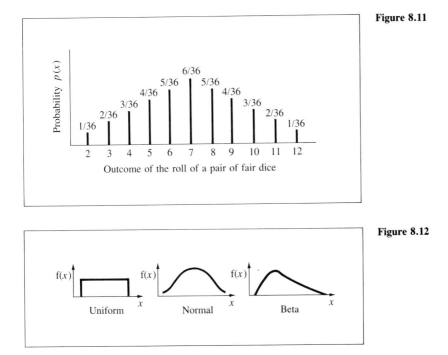

Outcome of the roll of a pair of fair dice

Figure 8.12

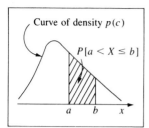

Figure 8.13

Consider a continuous random variable X, and let x represent a specific value of this random variable. The probability of x occurring between points a and b is illustrated in Figure 8.13. Strictly speaking, the probability of x occurring is zero, since X is a continous random variable. For a short interval of length Δx with midpoint $x = c$, the corresponding probability is approximately equal to $p(c)(\Delta x)$, because this is the area of the rectangle of width Δx and height $p(c)$. If Δx approaches 0, then $P[X = c] = 0$.

Expectation

The concept of **mathematical expectation** is useful in analyses involving random variables. The *expected value* of a *discrete* random variable X is given by

$$E[X] = \sum x \cdot p(x) \tag{8.6}$$

where the probability mass function, $p(x)$, is defined for every number x by $p(x) = P[X = x]$. The expected value of a *continuous* random variable X is given by

$$E[X] = \int_{-\infty}^{\infty} x \cdot f(x)dx \tag{8.7}$$

where $f(x)$ is the associated probability density function. The expected value is frequently called the **mean,** or **mean value,** of the distribution and is denoted by the Greek letter μ.

To illustrate, let's compute the expected value of the discrete random variable X, where X represents the sum of the upturned faces of a pair of fair dice when rolled, or $x = 2, 3, \ldots, 12$. (It may be assumed that rolling the dice has the effect of randomizing the variable.) Then,

$$E[X] = \sum x \cdot p(x)$$

$$= \frac{1}{36}(2) + \frac{2}{36}(3) + \cdots + \frac{1}{36}(12)$$

$$= 252/36 = 7$$

In this example, the mean value is the same as the most likely value but this is not always the case.

Another example may be instructive. Suppose that John and Mary are gambling on the flip of a fair coin. They agree to wager $1 on each flip: Heads, John wins $1 from Mary, and tails, Mary wins $1 from John. (See Table 8.1.) John's expected value may be easily computed:

EIGHT   RISK AND UNCERTAINTY

Table 8.1

Outcome	Probability	John wins	Mary wins
Heads	0.5	+1	−1
Tails	0.5	−1	+1

$$E[\text{winnings}] = \frac{1}{2}(\$1) + \frac{1}{2}(-\$1) = 0$$

Put somewhat differently, if John and Mary continue to play this game many times, then the *long-run* average winnings *per play* approaches zero. Note that on no single play is it possible to win exactly zero; John must either win $1 or lose $1 on each play. Nevertheless, the expected value is zero. This example clearly demonstrates that expectation is a mathematical concept and is not directly related to "anticipated" values.

As a final example, consider the numerical example discussed in Section 8.1. Recall that we were considering a certain random variable, namely annual operating savings, ranging from $2,400 to $3,300. Let's suppose that the following additional information is now available:

Annual savings $[A_i]$	Probability $P[A_i]$
less than $2,400	0
$2,400	0.04
2,500	0.08
2,600	0.10
2,700	0.12
2,800	0.14
2,900	0.16
3,000	0.16
3,100	0.10
3,200	0.08
3,300	0.02
more than $3,300	0
	1.00

The probability mass function is shown in Figure 8.14. The expected value of the annual operating savings is simply the sum of the products of the individual values and their respective probabilities:

$$E[\text{Savings}] = \sum A_i P[A_i]$$
$$= \$2,400(0.04) + \$2,500(0.08) + \cdots$$
$$+ \$3,300(0.02)$$
$$= \$2,848$$

Figure 8.14

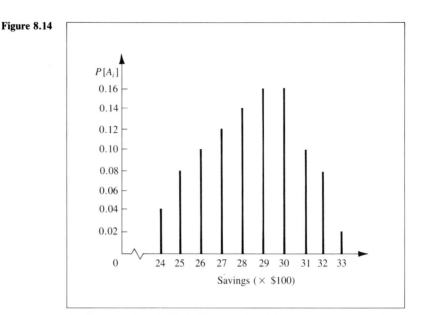

Again, note that $P[\$2,848] = 0$. Yet if we were to sample from this distribution a large number of times, the long-term average value per sample would approach $2,848.

As noted earlier, decisions under risk are those in which the probability distributions underlying one or more parameters of interest are either known or can be assumed. Given these probability distributions, analysts can use the theory of probability to compute either (1) the expected values of figures of merit, such as present worth, annual cost, rate of return, and so on, and/or (2) the resulting distributions of one or more of these random variables. (A function of random variables is itself a random variable.)

To illustrate let's return to the numerical example presented in Section 8.1. By spending $25,000 on automatic controls that have a thirty-year life, the firm can effect annual operating savings of $3,000 during each year over thirty years. If the minimum attractive rate of return is 10 percent and if all these input data are *certainty estimates* (assumed to be known with certainty), then, as shown previously,

$$PW = -\$25,000 + \$3,000(P/\overline{A}, 10\%, 30)$$

$$= -\$25,000 + \$3,000(9.891)$$

$$= \$4,673$$

Now suppose that the parameter, annual operating savings (A), is a random variable having the probability distribution given in Section 8.2. The distribution, as well as the expected value, of the PW may be computed as

$$PW = -\$25,000 + 9.891A$$

The expected present worth is given by

$$E[PW] = -\$25,000 + 9.891E[A] \quad \overset{\textstyle \Sigma A_i P[A_i]}{}$$

or, alternatively, by

$$E[PW] = \Sigma_i(-\$25,000 + 9.891A_i)P[A_i]$$

The solution is shown in Table 8.2.

The information now available to the analyst, and hence to the decision maker, is greatly enhanced. The probability distribution for the PW has been determined statistically and can be illustrated graphically, as in Figure 8.15. The minimum ($-\$1,262$), maximum ($\$7,640$), and expected ($\$3,170$) values are now known. Additional types of probability statements can also be made: What is the probability that the PW will exceed $\$P$? Here, for example, it is evident that the probability of losing money with this investment proposal is $0.04 + 0.08 = 0.12$. The probability that the return will exceed the expected value, $\$3,170$, is 0.52. (Note that the probability that PW will exceed the expected PW is not exactly 0.50, as might be anticipated. That is, the distribution of PW is not symmetrical about the mean.)

Before leaving this section, note that, in the literature of economic analysis, *risk* is frequently defined as variability of the

Table 8.2
Determining the Present Worth Distribution

Event i (1)	Annual savings A_i (2)	Probability $P[A_i]$ (3)	Present worth $-\$25,000 + 9.891A_i$ (4)	$(3) \times (4)$ (5)
1	less than $2,400	0.00	$ 0	$ 0
2	2,400	0.04	− 1,262	− 50
3	2,500	0.08	− 273	− 22
4	2,600	0.10	717	72
5	2,700	0.12	1,706	205
6	2,800	0.14	2,695	377
7	2,900	0.16	3,684	589
8	3,000	0.16	4,673	748
9	3,100	0.10	5,662	566
10	3,200	0.08	6,651	532
11	3,300	0.02	7,640	153
12	more than $3,300	0.00	0	0
		1.00		$3,170

Figure 8.15. Probability Mass Function for Present Worth

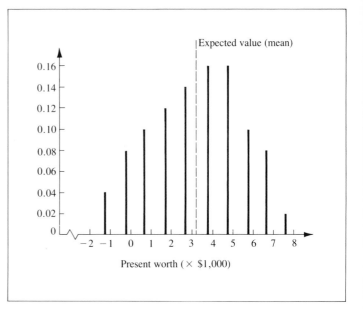

figure of merit. More precisely, risk is measured by a statistic that describes the extent to which the outcome may differ from the expected value, μ. Normally, the descriptive statistic is the variance (σ^2) of the probability distribution for the figure of merit. If one assumes that risk is to be avoided, then given two alternatives, A and B, with equal expected returns ($\mu_A = \mu_B$), the alternative having the smaller risk will be preferred. (See Figure 8.16.) Similarly, given two alternatives with equal risk ($\sigma_A^2 = \sigma_B^2$), the alternative with the larger expected return should be preferred. (See Figure 8.17.) If, as is the usual case, both the returns and risks of the alternatives are unequal, then Alternative B is clearly preferable to A only if $\mu_B > \mu_A$ and $\sigma_B^2 < \sigma_A^2$, that is, if Alternative B has both the higher return and the lower risk. (See Figure 8.18.) If the comparison indicates that $\mu_B > \mu_A$ and $\sigma_B^2 > \sigma_A^2$, that is, Project B has a higher expected return but is also more risky, then the decision maker must trade off the desire for maximum return against the propensity to avoid risk. (See Figure 8.19.) There are certain approaches to the "risk versus return" problem, but further discussion is beyond the scope of this book.

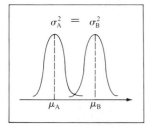

Figure 8.16

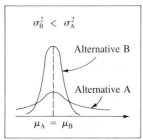

Figure 8.17

8.4
ADDITIONAL PROCEDURES FOR ANALYZING DECISIONS UNDER RISK

The approach to risk analysis outlined in Section 8.3 is based on the premise that the decision maker desires to (a) maximize expected return and (b) minimize risk. This section presents some additional principles of choice that may be appealing under certain conditions. A simple numerical example is used as a basis for the discussion.

Problem Statement

The International Investment Corporation is considering five mutually exclusive alternatives for constructing a new manufacturing plant in a certain Latin American country. The costs of each alternative, stated in terms of equivalent uniform annual costs, depend on the outcome of negotiations that are currently under way between International, lending agencies, and the government of the host country. International's analysts have concluded that four specific mutually exclusive outcomes are possible, and they have computed the equivalent uniform annual cost for each alternative-outcome combination. These are shown as cell values in the following *cost matrix*.

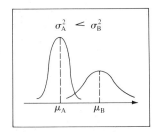

Figure 8.18

Possible outcomes

	s_1	s_2	s_3	s_4
a_1	18	11	11	10
a_2	16	16	16	16
Alternatives a_3	17	20	8	17
a_4	9	12	17	16
a_5	10	13	17	18

(cell entries are multiples of $10,000)

If the future is known with certainty, then the least costly alternative may be selected by any of the methods presented in Chapter 3. For example, if it is known that outcome s_3 will definitely occur, then a_3 should be selected, because it will result in the lowest equivalent uniform annual cost. On the other hand, if a_3 is selected and s_2 perversely occurs, choosing a_3 will have resulted in the most costly event.

Let's assume that sufficient information exists to warrant statements about the relative probabilities of the possible future outcomes. Specifically, these probabilities (expected relative frequencies) are

$P[s_1] = 0.3$ $P[s_3] = 0.2$

$P[s_2] = 0.4$ $P[s_4] = 0.1$

Given this additional information, which alternative should be selected? A number of principles that may be applied in this situation are discussed below.

The Dominance Principle

Before applying *any* of the principles of choice, it is first desirable (although not absolutely necessary) to apply the **dominance prin-**

ciple to determine which alternatives, if any, are dominated. If, of two alternatives, one would never be preferred no matter what future occurs, it is said to be dominated and may be removed from any further consideration. From the example, consider a_4 and a_5:

	s_1	s_2	s_3	s_4
a_4	9	12	17	16
a_5	10	13	17	18

Since a_5 is always at least as costly as a_4, irrespective of which future outcome occurs, a_5 may be ignored in the remaining discussion.

If one alternative dominates all others, it is said to be *globally dominant*, and the decision maker need look no further; the optimal solution has been found. Unfortunately, globally dominant alternatives are rare. But in any event, the dominance principle is frequently effective in reducing the number of alternatives to be considered. In our example problem, we are able to reduce the number of alternatives from five to four.

The Principle of Expectation

The **principle of expectation** states that the alternative to be selected is the one that has the minimum expected cost (or maximum expected profit or revenue). From the example:

$$E[C(a_1)] = 0.3(18) + 0.4(11) + 0.2(11) + 0.1(10) = 13.0$$
$$E[C(a_2)] = 0.3(16) + 0.4(16) + 0.2(16) + 0.1(16) = 16.0$$
$$E[C(a_3)] = 0.3(17) + 0.4(20) + 0.2(8) + 0.1(17) = 16.4$$
$$E[C(a_4)] = 0.3(9) + 0.4(12) + 0.2(17) + 0.1(16) = 12.5$$

where $E[C(a_i)]$ is the expected cost of alternative a_i. Here, a_4 should be selected, because it yields the minimum expected cost. (The entries in the cost matrix are multiples of $10,000; hence the minimum expected cost is $125,000.)

Principles that depend on determination of expected values by the mathematics of probability theory are frequently criticized on the grounds that the theory holds only when trials are repeated many times. That is, the relative frequency of tails resulting from the flip of a fair coin approaches 0.5 as the number of flips (trials) gets larger and larger. If the number of flips is extremely large, one can be reasonably confident that the relative frequency of tails will indeed be 0.5. On the other hand, this confidence is unwarranted if the number of flips is small. It is argued that, for certain types of decisions—for example whether to finance a major expansion—expectation is meaningless since this type of decision is not made very often. According to the counter-argument, even if the firm is not faced with a large number of repetitive decisions, it should apply the principle to many different decisions and thus realize the long-run effects. Moreover, even if the decision is unique, the only way to approach decisions for which probabilites are known is to behave as if the decision were a repetitive one and thus minimize expected cost or maximize expected revenue or profit.

The Expectation-Variance Principle

The **expectation-variance principle** focuses attention on the variability of the possible results for each alternative. It formally allows for the view that variability is of great interest to the decision maker and that, if feasible, it is desirable to minimize this measure of uncertainty. Thus the expectation-variance principle states that, if two or more alternatives have the same expected cost or return, the one with the smallest variance of cost or return should be chosen; if two or more alternatives have the same variance, the one with the smallest expected cost or largest expected return should be chosen. This view was expressed earlier in Section 8.3. (See Figures 8.16–8.19.)

The variance of a random variable, $\text{Var}[x]$, is defined as

$$\text{Var}[x] = E[x^2] - \{E[x]\}^2 \qquad (8.8)$$

where

$$E[x^2] = \sum_x x^2 P[x]$$

and

$$\{E[x]\}^2 = \{\sum_x x P[x]\}^2$$

Returning to the example, the variances of the random variable (cost) for each alternative are

$$\text{Var}[C(a_1)] = [0.3(18)^2 + 0.4(11)^2 + 0.2(11)^2$$
$$+ 0.1(10)^2] - (13)^2$$
$$= 179.8 - 169$$
$$= 10.8, \text{ or } \$108,000$$

$$\text{Var}[C(a_2)] = [0.3(16)^2 + 0.4(16)^2 + 0.2(16)^2$$
$$+ 0.1(16)^2] - (16)^2$$
$$= 256 - 256$$
$$= 0$$

$$\text{Var}[C(a_3)] = [0.3(17)^2 + 0.4(20)^2 + 0.2(8)^2$$
$$+ 0.1(17)^2] - (16.4)^2$$
$$= 288.4 - 269.0$$
$$= 19.4, \text{ or } \$194,000$$

$$\text{Var}[C(a_4)] = [0.3(9)^2 + 0.4(12)^2 + 0.2(17)^2$$
$$+ 0.1(16)^2] - (12.5)^2$$
$$= 165.3 - 156.3$$
$$= 9.0, \text{ or } \$90,000$$

From these results it is clear that variability is minimized if a_2 is selected; there is zero dispersion of possible results. But the expected cost of a_2 ($160,000) is considerably greater than the expected cost of either a_1 ($130,000) or a_4 ($125,000), and thus a_2 should not be chosen unless a large implicit cost is associated with variability of possible future events.

Since expected values and variances are rarely exactly equal for two or more alternatives, the expectation-variance principle in effect calls for the maximization of some function of the expectation and the variance. To illustrate, consider alternatives a_2 and a_4:

$$E[C(a_2)] = \$160,000 \qquad \text{Var}[C(a_2)] = 0$$
$$E[C(a_4)] = \$125,000 \qquad \text{Var}[C(a_4)] = \$90,000$$

To determine which alternative is preferable, it is first necessary to decide the relative value, or utility, of a dollar of expected cost versus that of a dollar of variance. Suppose, for example, that an increase of $100 in expected cost is equivalent to an increase of $50 in variance. Then the difference in expected costs between a_2 and a_4, $35,000, is more than offset by the corresponding value of the adjusted difference in variances, $45,000. Computing the total implicit cost for all four alternatives:

Implicit cost $(a_1) = \$130,000 + 0.5(\$108,000) = \$184,000$

Implicit cost $(a_2) = \$160,000 + 0.5(0) = \$160,000$

Implicit cost $(a_3) = \$164,000 + 0.5(\$194,000) = \$261,000$

Implicit cost $(a_4) = \$125,000 + 0.5(\$90,000) = \$170,000$

Consequently, if we assume a $2:1$ relationship between equivalent uniform annual cost and variance, the total cost is minimized if a_2 is selected.

The *range* of values is frequently used as an *estimate* of the variance, where

$$\text{Range}(x) = x_{\text{max}} - x_{\text{min}} \qquad (8.9)$$

To illustrate, consider the data from the example:

Alternative	Maximum	Minimum	Range	Variance
a_1	$180,000	$100,000	$ 80,000	$108,000
a_2	160,000	160,000	0	0
a_3	200,000	80,000	120,000	194,000
a_4	170,000	90,000	80,000	90,000

Although the range is only an approximate measure of dispersion, it may be especially useful when the analyst is unfamiliar with the mathematics required to calculate variance. The ability of the range to estimate the variance is improved as the number of values (possible outcomes) is increased.

The Most Probable Future Principle

Let us view the future somewhat myopically, using the **most probable future principle.** In particular, assume that the future event to expect is precisely the most likely event. Thus, observing that s_2 has the highest probability of occurring, assume that it will in fact occur. In this case, a_1 (with a cost of $110,000) is the least costly of the four available alternatives.

It would appear that this principle is particularly appealing in cases in which one future is significantly more probable than all other possibilites. To illustrate, we may admit to the finite probability that we may not survive a trip on the freeway, but since safe transit is significantly more likely than death or injury in an automobile accident, we do drive!

The Aspiration Level Principle

The **aspiration level principle** requires the establishment of a goal, or level of aspiration. Thus the alternative that maximizes the probability that the goal will be met or exceeded should be selected. To illustrate, suppose that the management of the International Investment Corporation wishes to minimize the probability that equivalent uniform annual costs will exceed $150,000. (This is identical to the requirement that it maximize the probability that costs will *not* exceed $150,000.) The probabilities are

$$Pr[C(a_1) > \$150,000] = 0.3$$

$$Pr[C(a_2) > \$150,000] = 0.3 + 0.4 + 0.2 + 0.1 = 1.0$$

$$Pr[C(a_3) > \$150,000] = 0.3 + 0.4 \qquad + 0.1 = 0.8$$

$$Pr[C(a_4) > \$150,000] = \qquad\qquad + 0.2 + 0.1 = 0.3$$

Thus the aspiration level will be met if either a_1 or a_4 is selected.

Summary of Results for Example Problem
Clearly, the selection from mutually exclusive alternatives is a matter of which principle is used to guide the decision. From the example:

Principle of choice	Alternative recommended
Expectation	a_4
Expectation-variance	a_2
Most probable future	a_1
Aspiration level	a_1 or a_4

**8.5
DECISIONS
UNDER UNCERTAINTY**

This section examines a number of principles of choice that may be used when the relative likelihoods of future states of nature *cannot* be estimated. These principles will be demonstrated by using the example problem introduced in Section 8.4.

The Minimax (Maximin) Principle
The **minimax principle** is pessimistic in the extreme. It assumes that, if an alternative is selected, the worst possible outcome will occur. The maximum cost associated with each alternative is examined, and the alternative that *minimizes* the *maximum* cost is selected. In general, the mathematical formulation of the minimax principle is

$$\underset{i}{\text{Min}}\left[\underset{j}{\text{Max}}\,(C_{ij})\right] \tag{8.10}$$

where C_{ij} is the cost that results when alternative i is selected and state of nature j occurs. From the example:

Alternative		State of nature j				$\underset{j}{\text{Max}}\,C_{ij}$
i	a_i	s_1	s_2	s_3	s_4	
1	a_1	18	11	11	10	18
2	a_2	16	16	16	16	16
3	a_3	17	20	8	17	20
4	a_4	9	12	17	16	17

The numbers in the cells represent costs (C_{ij}) in thousands of dollars. If the maximin principle is adopted, a_2 is indicated because it results in minimum costs, assuming the worst possible conditions.

The mirror image of the minimax principle, the **maximin**

principle, may be applied when the matrix contains *profits* or *revenue* measures. In this case the most pessimistic view suggests that the alternative to select is the one that *maximizes* the *minimum* profit or revenue associated with each alternative. The mathematical formulation of the maximin principle is

$$\text{Max}_i\left[\text{Min}_j\ (R_{ij})\right] \tag{8.11}$$

where R_{ij} is the revenue or profit resulting from the combination of a_i and s_j.

The Minimin (Maximax) Principle
The **minimin principle** is based on the view that the best possible outcome occurs when a given alternative is selected. It is optimistic in the extreme. The minimum cost associated with each alternative is examined, and the alternative that *minimizes* the *minimum* cost is selected. The mathematical formulation is

$$\text{Min}_i\left[\text{Min}_j\ (C_{ij})\right] \tag{8.12}$$

From the example,

a_i	$\underset{j}{\text{Min}}\ C_{ij}$
a_1	$100,000
a_2	160,000
a_3	80,000
a_4	90,000

Alternative a_3 minimizes the minimum cost.

As a corollary to the minimin principle, the **maximax principle** is appropriate when the decision maker is extremely optimistic and the matrix contains measures of profit or revenue. The maximum profit (or revenue) associated with each alternative is examined, and the alternative that *maximizes* the *maximum* profit (or revenue) is selected. The mathematical formulation is

$$\text{Max}_i\left[\text{Max}_j\ (R_{ij})\right] \tag{8.13}$$

The Hurwicz Principle
It may be argued that decision makers need not be either optimistic or pessimistic in the extreme, in which case the **Hurwicz Principle** permits selection of a position between the two limits.[4] When evaluating costs, C_{ij}, the *Hurwicz criterion* for alternative a_i is given by

4. This principle is named after the econometrician Leonid Hurwicz.

$$H(a_i) = \alpha \left[\underset{j}{\text{Min}} \ (C_{ij}) \right] + (1 - \alpha) \left[\underset{j}{\text{Max}} \ (C_{ij}) \right] \qquad \textbf{(8.14)}$$

where α is the "index of optimism" such that $0 \leqslant \alpha \leqslant 1$. Extreme pessimism is defined by $\alpha = 0$; extreme optimism is defined by $\alpha = 1$. The value of α used in any particular analysis is selectd by the decision maker based on subjective judgment. The alternative that *minimizes* the quantity $H(a_i)$ is the alternative to select.

When evaluating profits or revenues, R_{ij}, the expression for the Hurwicz criterion is

$$H(a_i) = \alpha \left[\underset{j}{\text{Max}} \ (R_{ij}) \right] + (1 - \alpha) \left[\underset{j}{\text{Min}} \ (R_{ij}) \right] \qquad \textbf{(8.15)}$$

From the example, the Hurwicz criteria are determined using Equation 8.14 as follows:

$$H(a_1) = \alpha(\$100{,}000) + (1 - \alpha)(\$180{,}000)$$

$$H(a_2) = \alpha(\$160{,}000) + (1 - \alpha)(\$160{,}000)$$

$$H(a_3) = \alpha(\$ \ 80{,}000) + (1 - \alpha)(\$200{,}000)$$

$$H(a_4) = \alpha(\$ \ 90{,}000) + (1 - \alpha)(\$170{,}000)$$

The values of $H(a_i)$ are plotted in Figure 8.20. We may determine, either graphically or algebraically, that a_2 will be chosen for $0 \leqslant \alpha \leqslant 0.125$, a_4 will be selected for $0.125 \leqslant \alpha \leqslant 0.75$, and a_3 is least costly for $0.75 \leqslant \alpha \leqslant 1.00$. Suppose that the personnel of the International Investment Corporation concerned with this decision agree that there is no particular reason for being either optimistic or pessimistic. Thus their value for α is approximately 0.5. Under this assumption the Hurwicz principle indicates that a_4 should be selected.

The Laplace Principle

The **Laplace principle,** sometimes known as the **principle of insufficient reason,** assumes that the probabilities of future events occurring are equal.[5] That is, in the absence of any information to the contrary, it is assumed that all future outcomes are equally likely to occur. The expected cost (or profit/revenue) of each alternative is then computed, and the alternative that yields the minimum expected cost (or maximum expected profit/revenue) is selected. The mathematical expression for this principle is

$$\underset{i}{\text{Min}} \left\{ \frac{1}{k} \sum_{j=1}^{k} C_{ij} \right\} \qquad \textbf{(8.16)}$$

5. This principle is named after the eighteenth-century French mathematician Pierre Simon de Laplace.

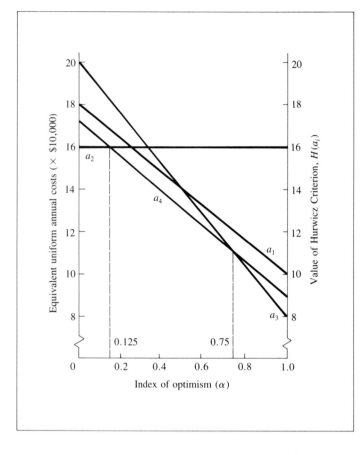

Figure 8.20. Hurwicz Criteria as Functions of α

when the figure of merit is expressed as a cost or as

$$\underset{i}{\text{Max}}\left\{\frac{1}{k}\sum_{j=1}^{k} R_{ij}\right\} \qquad (8.17)$$

when the figure of merit is expressed as revenue or profit.

Returning to our example, the insufficient reason assumption yields

$$P[s_1] = P[s_2] = P[s_3] = P[s_4] = \tfrac{1}{4}$$

Therefore,

$$E[C(a_1)] = \tfrac{1}{4}(18 + 11 + 11 + 10) = 12.5$$
$$E[C(a_2)] = \tfrac{1}{4}(16 + 16 + 16 + 16) = 16.0$$
$$E[C(a_3)] = \tfrac{1}{4}(17 + 20 + 8 + 17) = 15.5$$
$$E[C(a_4)] = \tfrac{1}{4}(9 + 12 + 17 + 16) = 13.5$$

Alternative a_1 should therefore be selected because it results in the minimum expected equivalent uniform annual cost.

The Savage Principle of Minimax Regret

The **Savage principle,** or **principle of minimax regret,** is based on the assumption that the decision maker's primary interest is the *difference* between the actual outcome and the outcome that would have occurred had he been able to accurately predict the future.[6] Given these differences, or *regrets,* the decision maker then adopts a conservative position and selects the alternative that minimizes the maximum potential regret for each alternative.

A *regret matrix* is then constructed, having for its cell values either

$$C_{ij} - \left[\begin{array}{c} \text{Min} \\ i \end{array} (C_{ij}) \right] \tag{8.18}$$

for cost data or

$$\left[\begin{array}{c} \text{Max} \\ i \end{array} (R_{ij}) \right] - R_{ij} \tag{8.19}$$

for revenue or profit data. In either case, these cell values, or regrets, represent the differences between (a) the outcome if alternative a_i is selected and state of nature s_j subsequently occurs and (b) the outcome that would have been achieved had it been known in advance which state of nature would occur, so that the best alternative could have been selected. To illustrate, consider alternative a_1 and state of nature s_1: C_{11} = \$18,000. However, if we had known *a priori* that state s_1 would in fact occur, we would have selected a_4, incurring a cost of only \$9,000. The difference (\$18,000 − \$9,000) is a measure of "regret" about selecting a_1 when we could have selected a_4 (had we known the state of nature in advance). The regret matrix for the example is

Possible outcomes

	s_1	s_2	s_3	s_4
a_1	18 − 9 = 9	11 − 11 = 0	11 − 8 = 3	10 − 10 = 0
a_2	16 − 9 = 7	16 − 11 = 5	16 − 8 = 8	16 − 10 = 6
a_3	17 − 9 = 8	20 − 11 = 9	8 − 8 = 0	17 − 10 = 7
a_4	9 − 9 = 0	12 − 11 = 1	17 − 8 = 9	16 − 10 = 6

Alternatives

Having completed the regret matrix, the alternative that *minimizes* the *maximum* regret is selected. That is,

6. This principle is named after statistician L. J. Savage.

$$\underset{i}{\text{Min}} \underset{j}{\text{Max}} \left\{ C_{ij} - \left[\underset{i}{\text{Min}} (C_{ij}) \right] \right\} \qquad \textbf{(8.20)}$$

for cost data or

$$\underset{i}{\text{Min}} \underset{j}{\text{Max}} \left\{ \left[\underset{i}{\text{Max}} (R_{ij}) \right] - R_{ij} \right\} \qquad \textbf{(8.21)}$$

for revenue or profit data. From the example,

Alternative	Maximum regret
a_1	9
a_2	⑧
a_3	9
a_4	9

Thus, according to the Savage principle of minimax regret, alternative a_2 should be selected.

A major inconsistency in using this principle is that the solution can be altered by adding still another alternative that itself is not selected as the most desirable. To illustrate, suppose that we are considering another alternative, a_6, with the following equivalent uniform annual costs ($\times$ \$1,000):

	s_1	s_2	s_3	s_4
a_6	19	18	6	18

Including this new alternative results in a revised regret matrix:

	s_1	s_2	s_3	s_4
a_1	9	0	5	0
a_2	7	5	10	6
a_3	8	9	2	7
a_4	0	1	11	6
a_6	10	7	0	8

Now, minimizing the maximum regret:

a_1	⑨
a_2	10
a_3	⑨
a_4	11
a_6	10

In this instance, the addition of alternative a_6 has shifted the solution from a_2 to a_1 or a_3—an unreasonable result. Thus by adding an alternative that is not selected under the Savage principle of minimax regret, the solution has been changed.

Summary of Example Problem Results

There is no special reason why the principles discussed above should yield the same solution. In fact, in our example all alternatives were selected at least once:

Principle of choice	Alternative recommended
Minimax	a_2
Minimin	a_3
Hurwicz ($\alpha = 0.5$)	a_4
Laplace (insufficient reason)	a_1
Savage (minimax regret)	a_2

Is one principle more "correct" than any other? There is no simple answer to this question, since the choice of principle largely depends on the predisposition of the decision maker. Each principle has certain obvious advantages, and each is deficient in one or more desirable characteristics.[7] Nevertheless, the principles in this section are useful if for no other reason than they shed some light on the subjective decision process.

8.6 DIGITAL COMPUTER (MONTE CARLO) SIMULATION

The statistical procedures related to risk analysis described in Section 8.3 suffer from at least one important drawback: The analytical techniques necessary to derive the mean, variance, and possibly the probability distribution of the figure of merit may be extremely difficult to implement. Indeed, the complexity of many real-world problems precludes the use of these computational techniques altogether; computations may be intractable, or the necessary underlying assumptions may not be met. Under these conditions, analysts may find **digital computer (Monte Carlo) simulation** especially useful.[8]

Digital computer simulation was first developed in connection with scientific research during World War II. *Monte Carlo* was the code name given by two scientists, John von Neumann and S. M. Ulam, to the mathematical technique they applied to a category of nuclear shielding problems that were too expensive for experimental solution and too complicated for analytical treatment. Originally, the concept referred to a situation in which there is a difficult nonprobabilistic problem to be solved for which a stochastic process may be invented having moments or a distribution that satisfies the relations of the problem. The application of this

---

7. Our discussion has primarily focused on simple structures and areas of application; little has been said of the major deficiencies of the principles of choice. For a more extensive discussion, see the references cited in the Bibliography.

8. Strictly speaking, *Monte Carlo* simulation and *digital computer* simulation are not synonymous. Monte Carlo simulation is a technique used in the digital computer simulation of systems behavior. However, in recent years, practitioners have tended to blur this semantic distinction, using the terms interchangeably.

simulation technique to economic analysis, using a digital computer, was subsequently perceived by a number of people. David B. Hertz, writing in the *Harvard Business Review,* was chiefly responsible for bringing the technique to the general attention of analysts, businesspeople, and other potential users.[9]

The objective of digital computer simulation is to generate a probability distribution for the figure of merit, generally present worth or rate of return, given the probability distributions for the various components of the analysis. The decision maker can thus compare expected returns as well as the variability of returns for two or more alternatives. Moreover, probability statements can be made, in this form: The probability is x that project y will result in a profit in excess of z.

Sampling from a Discrete Distribution

Recall the probability distribution for annual operating savings given in Section 8.3. Figure 8.21 shows the probabilities of the events $A = \$2,400, \ldots, \$3,300$, as well as the **cumulative distribution function** (CDF). The CDF represents the probability that the annual operating savings will be less than or equal to some given value. For example, the density function shows that

$$P[A \leq \$2,400] = 0.04$$

$$P[A \leq \$2,500] = 0.04 + 0.08 = 0.12$$

$$P[A \leq \$2,600] = 0.12 + 0.10 = 0.22$$

$$\vdots$$

$$P[A \leq \$3,300] = 0.98 + 0.02 = 1.00$$

Our problem now is one of sampling from this distribution, using either the probability mass function or its associated CDF in order to preserve precisely all the characteristics of the original distribution. To solve this problem, suppose that the ordinate of the CDF, the vertical axis labeled from 0 to 1.0 (as in Figure 8.22) is bent around to form a perfect circle, or wheel. Radii can be drawn from the center of the circle to divide it into n equal parts, or sectors. (It would be useful if n were a multiple of 10, say, 100 or 1,000.) Now let's place a balance point at the center of the wheel and a pointer at the edge. If the wheel is perfectly balanced, each of the n sectors has an equal chance of being selected if the

9. David B. Hertz, "Risk Analysis in Capital Investment," *Harvard Business Review* 42 (Jan./Feb. 1964): 95–106.

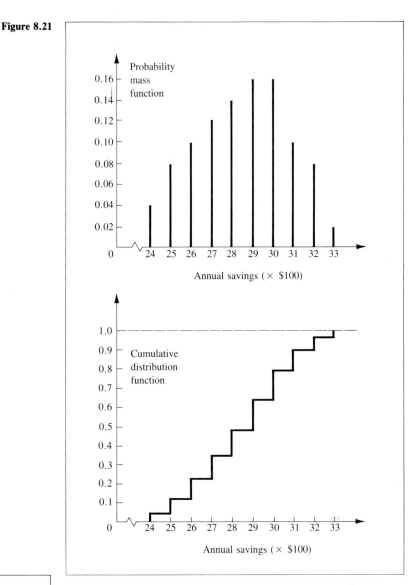

Figure 8.21

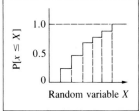

Figure 8.22

wheel were spun. As illustrated in Figure 8.23, the value selected by the pointer is a random number that can be used to enter the CDF to identify the simulated value of the random variable.

An alternative explanation of the random sampling process may be helpful at this point. Bear in mind that our problem is one of sampling from the probability distribution so that the characteristics of the distribution are retained. We can do this by obtaining, say, 100 perfectly matched balls, numbered from 00 to 99. Using the operating savings problem as an example, we want to label four of the balls "$2,400," eight of the balls "$2,500," and so on. The number of balls labeled with a particular amount is proportional to their relative probability in the original distribution:

Ball numbers	Number of balls	Label	Ball numbers	Number of balls	Label
01–04	4	$2,400	49–64	16	$2,900
05–12	8	2,500	65–80	16	3,000
13–22	10	2,600	81–90	10	3,100
23–34	12	2,700	91–98	8	3,200
35–48	14	2,800	99–00	2	3,300

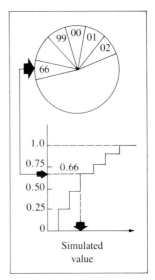

Figure 8.23

Now we can put all the balls into a large jar, shake it thoroughly so that the balls are completely mixed, and then draw out a single ball. Then, the result of this "random sample" is recorded and the ball is placed back in the jar to select our next sample. As we continue this process through a large number of samples, or trials, we can expect the resulting frequency distribution to approximate that of the original population.

Of course, in practical applications, the sampling process does not consist of spinning wheels or drawing balls from a jar. There are a variety of more elegant procedures, generally based on successive iterations of a predetermined formula. An alternative approach that is useful when the number of samples to be drawn is relatively small is to reference a **table of random numbers.** Such tables have been developed and the results recorded in tabular format. (A table of three-digit random numbers appears in Appendix C.) Inasmuch as the numbers in the table are randomly generated, users may enter the table at any point and proceed in any direction.

Sampling from a Normal Distribution

The **normal distribution** is frequently used to describe the probabilities of certain continuous random variables. The probability distribution function is given by

$$p(x) = \frac{1}{\sigma\sqrt{2\pi}}\left\{exp\left[\frac{-1}{2}\left(\frac{x-\mu}{\sigma}\right)^2\right]\right\} \qquad (8.22)$$

where μ = mean and σ = standard deviation of the random variable X and x is the particular value of the random variable. A particular normal distribution is fully described by the parameters μ and σ, where μ is a measure of central tendency and σ is a measure of dispersion.

The **standardized normal distribution** results from the special case wherein $\mu = 0$ and $\sigma = 1$. (See Figure 8.24.) The area under the curve from $-\infty$ to $+\infty$ is exactly 1.0. If one can develop a table of random numbers for a uniform distribution over the interval 0–1, it is possible to map a set of equivalent values for the

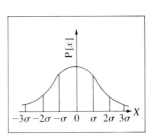

Figure 8.24

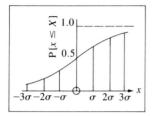

Figure 8.25

standardized normal distribution, as in Figure 8.25. The value along the ordinate represents the probability that the random variable X lies in the interval $-\infty$ to x. For any random number we can compute the equivalent value x. This latter value is called the **random normal deviate.**

Tables of random normal deviates exist that contain values from which one may generate a random sample from *any* normal distribution with known parameters μ and σ. (See Appendix D.) To illustrate, suppose that we wish to sample from a normal distribution with $\mu = 200$ and $\sigma = 25$. Consulting a table of random normal deviates, we "draw" the numbers $+0.289$, $+0.861$, -1.454, and so on. The corresponding values of the random variable are

Sample	Random normal deviate	Values of random variable
1	+0.289	200 + 0.289(25) = 207.2
2	+0.861	200 + 0.861(25) = 221.5
3	−1.454	200 − 1.454(25) = 163.6
⋮	⋮	⋮

General Framework

The previous sections discussed the process whereby random samples are drawn from an underlying probability distribution. This process, of course, is fundamental in the overall simulation. The general procedure can be described in four steps:

Step 1 Determine the probability distribution(s) for the significant factors, as illustrated in Figure 8.26.

Step 2 Using Monte Carlo simulation, select random samples from these factors according to their relative probabilities of occurring in the future. (See Figure 8.27.) Note that the selection of one factor (price, for example) may determine the probability distribution of another factor (total amount demanded, for example).

Step 3 Determine the figure of merit (rate of return or present worth, for example) for each combination of factors. One trial consists of one calculation of the figure of merit.

Figure 8.26

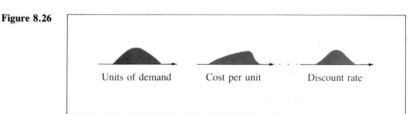

Units of demand   Cost per unit   Discount rate

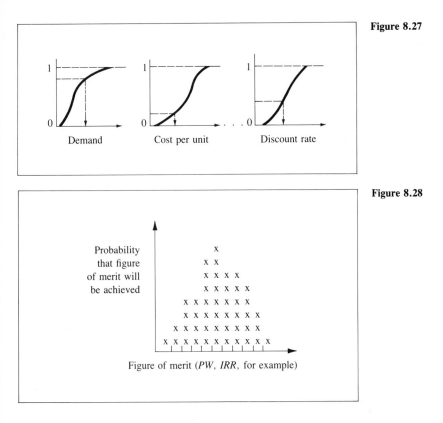

Figure 8.27

Figure 8.28

Step 4   Repeat the process, that is, conduct a series of trials, building a frequency histogram with the results, as in Figure 8.28. Continue until you are reasonably satisfied that the histogram yields a clear portrayal of the investment risk.[10]

A Numerical Example

A certain investment, if purchased, will result in annual operating savings described by the probability distribution shown in Column b of Table 8.3. The project life is described by the probability distribution shown in Column e, and the initial cost is a normally distributed random variable with $\mu = \$25,000$ and $\sigma = \$1,000$. The discount rate to be used in the analysis, a certainty estimate, is 0.10. It is assumed that the annual savings, project life, and initial cost are independent random variables. If there are no other relevant consequences of the proposed investment, the expected net present value is given by the equation

10. There is no universally accepted rule for determining the optimum number of trials. It is clearly less expensive to produce a smaller number of trials, yet a larger number of trials yields more information. Substantial literature is addressed to this interesting problem, but additional discussion is not warranted here.

Table 8.3

Example Simulation Problem: Input Data

Annual operating savings (a)	Probability (b)	Corresponding random numbers (c)	Project life (d)	Probability (e)	Corresponding random numbers (f)
$2,400	0.04	01–04	25	0.05	01–05
2,500	0.08	05–12	26	0.10	06–15
2,600	0.10	13–22	27	0.10	16–25
2,700	0.12	23–34	28	0.10	26–35
2,800	0.14	35–48	29	0.10	36–45
2,900	0.16	49–64	30	0.10	46–55
3,000	0.16	65–80	31	0.10	56–65
3,100	0.10	81–90	32	0.10	66–75
3,200	0.08	91–98	33	0.10	76–85
3,300	0.02	99–00	34	0.10	86–95
			35	0.05	96–00
	1.00			1.00	

$$PW = A(P/\bar{A}, 10\%, N) - P \qquad (8.23)$$

where

A = annual operating savings

N = project life

P = initial cost

Columns c and f of Table 8.3 contain the random numbers corresponding to the relative probabilities in Columns b and e, respectively. Note that two-digit random numbers are used. Inasmuch as the specified accuracy of the probability distributions is two significant digits, the corresponding random numbers must be specified by *at least* two digits. (A three-digit random number, say 843, could be rounded to 84, or simply truncated after the first two digits.) For the first variable, annual operating savings (A), there are 100 random numbers, the first four of which correspond to the event A = $2,400. The next eight numbers, 05 through 12, correspond to the event A = $2,500, and so on.

The results of ten simulated trials are shown in Table 8.4. Consider the first trial, for example. A random number, 09, is drawn from the table of random numbers in Appendix C. As shown in Table 8.3, this corresponds to the event A = $2,500. Next, a new random number, 52, is drawn, which corresponds to the event N = 30. Note that *the same random number cannot be used for both random variables, because they are independent.* The third random variable, P, is normally distributed, so a random normal deviate is drawn from the table of random normal deviates

Table 8.4

Example Simulation Problem: Simulated Trials

Trial	Random number	Operating savings	Random number	Project life (years)	Random normal deviate	Initial cost	Present worth	Cumulative average PW
1	09	$2,500	52	30	0.464	$25,464	−$ 737	−$ 737
2	54	2,900	80	33	0.137	25,137	3,979	1,621
3	42	2,800	45	29	2.455	27,455	72	1,105
4	01	2,400	68	32	−0.323	24,677	− 689	656
5	80	3,000	59	31	−0.068	24,932	4,903	1,506
6	06	2,500	48	30	0.296	25,269	− 569	1,160
7	06	2,500	12	26	−0.288	24,712	− 682	897
8	26	2,700	35	28	0.060	24,940	1,426	963
9	57	2,900	91	34	−2.526	22,474	6,761	1,607
10	79	3,000	89	34	−0.531	24,469	5,774	2,023

in Appendix D. This number, 0.464, indicates a simulated value for

$$P = \mu + (RND)\sigma \qquad (8.24)$$

$$= \$25,000 + (0.464)(\$1,000)$$

$$= \$25,464$$

The present worth for the first trial can now be computed:

$$PW = \$2,500(P/\overline{A}, 10\%, 30) - \$25,464$$

$$= -\$737$$

A frequency distribution can be developed from the resulting *PW* values, and relevant statistics can be computed. In this example, the cumulative average *PW* after ten trials is $2,023; 40 percent of the trials result in a negative *PW*. The minimum value simulated was −$737; the maximum value simulated was $6,761. Of course, if this were an actual application, the number of trials would be much larger, perhaps several thousand or more, and we would have considerably greater confidence in the resulting statistics.

In this particular example there are relatively few random variables, the relationships are not complex, and the variables are independent. Hence it is possible to compute the theoretical expected value for the net present value. That is,

$$PW = A(P/\overline{A}, 10\%, N) - P$$

$$E[PW] = E[A](P/\overline{A}, 10\%, E[N]) - E[P]$$

where

$$E[A] = 0.04(\$2,400) + 0.08(\$2,500)$$
$$+ \cdots + 0.02(\$3,300)$$
$$= \$2,848$$
$$E[N] = 0.05(25) + 0.10(26) + \cdots + 0.05(35)$$
$$= 30$$
$$E[P] = \$25,000$$

Thus

$$E[PW] = \$2,848(P/\overline{A}, 10\% \ 30) - \$25,000$$
$$= \$3,170$$

We can expect, therefore, that the cumulative average *PW* will approach \$3,170 as the number of trials increases. Although we can compute the theoretical *mean,* it is not possible in this case to compute the theoretical *distribution* for the *PW*. However an approximate distribution can be developed through simulation.

8.7 SOME ADDITIONAL APPROACHES

As indicated at the beginning of this chapter, risk and uncertainty are inherent in the general problem of resource allocation because all decisions depend on estimates about the noncertain future. Thus risk and uncertainty have occupied the attention of a great many theoreticians and practitioners. A substantial number of approaches have been proposed, several of which are summarized in this chapter. Now, four additional approaches are briefly identified. The first three are widely used in industry, despite certain important shortcomings; the fourth, as you will see, requires detailed discussion beyond the scope of this text.

Increasing the Minimum Attractive Rate of Return
Some analysts advocate adjusting the minimum attractive rate of return to compensate for risky investments, suggesting that, since the future is uncertain, stipulation of a minimum attractive rate of return of, say, $i + \Delta i$ will insure that i will be earned in the long run. Since some investments will not turn out as well as expected, they will be compensated for by the incremental "safety margin," Δi. This approach, however, fails to come to grips with the risk or uncertainty associated with estimates for specific alternatives, and thus an element Δi in the minimum attractive rate of return penalizes all alternatives equally.

Differentiating Rates of Return by Risk Class

Rather than building a "safety margin" into a single minimum attractive rate of return, some firms establish several risk classes with separate standards for each class. For example, a firm may require low-risk investments to yield at least 15 percent and medium-risk investments to yield at least 20 percent, and it may define a minimum attractive rate of return of 25 percent for high-risk proposals. The analyst then judges which class a specific proposal belongs in, and the relevant minimum attractive rate of return is used in the analysis. Although this approach is a step away from treating all alternatives equally, it is less than satisfactory in that it fails to focus attention on the uncertainty associated with the individual proposals. No two proposals have precisely the same degree of risk, and grouping alternatives by class obscures this point. Moreover, the attention of the decision maker should be directed to the causes of uncertainty, that is, to the individual estimates.

Decreasing the Expected Project Life

Still another procedure frequently employed to compensate for uncertainty is to decrease the expected project life. It is argued that estimates become less and less reliable as they occur further and further into the future; thus shortening project life is equivalent to ignoring those distant, unreliable estimates. Furthermore, distant consequences are more likely to be favorable than unfavorable; that is, distant estimated cash flows are generally positive (resulting from net revenues) and estimated cash flows near date zero are more likely to be negative (resulting from startup costs). Reducing expected project life, however, has the effect of penalizing the proposal by precluding possible future benefits, thereby allowing for risk in much the same way that increasing the minimum attractive rate of return penalizes marginally attractive proposals. Again, this procedure is to be criticized on the basis that it obscures uncertain estimates.

Statistical Techniques

A number of procedures have been advocated wherein mathematical statistics and probability theory are applied to problems of risk and uncertainty. Many of these procedures provide management with incomplete information on which to base decisions; others assume availability of input data that are extremely difficult to obtain. Nevertheless, some of these techniques appear to be quite useful, and they have been applied with increasing frequency in recent years. Neither space nor time allow a complete explanation of statistical techniques here. Moreover, the knowledge of statistics and probability theory required for reasonable comprehension of these techniques is probably beyond that of most readers. For those of you who wish to search the literature independently, several relevant references are cited in the Bibliography.

<table>
<tr><td>

━━━━━━━━━━

8.8
SUMMARY

</td><td>

Limited capital is allocated to competing investment proposals on the basis of estimates of future consequences. Since the future can never be known with absolute certainty, it follows that procedures must be developed that analysts can use when dealing with risk and uncertainty in capital budgeting problems. Several of those procedures are discussed in this chapter.

</td></tr>
</table>

Sensitivity (or break-even) analysis is widely used because of its simplicity and ability to focus on particular estimates. It generally treats only one estimate at a time, however, and is difficult to apply when determining possible effects of combinations of errors in the estimates.

Decision theory provides a number of principles that may be used to select from mutually exclusive alternatives when various future outcomes are possible. They may be grouped into two classes: those that are applicable when nothing at all is known about the relative probabilities of future events (decisions under uncertainty) and those that are applicable when these probabilities can be defined with reasonable precision (decisions under risk). There are significant differences between the principles of choice employed, and in some instances principles are mutually contradictory. In any event, they are helpful in formalizing the decision process and thereby exposing the relative strengths and weaknesses of the various viewpoints.

Digital computer (Monte Carlo) simulation is a technique that is becoming more widely used, especially in view of the recent dramatic increase in computing power and the corresponding decrease in the cost of simulation. However, this approach requires the analyst to provide estimates of the probability distributions of all the relevant parameters. This requirement may be difficult to fulfill in many applications.

Some firms adjust for risk and uncertainty by requiring prospective investments to promise somewhat higher returns than would normally be expected if future consequences were known with perfect certainty. In some cases the minimum attractive rate of return is increased above the risk-free rate; in others expected project life is truncated. These approaches are criticized primarily on the grounds that they penalize all alternatives equally and that they tend to obscure the degree of uncertainty associated with specific estimates.

There are no simple, widely accepted techniques for effectively dealing with risk and uncertainty in capital allocation decisions. Although some promising advances have been made in recent years in the application of probability theory and decision theory, practical application remains as much an art as a science.

Sensitivity Analysis

8.1 A phased program of plant expansion is being compared with a program wherein full expansion will be undertaken immediately with a total initial investment of $1,400,000. The phased program requires $800,000 now, $600,000 in five years, and another $600,000 in ten years. If the full program is selected, estimated annual disbursements will be an additional $40,000 during the first five years and an additional $20,000 during the second five years. There are no other differences between the alternatives.
 a. If the pretax minimum attractive rate of return is 10 percent, which alternative should be selected?
 b. At what range of values for the minimum attractive rate of return is the phased program economically superior?
 (*Answers:* (a) phased program; (b) greater than 4.9 percent)

8.2 The research director and the controller of the Truline Trucking Company are discussing the merits of a proposed labor-saving device designed to eliminate a certain freight-handling operation. The proposal has a first cost of $25,000, an estimated life of ten years, and zero net salvage value. The manual operation that it will replace currently requires 500 labor hours per year at a cost of $10 per hour. There are no other differences between alternatives. The minimum attractive rate of return is 20 percent before taxes.
 a. The controller believes that the investment should not be made. Is he correct? Why or why not?
 b. The research director claims that labor costs may be expected to increase by about 10 percent per year. That is, the cost per man-hour will be $10 the first year, $11 the second, $12.10 the third, and so on. Under these conditions, she says, the new device should be purchased. Is she correct? Why or why not?

8.3 An industrial sales firm is considering the purchase of a personal computer for $10,000. If this computer is purchased, clerical costs will be reduced by $1,000 per year. The computer will be used for ten years, at the end of which time the net salvage value will be zero. The firm requires a minimum attractive rate of return of 15 percent before taxes.
 a. Assuming no other relevant differences between alternatives, should the computer be purchased?
 b. Test the sensitivity of the solution in (a) to the assumption concerning annual labor costs by assuming that these costs will increase by $50 per year, that is, $1,000 the

first year, $1,050 the second, $1,100 the third, and so on.
c. How much must the cost of labor increase per year for the proposed investment to be economically sound?

(*Answers:* (a) No; (b) $PW = -\$4,132$; (c) $G = \$293$)

8.4 A company has fixed costs of $80,000 with variable costs equal to 60 percent of net sales. The company is planning to increase its present capacity of $400,000 net sales by 30 percent, with a 20 percent increase in fixed costs. The tax rate on profits is 50 percent.
a. What would be the net profit after taxes if the *present plant* were operated at *full capacity?*
b. Assume that the present plant is enlarged as described above. What must the new sales revenue be if the company is to "break even," that is, with revenues equal to expenses?

Decisions under Risk

8.5 The following table gives the costs that would result from alternative testing sequences required in connection with a certain research program. All costs are in thousands of dollars:

Possible outcomes

		s_1	s_2	s_3	s_4	s_5
	a_1	18	16	10	14	15
Mutually	a_2	14	15	15	14	15
exclusive	a_3	5	16	12	10	15
alternatives	a_4	14	22	10	12	15
	a_5	10	12	15	10	15

Suppose that information is available suggesting that failure patterns will occur with the probabilities $P[s_1] = 0.20$, $P[s_2] = 0.15$, $P[s_3] = 0.40$, $P[s_4] = 0.10$, and $P[s_5] = 0.15$. Apply criteria for decision making when probabilities are known or estimated and comment on the results. Use the various principles of choice discussed in Chapter 8: (a) expectation, (b) expectation-variance, and (c) most probable future.

(*Answers:* (a) a_3; (b) a_5; (c) a_1 or a_4)

8.6 Managers of a certain company are attempting to determine which cleaning equipment they will purchase for a California plant. The decision depends on wage rates used in the analysis, but these rates are uncertain because they are the subject of current negotiations with the union. Unfortunately, the investment decision must be made immediately. The following table represents the annual benefits in millions of dollars (reduction in labor and other operating costs) that may result from four different alternatives coupled with the four possible outcomes of the current wage negotiations.

Possible outcomes

		s_1	s_2	s_3	s_4
Mutually	a_1	1	2	3	4
exclusive	a_2	3	2	1	0
alternatives	a_3	2	3	4	2
	a_4	3	4	2	1

Suppose there is reason to believe that $P[s_1] = 1/8$, $P[s_2] = 1/8$, $P[s_3] = 1/4$, and $P[s_4] = 1/2$. Determine which alternative is indicated by the following decision rules: (a) expectation, (b) most probable future, and (c) expectation-variance.

8.7 A heat exchanger is being installed as part of a plant modernization program. It costs $80,000, will last for four years, and is expected to reduce the overall plant fuel cost by x dollars per year, where

x	Prob $[x]$
$18,000	0.2
19,000	0.4
20,000	0.3
21,000	0.1

Estimates of the salvage value range from an optimistic $25,000 to a pessimistic $5,000. The most likely estimate is $10,000.
a. Find the expected annual reduction in fuel costs.
b. Find the Beta estimate expected salvage value. (*Hint:* An estimate of mean (expected) value of a *Beta-distributed random variable* is

$$m = \frac{a + 4b + c}{6}$$

where a is the minimum anticipated value, b is the most likely value, c is the maximum anticipated value, and m is the estimate of the mean value.)
c. Using the expected values of (a) and (b) and assuming a 10 percent discount rate, find the expected present worth for this proposal.
(*Answers:* (a) $19,300; (b) $11,667; (c) −$10,850)

8.8 A manufacturer is considering a new product:

Initial cost = $100,000

Cost per unit sold = $10

Discount rate = 20% before taxes

Service life = 15 years

Suppose that there are three estimates for number of units sold, revenue per unit sold, and salvage value:

	Optimistic	Most likely	Pessimistic
Number of units sold per year (x_1)	7,500	7,000	6,000
Revenue per unit sold (x_2)	$13	$12	$10
Salvage value at end of 15 years (x_3)	$22,000	$20,000	$15,000

a. Use the Beta distribution to provide estimates of the mean (expected) values for x_1, x_2, and x_3. (See problem 8.7b.)
b. Using your estimates from (a), find the expected equivalent uniform annual benefits.

8.9 Engineering economy can be used to evaluate measures designed to reduce the cost of risk, as illustrated by this problem.*

A firm currently pays a $550 per year fire insurance premium to provide $100,000 coverage. If a fire were to occur, however, the uninsured loss would be $50,000. That is, the *total* cost of fire damage would be $150,000, of which $100,000 would be reimbursed by the insurance company and $50,000 would be a cost to the firm.

The firm is considering the installation of a sprinkler system that would reduce the premium to $300 annually. The sprinkler system costs $4,500, has a fifteen-year life, and will have no salvage value at the end of fifteen years. Coverage would remain at $100,000, but the expected uninsured loss is only $20,000 with the sprinkler system in place.

The key to this problem is the question of probability that a loss will occur. Assume that the insurance premiums directly reflect this probability: annual insurance premium = (probability of loss)(expected damage). For example, the implied probability of loss with no sprinkler system is $550/$100,000 = 0.0055$. Assume that this same probability applies to uninsured damage. Let $i = 0.10$.

Complete the analysis. Determine the equivalent uniform annual cost both *with* and *without* the sprinkler. (*Answers: EUAC* (with) $= 952; *EUAC* (without) $= 825)

8.10 A certain retail firm estimates that the probability that a theft will occur during any one-month period is approximately 0.02. The probability that more than one theft will occur during any one-month period is zero. If a theft occurs, the firm will incur a cost of $10,000. Theft insurance may be purchased at a cost of x at the start of each month. The firm's effective minimum attractive rate of return is 1 percent per month before income taxes.

a. Based on the above assumptions, how much can the firm afford to pay for theft insurance?
b. Suppose that a more rigorous analysis of theft likelihood reveals the following estimates of probabilities:

*Adapted from G. A. Taylor, *Managerial and Engineering Economy*, 3rd ed. (New York: D. Van Nostrand, 1980), pp. 312–313.

Optimistic	0.018
Most likely	0.020
Pessimistic	0.025

Using the "Beta distribution assumption," how much can the firm afford to pay for theft insurance? (See Problem 8.7 for the mean value of a Beta-distributed random variable.)

8.11 A firm is planning to keep a certain old machine for another two years. The maintenance department advises that the machine will probably break down and require repair sometime during the two-year period unless an immediate overhaul is performed. It estimates that, without an overhaul, the probability is 0.5 that the machine will break down the first year and 0.3 that the machine will survive the first year but break down during the second year.

Repairing the machine after a breakdown would (a) cost an estimated $1,000, (b) guarantee against further breakdowns in the next two years, and (c) add $500 to the salvage value. Overhauling the machine now would (a) cost $800, (b) guarantee against further breakdown in the next two years, and (c) add $500 to the salvage value. The minimum attractive rate of return is 10 percent before taxes.

Use "expected cost" calculations and the end-of-year convention to determine whether the machine should be repaired now. Assume that cash expenses during any year will be concentrated at the end of that year.
(*Answers:* E[cost given overhaul now] = $387; E[cost given do nothing] = $372)

8.12 The Military Air Transport Service (MATS) operates a fleet of transport aircraft that is periodically overhauled after so many hours of operation at three different facilities: Upton Park, Flyby Service, and Mitchel Air Force Base.*

Major John Hardesty, a recently assigned cost analyst with MATS, notes while reviewing the maintenance plans for the three facilities that each facility follows a different procedure in overhauling the transports' pressurization system. Upton Park only ascertains whether the system is functioning properly. Flyby Service overhauls the system on each transport at an average unit cost of $24,000. Mitchel AFB replaces the entire system at an average unit cost of $42,000. Further examination by Major Hardesty discloses that, in a majority of cases, there are no major problems with the pressurization system between overhauls. Sometimes, however, there is a normal component malfunction during line operations that costs an average of $67,000 in repair and lost aircraft time. Occasionally, a system failure is more severe, causing damage not only to the pressurization system but

*This problem was prepared by R. J. Borntraeger, U.S. Civil Service Commission, for use in economic analysis training by the Financial Management and FPB Training Center, Bureau of Training, U.S. Civil Service Commission.

also to the aircraft structure and to other equipment housed near the pressurization components. When these malfunctions occur, special crews must be dispatched to make the repairs, and more aircraft time is lost. Historically, costs in such instances have averaged about $330,000.

Major Hardesty questions the incremental costs of overhauling and replacing the systems and suspects that the present practice employed by Upton may be the cheapest alternative. He directs his assistant, Lieutenant B. G. Hotshot, to compile past service data on systems serviced by each facility. Lt. Hotshot's sample investigation provided the following information:

	Upton	Flyby	Mitchel
Aircraft overhauled	87	160	100
Subsequent failures	48	48	10
Subsequent damage to other components	6	8	3
No subsequent problems	33	104	87

Based on the information available, is Major Hardesty's position sound? (*Hint:* Assume that the historical data provide the best estimates of future performance. Determine the expected cost per aircraft for each alternative.)

Decisions under Uncertainty

8.13 Suppose that nothing at all is known or can be estimated about the relative likelihoods of the future states of nature given in Problem 8.5. Which alternative(s) would be selected under each of the following principles of choice: (a) minimax, (b) minimin, (c) Hurwicz with $\alpha = 0.25$, (d) Savage, and (e) Laplace?
(*Answers:* (a) a_5; (b) a_3; (c) a_3; (d) a_3; (e) a_3)

8.14 Suppose that nothing at all is known or can be estimated about the probabilities of the future states of nature given in Problem 8.6. Which alternative(s) would be selected under each of the following principles of choice: (a) maximin, (b) maximax, (c) Hurwicz with $\alpha = 0.5$, (d) Savage, and (e) Laplace?

8.15 The following questions refer to the maximin principle.
a. Consider the following profit matrix:

	s_1	s_2
a_1	$1	$100
a_2	$2	$2

Alternative a_2 is chosen under the maximin principle of choice. What in general may be inferred from this problem?

b. Consider the following profit matrix (Case I):

	s_1	s_2
a_1	$1	$4
a_2	$3	$2

Suppose that additional information has been developed indicating that, if s_1 occurs, profits will increase by $3 over the original values. Moreover, profits will increase by $3 regardless of which alternative is selected. The revised profit matrix (Case II) is

	s_1	s_2
a_1	$4	$4
a_2	$6	$2

Comment on the maximin solutions for these two cases. (*Answers (partial):* (a) a_2; (b) a_2 under Case I and a_1 under Case II)

8.16 A manufacturing firm is planning to construct a large industrial smoke stack that, because of certain design characteristics, may be either 80 feet or 150 feet high. The tall stack can be accommodated only by a large footing, whereas the short stack can be built with either a large or a small footing. The initial costs are

Large footing—$40,000   Tall stack—$140,000
Small footing—$20,000   Short stack—$ 40,000

At first glance it appears that the company should build the short stack with the small footing with a total first cost of $60,000. Unfortunately, life is not quite so simple. If a short stack is built, the company risks being cited by the local air pollution control board and having to build a tall stack.

The tall stack and the footing are permanent, but the short stack will have to be replaced by a tall one in ten years. Using a minimum attractive rate of return of 10 percent before taxes, analyze this problem so that management will have a reasonable basis for decision. Assume that cash expenses during any year will be concentrated at the end of that year. (*Hint:* Construct a matrix of expected costs for the various combinations of alternatives and states of nature.)

8.17 A plant manager is considering two mutually exclusive alternatives: expand the existing warehouse (a_1) or build an additional warehouse (a_2). Three possible future events are likely: there will be no change in output other than normal growth (s_1), a new product, A, will be manufactured at the plant (s_2), or a new product, B, will be manufactured at the plant (s_3).

a. The matrix of equivalent uniform annual costs to be considered by the plant manager is

	s_1	s_2	s_3
a_1	$140,000	$300,000	$300,000
a_2	$200,000	$250,000	$250,000

Using the Laplace principle of insufficient reason, decide which alternative should be selected.

b. Suppose that the decision is reviewed by the comptroller at corporate headquarters. Noting that costs depend only on whether or not a new product line is manufactured at the plant, he simplifies the cost matrix as follows:

	s_1	s_2
a_1	$140,000	$300,000
a_2	$200,000	$250,000

In the revised matrix, s_2 represents the outcome "A new product line will be manufactured at the plant." Applying the Laplace principle, decide which alternative should be selected.

c. Comment on the results for (a) and (b).

(*Answers (partial)*: (a) a_2; (b) a_1)

Digital Computer (Monte Carlo) Simulation

8.18 A certain economic analysis consists of four parameters: initial cost, annual cash flows, service life, and discount rate. The analyst wants to account for the uncertainties inherent in the estimates.

a. First, consider the discount rate, i. It is estimated that the most likely value is 8 percent, with the actual value ranging somewhere from 6 percent to 13 percent. Assuming that this random variable is "Beta distributed"—drawn from a Beta distribution as described in Problem 8.7—find the mean value.

b. Now assume that the analyst has some additional information about the probability function for the discount rate:

Discount rate (i)	6%	7%	8%	9%	10%	11%	12%	13%
Probability $P(i)$	.05	.10	.25	.20	.15	.10	.10	.05

Find the *expected value*.

c. Assume that the cash flow at time t is a linear function of t and that t is a normally distributed random variable with a mean of ten years and standard deviation of one year. That is,

$$A(t) = 100t - 200$$

where

$$t = N(\mu = 10, \sigma = 1)$$

A digital computer (Monte Carlo) simulation process is used to generate random samples. On the first trial, the random normal deviate is -1.32. What is the resulting $C(t)$? That is, what value of $A(t)$ corresponds to a random normal deviate of -1.32?

d. Draw twenty-five random samples using the probability distribution in (c), then plot your results in the form of a frequency histogram. Compute the mean of your sample.

8.19 Suppose that you want to use digital computer simulation to simulate the present worth of a certain proposed project, where

$$PW = B - C$$

Both the benefits (B) and costs (C) are independent random variables with the following distributions:

B_i ($\times$ $1,000)	$P(B_i)$	C_i ($\times$ $1,000)	$P(C_i)$
195–205	0.1	195–205	0.1
205–215	0.1	205–215	0.3
215–225	0.2	215–225	0.2
225–235	0.3	225–235	0.2
235–245	0.2	235–245	0.1
245–255	0.1	245–255	0.1

In your simulation, use the midpoints of the ranges to represent the events. For example, if $195 \leq x \leq 205$, assume that $x = 200$.

a. What are the minimum and maximum possible values of present worth?
b. What are the expected values of benefits, costs, and present worth?
c. Draw twenty-five independent random samples (trials) and compute the average PW as the result of these trials. (If an electronic digital computer is available, conduct a minimum of 2,500 trials.)

(*Answers (partial):* (a) $-$$50,000, $50,000; (b) $E[PW]$ = $5,000)

8.20 A set of future cash flows are random variables. The resulting present worth is a normally distributed random variable with mean (μ) = $1,000 and standard deviation (σ) = $500. What is the probability that this project will result in a loss?

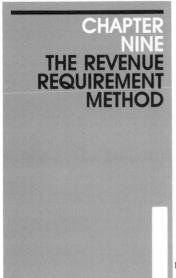

CHAPTER NINE
THE REVENUE REQUIREMENT METHOD

n the private sector of the economy, business firms are generally free to price goods and services so as to meet the owner's objectives. Revenues are affected by the interaction of market forces and the firm's investment strategies. However an important exception arises in the case of regulated utilities, whether owned by private shareholders or public entities. Because utilities hold monopolistic or oligopolistic positions within a given community, the community protects its interests by regulating the behavior of the utility through a "utilities commission" or a similar regulatory body.[1] Prices, or rates, charged by the utility are normally subject to review and approval by the regulatory agency.

The philosophy underlying the regulation of prices (rates) is that utilities are entitled to price their products and services so that all costs are recovered, including a *fair return* on the rate base. (There is no guarantee that a utility will in fact earn a fair return for a permitted rate structure. Operational experience that occurs after the authorization may well prove to be different from what was predicted, or forecasted, when authorization to charge certain rates was originally sought.) Because regulatory agencies act on behalf of the consumers of the utility's products and services, investment decisions should be made in such a way that revenue requirements, while meeting the costs and fair return of the utility, are minimized.

The **revenue requirement method,** outlined in the following sections, has been used for many years by utilities whose rates are subject to government regulation. Since there is an equivalence between this technique and the present worth method, the revenue requirement method could be used by nonregulated industries as well.

1. In California, the regulatory body is the State Public Utilities Commission (PUC).

As illustrated in Figure 9.1, the permitted revenue (R), or **revenue requirement,** may, in general, be characterized as the sum of six elements:

(K) Current operating disbursements (labor and material costs directly associated with the investment, but also other expenses such as property taxes and insurance)

(D) Depreciation expense

(I) Interest on that portion of the investment representing debt capital

(T) Income taxes paid on taxable income

(P) Dividends paid to preferred stockholders based on the portion of the investment that represents funds raised through preferred stock

(S) Return to owners, that is, return to the utility's common stockholders

Figure 9.1. Elements of Permitted Revenue (Revenue Requirement)

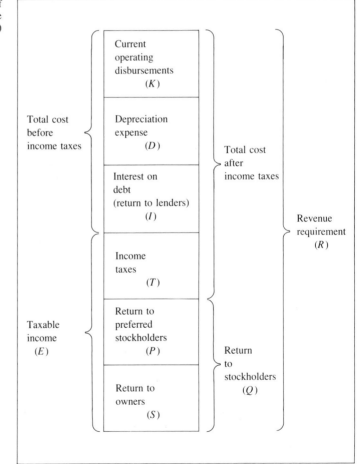

It may be shown that there are several intermediate relationships. The total cost before income taxes is $K + D + I$, and the *taxable income* (E) is given by

$$E = R - (K + D + I) \qquad \textbf{(9.1)}$$

Income taxes (T) are equal to tE, where t is the effective income tax rate.[2]

At this point it will be helpful to review the concepts underlying the "costs" of the various types of capital invested by the utility.[3] These concepts, as you will see, are essential to the development of the revenue requirement models.

Broadly speaking, a utility's investment funds consist of debt and equity capital. **Debt** is simply borrowed money, the cost of which is reflected by interest payments to lenders. **Equity capital** is obtained through investments by the owners—taxpayers, in the case of publicly owned utilities—and through retained earnings. There is, of course, an opportunity cost associated with equity capital in the sense that these funds could be employed elsewhere if not used for investment by the utility. In addition to debt and equity, utilities (and other firms) frequently obtain needed investment funds by issuing **preferred stock.** Dividends on preferred stock are paid at a specified rate after all other expenses, including interest on debt, are paid, but before returns to common stockholders are paid.[4] In summary, then, if investment funds are made available by lenders as well as by common and preferred stockholders, the investment must provide sufficient revenues to reward lenders as well as common and preferred stockholders.

The **weighted average cost of capital** (i) after income taxes is given by

$$i = w_b k_b + w_p k_p + w_e k_e \qquad \textbf{(9.2)}$$

where

$w_b =$ proportion of debt (borrowing)

$w_p =$ proportion of preferred stock

$w_e =$ proportion of common stock (equity)

$\left.\begin{array}{l} \end{array}\right\} \; \begin{array}{l} w_b + w_p \\ + w_e = 1 \end{array}$

$k_b =$ cost of debt

2. Inasmuch as the notation introduced in this chapter differs somewhat from the notation used in other chapters, a special glossary has been included as Appendix 9A.

3. The cost of capital is discussed at length in Chapter 11.

4. Dividends to preferred stockholders are not guaranteed, of course. These dividends can be paid only after more senior obligations (debt, for example) have been met.

$$k_p = \text{cost of preferred stock}$$

$$k_e = \text{cost of equity}$$

The cost of debt (k_b) is the average interest rate paid for the borrowed capital committed for investment. Similarly, the cost of preferred stock (k_p) is the average dividend rate paid on investment funds raised through issuing preferred stock. The cost of equity (k_e) is that **fair return** necessary to attract investors (in the case of investor-owned utilities) or to compensate taxpayers (in the case of publicly owned utilities).

To illustrate, suppose that a utility's investment capital consists of the following:

Source	Proportion (w)	Cost (k)
Debt	0.55	0.10
Preferred stock	0.10	0.10
Common stock	0.35	0.14

The weighted average cost of capital is

$$i = 0.55(0.10) + 0.10(0.10) + 0.35(0.14)$$

$$= 0.055 + 0.010 + 0.049$$

$$= 0.114$$

9.3 DETERMINING INCOME TAXES

The income taxes paid by the utility in the jth year (T_j) are found by multiplying the effective tax rate for that year (t_j) by taxable income in that year (E_j). From Equation 9.1,

$$T_j = t_j E_j$$
$$= t_j [R_j - (K_j + D_j + I_j)] \qquad (9.3)$$

where

$$R_j = K_j + D_j + I_j + T_j + P_j + S_j \qquad (9.4)$$

Combining Equations 9.3 and 9.4 and solving for T_j,

$$T_j = t_j [(K_j + D_j + I_j + T_j + P_j + S_j) - (K_j + D_j + I_j)]$$
$$= t_j T_j + t_j (P_j + S_j)$$
$$= \left[\frac{t_j}{1 - t_j} \right] (P_j + S_j) \qquad (9.5)$$

Since P_j and S_j are the returns to the preferred and common stockholders, Equation 9.5 can be formulated somewhat differently by noting that $P = w_p k_p \hat{B}_j$ and $S = w_e k_e \hat{B}_j$, where $\hat{B}_j$ is the unrecovered investment in year j.

From Equation 9.2,

$$i_j = w_{bj} k_{bj} + w_{pj} k_{pj} + w_{ej} k_{ej}$$
$$= w_{bj} k_{bj} + (P_j + S_j)/\hat{B}_j$$

Thus Equation 9.5 can be rewritten

$$T_j = \left[\frac{t_j}{1 - t_j}\right][i_j - w_{bj} k_{bj}]\hat{B}_j \qquad (9.6)$$

A basic model can be developed adopting certain simplifying assumptions:

1. The tax rate (t_j) remains constant throughout the planning horizon $(j = 1, 2, \ldots, N)$.

2. The costs of debt, preferred stock, and equity $(k_b, k_p,$ and $k_e)$ remain constant throughout the life of the investment. Furthermore, the proportions of debt, preferred stock, and equity (w_b, w_p, w_e) also remain constant for all values of j.

3. The amount of unrecovered invested capital remaining in year j $(\hat{B}_j)$ is equal to the book value of the investment at the *start* of the year (B_j).

4. Let $\hat{D}_j = \hat{B}_j - \hat{B}_{j+1}$, as shown in Figure 9.2. That is, the reduction in unrecovered invested capital in year j $(\hat{D}_j)$ is the change in book value from the start of year j to the start of year $j + 1$. Let D_j represent the depreciation expense affecting taxable income in year j. *Assume here that* $D_j = \hat{D}_j$. Put somewhat differently, assume that the book value of the investment (B_j) represents the unrecovered investment $(\hat{B}_j)$ at that point in time. The importance of this assumption stems from the fact that several of the elements of the permitted revenue are functions of $\hat{B}_j$: $I_j = w_b k_b \hat{B}_j$, $P_j = w_p k_p \hat{B}_j$, $S_j = w_e k_e \hat{B}_j$, and $T_j = \left[\frac{t}{1 - t}\right][i - w_b k_b]\hat{B}_j$. Calculations are greatly simplified if it is assumed that $\hat{B}_j = B_j$.

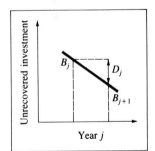

Figure 9.2

5. The straight line method is used for computing annual depreciation expenses. Thus the depreciation expense (D) remains constant over the depreciable life of the investment, and

$$D = \frac{C - L}{N} \qquad (9.7)$$

where C is the amount of the initial investment (cost basis), L is the expected salvage value, and N is the depreciable life. It also follows that the book value at the *start* of year j is given by

$$B_j = C - (j - 1)D \qquad (9.8)$$
$$= C + D - jD$$

Given these assumptions, Equation 9.6 can be rewritten as

$$T_j = \left[\frac{t}{1-t}\right][i - w_b k_b][C + D - jD]$$
$$= \alpha[C + D - jD] \qquad (9.9)$$
$$= \alpha[C + D] - j(\alpha D)$$

where

$$\alpha = \left[\frac{t}{1-t}\right](i - w_b k_b)$$

As we examine the year-by-year taxes, it may be seen that the taxes decrease by a uniform amount, αD, each year:

$$
\left.
\begin{aligned}
T_1 &= \alpha(C + D) - \alpha D = \alpha C \\
T_2 &= \alpha(C + D) - 2\alpha D = \alpha C - \alpha D \\
T_3 &= \alpha(C + D) - 3\alpha D = \alpha C - 2\alpha D \\
&\;\;\vdots \\
T_N &= \alpha(C + D) - N\alpha D = \alpha C - (N - 1)\alpha D
\end{aligned}
\right\}
\begin{aligned}
&\text{gradient} \\
&= \alpha D
\end{aligned}
$$

Thus the uniform annual series, T, equivalent to the series of income taxes $(T_1, T_2, \ldots, T_N)$, can be found from

$$T = \alpha C - \alpha D(A/G, i, N)$$
$$= \alpha[C - D(A/G, i, N)] \qquad (9.10)$$

(The annualized income taxes are generally known as **levelized taxes** in the literature of utility economics.)

It is frequently useful to express income taxes as a percentage of the initial investment. Let this ratio be defined as $\tau = T/C$. Then

$$\tau = \alpha\left[1 - \frac{D}{C}(A/G, i, N)\right] \qquad (9.11)$$

But $D/C = (1 - c)/N$, where $c = L/C$. Thus

$$\tau = \alpha\left[1 - \left(\frac{1 - c}{N}\right)(A/G, i, N)\right] \qquad (9.12)$$

After additional algebraic manipulation, it may be shown that

$$\tau = \left[\frac{t}{1 - t}\right]\left[1 - w_b k_b/i\right]$$
$$\times \left[i - \left(\frac{1 - c}{N}\right) + (1 - c)(A/F, i, N)\right] \qquad (9.13)$$

Both Equations 9.12 and 9.13 may be found in the literature of utility economics. Of course, they give identical results.

State income taxes are generally treated as a deductible business expense on federal income tax returns, whereas federal income taxes are not deductible on state returns. When a utility pays income taxes to both the federal and state governments and procedures for computing taxable income are the same in both instances, the appropriate *combined* rate to use in the analysis is

9.5 COMBINED INCREMENTAL INCOME TAX RATE

$$t = t_s + t_f(1 - t_s) \qquad (9.14)$$

where t_s = incremental state income tax rate and t_f = incremental federal income tax rate. To illustrate, suppose that the state income tax rate is 0.095 and the federal income tax rate is 0.46. Then the combined incremental income tax rate is

$$t = 0.095 + 0.46(1 - 0.095) = 0.511$$

This is the rate that should be used in Equations 9.12 and 9.13.

The cost of capital recovery is an important element in evaluating proposed investments by public utilities. Given an initial investment (C), expected terminal salvage value (L), and service life (N), as portrayed in Figure 9.3, the cost of capital recovery (CR) is the equivalent uniform annual series of the two cash flows:

9.6 THE COST OF CAPITAL RECOVERY

$$CR = Ci + (C - L)(A/F, i, N) \qquad (9.15)$$

Expressing the cost of capital recovery as a fraction of the original investment

$$\chi = (CR)/C = i + (1 - c)(A/F, i, N) \qquad (9.16)$$

where $c = L/C$ as before.

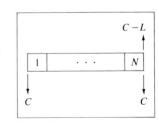

Figure 9.3

It is often useful to express the fixed costs of a proposed investment or class of investments as a constant percentage of the investment. To illustrate this procedure, consider the following example:

Capital Structure

Source	Proportion (w)	Cost (k)
Debt	0.55	0.10
Preferred stock	0.10	0.10
Common stock	0.35	0.14

Income tax rates: State $t_s = 0.095$
                    Federal $t_f = 0.460$

The investment under consideration has a depreciable life (N) of twenty-eight years and an expected salvage value (L) of zero. *Ad valorem* taxes are assumed to be 1 percent of the initial cost each year; annual maintenance costs are assumed to be 2 percent of the initial cost. Adopt all other assumptions as specified in Section 9.4.

The capital structure for this problem is the same as that used earlier to illustrate the computation of the weighted average cost of capital. From Equation 9.2, $i = 0.114$. Compound interest factors for $i = 0.114$ and $N = 28$ are summarized in Appendix 9C. The appropriate income tax rate was also computed in an earlier illustration: $t = 0.095 + 0.46(1 - 0.095) = 0.511$. Because the salvage value (L) is zero, the factor $c = L/C = 0$.

The annual **fixed charge rate** (FCR) is now determined as follows:

Income taxes (Equation 9.13)

$$\tau = \left[\frac{0.511}{1 - 0.511}\right]\left[1 - \frac{0.55(0.10)}{0.114}\right]\left[0.114 - \left(\frac{1 - 0}{28}\right)\right.$$
$$\left. + (1 - 0)(A/F, 11.4\%, 28)\right] = 0.0455$$

Capital recovery (Equation 9.16)

$\chi = 0.114 - (1 - 0)(A/F, 11.4\%, 28) =$	0.1198
Ad valorem taxes	$= 0.01$
Maintenance	$= 0.02$
Total fixed charge rate	$= 0.1953$

That is, for each dollar invested, 19.53 cents of revenue is needed each year for twenty-eight years to cover these fixed charges.

In general, the fixed charge rate may be used to determine the **levelized annual revenue requirement** for proposed investments that meet the assumptions as outlined previously. Suppose, for example, that the utility is considering the construction and operation of a steam-powered generating plant that will produce 300,000 kilowatts (kw) of electric energy.[5] The investment cost per kw is $350; operating costs, other than fixed charges, are anticipated to be $17 per kw per year. If the plant is constructed, it is expected that it will be operated for twenty-eight years with no residual salvage value. The fixed charge rate, as outlined in Section 9.7, is 0.1953. (The FCR includes income taxes, capital recovery, *ad valorem* taxes, and maintenance.) Thus the total equivalent uniform annual cost per kw for the proposed investment is

$$(\$350 \times 0.1953) + \$17 = \$85.355.$$

The toal cost for 300,000 kw is $25,607,000. This is the levelized annual revenue that the utility requires if it is to recover all relevant costs of the investment.

Once we recognize that the levelized annual revenue requirement (for a given investment proposal) is the equivalent uniform annual cost, it follows that alternatives may be rank-ordered by simply minimizing the revenue requirement. Thus the annual worth method or the present worth method may be applied as outlined in Chapter 3.

To illustrate, suppose that the utility is considering an alternative to the steam-powered plant described above. The alternative plant is powered by internal combustion, has an initial cost of $150 per kw, and has an operating life of fourteen years. The residual value at the end of its service life is expected to be negligible (zero). The fixed charge rate is

Income taxes (τ)	0.0405
Capital recovery (χ)	0.0817
Ad valorem taxes	0.0100
Maintenance	0.0200
Total FCR	0.1522

If it is assumed that total operating costs, other than fixed charges, will be $70 per kw each and every year over the fourteen-year

---

5. The numerical illustration in this chapter is based on an example in: J. R. Canada and J. A. White, *Capital Investment Decision Analysis for Management and Engineering*, (Englewood Cliffs, N.J.: Prentice-Hall, 1980).

service life, then the total equivalent uniform annual cost, or revenue requirement, is

$$(\$150 \times 0.1522) + \$70 = \$92.83.$$

If the plant is sized to generate 300,000 kw annually, then the total annual revenue requirement would be $27,849,000.

It should be emphasized that direct comparison of the annualized (levelized) revenue requirements—$25.61 million for the steam plant versus $27.85 million for the internal combusion plant—is valid *only* under two critical assumptions. First, it must be assumed that there will be a need for this power output for at least twenty-eight years, the least common multiple of the two alternatives. Second, it must be assumed that the cash flow consequences during the first life cycle of the internal combustion plant will also occur during the second life cycle. (See Figure 9.4.) If these assumptions hold, then annual worths (costs) may be compared directly. Otherwise, it is necessary to specify the likely consequences after the shorter-lived alternative completes its first life cycle.

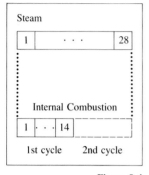

Figure 9.4

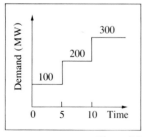

Figure 9.5

9.10 STAGGERED DEVELOPMENT

In the event that the utility is faced with growing demand over time, it may be worthwhile to stagger, or phase, the construction of a new plant so as to bring capacity on-line when it is needed. To illustrate, suppose that the electricity generating plants described previously are to be used to meet a demand of 100 megawatts (mw) annually during the first five years, 200 mw annually during years 6 through 10, and 300 mw annually thereafter. (See Figure 9.5.) Assume here that the utility may choose between *either* building a 300 mw steam plant now *or* constructing internal combustion plants in phases to meet demand when it occurs. There are two methods that are generally used when addressing this type of problem: the *repeated plant method* and the *co-terminated plant method*. Each is appropriate under somewhat different assumptions.

The Repeated Plant Method

The **repeated plant method** assumes an infinite series of identical replacements. For the steam-powered plant, for example, it is assumed that the plant will be replaced at the end of its first cycle, that is, after twenty-eight years; with a new facility that will be identical in every respect to the first; there will be another identical replacement after the second twenty-eight years; and so on. We have determined that the revenue requirement (R) is $85.355 per kw for the steam plant. Since there are an "infinite" series of replacements, we find that the equivalent present worth of costs

$(PWOC)$ of this infintie series is[6]

$$PWOC = R/i$$
$$= \$85.355/0.114$$
$$= \$748.73 \text{ per kw}$$

To produce 300 mw, the total cost is $\$748.73 \times 300,000 = \$224,619,000$. The calculations leading to this result are summarized in Table 9.1.

The total equivalent present worth of costs may be determined in a similar manner for the internal combustion alternative. In this instance, however, only one 100 mw plant will be built initially. Phase B, the second plant, will be built after five years, and Phase C, the third internal combustion plant, will not be available until the start of the eleventh year (end of year 10).[7] Thus the calculations of the present worths of infinite series for the three phases, as indicated in Table 9.1, are end of year 0 for Phase A, end of year 5 for Phase B, and end of year 10 for Phase C.

Table 9.1
Determination of Capitalized Cost
under the Repeated Plant Method

Years of operation over planning horizon

| Cycle | Steam | Internal combustion | | |
		Phase A	Phase B	Phase C
1st	1–28	1–14	6–19	11–24
2nd	29–56	15–28	20–33	25–38
3rd	57–84	29–42	34–47	34–52
⋮	⋮	⋮	⋮	⋮
Revenue Requirement (R) during each year of life cycle	$ 85.355	$814.30	$814.30	$814.30
Equivalent *PWOC* at 1st year of 1st cycle	$748.73	$814.30	$814.30	$814.30
Equivalent *PWOC* at 1st year of planning horizon	$748.73	$814.30	$747.63	$276.65
Capacity (kw)	300,000	100,000	100,000	100,000
Total equivalent *PWOC* (in millions)	$224.619	$ 81.430	$ 47.463	$ 27.665
			$156.558	

6. Recall that $(P/A, i, \infty) = 1/i$

7. The artificiality of this illustration is obvious: Clearly, it is impossible to build 100 mw of power-generating capacity instantaneously, that is, at the end of a given year. The construction would almost certainly require a substantial period of time, perhaps a year or longer. Literary license is exercised in this case so that numerical complexity will be minimized and attention more sharply focused on other elements of the analysis.

The equivalent amounts for Phase B and Phase C must be discounted to the beginning of the planning horizon, that is, to the end of year zero.

Phase B   ($814.30/kw)($P/F$, 11.4%, 5)   = $474.63/kw
Phase C   ($814.30/kw)($P/F$, 11.4%, 10) = $276.65/kw

The capacity of each phase is 100,000 kilowatts. The total equivalent present worth of costs, then, is

Phase A   100,000 kw × $814.30/kw = $ 81,430,000
Phase B   100,000 kw × $474.63/kw =   47,463,000
Phase C   100,000 kw × $276.65/kw =   27,665,000
                Total cost for 300 mw $156,558,000

This is approximately $68 million lower than the cost that results from the steam plant built to generate 300 megawatts immediately: $224,619,000 − $156,558,000 = $68,061,000.

The Co-Terminated Plant Method
The **co-terminated plant method** assumes that all alternatives are terminated at the end of some specified planning horizon. It requires estimates of the residual values of any assets that will (or could) remain in service at the end of the planning horizon. To illustrate this method, suppose that the internal combustion plant will be built in phases to meet demand, as before, but that the planning horizon is only twenty-eight years. The economic analysis is summarized in Table 9.2.

Phases A, B, and C are constructed at the end of years 0, 5, and 10 respectively. Since each facility has a life of fourteen years, there will be two cycles for each phase. At the end of year 28, Phase A has no remaining life, Phase B has a remaining life of five years, and Phase C has a remaining life of ten years. (See Figure 9.6.) The ages of Phases A, B, and C at time of termination are 14, 9, and 4 years, respectively.

Assume here that the residual value of Phases B and C will be equal to the book values of these assets at the time of termination. (This assumption may not be suitable for all applications. The analyst should adopt the assumption that appears most reasonable under any given set of circumstances.) Using straight line depreciation, the book value at the start of the jth year is:

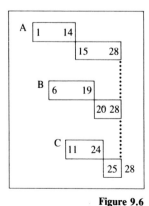

Figure 9.6

$$B_j = C - (j - 1)D$$

Table 9.2
Determination of Revenue Requirement under the Co-Terminated Plant Method

Years of operation over 28-year planning horizon

Cycle	Phase A	Phase B	Phase C
1st	1–14	6–19	11–24
2nd	15–28	20–28	25–28
Age of plant at time of co-termination	14 years	9 years	4 years
Estimated residual value at end of planning horizon (at end of year 28)	0	$5,357,000	$10,714,000
(% of initial cost)	(0.000)	(0.357)	(0.714)
Fixed charge rate during 2nd cycle			
Income taxes (τ)	0.0405	0.0472	0.0556
Capital recovery (χ)	0.1463	0.1586	0.1744
Ad valorem taxes	0.0100	0.0100	0.0100
Maintenance	0.0200	0.0200	0.0200
Total FCR	0.2168	0.2358	0.2600

where

$$C = 100,000 \text{ kw} \times \$150/\text{kw} = \$15,000,000$$

$$D = (C - L)/N = \$15,000,000/14 = \$1,071,429$$

Thus, for Phase B, $B_{10} = \$15,000,000 - (10 - 1)(\$1,071,429)$ = $5,357,000, and, for Phase C, B_5 = $15,000,000 - $(5 - 1)(\$1,071,429) = \$10,714,000$. (See Figure 9.7.) Using

Figure 9.7

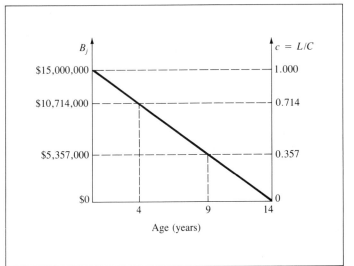

these estimated residual values, the ratio of residual value to initial cost ($c = L/C$) may be shown to be 0, 0.357, and 0.714 for Phases A, B, and C, respectively.

The fixed charge rate of income taxes (τ) for the three phases during each of their two cycles may be determined by Equation 9.13. For the *first* cycle, $N = 14$ and $c = 0$ in all three phases. Thus

$$\tau_1 = \left\{\left[\frac{0.511}{1 - 0.511}\right]\left[1 - \frac{0.55(0.10)}{0.114}\right]\right\}$$
$$\times \left[0.114 - \left(\frac{1 - 0}{14}\right) + (1 - 0)(A/F, 11.4\%, 14)\right]$$

$$= \{0.540828\}[0.07484]$$

$$= 0.0405 \text{ for Phases, A, B, and C}$$

Note that the value in {braces} is α and is independent of service life (N) and terminal salvage value (L). For the *second* cycle, the τ values are:

$$\tau_2(A) = 0.0405$$

$$\tau_2(B) = \{0.540828\}\left[0.114 - \left(\frac{1 - 0.357}{9}\right)\right.$$
$$\left. + (1 - 0.357)(A/F, 11.4\%, 9)\right]$$

$$= 0.0472$$

$$\tau_2(C) = \{0.540828\}\left[0.114 - \left(\frac{1 - 0.714}{4}\right)\right.$$
$$\left. + (1 - 0.714)(A/F, 11.4\%, 4)\right]$$

$$= 0.0556$$

In an analogous manner, the fixed charge rate for capital recovery during the first cycle ($N = 14$, $L = 0$) is found from Equation 9.16:

$$\chi_1 = 0.114 + (1 - 0)(A/F, 11.4\%, 14)$$
$$= 0.1463 \text{ for Phases A, B, and C}$$

For the second cycle, the χ values are:

$$\chi_2(A) = 0.1463$$

$$\chi_2(B) = 0.114 + (1 - 0.357)(A/F, 11.4\%, 9) = 0.1586$$

$$\chi_2(C) = 0.114 + (1 - 0.714)(A/F, 11.4\%, 4) = 0.1744$$

Assuming that the fixed charge rates for *ad valorem* taxes and routine maintenance remain constant during the second cycles of Phases A, B, and C, the total FCR's are as shown in Table 9.2. Note that the FCR for Phase A in the second cycle (0.2168) is the FCR for all phases in their first cycles.

The equivalent uniform annual costs (revenue requirements) during the relevant years of the twenty-eight-year planning horizon are shown schematically in Figure 9.8. The revenue requirements during each cycle are found by summing the fixed cost and other costs. The fixed cost is the product of the fixed charge rate and the initial cost. Thus

$$R_1(A) = R_2(A) = 0.2168(\$150/\text{kw}) + \$70/\text{kw}$$
$$= \$102.52/\text{kw} = \$10,252,000 \text{ for } 100 \text{ mw}$$
$$R_1(B) = \$102.52/\text{kw} = \$10,252,000 \text{ for } 100 \text{ mw}$$
$$R_2(B) = 0.2358(\$150/\text{kw}) + \$70/\text{kw}$$
$$= \$105.37/\text{kw} = \$10,537,000 \text{ for } 100 \text{ mw}$$
$$R_1(C) = \$102.52/\text{kw} = \$10,252,000 \text{ for } 100 \text{ mw}$$
$$R_2(C) = 0.26(\$150/\text{kw}) + \$70$$
$$= \$109.00/\text{kw} = \$10,900,000 \text{ for } 100 \text{ mw}$$

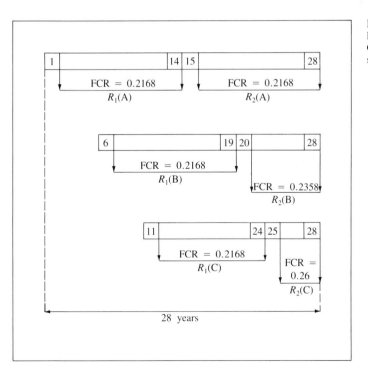

Figure 9.8. Equivalent Revenue Requirements: Internal Combustion Plant (note that strategy C is not drawn to scale)

The equivalent present values at the beginning of the twenty-eight-year planning horizon are[8]

$$P(A) = [R(A)](P/A, 11.4\%, 28)$$

$$= \$10,252,000(8.3451)$$

$$= \$85,554,000$$

$$P(B) = \{[R_2(B)](P/A, 11.4\%, 23) - [R_2(B) - R_1(B)]$$
$$\times \cdots \times (P/A, 11.4\%, 14)\}(P/F, 11.4\%, 5)$$

$$= \{\$10,537,000(8.03955)$$
$$- \$285,000(6.8368)\}(0.58287)$$

$$= \$48,241,000$$

$$P(C) = \{[R_2(C)](P/A, 11.4\%, 28) - [R_2(C) - R_2(B)]$$
$$\times \cdots \times (P/A, 11.4\%, 14)\}(P/F, 11.4\%, 10)$$

$$= \{\$10,900,000(7.51543)$$
$$- \$648,000(6.8368)\}(0.33974)$$

$$= \$26,326,000$$

Thus the equivalent present value of the revenue requirements for the internal combustion plant constructed in three 100 mw phases is $P(A) + P(B) + P(C) = \$160,120,000$. Expressed as an equivalent annual (levelized) revenue requirement:

$$R = \$160,120,000(A/P, 11.4\%, 28)$$

$$= \$19,187,000$$

8. See Appendix 9C for interest factors when $i = 11.4\%$.

Table 9.3
Accounting and Cash Flow Data for Steam Plant Example

			Participation			Return to participants		
Year j (1)	Depreciation expense D_j (2)	Book value at start of year B_j (3)	Debt (4)	Preferred stock (5)	Common stock (6)	Lenders I_j (7)	Preferred stockholders P_j (8)	Owners S_j (9)
1	$12.5	$350.0	$192.500	$35.00	$122.500	$19.2500	$3.500	$17.1500
2	12.5	337.5	185.625	33.75	118.125	18.5625	3.375	16.5373
3	12.5	325.5	178.750	32.50	113.750	17.8750	3.250	15.9250
⋮	⋮	⋮	⋮	⋮	⋮	⋮	⋮	⋮
27	12.5	25.0	13.750	2.50	8.750	1.3750	0.250	1.2250
28	12.5	12.5	6.875	1.25	4.375	0.6875	0.125	0.6125
Increment	0	−12.5	−6.875	−1.25	−4.375	−0.6875	−0.125	−0.6125

It may be shown that the revenue requirement method is directly related to the more traditional approaches widely used in non-utility industries: the present worth method, the (internal) rate of return method, and the like. Recall that these more "conventional" techniques depend on the determination of after-tax cash flows, from which the appropriate figure of merit (PW, AW, IRR, and so on) may be computed. We will examine the steam-plant example presented earlier to illustrate these relationships.

The relevant calculations for this example are summarized in Table 9.3. All costs and revenue figures in the table are per kilowatt. The various columns are described as follows:

Column (1)   Planning horizon (life of investment)
 = 28 years

Column (2) $D_j = (C - L)/N = \$350/28$
 = \$12.5 for all j

Column (3) $B_j = C + D - jD = C - D(j - 1)$
 = \$350 - \$12.5$(j - 1)$

Column (4)   Lenders' share = $w_b B_j = 0.55 B_j$

Column (5)   Preferred stockholders' share = $w_p B_j$
 = $0.10 B_j$

Column (6)   Common stockholders' (owners') share
 = $w_e B_j = 0.35 B_j$

Column (7) $I_j = k_b w_b B_j = 0.10 \times$ Column 4
 = $0.055 B_j$

Column (8) $P_j = k_p w_p B_j = 0.10 \times$ Column 5
 = $0.010 B_j$

Column (9) $S_j = k_e w_e B_j = 0.14 \times$ Column 6
 = $0.049 B_j$

Return to all participants is $0.114 B_j$

Table 9.3 (continued)

End of year j (1)	Revenue R_j (10)	Miscellaneous pretax cash flows K_j (11)	Taxable income (\$) E_j (12)	Income taxes (\$) T_j (13)	Permitted revenue (\$) R_j (14)
1	R_1	\$27.5	R_1 − 59.2500	$0.511R_1$ − 30.27675	$0.511R_1$ + 49.62325
2	R_2	27.5	R_2 − 58.5625	$0.511R_2$ − 29.9254375	$0.511R_2$ + 48.5495625
3	R_3	27.5	R_3 − 57.8750	$0.511R_3$ − 29.574125	$0.511R_3$ + 47.475875
⋮	⋮	⋮	⋮	⋮	⋮
27	R_{27}	27.5	R_{27} − 41.3750	$0.511R_{27}$ − 21.142625	$0.511R_{27}$ + 21.707375
28	R_{28}	27.5	R_{28} − 40.6875	$0.511R_{28}$ − 20.7913125	$0.511R_{28}$ + 20.6336875
Increment	0	0	0.6875	0.3513125	1.0736875

Column (10) R_j is to be determined

Column (11) K_j = *ad valorem* taxes, routine maintenance, and other operating costs

$$= (0.01 + 0.02)C + \$17 = \$10.5$$
$$+ \$17 = \$27.5$$

Column (12) $E_j = R_j - (D_j + I_j + K_j)$

$$= R_j$$
$$- [\text{Column 2} + \text{Column 7} + \text{Column 11}]$$
$$= R - [\$12.5 + 0.055B_j + \$27.5]$$
$$= R - \$40 - 0.055B_j$$

Column (13) $T_j = tE_j = 0.511\,E_j = 0.511(\text{Column 12})$

Column (14) $R_j = D_j + I_j + P_j + S_j + K_j + T_j$

$$= \text{Column 2} + \text{Column 7} + \text{Column 8}$$
$$+ \text{Column 9} + \text{Column 11} + \text{Column 13}$$

Note here that the *levelized* annual requirements may be determined directly from the values given in Column 14 of Table 9.3.
That is, we may find the uniform annual amount (R) that is
equivalent to the series $R_1, R_2, \ldots, R_{28}$:

$$R = 0.511R + \$49.62325$$
$$- \$1.0736875(A/G, 11.4\%, 28)$$

$$R - 0.511R = \$49.62325 - \$1.0736875(7.3396)$$

$$R = \$85.36$$

This, of course, is the same value as that obtained in the earlier
computation.

A somewhat different, but equivalent, approach is illustrated
in Table 9.4. Here, the cash flows for income taxes are computed
directly through application of Equation 9.6.

$$T_j = \left[\frac{t}{1-t}\right][i - w_b k_b]B_j$$

$$= \left[\frac{0.511}{1-0.511}\right][0.114 - 0.55(0.10)]B_j$$

$$= 0.06165B_j$$

The year-by-year revenue requirements, then, are obtained by
using Equation 9.4.

$$R_j = D_j + I_j + P_j + S_j + K_j + T_j$$

$$= [\text{Columns 2, 7, 8, 9, and 11 from Table 9.3}]$$
$$\text{plus [Column 3 from Table 9.4]}$$

Table 9.4

Table 9.4
Income Taxes and Permitted Revenue:
Steam Plant Example

Year j (1)	Book value at start of year B_j (2)	Income taxes T_j (3)	Permitted revenue R_j (4)
1	$350.0	$21.57904	$101.47904
2	337.5	20.80836	99.28336
3	325.0	20.03768	97.08768
⋮	⋮	⋮	⋮
27	25.0	1.54136	44.39136
28	12.5	0.77068	42.19568
Increment	−12.5	−0.77068	−2.19568

Note also that

$$R_j = D_j + I_j + P_j + S_j + K_j + T_j$$

$$= \left[\frac{C-L}{N}\right] + k_b w_b B_j + k_p w_b B_j + k_e w_e B_j + K$$

$$+ \left[\frac{t}{1-t}\right](i - w_b k_b)B_j$$

$$= D + iB_j + K + \left[\frac{t}{1-t}\right](w_p k_p + w_e k_e)B_j$$

$$= \$12.5 + 0.114B_j + \$27.5$$

$$+ \left[\frac{0.511}{0.489}\right](0.010 + 0.049)B_j$$

$$= \$40 + 0.17565B_j$$

This equation, of course, is the easiest way to compute the year-by-year revenue requirements. But however these values are determined, the levelized annual revenue requirement (R) may be calculated by noting the arithmetic gradient series in Column 4 of Table 9.4.

$$R = \$101.47904 - \$2.19568(A/G, 11.4\%, 28)$$

$$= \$85.36 \text{ per kw}$$

Again, this is the same result as obtained previously.

The *cash flows* expected to result from the proposed investment in the steam plant are summarized in Table 9.5. For years 1 through 28 the net after-tax cash flow in year j (Y_j) is equal to the permitted revenue (R_j) less the negative cash flows: principal (U_j) and interest (I_j) payments on that part of the investment representing borrowed funds, preferred stock dividends (P_j), and

Table 9.5

Cash Flow Summary: Steam Plant Example

End of year j (1)	Revenue R_j (2)	Loan Principal U_j (3)	Loan Interest I_j (4)	Preferred stock Payments V_j (5)	Preferred stock Dividends P_j (6)	Cost of asset (7)	Miscellaneous pretax cash flows K_j (8)	Income taxes T_j (9)	Net cash flow after taxes Y_j (10)
0	$ ——	$191.400	$ ——	$35.00	$ ——	− $350	$ ——	$ ——	− $122.5
1	101.479	− 6.875	− 19.25	− 1.25	− 3.5		− 27.5	− 21.579	21.525
2	99.283	− 6.875	− 18.5625	− 1.25	− 3.375		− 27.5	− 20.808	20.9125
3	97.088	− 6.875	− 17.875	− 1.25	− 3.25		− 27.5	− 20.038	20.3
⋮	⋮	⋮	⋮	⋮	⋮		⋮	⋮	⋮
27	44.391	− 6.875	− 11.375	− 1.25	− 0.25		− 27.5	− 11.541	5.6
28	42.196	− 6.875	− 0.6875	− 1.25	− 0.125	0	− 27.5	− 0.771	4.9875
Increment	− 2.196	0	0.6875	0	0.125		0	0.771	− 0.6125

retirement of preferred stock (V_j) on that part of the investment funded by preferred stock, miscellaneous other pretax cash flows (K_j), and income taxes (T_j). That is,

$$Y_0 = \text{(loan + preferred stock contribution)} - \text{(initial investment)}$$

$$= \text{owners' contribution}$$

$$Y_j = R_j - (U_j + I_j + V_j + P_j + K_j + T_j),$$
$$\text{for } j = 1, 2, \ldots, 28 \quad (9.17)$$

where

$$R_j = \$40 + \$0.17565B_j \text{ (from before)}$$

$$U_j = w_bC/N = 0.55(\$350)/28 = \$6.875$$

$$I_j = w_bk_bB_j = 0.55(0.10)B_j = 0.055B_j$$

$$V_j = w_pC/N = 0.10(\$350)/28 = \$1.25$$

$$K_j = (0.01 + 0.02)C + \$17 = \$27.5$$

$$T_j = 0.06165B_j \text{ (from before)}$$

Since all of the above factors are functions of B_j, Equation 9.17 can be rewritten

$$Y_j = \$4.375 + 0.049B_j, \text{ for } j = 1, 2, \ldots, 28$$

The internal rate of return from this sequence of cash flows is precisely 0.14, the cost of equity capital. That is, the present worth of all future positive cash flows, when discounted at $k_e = 0.14$, is equal to the initial investment by owners, or

$$\$21.525 - \$0.6125(P/G, 14\%, 28) = \$122.5.$$

This is the expected result, of course, inasmuch as the permitted revenues have been determined so as to provide the 15 percent return to owners.

9.12 SUMMARY

Because of the relatively large capital investment in plant and equipment required of companies furnishing utility services to the general public, and because most of these investments are incurred well in advance of the period during which the costs are to be recovered through revenues, public utility companies are often allowed to operate as a monopoly within a political jurisdiction. In return for this special treatment, certain aspects of the firm's behavior are regulated by some regulatory body, generally a "utilities commission" or the like. Of particular concern are the rates, or prices, that the firm is permitted to charge customers for its services.

A fundamental principle underlying the regulation of rates is that they be sufficient to cover all costs, including a "fair return" on invested capital. Consequently, decisions between alternative investment opportunities may be judged on the basis of their respective effects on revenue requirements, or the rates charged to customers: The preferable alternative is the one that minimizes revenue requirements.

As demonstrated in Section 9.11, if revenue requirements include the cost of capital, then minimizing revenue requirements is equivalent to maximizing present worth or annual worth. The choices resulting from these methods of analysis are consistent. Only the perspective is different. Nevertheless, it follows that decisions taken from the viewpoint of stockholders (the present worth or annual worth method) will be consistent with the interests of ratepayers (the revenue requirement method) when rates are regulated.

B_j Book value for tax purposes at start of year j

$\hat{B}_j$ Value of the unrecovered investment at the start of year j

C Initial investment (cost basis)

c Ratio of salvage value to initial investment

D Depreciation for tax purposes in year j (affecting taxable income)

$\hat{D}_j$ Reduction in unrecovered invested capital in year j

E Taxable income

I Interest (return to lenders)

i Weighted average cost of capital

j Index for year (shown as subscript for appropriate cost/revenue element)

K Total pretax cash flows, other than debt service and preferred stock dividends

k_b Cost of debt (interest rate paid on borrowed capital)

k_e Cost of equity (return to common stockholders)

k_p Cost of preferred stock (return to preferred stockholders)

L Terminal salvage value of investment

N Life of investment

P Preferred stock dividends

Q Net profit after income taxes

R Operating revenue (permitted revenue)

S Return to common stockholders

T Income taxes

t_s State income tax rate (incremental)

t_f Federal income tax rate (incremental)

t Combined (federal and state) effective income tax rate

U Repayment of loan principal

V Retirement of preferred stock

w_b Proportion of debt (borrowing)

w_e Proportion of equity (common stock)

w_p Proportion of preferred stock

Y Cash flow after income taxes and after payment of preferred stock dividends

$$\alpha = \left[\frac{t}{1-t}\right](i - w_b k_b) \qquad \text{(See Equation 9.9)}$$

$$\tau = T/C \qquad \text{(See Equation 9.11)}$$

χ Ratio of capital recovery cost to the original investment

(9.1) $E = R - (K + D + I)$

(9.2) $i = w_b k_b + w_p k_p + w_e k_e$

(9.3) $T_j = t_j [R_j - (K_j + D_j + I_j)]$

(9.4) $R_j = K_j + D_j + I_j + T_j + P_j + S_j$

(9.5) $T_j = \left[\dfrac{t_j}{1 - t_j}\right](P_j + S_j)$

(9.6) $T_j = \left[\dfrac{t_j}{1 - t_j}\right][i_j - (w_{bj} \cdot k_{bj})]\hat{B}_j = \left[\dfrac{t}{1 - t}\right][i - w_b k_b]B_j$

(9.7) $D = \dfrac{C - L}{N}$

(9.8) $B_j = C + D - jD$

(9.9) $T_j = \left[\dfrac{t}{1 - t}\right][i - w_b k_b][C + D - jD]$

$\qquad = \alpha(C + D) - j(\alpha D)$

$\qquad$ where $\quad \alpha = \left[\dfrac{t}{1 - t}\right](i - w_b k_b)$

(9.10) $T = \alpha[C - D(A/G, i, N)]$

(9.11) $\tau = \alpha\left[1 - \dfrac{D}{C}(A/G, i, N)\right]$

(9.12) $\tau = \alpha\left[1 - \left(\dfrac{1 - c}{N}\right)(A/G, i, N)\right]$

(9.13) $\tau = \left[\dfrac{t}{1 - t}\right]\left[1 - w_b k_b/i\right]\left[i - \left(\dfrac{1 - c}{N}\right) + (1 - c)(A/F, i, N)\right]$

(9.14) $t = t_s + t_f(1 - t_s)$

(9.15) $CR = Ci + (C - L)(A/F, i, N)$

(9.16) $\chi = i + (1 - c)(A/F, i, N)$, where $c = L/C$

(9.17) $Y_j = R_j - (U_j + I_j + V_j + P_j + K_j + T_j)$,
          for $j = 1, 2, \ldots, 28$

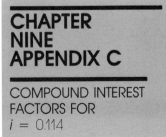

CHAPTER
NINE
APPENDIX B

SUMMARY OF
PRINCIPAL EQUATIONS

CHAPTER
NINE
APPENDIX C

COMPOUND INTEREST
FACTORS FOR
$i = 0.114$

	$N = 14$	$N = 28$
$(A/P, 11.4\%, N) = \dfrac{0.114(1.114)^N}{(1.114)^N - 1}$	0.14627	0.11983
$(P/A, 11.4\%, N) = (A/P, 11.4\%, N)^{-1}$	6.8368	8.3451
$(A/F, 11.4\%, N) = \dfrac{0.114}{(1.114)^N - 1}$	0.03227	0.00583
$(A/G, 11.4\%, N) = (0.114)^{-1} - N[(1.114)^N - 1]^{-1}$	4.8094	7.3396

9.1 Recall the numerical example introduced in Section 9.7. We assumed that *straight line* depreciation would be used for tax purposes as well as for determining the unrecovered invested capital. Using Equation 9.13, we determined that the fixed charge for income taxes, τ, is 0.0455 when $i = 0.114$, $t = 0.511$, $w_b = 0.55$, $k_b = 0.10$, $N = 28$, and $c = 0$. Now, we will change the assumptions in only one respect: Assume that the property will be depreciated under ACRS as a 15-year public utility property, with annual depreciation percentages as given in Table 7.3, Chapter 7. Compute the annual fixed charge rate for income taxes, τ, with this modification.
(*Answer:* 0.032)

9.2 Recall the numerical example introduced in Section 9.7. Here, all assumptions are as originally stated, except that we now assume that the capital structure consists of 60 percent debt with $k_b = 0.11$ and 40 percent common stock with $k_e = 0.15$.
a. Find the weighted average cost of capital (i)
b. Determine the sinking fund factor for this value of i and $N = 28$: $(A/F, i, 28)$.
c. Determine the revised total fixed charge rate with these revised data.

9.3 In Section 9.10 we discussed the issue of staggered development in the context of a numerical example. To meet forecasted demand, we assumed that a new 100 mw plant would be built at the start of years 1, 6, and 11, with each plant having a twenty-eight-year service life. Now, let us assume that the demand forecast has been revised and the firm is considering building one 150 mw plant now and a second 150 mw plant after ten years. With all other assumptions as originally stated, use the *repeated plant method* to determine the capitalized cost of this two-stage construction plan.
(*Answer:* $163.6 million)

9.4 Use the co-terminated plant method to determine the capitalized cost of the two-stage construction plan given in Problem 9.3.

9.5 In Table 9.5, the net cash flows after taxes, Y_j, are determined, with the assumptions reflected in columns 2 through 9. Reconstruct Table 9.5 for the steam plant example, assuming that the miscellaneous pretax cash flows, K_j, are expected to be $15,000 the first year, increasing by $1,000 each year over the twenty-eight-year life of the plant. All other assumptions are as originally stated. With the revised values of Y_j in column 10, determine the levelized annual revenue requirements.
(*Answer:* $57.938 per kilowatt)

9.6 A public utility is considering investment in an asset that has a first cost of $130,000, an expected service life of three years, and an estimated salvage value of $40,000. Recurring costs of operation and administration are expected to be $52,000 the first year, $59,000 the second year, and $66,000 the third year. The company's before-tax rate of return requirement is 20 percent per year. Determine the (pretax) levelized annual revenue requirement. (This pretax analysis would be suitable for publicly owned utilities that do not pay income taxes.)*

9.7 (This is an after-tax analysis of the investment described in Problem 9.6.) The company's combined federal and state income tax rate is 50 percent, financing will be by 30 percent debt and 70 percent equity, the cost of debt financing is 8 percent per year, and the after-tax rate of return requirement, the overall cost of capital, is 10 percent per year. The asset will be depreciated by the sum of years digits method with a three-year life and $40,000 salvage value. Assume that the unrecovered investment at any time is equal to book value and that the amount of debt at any time (30 percent) remains constant throughout the asset life.
a. Determine the levelized annual income taxes.
b. Determine the (after-tax) levelized annual revenue requirements.
    (*Answers:* (a) $7,022; (b) $105,770)

9.8 A publicly owned gas company is considering the construction of a small warehouse on land that it currently owns. The need for this warehouse is not pressing; it may be built now or, alternatively, three years from now. The issue is whether it is economically desirable to delay construction.

The cost of the warehouse is $1,000,000; it has an expected useful life of thirty years. These estimates are independent of when the warehouse is built.

The company plans to sell the land and the warehouse in twenty years. If the warehouse is built now, it will be twenty years old at the time of sale, at which point its residual value is estimated to be $200,000; if the company waits three years to build, the warehouse will be seventeen years old at the time of sale, at which point its residual value is estimated to be $225,000.

If construction of the warehouse is delayed, the company will rent space nearby at a cost of $10,000 per month over the three-year period.

The company uses 30 percent debt at a cost of 7 percent. The regulatory body permits a "fair return" of 9 percent on the remaining 70 percent equity capital. This publicly owned company pays no income taxes, but straight line depreciation is used for book purposes. (Here, use a thirty-year life with zero salvage value.)

Determine the levelized annual revenue requirements for both alternatives: (a) build now and (b) delay construction three years.

*Adapted from R. R. Mayer, "Finding Your Minimum Revenue Requirements," *Industrial Engineering* (April 1977): 16–22.

9.9 An investor-owned power company is considering the acquisition of a small coal-fired boiler.* The initial cost is $2,500,000, the expected useful life is four years, and estimated salvage value at the end of four years is $500,000. The boiler will be depreciated by the straight line method for book purposes; the sum of years digits method will be used for depreciation purposes. A four-year life and a $500,000 salvage value will be used for both book and tax purposes. (Note here that the value of unrecovered investment at any point in time is based on straight line depreciation; income taxes are based on SYD depreciation.)

Operating costs are expected to be $300,000 per year over the four-year useful life. The thermal output of the boiler is 300×10^9 Btu per year.

The boiler will be purchased entirely with equity funds that cost 20 percent per year. The effective income tax rate is 60 percent.

a. Determine the levelized equivalent revenue requirement for the proposed investment.

b. Determine the revenue requirement per 10^6 Btu.

   (*Answers:* (a) $1,663,700; (b) $5.55/$10^6$ Btu)

9.10 The Admiral Telephone Company is a regulated, investor-owned utility. Two plans, Plan A and Plan B, are under consideration for providing telephone service over the next twenty years to a rapidly growing portion of its service area.

Plan A, a full-development plan, calls for an initial investment of $400,000 in facilities with an estimated service life of twenty years and salvage value of $80,000 at that time. Costs of operations and maintenance (O and M) are expected to be $30,000 per year with *ad valorem* taxes of $8,000 per year.

Plan B, phased (stepped) development, calls for an initial investment of only $300,000, with a second investment of $300,000 after seven years. Operations and maintenance (O and M) costs will be $20,000 a year during the first seven years, increasing to $35,000 a year thereafter. Each installation will have a twenty-year service life with estimated salvage value of $60,000. *Ad valorem* taxes in any year will be 2 percent of the intial cost of the plant in place at that time.

The capital structure of Admiral Telephone includes 50 percent debt at a cost of 9 percent before income taxes. A "fair return" of 10 percent is permitted by the regulatory agency. The applicable combined federal and state income tax rate is 55 percent. The company uses straight line depreciation for book and tax purposes.

Assuming the repeated plant method as discussed in Section 9.10, determine the levelized annual revenue requirements for the two plans.

---

*Adapted from T. L. Ward and W. G. Sullivan, "Equivalence of the Present Worth and Revenue Requirement Methods of Capital Investment Analysis," *AIIE Transactions* (March 1981): 29–40.

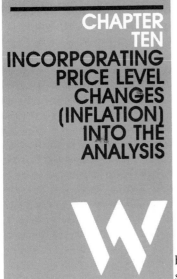

CHAPTER
TEN
INCORPORATING PRICE LEVEL CHANGES (INFLATION) INTO THE ANALYSIS

hen allocating limited capital resources among competing investment alternatives, the effects of price level changes can be significant to the analysis. Cash flows, proxy measures of goods and services received and expended, are affected by both the *quantities* of goods and services as well as their *prices*. Thus, to the extent that changes in price levels affect cash flows, these changes must be incorporated into the analysis.

Prior to the mid-1970s, the traditional literature of economic analysis in the United States has largely ignored price level changes because they have not been historically significant. As measured by the Consumer Price Index (CPI), price levels were reasonably constant during the twenty-year period prior to World War II, rose somewhat during the World War II and Korean War years, and then were flat (about $1\frac{1}{2}$ percent per year) from 1953 to 1965. Only in recent years have price level changes become so pronounced as to demand attention. During the 1970s, for example, the general rate of inflation as measured by the CPI has been about 8.6 percent per year.[1] (See Figure 10.1.) The annualized rate of inflation was "double-digit" in the period 1979–1982, but as this is being written (mid-1983), the rate has declined substantially. Our inability to see into the future notwithstanding, it is clear that the topic of inflation merits close attention.

A number of inflation-related issues are explored in this chapter. First, a variety of techniques for measuring price level changes are described, emphasizing the most popular of these measures, the Consumer Price Index. (A general discussion of

1. *Inflation* is a term commonly used to describe price level changes. Although it generally implies price *increases*, we can use *inflation* synonymously with *price level changes* to describe both increases and decreases in relative prices. (Some writers refer to price level decreases as *disinflation*.)

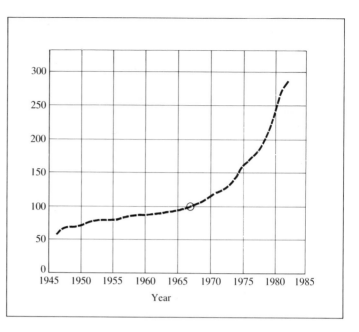

Figure 10.1. Consumer Price Index U.S. City Average, All Urban Consumers, All Items (1967 = 100)

index numbers is included in Appendix 10A.) Then the effect of inflation on two measures of effectiveness, present worth and internal rate of return, are discussed, and appropriate steps for incorporating price level changes directly into the analysis are presented. The effect of inflation on after-tax economy studies is also discussed.

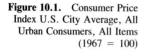

10.1
THE CONSUMER PRICE INDEX AS A MEASURE OF INFLATION

A price index, in general, is simply a relative measure of prices, as measured over time, of specific goods and/or services. The **Consumer Price Index** (CPI), in particular, is widely used to measure a given set (*market basket*) of goods and services consumed by "typical" American consumers. It is a highly visible measure of general inflation, widely quoted in print and broadcast media as well as in the investment community. Its importance is further emphasized because it is used to adjust income payments for a great many Americans, including, as of 1977, 8.5 million wage earners, 20 million food-stamp recipients, and 2.5 million retired military/federal employees and their survivors. The CPI also influences the distribution of revenue-sharing funds, the basis of eligibility in many health and welfare programs, and various school lunch programs funded by the federal government. In all, about half the population of the United States may be directly affected by changes in the CPI.

The CPI was first published by the U.S. Bureau of Labor Statistics (BLS) in 1919 to help set new wage levels for workers in shipbuilding yards. The first major revision occurred in 1940,

340   TEN   INCORPORATING PRICE LEVEL CHANGES (INFLATION) INTO THE ANALYSIS

at which time the reference base period was updated from the years 1917–1919 to 1935–1939.

A second major revision occurred in 1953, when the base period was updated to the years 1947–1949; a third major revision was completed in 1964, based on a survey of consumer expenditures conducted in the period 1960–1961. The most recent major revision, completed in 1978, included the following changes:

1. A new index, the **CPI-U,** was issued, representing all urban residents, about 80 percent of the nation's total noninstitutional civilian population.

2. The original index was revised. This index, now known as the **CPI-W,** includes only wage earners and clerical workers, roughly half the population covered by the CPI-U.

3. The distribution of individual items making up the "market basket" was modified to reflect changing consumption patterns as measured during the 1972–1973 survey period. (See Table 10.1.)

4. Monthly or bimonthly indexes are now published for twenty-eight cities; regional indexes are available for urban areas of different population sizes.

Prices are collected periodically (monthly or bimonthly) in eighty-five urban areas across the country. The prices of most goods and services are collected by BLS representatives, but mail questionnaires are also used to obtain public utility rates, some fuel prices, and certain other items. All taxes directly associated with the purchase and use of items are included in the index.

Price changes for the various items in the various locations are

Table 10.1

Percent Distribution of the Consumer Price Index
by Major Expenditure Group*

Major group	Wage earners and clerical workers				All urban consumers 1972–1973
	1935–1939	1952	1963	1972–1973	
Food and alcoholic beverages	35.4	32.2	25.2	20.4	18.8
Housing	33.7	33.5	34.9	39.8	42.9
Apparel	11.0	9.4	10.6	7.0	7.0
Transportation	8.1	11.3	14.0	19.8	17.7
Medical care	4.1	4.8	5.7	4.2	4.6
Entertainment	2.8	4.0	3.9	4.3	4.5
Personal care	2.5	2.1	2.8	1.8	1.7
Other goods and services	2.4	2.7	2.9	2.7	2.8
	100.0	100.0	100.0	100.0	100.0

*Source: U.S. Dept. of Labor, Bureau of Labor Statistics, *The Consumer Price Index: Concepts and Content Over the Years,* Report 517 (Washington, DC: U.S. Government Printing Office, 1978).

averaged together using weights that represent their importance in the consumption pattern of the appropriate population group. The formula for computing both the CPI-U and CPI-W is the **modified Laspeyres index:**

$$I_j = I_{j-1} \left\{ \frac{\sum_n [p(n,j-1)q(n,0)] \times [p(n,j)/p(n,j-1)]}{\sum_n [p(n,j-1)q(n,0)]} \right\} \quad (10.1)$$

where

$$I_j = \text{index at end of period } j$$

$$I_{j-1} = \text{index at end of period } j-1$$

$$p(n,j) = \text{price of good or service } n \text{ at end of period } j$$

$$p(n,j-1) = \text{price of good or service } n \text{ at end of period } j-1$$

$$q(n,0) = \text{quantity of good or service } n \text{ at base period}$$

The index measures changes from the designated reference date, 1967, for which the CPI was 100.

As a measure of the "cost of living," the CPI is not without its conceptual problems. One problem has to do with who, exactly, should make up the index population. Ideally, the population should be large enough to encompass most current uses of the index and narrow enough to be considered homogeneous in reflecting price experience. The BLS is considering using a family of indexes, but this approach is not feasible at present, because of time and money constraints. A second problem involves accurate measurement and treatment of quality differences, given the fact that products and consumption patterns are constantly changing. A third problem has to do with owner-occupied housing: Under what concept should housing be priced, and what is the best way to measure price changes under that concept? These issues are discussed at length in BLS Report 517, *The Consumer Price Index: Concepts and Content Over the Years.*

**10.2
OTHER STANDARD
INDEXES OF
RELATIVE PRICE
CHANGES**

The Consumer Price Index is but one of a large number of indexes that are regularly used to monitor and report relative price changes. The CPI is by far the most well known, yet it is not particularly useful for specific economic analyses. Analysts should be interested in *relative price changes of goods and services that are germane to the particular investment alternatives under consideration.* The appropriate indexes are those that are related, say, to construction materials, costs of certain labor skills, energy, and other cost and revenue factors.

Government agencies regularly produce a number of indexes and other price level data that analysts may find useful. Some examples are the **Wholesale Price Index,** the **Implicit Price Index,** and the **Industrial Production Indexes.** Industry trade associations are also excellent sources for price level data, and banks and other lending institutions often maintain useful data bases for economic activity in their areas of operation. Still other indexes are available from private firms that specialize in econometric data, such as Data Resources, Inc. and Standard & Poor's. Finally, if analysts can find no readily available indexes relevant to a specific investment problem or class of problems, they may develop their own through continual monitoring of actual prices over time. In fact, it is sound practice for analysts to maintain a current file of appropriate indexes.

Let i represent the minimum attractive rate of return as specified by the opportunity cost of marginal investment opportunities. That is, funds considered for investment in candidate projects or programs could be invested "elsewhere" at rate of return i. In the absence of inflation, it is assumed that all net cash receipts may be reinvested at any time during the planning horizon at rate i per period. Thus the *inflation-free* measures of equivalent present worth (PW), annual worth (AW), and future worth (FW) may be specified as

10.3 EFFECTS OF RELATIVE PRICE CHANGES IN PRETAX ECONOMY STUDIES

$$PW = \sum_{j=0}^{N} A_j (1 + i)^{-j} \qquad \textbf{(10.2)}$$

$$AW = (PW)(A/P, i, N) = (PW)\left[\frac{i(1 + i)^N}{(1 + i)^N - 1}\right] \qquad \textbf{(10.3)}$$

$$FW = \sum_{j=0}^{N} A_j (1 + i)^{N-j} \qquad \textbf{(10.4)}$$

where A_j is the cash flow occurring at the end of period j and N is the number of periods in the planning horizon.

Again, in these classical inflation-free models, it is assumed that investment and reinvestment at any point in time earns interest at rate i. The compounding rate equals the discounting rate. Thus it follows that competing investment alternatives may be rank-ordered by either PW, AW, or FW. The following section examines the effects of inflation on these figures of merit.

Under the assumption of general inflation, the relative prices of all factors of the analysis are assumed to increase at rate f per period.[2]

10.4 GENERAL INFLATION

2. Given price levels as measured by index numbers at the beginning (I_0) and end (I_1) of a period of time, the inflation rate for that period is found from

$$f = \frac{I_1 - I_0}{I_0}$$

Note that f is unconstrained in sign. Negative f implies what is generally known as disinflation.

If a cash flow at the end of j periods from now, measured in today's ("now") dollars, is A_j, then the inflated value in equivalent current ("then") dollars, A_j^*, is

$$A_j^* = A_j(1 + f)^j \tag{10.5}$$

(This formulation assumes that the inflation rate f is constant over all j.) For example, suppose that the current cost of one man-hour of labor is \$20. If the increase in relative prices is, say, 5 percent per period, the equivalent amount at the end of ten periods is $\$10(1.05)^{10}$, or \$16.29.

To determine the relationships between price level changes (inflation) and the various figures of merit used in economic analyses, let's examine a simple numerical example. Consider results in returns of \$576.20, measured in "now" dollars, at the ends of both periods 1 and 2. If the inflation rate f is 5 percent per period, the cash flows are

End of period j	Now dollars A_j	Then dollars A_j^*
0	$-\$1,000.00$	$-\$1,000.00$
1	576.20	$605.01 = \$576.20(1.05)$
2	576.20	$635.26 = \$576.20(1.05)^2$

Suppose further that the firm's inflation-free minimum attractive rate of return (i) is 10 percent per period. Is this proposed investment justified?

Clearly, *if inflation is ignored,* we are indifferent between Proposal W and the "do nothing" alternative, because from Equation 10.2 (or 10.4) the present worth (or future worth) is approximately zero.

$$PW = -\$1,000 + \$576.20(1.10)^{-1} + \$576.20(1.10)^{-2}$$
$$= \$0.0165,$$

or

$$FW = -\$1,000(1.10)^2 + \$576.20(1.10) + \$576.20$$
$$= \$0.02$$

That is, the investor is about as well off investing \$1,000 in the proposal as investing the \$1,000 elsewhere at a 10 percent rate of return. But is the same result obtained in the presence of inflation? The answer, as you will see, depends on the reinvestment assumption.

If the investment/reinvestment rate is i, then the value at the end of the second period of $1,000 invested elsewhere is $1,000(1.10)^2$, or $1,210. Similarly, the positive cash flow of $605.01 at the end of period 1 may be reinvested at 10 percent for period 2, resulting in $665.51 at the end of period 2. Thus the total future worth of the proposal is $665.51 + $635.26 = $1,300.77. The net advantage of investing in Proposal W rather than investing elsewhere at 10 percent is $90.77. In summary,

Proposal, 1st period	$605.01(1.10) =	$ 665.51
Proposal, 2nd period		635.26
Net positive cash flows,		
end of planning horizon		$1,300.77
Opportunity forgone	$1,000(1.10)^2 = -	1,210.00
		$ 90.77

The general formulation for this approach is given by

$$FW = \sum_{j=0}^{N} A_j^*(1 + i)^{N-j} \qquad (10.6)$$

Note that this result is very different from the one obtained using Equation 10.4, in which the inflationary effects were simply ignored. The investor should *not* be indifferent to this proposal; its equivalent future worth is positive. The difference lies in the fact that it is the inflated cash flows (A_j^*) that are to be reinvested, not the equivalent "now" dollars (A_j).

It may be argued that, in the presence of general inflation, it is unreasonable to expect the investment/reinvestment rate to remain at i, the inflation-free rate. The return on invested capital must cover the implicit cost of inflation as well as provide a return to owners. Thus the required minimum attractive rate of return, *including* general inflation, is

$$i^* = (1 + i)(1 + f) - 1$$
$$= i + f + if \qquad (10.7)$$

(This composite rate is not simply the sum of the inflation-free *MARR* and the general inflation rate. If both i and f are small, the error introduced by ignoring the product term if may be insignificant, but this is not necessarily the case.)

Returning to the example, *if all cash flows may be invested or reinvested at* rate i^*, where $i^* = (1.10)(1.05) - 1 = 0.155$, the future worth of the proposed investment is as shown in column 5 of Table 10.2. The general formulation is given by

$$FW = \sum_{j=0}^{N} A_j^*(1 + i^*)^{N-j} \qquad (10.8)$$

Table 10.2
Determining the Future Worth of Proposal W under
Alternative Reinvestment Assumptions

End of period	Cash flows			Future value (end of period 2)		
	Now dollars	Then dollars	Reinvestment at $i = 0.10$	Reinvestment at $i^* = 0.155$		Difference
j	A_j	A_j^*	$A_j^*(1.10)^{2-j}$	$A_j^*(1.155)^{2-j}$		Column 5 − Column 4
(1)	(2)	(3)	(4)	(5)		(6)
0	− $1,000.00	− $1,000.00	− $1,210.00	− $1,334.03		− $124.03
1	576.20	605.01	665.51	698.79		33.28
2	576.20	635.26	635.26	635.26		0
		Future worth (FW)	$ 90.77	$ 0.02		− $ 90.75

Note: Assumes $i = 0.10$ and $f = 0.05$, or $i^* = 0.155$.

Note that the *FW* is almost zero ($0.02) in our sample problem, primarily as a result of the greater opportunity cost. If the initial $1,000 were to be invested elsewhere at return i^* per period, rather than in the proposed project, then $1,334.03 would be obtained at the end of two periods. The additional cost ($124.03) more than offsets the additional return from the reinvestment of the positive cash flow at the end of the first period ($33.28).

Given the alternative formulations as described by Equations 10.6 and 10.8, it is possible to determine a figure of merit, future worth (*FW*), that is a measure of the equivalent value of the investment measured at the end of the planning horizon. That is, the investor's wealth is maximized by choosing the alternative for which *FW* is maximum. Although the numerical example assumed only two investment alternatives, do nothing or choose Proposal W, the argument extends to any number of mutually exclusive alternatives.

In the event that present worth (*PW*) is the desired figure of merit, the future worth models described by Equations 10.6 and 10.8 must be modified appropriately. Specifically, if investment or reinvestment takes place at rate i per period, then

$$PW = \sum_{j=0}^{N} A_j^*(1 + i)^{-j} \qquad (10.9)$$

Alternatively, if this "opportunity cost" rate takes place at rate i^* per period, then

$$PW = \sum_{j=0}^{N} A_j^*(1 + i^*)^{-j} \qquad (10.10)$$

From the example, the equivalent present worth of Proposal W, using Equation 10.10, is

$$PW = -\$1,000 + \$605.01(1.155)^{-1} + \$635.26(1.155)^{-2}$$

$$= \$0.016$$

It is instructive at this point to take a closer look at the future worth model with reinvestment at i^*. Since $A_j^* = A_j(1 + f)^j$ and $i^* = (1 + i)(1 + f) - 1$, Equation 10.8 may be rewritten as

$$FW = (1 + f)^N \left[\sum_{j=0}^{N} A_j(1 + i)^{N-j} \right]$$

The term in [brackets] in this modified equation is the future worth of the uninflated cash flows, assuming reinvestment at rate i (Equation 10.4). Since f is the general inflation rate and affects all factors equally, it follows that rank-ordering alternatives by Equation 10.4 is equivalent to rank-ordering by Equation 10.8. In other words, in the presence of general inflation at rate f, and assuming that inflation affects the desired minimum attractive rate of return i so that investment or reinvestment takes place at rate i^*, the appropriate rank-ordering of alternatives results from either (a) finding the future values of A_j using interest rate i or (b) finding the future values of A_j^* using interest rate i^*.

Similarly, the present worth model shown in Equation 10.10 may be rewritten as

$$PW = \sum_{j=0}^{N} A_j^*(1 + i^*)^{-j}$$

$$= \sum_{j=0}^{N} A_j(1 + f)^j[(1 + i)(1 + f)]^{-j}$$

$$= \sum_{j=0}^{N} A_j(1 + i)^{-j}$$

Thus, under the assumptions of general inflation and investment or reinvestment (discounting) at rate i^*, the appropriate rank-ordering of alternatives results form either (a) finding the present values of A_j using discount rate i or (b) finding the present values of A_j^* using discount rate i^*.

Consider a new investment, Proposal X, which is competing with both Proposal W and the do-nothing alternative. Assume that the cash flows for Proposal W are affected only by general inflation, with $f = 0.05$, as before, but that the cost and revenue factors for Proposal X are influenced by an inflation rate, h, of 8 percent per period. Assume also that investment or reinvestment of funds

**10.5
DIFFERENTIAL
INFLATION**

invested "elsewhere" takes place at effective interest rate $i*$ = 0.155. In summary,

The inflation-free *MARR* (i) is 0.10.

Returns from all funds invested "elsewhere" increase at rate ($i*$) = 0.155 per period.

Prices of all elements of Proposal W increase at rate (f) = 0.05 per period.

Prices of all elements of Proposal X increase at rate (h) = 0.08 per period.

The inflation-free cash flows ("now" dollars) for Proposal X are shown in column 5 of Table 10.3; the inflated values ("then" dollars) are shown in column 6; and the equivalent future values, resulting from reinvestment at $i*$ = 0.155, are given in column 7. The total *future worth* for Proposal X, $17.70, is the solution to the equation

$$FW = \sum_{j=0}^{N} A_j (1 + h)^j (1 + i*)^{N-j} \qquad (10.11)$$

The comparable equation for equivalent *present worth* of the cash flow for Proposal X is

$$PW = \sum_{j=0}^{N} A_j (1 + h)^j (1 + i*)^{-j} \qquad (10.12)$$

Note that it is *incorrect* to compare the alternatives on the basis of their uninflated ("now" dollar) cash flows only, using the inflation-free interest rate i. To do so would result in $PW(W) >$

Table 10.3

Determining the Future Worth of Proposals W and X

Assuming Reinvestment at Rate $i*$ = 0.155

	Proposal W			Proposal X		
	Cash flows			Cash flows		
End of period j (1)	Now dollars A_{jw} (2)	Then dollars A_{jw}^* (3)	Equivalent future value $A_{jw}^*(1.155)^{2-j}$ (4)	Now dollars A_{jx} (5)	Then dollars A_{jx}^* (6)	Equivalent future value $A_{jx}^*(1.155)^{2-j}$ (7)
0	−$1,000.00	−$1,000.00	−$1,334.03	−$1,000.00	−$1,000.00	−$1,334.03
1	576.20	605.01	698.79	560.00	604.80	698.54
2	576.20	635.26	635.26	560.00	653.18	653.18
	Future worth		$ 0.02			$ 17.70

Column 3: $A_{jw}(1 + f)^j = A_{jw}(1.05)^j$
Column 6: $A_{jx}(1 + h)^j = A_{jx}(1.08)^j$

$PW(X)$ and $FW(W) > FW(X)$. This incorrect approach disregards the *differential* rates of price increases (inflation) for the two proposals.

An alternative solution to the problem of differential inflation rates requires the determination of the *incremental rate of inflation* (d), as given by

$$d = \left[\frac{1 + h}{1 + f}\right] - 1 \tag{10.13}$$

This result stems from Equation 10.12, the *present worth* of a proposal in which the inflation rate, h, is different from the general inflation rate, f. Since $1 + i^* = (1 + f)(1 + i)$, Equation 10.12 can be rewritten

$$
\begin{aligned}
PW &= \sum_{j=0}^{N} A_j \left[\frac{1 + h}{1 + f}\right]^j (1 + i)^{-j} \\
&= \sum_{j=0}^{N} A_j (1 + d)^j (1 + i)^{-j} \tag{10.14}
\end{aligned}
$$

That is, the prospective cash flows in "now" dollars are inflated at rate d, then discounted at the uninflated $MARR = i$. The comparable model for *future worth* is

$$FW = (1 + f)^N \sum_{j=0}^{N} A_j (1 + d)^j (1 + i)^{N-j} \tag{10.15}$$

To illustrate the use of Equations 10.14 and 10.15, consider Proposal X:

$$d = \frac{1.08}{1.05} - 1 = 0.02857$$

$$
\begin{aligned}
PW &= \sum_{j=0}^{2} A_{jx} (1.02857)^j (1.10)^{-j} \\
&= -\$1{,}000 + \$560(1.02857)(1.10)^{-1} \\
&\quad + \$560(1.02857)^2 (1.10)^{-2} \\
&= \$13.27
\end{aligned}
$$

and

$$
\begin{aligned}
FW &= (1.05)^2 \sum_{j=0}^{2} A_j (1.02857)^j (1.10)^{2-j} \\
&= (1.05)^2 [-\$1{,}000(1.10)^2] + \$560(1.02857)^2 \\
&= \$17.70
\end{aligned}
$$

Of course, the future worth may also be computed directly if the *PW* is known:

$$FW = (PW)(1 + i*)^N$$
$$= \$13.27(1.155)^2$$
$$= \$17.70$$

To summarize, given the assumptions in this section, Equation 10.12 equals Equation 10.14 and Equation 10.11 equals Equation 10.15.

10.6 EFFECTS OF INFLATION ON AFTER-TAX ECONOMY STUDIES

Relative price level changes have an important effect on economy studies in which income taxes are considered. This effect may be illustrated by a numerical example. Assume here that the effective income tax rate (t_j) is constant for all periods over the life of the asset $(j = 1, 2, \ldots, N)$. For simplicity, assume also that allowances for depreciation are computed using the straight line method. The inferences to be drawn from this example do not strictly depend on either assumption, however.

Consider an investment opportunity, Alternative I, with an initial cost (P) of \$100, a life (N) of five years, and an expected salvage value (S) of \$20 at the end of its depreciable life. Thus the annual depreciation expense is

$$D_j = \frac{P - S}{N} = \frac{\$100 - \$20}{5}$$
$$= \$16 \text{ for } j = 1, 2, \ldots, 6$$

If purchased, this asset will result in savings (R_j) of \$30 each year over the five-year planning horizon. (For convenience, assume end-of-year cash flows.) The cash flows before taxes and the annual depreciation expenses are shown in columns 2 and 3 of Table 10.4.

Table 10.4
Cash Flow Table for Proposal I in the Absence of Inflation

End of year j (1)	Cash flow before taxes $A_j = (P,R_j,S)$ (2)	Depreciation D_j (3)	Taxable income $R_j - D_j$ (4)	Cash flow for taxes $-t_j(R_j - D_j)$ (5)	Cash flow after taxes $\hat{A}_j$ (6)	Notes
0	−\$100	\$ —	\$ —	\$ —	−\$100	First cost
1	30	16	14	−7	23	
2	30	16	14	−7	23	
3	30	16	14	−7	23	Annual operations
4	30	16	14	−7	23	
5	30	16	14	−7	23	
5	20	<20>	—	—	20	Salvage value

Note: <Book value> in column 3 at $j = 5$

Inflation Negligible (or Ignored)

The *internal rate of return before income taxes* (ρ_0) is that interest rate at which the net present value of all cash flows is equal to zero.[3] That is,

$$PW\,(\rho_0) \equiv 0 = \sum_{j=0}^{N} A_j (1 + \rho_0)^{-j} \qquad \textbf{(10.16)}$$

where

$$
\left.
\begin{array}{l}
A_0 = -P \\[4pt]
A_j = R_j \quad \text{for } j = 1, 2, \ldots, N - 1 \\[4pt]
A_N = R_N + S
\end{array}
\right\}
\begin{array}{l}
\text{Cash flows} \\
\text{before income} \\
\text{taxes}
\end{array}
$$

The pretax *IRR* (ρ_0) may be shown to be approximately 19.04 percent for Alternative I.

In the absence of inflation, the cash flows *after* income taxes ($\hat{A}_j$) are shown in column 6 of Table 10.4. The effective income tax rate (t) is assumed to be 0.50. Note that the after-tax revenues are found from the equation

$$
\begin{aligned}
\hat{R}_j &= R_j - t_j(R_j - D_j) \\[4pt]
&= R_j(1 - t_j) + t_j D_j \qquad \text{for } j = 1, 2, \ldots, N \quad \textbf{(10.17)}
\end{aligned}
$$

The *after-tax* internal rate of return is that value of ρ_1 such that[4]

$$PW\,(\rho_1) \equiv 0 = \sum_{j=0}^{N} \hat{A}_j (1 + \rho_1)^{-j} \qquad \textbf{(10.18)}$$

where

$$
\left.
\begin{array}{l}
\hat{A}_0 = -P \\[4pt]
\hat{A}_j = \hat{R}_j \quad \text{for } j = 1, 2, \ldots, N - 1 \\[4pt]
\hat{A}_N = \hat{R}_N + S
\end{array}
\right\}
\begin{array}{l}
\text{Cash flows after} \\
\text{income taxes}
\end{array}
$$

When R_j, t_j, and D_j are constant for $j = 1, 2, \ldots, N$, as in the example problem, Equation 10.18 may be simplified to

---

3. Earlier chapters used the symbol i^* to represent the internal rate of return. The change in notation—the Greek letter ρ is used here—is due to the possible confusion about the asterisk (*), which denotes that cash flows reflect the effect of inflation. In other words, ρ in this chapter is the same as i^* in previous chapters.

4. The subscript 0 denotes the *pretax* internal rate of return and 1 denotes the *after-tax* internal rate of return. That is, ρ_0 is the pretax *IRR*; ρ_1 is the after-tax *IRR*. Moreover, the circumflex, or "hat" ($\hat{A}$, for example) denotes *after-tax cash flows*.

$$PW(\rho_1) \equiv 0 = [R(1 - t) + tD](P/A, \rho_1, N)$$
$$- P + S(P/F, \rho_1, N) \tag{10.19}$$

Solving the equation, $\rho_1 = 9.85\%$ for Alternative I.

Inflation at Rate $f > 0$: Discounting Inflated Cash Flows
Assume that all future cash flows are affected by price level changes at the rate (f) of 10 percent per year. That is, future cash flows, as measured in "then" (current) dollars, are found from

$$A_j^* = A_j(1.10)^j$$

The cash flow table is given in Table 10.5. The after-tax cash flows for the annual revenues are found from

$$\hat{R}_j^* = R_j^* - t_j(R_j^* - D_j)$$
$$= R_j(1 + f)^j(1 - t_j) + t_jD_j \tag{10.20}$$
$$\text{for } j = 1, 2, \ldots, N$$

If inflation persists at the rate of 10 percent per year over five years, the expected salvage value will be $\$20(1.10)^5 = \$32.21$. The book value, however, will be only $20, resulting in a gain on disposal of $12.21. This gain will be taxed as ordinary income, resulting in an increase in taxes of $6.11. Thus the after-tax cash flow will be $\$32.21 - \$6.11 = \$26.10$. In general,

$$\hat{S}^* = S^* - t_N(S^* - S)$$
$$= S[(1 + f)^N(1 - t_N) + t_N] \tag{10.21}$$

where $S^* = S(1 + f)^N$.

The "inflated" internal rate of return before income taxes is that value of α_0 such that

Table 10.5
Cash Flow Table for Proposal I Assuming 10 Percent Inflation Per Year

End of year j (1)	Cash flow before taxes $A_j^* = (P^*,R_j^*,S^*)$ (2)	Depreciation D_j (3)	Taxable income $A_j^* - D_j$ (4)	Cash flow for taxes $-t_j(R_j^* - D_j)$ (5)	Cash flow after taxes $\hat{A}_j^*$ (6)	Notes
0	− $100.00	$———	$———	$———	− $100.00	First cost
1	33.00	16.00	17.00	− 8.50	24.50	
2	36.30	16.00	20.30	− 10.15	26.15	
3	39.93	16.00	23.93	− 11.97	27.97	Annual operations
4	43.92	16.00	27.92	− 13.96	29.96	
5	48.32	16.00	32.32	− 16.16	32.16	
5	32.21	<20.00>	12.21	− 6.11	26.10	Salvage value

Note: <Book value> in column 3 at $j = 5$

$$PW(\alpha_0) \equiv 0 = \sum_{j=0}^{N} A_j^*(1 + \alpha_0)^{-j} \qquad \textbf{(10.22)}$$

where

$$A_0^* = -P$$

$$A_j^* = R_j^* = R_j(1 + f)^j \qquad \text{for } j = 1, 2, \ldots, N - 1$$

$$A_N^* = R_N(1 + f)^N + S$$

In Equation 10.22, A_j^* is the pretax cash flow in "then" (inflated) dollars at the end of period j. Using the data from the example:

$$PW \equiv 0 = -\$100 + \$33.00(1 + \alpha_0)^{-1} + \cdots$$
$$+ (\$48.32 + \$32.21)(1 + \alpha_0)^{-5}$$

Solving the equation, $\alpha_0 = 30.95\%$.

The "inflated" after-tax internal rate of return is that value of α_1 such that

$$PW(\alpha_1) \equiv 0 = \sum_{j=0}^{N} \hat{A}_j^*(1 + \alpha_1)^{-j} \qquad \textbf{(10.23)}$$

where

$$\hat{A}_0^* = -P$$

$$\hat{A}_j^* = R_j(1 + f)^j(1 - t_j) + t_j D_j$$
$$\text{for } j = 1, 2, \ldots, N - 1$$

$$\hat{A}_N^* = R_N(1 + f)^N(1 - t_N) + t_N D_N$$
$$+ S[(1 + f)^N(1 - t_N) + t_N]$$

In Equation 10.23, $\hat{A}_j^*$ is the after-tax cash flow in "then" (inflated) dollars at the end of the period j. Returning to the example,

$$PW \equiv 0 = -\$100.00 + \$24.50(1 + \alpha_1)^{-1} + \cdots$$
$$+ (\$32.16 + \$26.10)(1 + \alpha_1)^{-5}$$

Solving the equation, $\alpha_1 = 17.02\%$.

The presence of inflation at the rate of 10 percent annually has increased the apparent rate of return from 9.85 percent to 17.02 percent. This increase is due entirely to growth in the pretax cash flows.

Inflation at Rate $f > 0$: Discounting Deflated Cash Flows

The apparent inflated rates of return, either before or after income taxes, may be adjusted by converting cash flows from current (inflated) dollars to "now" (deflated) dollars. For the pretax case, the internal rate of return after correcting for inflation is clearly the same as the one that was obtained when inflation was ignored. The deflated pretax cash flows are equal to $A_j^*(1 + f)^{-j}$, which is the same as the inflation-free cash flows (A_j). Thus the pretax internal rate of return, after applying the adjustment for inflation, is approximately 19.04 percent, the same as that determined previously. Letting β_0 be defined as the pretax *IRR* using deflated cash flows, it follows that $\beta_0 = \rho_0$.

The after-tax internal rate of return, after adjusting the inflated cash flows, is determined from the following cash flows:

End of year j	Cash flow after taxes $\hat{A}_j^*$	Deflation factor $(1 + f)^{-j}$	Deflated cash flows after taxes $\hat{A}_j^*(1 + f)^{-j} = \hat{A}_{jd}^*$
0	−$100.00	1.100	−$100.00
1	24.50	0.909	22.27
2	26.15	0.826	21.61
3	27.97	0.751	21.01
4	29.96	0.683	20.46
5	32.16	0.621	19.97
5	26.10	0.621	16.21

In general, the corrected internal rate of return is that value of β_1 that satisfies the equation

$$PW(\beta_1) \equiv 0 = \sum_{j=0}^{N} \hat{A}_{jd}^*(1 + \beta_1)^{-j} \qquad (10.24)$$

where

$$\hat{A}_{0d}^* = -P$$

$$\hat{A}_{jd}^* = R_j(1 - t_j) + t_j D_j(1 + f)^{-j}$$
$$\text{for } j = 1, 2, \ldots, N - 1$$

$$\hat{A}_{Nd}^* = R_N(1 - t_N) + t_N D_N(1 + f)^{-N}$$
$$+ S[(1 - t_N) + t_N(1 + f)^{-N}]$$

In Equation 10.24, $\hat{A}_{jd}^*$ is the after-tax cash flow in "now" (deflated, or inflation-adjusted) dollars at the end of period j. Returning to the example,

$$PW \equiv 0 = -\$100.00 + \$22.70(1 + \beta_1)^{-1} + \cdots$$
$$+ (\$19.97 + \$16.21)(1 + \beta_1)^{-5}$$

Solving the equation, $\beta_1 = 6.39\%$.

It has frequently been argued that current tax laws penalize investors during periods of inflation, because allowable depreciation expenses are based on *historical* costs. If depreciation expenses could be *indexed* to inflation, that is, if the depreciation expenses used to determine taxable income could be allowed to increase with general price levels, income taxes would be reduced. The increased after-tax cash flows could then be used by investors to support the replacement of depreciated plant and equipment. The effect of indexed inflation on the example problem is illustrated in Table 10.6. Again, assume straight line depreciation with inflation (f) at the rate of 10 percent per year. The pretax inflated cash flows (A_j^*) are as before. However, the annual allowances for depreciation are affected as follows:

$$D_j^* = D_j(1 + f)^j \qquad (10.25)$$
$$= \$16(1.10)^j$$

These are the values shown in column 3 of Table 10.6. The apparent after-tax rate of return is that value of γ_1 such that

$$PW(\gamma_1) \equiv 0 = \sum_{j=0}^{N} \hat{A}_{ji}^*(1 + \gamma_1)^{-j} \qquad (10.26)$$

where

$$\hat{A}_{0i}^* = -P$$
$$\hat{A}_{ji}^* = (1 + f)^j[R_j(1 - t_j) + t_j D_j]$$
$$\text{for } j = 1, 2, \ldots, N - 1$$
$$\hat{A}_{Ni}^* = (1 + f)^N[R_N(1 - t_N) + t_N D_N] + S(1 + f)^N$$

Table 10.6
Cash Flow Table for Proposal I, Reflecting 10 Percent Inflation Per Year
and Depreciation Fully Indexed to Inflation

End of year j (1)	Cash flow before taxes A_j^* (2)	Indexed depreciation D_j^* (3)	Taxable income $A_j^* - D_j^*$ (4)	Cash flow for taxes $-t_j(A_j^* - D_j^*)$ (5)	Cash flow after taxes $\hat{A}_{ji}^*$ (6)	Notes
0	$-\$100.00$	$\$ \text{———}$	$\$ \text{———}$	$\$ \text{———}$	$-\$100.00$	First cost
1	33.00	17.60	15.40	$- 7.70$	25.30	
2	36.30	19.36	16.94	$- 8.47$	27.83	Annual
3	39.93	21.30	18.63	$- 9.32$	30.61	operations
4	43.92	23.43	20.49	-10.25	33.67	
5	48.32	25.77	22.55	-11.27	37.05	
5	32.21	$<32.21>$	———	———	32.21	Salvage value

Note: <Book value> in column 3 at $N = 5$

In Equation 10.26, $\hat{A}_{ji}^*$ is the after-tax cash flow in "then" (inflated) dollars at the end of period j, with depreciation expenses indexed to the inflation rate. As you can see by examining Table 10.6, the cash flows after taxes due to net operating revenues in years 1 through 5 are found from $[A_j^* - t_j(A_j^* - D_j^*)]$. Using the data from the example in Equation 10.26,

$$PW \equiv 0 = -\$100.00 + \$25.30(1 + \gamma_1)^{-1} + \cdots$$
$$+ (\$37.05 + \$32.21)(1 + \gamma_1)^{-5}$$

Solving the equation, $\gamma_1 = 20.84\%$.

Note that this solution for the *IRR* could also have been obtained by observing the after-tax rate of return in the absence of inflation: $\rho_1 = 9.85\%$. This result, as you may recall, was obtained from Equation 10.18, which can be written

$$PW(\rho_1) \equiv 0 = -P$$
$$+ \sum_{j=1}^{N} [R_j(1 - t_j) + t_j D_j](1 + \rho_1)^{-j}$$
$$+ S(1 + \rho_1)^{-N} \qquad (10.27)$$

Similarly, Equation 10.26 can be rewritten

$$PW(\gamma_1) \equiv 0 = -P$$
$$+ \sum_{j=1}^{N} [R_j(1 - t_j) + t_j D_j]\left(\frac{1 + f}{1 + \gamma_1}\right)^{j}$$
$$+ S\left(\frac{1 + f}{1 + \gamma_1}\right)^{N} \qquad (10.28)$$

The equations are equal to each other when $(1 + \gamma_1)/(1 + f) = 1 + \rho_1$. Solving, $\gamma_1 = (1 + \rho_1)(1 + f) - 1$. Given that f and ρ_1 are known, therefore, it is possible to solve for γ_1 directly. In the example, $\gamma_1 = (1.0985)(1.10) - 1 = 0.2084$. This is the same solution as that obtained by using Equation 10.26.

10.8 INTERPRETATION OF (INTERNAL) RATES OF RETURN UNDER INFLATION

Summarizing the results for this example, the (internal) rates of return for Alternative I are:

Condition	Before-tax rate of return	After-tax rate of return
No inflation	$\rho_0 = 19.04\%$	$\rho_1 = 9.85\%$
Inflation at uniform rate of 10 percent per year	$\alpha_0 = 30.95\%$	$\alpha_1 = 17.02\%$
Inflation at uniform rate of 10 percent per year but assuming "deflation" of current dollars to present dollars	$\beta_0 = 19.04\%$	$\beta_1 = 6.39\%$
Inflation at uniform rate of 10 percent per year but assuming that depreciation expenses are indexed to inflation	$\gamma_0 = 30.95\%$	$\gamma_1 = 20.84\%$

What is the significance of these results? How can they be used to guide the decision maker? What are the appropriate decision rules for these conditions? Let's now turn our attention to answering these questions.

Zero Inflation Rate: Inflation Negligible (or Ignored)
Let's begin by examining the case in which there is no inflation, or $f = 0$. First, consider Equation 10.2:

$$PW(i) = \sum_{j=0}^{N} A_j(1 + i)^{-j}$$

which represents the present worth of a series of end-of-period cash flows (A_j) when discounted at interest rate i per period. If $A_j \leq 0$ for $0 \leq j \leq n$ and $A_j \geq 0$ for $n \leq j \leq N$, for $n \leq N$, then $PW(i)$ has the general form shown in Figure 10.2. Observe that the present worth function intersects the abscissa at ρ, that is, $PW(\rho) = 0$. It is also evident that $PW(k) > 0$ for $k < \rho$ and $PW(k) < 0$ for $k > \rho$.

Now, consider two alternative investments: *either* invest P dollars and earn the (internal) rate of return *or* invest the same P dollars elsewhere at some predetermined minimum attractive rate of return (k). Let I represent the investment and $\emptyset$ represent the do-nothing (invest elsewhere) alternative. At the end of the planning horizon (at the end of N periods), the *future worth* of Alternative I, assuming reinvestment at rate k, is[5]

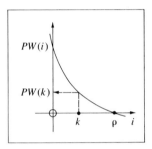

Figure 10.2

$$FW(\mathrm{I}:k) = \sum_{j=0}^{N} A_j(1 + k)^{N-j}$$
$$= (1 + k)^N \left[\sum_{j=0}^{N} A_j(1 + k)^{-j}\right] \qquad \textbf{(10.29)}$$
$$= (1 + k)^N[PW(\mathrm{I}:k)]$$

As you saw in Section 3.2, the present worth of the do-nothing alternative, Alternative $\emptyset$, is exactly zero, and thus the future worth is also zero. That is $PW(\emptyset:i) = 0$ and $FW(\emptyset:i) = 0$ for all i. Since the objective is to maximize the investor's wealth in the form of present worth or future worth, the investment in Alternative I is justified if, at the end of the planning horizon, its worth is greater than the worth that could be obtained if the do-nothing alternative were to be adopted instead. The decision rule, therefore, is to invest (choose I rather than $\emptyset$) if

$$FW(\mathrm{I}:k) > FW(\emptyset:i) = 0$$

5. In the equations that follow, the symbol $x:y$ is read x *given* y.

or

$$(1 + k)^N[PW(\mathrm{I}:k)] > PW(\emptyset:i) = 0$$

Since $FW(\emptyset:i) = 0$ for all i, it follows that $FW(\emptyset:\rho) = 0$ when $i = \rho$, the internal rate of return. Thus the decision rule can be restated as

$$FW(\mathrm{I}:k) > FW(\emptyset:\rho) = 0$$

or

$$PW(\mathrm{I}:k) > PW(\emptyset:\rho) = 0$$

If the cash flows are as described in the previous paragaph, that is, if there is a sequence of negative cash flows followed by a sequence of positive cash flows, then the decision rule is equivalent to selecting I over $\emptyset$ when $\rho > k$, or $IRR > MARR$.

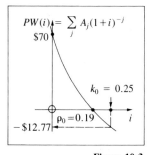

$$PW(i) = \sum_j A_j(1+i)^{-j}$$

$\$70$

$k_0 = 0.25$

$\rho_0 = 0.19$

$-\$12.77$

Figure 10.3

For the example, suppose that the before-tax minimum attractive rate of return (k_0) is 0.25 and the before-tax internal rate of return for the project (ρ_0) is 0.19, as illustrated in Figure 10.3. In this case the investment should be rejected, since $k_0 > \rho_0$. This is the same result, of course, as the one obtained using either the future worth or present worth methods. It may be shown that $FW(\mathrm{I}:k_0 = 0.25) = -\$38.97$, so the investment should be rejected, since $FW(\mathrm{I}:k) < 0$. Similarly, it may be shown that $PW(\mathrm{I}:k_0 = 0.25) = -\$12.77$; again, the investment should be rejected, since $PW(\mathrm{I}:k) < 0$.

Inflation at Rate $f > 0$: Reinvestment at Rate k

Consider the case wherein the rate of relative price changes remains constant at rate f throughout the planning horizon. All cash flows, it is assumed, are affected by this inflation rate. The cash flows in current ("then") dollars are, by Equation 10.5,

$$A_j^* = A_j(1 + f)^j$$

If these cash flows are reinvested at rate i, the future worth is given by Equation 10.6:

$$FW(i) = \sum_{j=0}^{N} A_j^*(1 + i)^{N-j}$$

$$= (1 + i)^N \sum_{j=0}^{N} A_j^*(1 + i)^{-j}$$

$$= (1 + i)^N[PW(i)]$$

Again, consider the two alternatives: *either* invest P dollars in Alternative I or invest the dollars elsewhere (Alternative $\emptyset$) earning a rate of return of k per period. The future worth of the investment is

$$FW(\text{I}:k) = \sum_{j=0}^{N} A_j^*(1 + k)^{N-j}$$
$$= (1 + k)^N[PW(\text{I}:k)] \qquad \textbf{(10.30)}$$

Since the (net) present worth of the do-nothing alternative is zero, it follows that

$$FW(\emptyset:k) = (1 + k)^N [PW(\emptyset:k)] = 0 \qquad \textbf{(10.31)}$$

Moreover, by definition α is that value of i such that the present worth (or future worth) of the investment is zero, so that $FW(\emptyset:k) = 0 = FW(\text{I}:\alpha)$.

The proposed investment is preferable to the do-nothing alternative if

$$FW(\text{I}:k) > [FW(\emptyset:k) = FW(\text{I}:\alpha)]$$
$$(1 + k)^N[PW(\text{I}:k)] > (1 + \alpha)^N[PW(\text{I}:\alpha)]$$
$$(1 + k)^N[PW(\text{I}:k)] > 0$$

Since $k > 0$, the condition becomes

$$PW(\text{I}:k) > 0$$

or

$$PW(\text{I}:k) > PW(\text{I}:\alpha)$$

If the cash flows are as described previously, the proposed investment should be accepted when $\alpha > k$.

Returning to the example, $k_0 = 0.25$, $f = 0.10$, and $\alpha_0 = 0.31$. The proposed investment should be *accepted*, since $\alpha_0 > k_0$. The same result is obtained when using either the future worth or present worth methods. Using the pretax cash flows (A_j^*) from Table 10.5, it may be shown that

$$FW(\text{I}:k_0 = 0.25) = \sum_{j=0}^{5} A_j^*(1.25)^{5-j} = \$44.11$$

$$PW(\text{I}:k_0 = 0.25) = \sum_{j=0}^{5} A_j^*(1.25)^{-j} = \$14.44$$

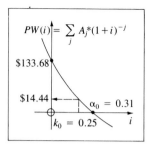

$$PW(i) = \sum_j A_j^*(1+i)^{-j}$$

$133.68

$14.44 — $\alpha_0 = 0.31$

$k_0 = 0.25$ — i

Figure 10.4

These results are summarized in Figure 10.4. The proposed investment is preferable to the do-nothing alternative, since $FW(I:k) > 0$ or, alternatively, since $PW(I:k) > 0$.

Inflation at Rate $f > 0$: Reinvestment at Rate k^*

All conditions are the same here as in the previous case, with one exception: It is now assumed that all cash flows may be reinvested at a rate that reflects both the "inflation-free" rate (k) as well as the general rate of inflation (f). Specifically, this reinvestment rate, from Equation 10.7, is $k^* = (1 + k)(1 + f) - 1$.

Consider the function

$$FW(i) = \sum_{j=0}^{N} A_j^*(1 + i)^{N-j} = (1 + i)^N[PW(i)]$$

As before, define α as the value of i such that $FW(\alpha) = 0$. The proposed investment is preferable to the do-nothing alternative when

$$FW(I:k^*) > [FW(\emptyset:k^*) = FW(I:\alpha)]$$

$$(1 + k^*)^N \sum_{j=0}^{N} A_j^*(1 + k^*)^{-j} > (1 + \alpha)^N \sum_{j=0}^{N} A_j^*(1 + \alpha)^{-j}$$

$$PW(I:k^*) > PW(I:\alpha)$$

If the cash flows (A_j^*) are as described previously, the acceptance condition is equivalent to $\alpha > k^*$.

Returning to the example, $k_0^* = (1.25)(1.10) - 1 = 0.375$ and $\alpha_0 = 0.31$. Thus the proposed investment should be *rejected*, since $\alpha_0 < k_0^*$. The future worth and present worth methods yield consistent results. It may be shown that

$$FW(I:k_0^* = 0.375) = \sum_{j=0}^{5} A_j^*(1.375)^{5-j} = -\$62.75$$

$$PW(I:k_0^* = 0.375) = \sum_{j=0}^{5} A_j^*(1.375)^{-j} = -\$12.77$$

These results are summarized in Figure 10.5. The proposed investment should be rejected, since $FW(I:k_0^*) < 0$, or $PW(I:k_0^*) < 0$.

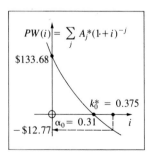

$$PW(i) = \sum_j A_j^*(1+i)^{-j}$$

$133.68

$k_0^* = 0.375$

$-\$12.77$ — $\alpha_0 = 0.31$ — i

Figure 10.5

Deflate before Compounding/Discounting at Rate k

We now consider the interpretation of the proposal's internal rate of return (β) resulting from deflating the current-period cost flows (A_j^*) prior to compounding or discounting. That is, cash flows in "then" dollars (A_j^*) will first be deflated to "now" dollars (A_j) prior to determining the internal rate of return. Assume that the rate of return to be earned on all reinvested cash flows is the inflation-free rate k.

The future worth of the proposed investment is given by Equation 10.30. By definition, the IRR is that rate (β) such that

$$PW(\mathrm{I}:\beta) = \sum_{j=0}^{N} A_j(1+\beta)^{-j} = 0$$

or

$$FW(\mathrm{I}:\beta) = \sum_{j=0}^{N} (\mathrm{I} + \beta)^{N-j} = 0$$

The proposed investment (I) is preferable to the do-nothing alternative (∅) when

$$FW(\mathrm{I}:k) > [FW(\varnothing:k) = FW(\mathrm{I}:\beta)]$$

$$(1+k)^N[PW(\mathrm{I}:k)] > (1+\beta)^N[PW(I:\beta)] = 0$$

$$PW(\mathrm{I}:k) > PW(\mathrm{I}:\beta)$$

$$\sum_{j=0}^{N} A_j^*(1+k)^{-j} > \sum_{j=0}^{N} A_j(1+\beta)^{-j}$$

$$\sum_{j=0}^{N} A_j\left[\frac{1+f}{1+k}\right]^j > \sum_{j=0}^{N} A_j\left[\frac{1}{1+\beta}\right]^j \qquad \textbf{(10.32)}$$

If the cash flows are as described previously, the proposed investment is preferable to the do-nothing alternative when

$$\frac{1}{1+\beta} < \frac{1+f}{1+k}$$

or

$$\beta > p = \left[\frac{1+f}{1+k}\right] - 1 \qquad \textbf{(10.33)}$$

Letting $p = [(1+k)/(1+f)] - 1$, note that

$$PW(\mathrm{I}:p) = \sum_{j=0}^{N} A_j(1+p)^{-j} \qquad \textbf{(10.34)}$$

Since $PW(\mathrm{I}:k) = \sum A_j^*(1+k)^{-j}$ and $PW(\mathrm{I}:k) = PW(\mathrm{I}:p)$, it

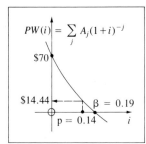

$$PW(i) = \sum_j A_j(1+i)^{-j}$$

$70

$14.44 ⊢--- β = 0.19

p = 0.14   i

Figure 10.6

follows that discounting *inflated* cash flows (A_j^*) at rate k is equivalent to discounting *deflated* cash flows (A_j) at rate p.

In the example, the pretax internal rate of return $\beta_0 = 0.19$. For $k_0 = 0.25$ and $f = 0.10$, $p = (1.25/1.10) - 1 = 0.14$. Thus the proposed investment should be *accepted*, since $p < \beta$. These results are summarized in Figure 10.6.

Deflate before Compounding/Discounting at Rate k^*

This case is identical to the previous one except that it is now assumed that the rate of return to be earned on all reinvestments, the minimum attractive rate of return, is $k^* = (1+k)(1+f) - 1$. This is the marginal cost of investment forgone.

The future worth of the proposed investment is given by

$$FW(I:k^*) = \sum_{j=0}^{N} A_j^*(1 + k^*)^{N-j}$$

$$= (1 + k^*)^N \sum_{j=0}^{N} A_j^*(1 + k^*)^{-j}$$

$$= (1 + k^*)^N[PW(I:k^*)] \qquad (10.35)$$

The proposed investment is preferable to the do-nothing alternative when

$$FW(I:k^*) > [FW(\emptyset:k) = FW(I:\beta)]$$

$$(1 + k^*)^N[PW(I:k^*)] > (1 + \beta)^N[PW(I:\beta)] = 0$$

$$\sum_{j=0}^{N} A_j^*(1 + k^*)^{-j} > \sum_{j=0}^{N} A_j(1 + \beta)^{-j}$$

$$\sum_{j=0}^{N} A_j(1 + f)^j[(1 + k)(1 + f)]^{-j} > \sum_{j=0}^{N} A_j(1 + \beta)^{-j}$$

$$\sum_{j=0}^{N} A_j(1 + k)^{-j} > \sum_{j=0}^{N} A_j(1 + \beta)^{-j}$$

Therefore, if the cash flows are as described previously, the proposed investment is preferable if $\beta > k$. Moreover, this is the same decision rule as shown in the case of no inflation. For before-tax cash flows, it may be seen that $\rho_0 = \beta_0$. (Because depreciation is not indexed to inflation, however, the after-tax rates of return are not identical. Indeed, it may be shown that, in general, $\rho_1 > \beta_1$.)

**10.9
SUMMARY**

The appropriate models for the three most-used figures of merit—future worth (*FW*), present worth (*PW*), and internal rate of return—are summarized in Table 10.7 for three basic cases under each of two fundamental, alternative assumptions. The basic cases are (a) no inflation, (b) inflation at rate f with discounting of inflated cash flows, and (c) inflation at rate f but with discounting of "deflated" cash flows. The two fundamental assumptions are (1) investment opportunities available at rate k, or (2) investment opportunities at rate $k^* = (1 + k)(1 + f) - 1$.

The associated decision rules for rates of return, however calculated, are also summarized in Table 10.7. (As before, a proposed investment should be accepted if $FW > 0$ or if $PW > 0$.)

Of critical importance is the assumption concerning the rate of return to be earned on reinvested capital during the study period. If all elements of the analysis are affected equally by inflation, that is, if the differential rates of inflation are zero, and if it can be assumed that the general rate of inflation affects the reinvestment rate as well as cash flows generated by the elements of the proposed projects, then the analyst may consider *either* the uninflated cash flows using the inflation-free discount rate *or* the inflated cash flows and the inflation-adjusted discount rate. If these assumptions do not hold, however, the differential effects must be considered explicitly, using the methods outlined in this chapter.

Table 10.7
Summary of Figures of Merit and Associated Decision Rules

	Reinvestment at rate k	Reinvestment at $k^* = (1 + f)(1 + k) - 1$
No inflation	$FW(\mathrm{I}:k) = (1 + k)^N[PW(\mathrm{I}:k)]$	
	$PW(\mathrm{I}:k) = \sum A_j(1 + k)^{-j}$	
	$FW(\mathrm{I}:\rho) = (1 + \rho)^N[PW(\mathrm{I}:\rho)]$	Not applicable
	$PW(\mathrm{I}:\rho) = \sum A_j(1 + \rho)^{-j} = 0$	
	Accept if $\rho > k$	
Inflation at rate f	$FW(\mathrm{I}:k) = \sum A_j^*(1 + k)^{N-j}$	$FW(\mathrm{I}:k^*) = \sum A_j^*(1 + k)^{N-j}$
Discount inflated cash flows	$PW(\mathrm{I}:k) = \sum A_j^*(1 + k)^{-j}$	$PW(\mathrm{I}:k^*) = \sum A_j^*(1 + k^*)^{N-j}$
	$FW(\mathrm{I}:\alpha) = \sum A_j^*(1 + \alpha)^{N-j} = 0$	$FW(\mathrm{I}:\alpha) = \sum A_j^*(1 + \alpha)^{N-j} = 0$
	$PW(\mathrm{I}:\alpha) = \sum A_j^*(1 + \alpha)^{-j} = 0$	$PW(\mathrm{I}:\alpha) = \sum A_j^*(1 + \alpha)^{-j} = 0$
	Accept if $\alpha > k$	*Accept if $\alpha > k^*$*
Inflation at rate f	$FW(\mathrm{I}:k) = \sum A_j^*(1 + k)^{N-j}$	$FW(\mathrm{I}:k^*) = \sum A_j^*(1 + k^*)^{N-j}$
Deflate cash flows before determining rate of return	$PW(\mathrm{I}:k) = \sum A_j^*(1 + k)^{-j}$	$PW(\mathrm{I}:k^*) = \sum A_j^*(1 + k^*)^{-j}$
	$FW(\mathrm{I}:\beta) = \sum A_j(1 + \beta)^{N-j} = 0$	$FW(\mathrm{I}:\beta) = \sum A_j(1 + \beta)^{N-j} = 0$
	$PW(\mathrm{I}:\beta) = \sum A_j(1 + \beta)^{-j} = 0$	$PW(\mathrm{I}:\beta) = \sum A_j(1 + \beta)^{-j} = 0$
	Accept if $\beta > p$ where $p =$ $\left[\dfrac{1 + k}{1 + f}\right] - 1$	*Accept if $\beta > k$ where $k =$* $\left[\dfrac{1 + k^*}{1 + f}\right] - 1$

The conclusions for the numerical example are summarized in Table 10.8 and shown graphically in Figure 10.7. Note the importance of the assumption concerning the reinvestment rate in the presence of inflation. If the rate of return on funds invested elsewhere is *not* affected by inflation, investment in Alternative I appears to be economically attractive: The future positive cash flows increase due to inflation, yet the inflation-free cost of capital remains at $k = 0.25$. Under these conditions, the *FW* and *PW* are positive. Using the rate of return method, the investment is attractive, since $\alpha > k$ (or, alternatively, $\beta > p$). The *IRR* may be determined either from the inflated cash flows (A_j^*), in which case α is compared to k, or from the deflated, inflation-free cash flows (A_j), in which case β is compared to $p = [(1 + k)/(1 + f)] - 1$. In the example,

$$(\alpha = 0.31) > (k = 0.25)$$

and

$$(\beta = 0.19) > (p = 0.136)$$

If it is assumed that the prospective reinvestment rate is also affected by the general rate of inflation, that is, if the discount rate is $k^* = (1 + k)(1 + f) - 1 = 0.375$, the results are quite

Table 10.8
Summary of Findings and Conclusions: Example Problem
(pretax cash flows as given in Table 10.4)

	Reinvestment at $k = 0.25$	Reinvestment at $k^* = 0.375$
No inflation	$PW = -\$12.77$ *Reject*, since $PW < 0$ $FW = -\$38.97$ *Reject*, since $FW < 0$ $IRR = \rho = 0.19$ *Reject*, since $\rho < k$	Not applicable
Inflation at $f = 0.10$ Discount inflated cash flows	$PW = \$14.44$ *Accept*, since $PW > 0$ $FW = \$44.11$ *Accept*, since $FW > 0$ $IRR = \alpha = 0.31$ *Accept*, since $\alpha > k$	$PW = -\$12.77$ *Reject*, since $PW < 0$ $FW = -\$62.75$ *Reject*, since $FW < 0$ $IRR = \alpha = 0.31$ *Reject*, since $\alpha < k^*$
Inflation at $f = 0.10$ Deflate cash flows before determining rate of return	$PW = \$14.44$ *Accept*, since $PW > 0$ $FW = \$44.11$ *Accept*, since $FW > 0$ $IRR = \beta = 0.19$ *Accept*, since $\beta > p$, where $p = \left[\dfrac{1 + k}{1 + f}\right] - 1 = 0.136$	$PW = -\$12.77$ *Reject*, since $PW < 0$ $FW = -\$62.75$ *Reject*, since $FW < 0$ $IRR = \beta = 0.19$ *Reject*, since $\beta < k$, where $k = \left[\dfrac{1 + k^*}{1 + f}\right] - 1 = 0.25$

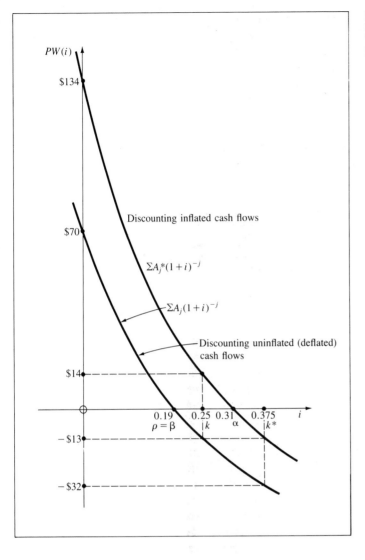

Figure 10.7. Graphic Representation of Present Worth Models for Example Problem

different. In this case Alternative I should be rejected. Both the FW and PW are negative because of the increased cost of capital. This conclusion is consistent with the rate of return method, since $\alpha < k^*$ (or, alternatively, $\beta < k$). The IRR may be determined either from the inflated cash flows, in which case α is compared to k^*, or from the deflated, inflation-free cash flows, in which case β is compared to k.

Now let's consider the after-tax investment problem, the results of which are summarized in Section 10.8. Recall that $\alpha_1 = 17.02\%$ and $\beta_1 = 6.39\%$. The interpretation of these results depends on the assumption concerning the effect of inflation on the reinvestment rate, that is, the cost of capital during the study period. Suppose that the inflation-free *MARR* is 15 percent after income taxes. If this rate is *not* affected by inflation, then the proposed investment is attractive in the presence of 10 percent per year inflation, since

$$(\alpha_1 = 0.1702) > (k_1 = 0.15)$$

or, alternatively, since

$$(\beta_1 = 0.0639) > (p_1 = (1.15/1.10) - 1 = 0.0455)$$

On the other hand, if the cost of capital is affected by inflation to the same extent as the cash flows, then the investment is not attractive, since

$$(\alpha_1 = 0.1702) < (k_1^* = (1.15)(1.10) - 1 = 0.265)$$

and

$$(\beta_1 = 0.0639) < (k = 0.15)$$

An **index number** is a statistic that describes the relative change in a variable (or group of variables) with respect to some reference characteristic, such as time. Index numbers are used in a variety of contexts, most notably in business and economics. Perhaps the most widely known application is the Consumer Price Index, described in Section 10.1. Other applications include the Wholesale Price Index, employment/unemployment indexes, production indexes for certain industries and/or commodities, and wage indexes. This appendix examines a variety of the most commonly used formulations.

One type of index number is the **price relative,** which describes the relative changes in price of a commodity (or group of commodities) over time. Specifically,

$$p(a/b) = p(b)/p(a) \qquad \textbf{(10A.1)}$$

where

$p(a/b)$ = index number describing the change in relative prices between periods a and b (price relative)

$p(a)$ = price of the commodity in period a

$p(b)$ = price of the commodity in period b

Normally, price indexes are developed so as to compare prices in a given period (n) with those in the base period (0). See Figure 10A.1. In this case, the price relative is written p_n/p_0. Note, in general, that $p(a/b)$ is the reciprocal of $p(b/a)$, and therefore

$$p(a/b)p(b/a) = 1 \qquad \textbf{(10A.2)}$$

The **quantity relative** is an index number that describes the relative changes in quantities (or volume) of commodities over time. Similar to the previous equation,

$$q(a/b) = q(b)/q(a) \qquad \textbf{(10A.3)}$$

where $q(a)$ and $q(b)$ are the quantities (or volumes) in periods a and b, respectively.

CHAPTER TEN APPENDIX A INDEX NUMBERS

10A.1 PRICE, QUANTITY, VALUE, AND LINK RELATIVES

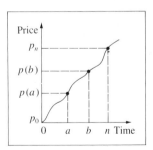

Figure 10A.1

The value in a given period (v) is the product of the price of the commodity in the period (p) times the quantity produced or sold in that same period (q). The **value relative,** then, is the index number given by

$$v(a/b) = v(b)/v(a)$$

$$= [p(b)q(b)] / [p(a)q(a)]$$

$$= [p(b)/p(a)] \times [q(b)/q(a)]$$

$$= p(a/b)q(a/b) \qquad \qquad \textbf{(10A.4)}$$

That is, the quantity relative is the product of the price relative times the quantity relative.

Link relatives are index numbers that relate relatives of each time period to those of the preceding time period. For example, if $p_1, p_2, p_3, \ldots$ represent prices during successive periods of time $1, 2, 3, \ldots$, then the link relatives are represented by $p(1/2)$, $p(2/3)$, $p(3/4)$, and so forth. Relatives for any two periods of time can always be expressed by link relatives. For example, $p(1/4) = p(1/2)p(2/3)p(3/4)$. Relatives with respect to a fixed base period are sometimes known as **chain relatives** for that base.

Properties of Relatives

There are four properties of relatives that follow directly from their definitions. (These are expressed in the following list with respect to price relatives, but the same properties may be associated with quantity and value relatives). The properties are

1. Identity	$p(a/a) = 1$
2. Time reversal	$p(a/b)p(b/a) = 1$
	or $p(a/b) = [p(b/a)]^{-1}$
3. Cyclical	$p(a/b)p(b/c)p(c/a) = 1$
	$p(a/b)p(b/c)p(c/d)p(d/a) = 1,$
	and so on
4. Modified cyclical	$p(a/b)p(b/c) = p(a/c)$
	$p(a/b)p(b/c)p(c/d) = p(a/d),$
	and so on

An additional property was noted earlier for value relatives. As shown in Equation 10A.4, $v(a/b) = p(a/b)q(a/b)$. This is known as the **factor reversal** property.

**10A.2
THEORETICAL TESTS**

The discussion of relatives in Section 10A.1 is based on the notion of relative changes in price/quantity/value of a single commodity or group of homogeneous commodities. As a practical matter, however, analysts are most often interested in measuring changes in groups of nonidentical commodities. A variety of index numbers can be used for this purpose, as outlined below. Theoretically, these should have properties that are satisfied by relatives. An index number having one of these properties is said to meet the test associated with that property. Nevertheless, as you will see, no index number meets all the tests, although some are approximately satisfied.

The methods for computing indexes outlined in this section are primarily related to prices. But the formulas shown here can easily be modified to obtain quantity or volume index numbers. This characteristic is demonstrated with respect to the weighted aggregate method, discussed in this section.

10A.3 METHODS FOR COMPUTING INDEX NUMBERS

Simple Aggregate Method

The **Simple Aggregate Price Index** (SAPI) is given by

$$SAPI = \sum_j p(j,n) \,/\, \sum_j p(j,0) \qquad \textbf{(10A.5)}$$

where

$p(j,0) =$ price of commodity j in the *base* period

$p(j,n) =$ price of commodity j in the *given* period n

The sample problem shown in Table 10A.1 illustrates this method of calculation. Here,

$$SAPI = \sum_{j=1}^{4} p(j,n) \,/\, \sum_{j=1}^{4} p(j,0)$$
$$= \$5.00/\$4.00$$
$$= 1.25$$

Table 10A.1
Prices for Example Problem

Commodity	Price		Price relatives
	Base period	Given period	$\dfrac{p(j,n)}{p(j,0)}$
j	$p(j,0)$	$p(j,n)$	
(1)	(2)	(3)	(4)
1	$0.50	$1.00	2.00
2	1.00	1.00	1.00
3	1.00	0.80	0.80
4	1.50	2.20	1.47
Totals	$4.00	$5.00	5.27

This method is simplistic in the extreme: Although easily applied, it assumes that *all commodities are equally important*. A second major disadvantage is that the index is especially *sensitive to the units of measurement used for the various commodities* in the index. To illustrate, consider the following two-commodity index:

Commodity	Units of measurement	Prices per unit	
		Base period	Given period
A	Metric tons	$1,000	$1,200
B	Liters	1	2
		$1,001	$1,202

Using these values, $SAPI = \$1,202/\$1,001 = 1.20$. But suppose that the prices of Commodity A are given in kilograms rather than metric tons. (1 MT $=$ 1,000 kg.) That is, the price of Commodity A is $1.00 per kg in the base period and $1.20 per kg in the given period. Thus, $SAPI = (\$1.00 + \$1.00)/(\$1.20 + \$2.00) = 1.60$, not 1.20 as before.

Simple Average of Relatives Method

In the simple average of relatives method, the price relatives of the individual commodities are averaged. The statistic used for averaging is usually, although not necessarily, the arithmetic mean, in which case the **Simple Average of Relatives Price Index** (SARPI) is given by[1]

$$SARPI = \left[\sum_j p(j,n)/p(j,0) \right]/N \qquad \text{(10A.6)}$$

where N is the total number of commodities in the index. Returning to the example given in Table 10A.1,

$$SARPI = \left[\sum_{j=1}^{4} p(j/n)/p(j,0) \right]/4$$

$$= 5.27/4$$

$$= 1.32$$

(As an exercise, you may recompute the *SARPI* using the geometric mean rather than the arithmetic mean.)

This method, unlike the simple aggregate method, is *not* sensitive to the units of measurement of the individual commodities. However, it is subject to the disadvantage that there is

---

1. Alternate formulations employ the geometric mean, the median, or the harmonic mean, among others.

no measure of the relative importance of the various commodities, a problem addressed next.

Weighted Aggregate Method

The problem of relative importance may be solved by multiplying each commodity by an appropriate weighting factor, usually the sales quantity, or volume, in (a) the base period, (b) the given period, or (c) the average of a set of periods representing some "typical" period.

The **Laspeyres Index** is a weighted aggregate price index (WAPI) with *base period* quantities used as the weights.

$$WAPI(B) = \frac{\sum_j p(j,n)q(j,0)}{\sum_j p(j,0)q(j,0)} \qquad \textbf{(10A.7)}$$

The base period and given period prices for our sample problem are repeated in Table 10A.2 (columns 2 and 4), along with a set of assumed quantities sold for each of the four commodities (columns 3 and 5). The numerator and denominator of Equation 10A.7 are tabulated in columns 6 and 7. Thus

$$WAPI(B) = \$800/\$650 = 1.23$$

The **Paasche Index** is an aggregate price index weighted with *given period* quantities.

$$WAPI(G) = \frac{\sum_j p(j,n)q(j,n)}{\sum_j p(j,0)q(j,n)} \qquad \textbf{(10A.8)}$$

The numerical values for this index are also tabulated in Table

Table 10A.2

Prices for Example Problem Weighted by Quantities

	Base period		Given period		Base period weights		Given period weights	
Commodity j (1)	Price $p(j,0)$ (2)	Quantity $q(j,0)$ (3)	Price $p(j,n)$ (4)	Quantity $q(j,n)$ (5)	Given period (6)	Base period (7)	Given period (8)	Base period (9)
1	$0.50	100	$1.00	100	$100	$ 50	$100	$ 50
2	1.00	100	1.00	150	100	100	150	150
3	1.00	200	0.80	100	160	200	80	100
4	1.50	200	2.20	150	440	300	330	225
				500	$800	$650	$660	$525

Column (6) = $p(j,n)q(j,0)$ = column (4) × column (3)
Column (7) = $p(j,0)q(j,0)$ = column (4) × column (3)
Column (8) = $p(j,n)q(j,n)$ = column (4) × column (5)
Column (9) = $p(j,0)q(j,n)$ = column (2) × column (5)

10A.2: the numerator of Equation 10A.8 is column 8, and the denominator is column 9. Thus, for the sample problem,

$$WAPI(G) = \$660/\$525 = 1.26$$

In the **typical period method,** prices are weighted by the quantities in some typical, or "average," period t:

$$WAPI(T) = \frac{\sum_j p(j,n)q(j,t)}{\sum_j p(j,0)q(j,t)} \qquad (10A.9)$$

If $t = 0$, then $WAPI(T) = WAPI(B)$; if $t = n$, then $WAPI(T) = WAPI(G)$.

Since these price indexes use *quantities* as weights, they are closely related to quantity, or volume, indexes. If *prices* are used as weights, then we have

Laspeyres Volume Index

$$WAVI(B) = \frac{\sum_j q(j,n)p(j,0)}{\sum_j q(j,0)p(j,0)} \qquad (10A.10)$$

Paasche Volume Index

$$WAVI(G) = \frac{\sum_j q(j,n)p(j,n)}{\sum_j q(j,0)p(j,n)} \qquad (10A.11)$$

Fisher's Ideal Index
Fisher's Ideal Price Index (FIPI) is the geometric mean of the Laspeyres and Paasche index numbers as given in Equations 10A.7 and 10A.8:

$$FIPI = \sqrt{[WAPI(B)][WAPI(G)]}$$

$$= \sqrt{\left[\frac{\sum_j p(j,n)q(j,0)}{\sum_j p(j,0)q(j,0)}\right]\left[\frac{\sum_j p(j,n)q(j,n)}{\sum_j p(j,0)q(j,n)}\right]} \qquad (10A.12)$$

This index is "ideal" because it comes closer than any other index to satisfying the important tests—both the time reversal and factor reversal tests.

Applying Fisher's ideal price index to our sample problem,

$$FIPI = \sqrt{(\$800/\$650)(\$660/\$525)}$$

$$= 1.244$$

The Marshall-Edgeworth Index

The **Marshall-Edgeworth Price Index** (MEPI) is a variant of the weighted aggregate method. The weight used for each commodity is the arithmetic mean of the base period and given period quantities. Thus

$$MEPI = \frac{\sum_j p(j,n)w(j)}{\sum_j p(j,0)w(j)} \qquad \textbf{(10A.13)}$$

where $w(j) = 0.5[q(j,0) + q(j,n)]$. It may be shown that the Marshall-Edgeworth index will always be between the Paasche index and the Laspeyres index.

Weighted Average of Relatives Method

As noted earlier, the simple average of relatives method suffers from the disadvantage that the relative importance of the various commodities are ignored. This problem may be addressed by weighting the average of relatives, usually the weighted arithmetic mean. The statistic used here for weighting the price relative is the total value of the commodity: $p(j)q(j)$. As was the case with the weighted aggregate method, the weight may be chosen for (a) the base period, (b) the given period, or (c) some specific typical period. The formulas are given below.

With the weight for commodity j defined as the total value of the commodity sold in the *base* period, the formula for computing the weighted average relative price index is

$$WARPI(\text{B}) = \frac{\sum_j [PR(j)][w(j,0)]}{\sum_j w(j,0)} \qquad \textbf{(10A.14)}$$

where

$$PR(j) = p(j,n)/p(j,0)$$

$$w(j,0) = p(j,0)q(j,0)$$

Substituting and simplifying, the equation may be rewritten as

$$WARPI(\text{B}) = \frac{\sum_j p(j,n)q(j,0)}{\sum_j p(j,0)q(j,0)}$$

(This is the same as the Laspeyres index as given in Equation 10A.7.)

This method is illustrated with respect to our sample problem in Table 10A.3. The numerator and denominator of Equation 10A.14 are shown to be \$800 and \$650, respectively. Thus

Table 10A.3

Weighted Price Relatives for Example Problems

Commodity j (1)	Base period		Given period		Unweighted price relatives (6)	Weights		Weighted price relatives	
	Price $p(j,0)$ (2)	Quantity $q(j,0)$ (3)	Price $p(j,n)$ (4)	Quantity $q(j,n)$ (5)		Base period (7)	Given period (8)	Base period weights (9)	Given period weights (10)
1	$0.50	100	$1.00	100	2.00	$ 50	$100	$100	$200
2	1.00	100	1.00	150	1.00	100	150	100	150
3	1.00	200	0.80	100	0.80	200	80	160	64
4	1.50	200	2.20	150	1.47	300	330	440	484
						$650	$660	$800	$898

Column (6) = $p(j,n)/p(j,0)$ = column (4) ÷ column (2)
Column (7) = $p(j,0)q(j,0)$ = column (2) × column (3)
Column (8) = $p(j,n)q(j,n)$ = column (4) × column (5)
Column (9) = column (6) × column (7)
Column (10) = column (6) × column (8)

$$WARPI(\text{B}) = \$800/\$650$$

$$= 1.231$$

Using *given* period weights, the weighted average relative price index is given by

$$WARPI(\text{G}) = \frac{\sum_j [PR(j)][w(j,n)]}{\sum_j w(j,n)} \qquad (10A.15)$$

where $w(j,n) = p(j,n)q(j,n)$.
The relevant calculations are summarized in Table 10A.3 for the sample problem.

$$WARPI(\text{G}) = \$898/\$660$$

$$= 1.361$$

When *typical* year weights are used,

$$WARPI(\text{T}) = \frac{\sum_j [PR(j)][w(j,t)]}{\sum_j w(j,t)} \qquad (10A.16)$$

where $w(j,t) = p(j,t)q(j,t)$.

10A.4 CHANGING THE BASE PERIOD

A practical problem in the calculation of index numbers is the choice of base period. There are a number of considerations, one of which is that the base period should not be too distant from the present. Thus it may be desirable to update the base period from time to time. Of course, all index numbers in the time sequence may be recomputed using the new base period, but this process may require extensive calculations. A simpler method is to use the relationship

$$I(n/k) = I(m/k)/I(m/n) \qquad\qquad \textbf{(10A.17)}$$

where

$I(n/k)$ = new index number for period k, using new period n as the base

$I(m/k)$ = old index number for period k, using old period m as the base

$I(m/n)$ = old index number for period n, using old period m as the base.

A simple numerical example is

Base	Period k	Old index number $I(a/k)$	New index number $I(b/k) = I(a/k)/I(b/k)$
Old	$m = 1$	$\boxed{1.00}$	$1.00/1.25 = 0.80$
	2	1.50	$1.50/1.25 = 1.20$
New	$n = 3$	1.25	$1.25/1.25 = \boxed{1.00}$

The general formulation of the problem of changing base period is illustrated in Table 10A.4 for price relatives as well as index numbers. The price relative in period k using the old period m as a base is

$$p(m/k) = p(k)/p(m) \qquad\qquad \textbf{(10A.18)}$$

Table 10A.4
The General Problem of Changing Base Periods

Period k (1)	Prices p_k (2)	Price relatives Period m as base $p(m,k)$ (3)	Price relatives Period n as base $p(n/k)$ (4)	Index numbers Period m as base $I(m/k)$ (5)	Index numbers Period n as base $I(n/k)$ (6)
1	p_1	$p(m/1)$	$p(n/1)$	$I(m/1)$	$I(n/1)$
2	p_2	$p(m/2)$	$p(n/2)$	$I(m/2)$	$I(n/2)$
⋮	⋮	⋮	⋮	⋮	⋮
Old m	p_m	$\boxed{1.00}$	$p(n/m)$	$\boxed{1.00}$	$I(n/m)$
⋮	⋮	⋮	⋮	⋮	⋮
New n	p_n	$p(m/n)$	$\boxed{1.00}$	$I(m/n)$	$\boxed{1.00}$
⋮	⋮	⋮	⋮	⋮	⋮
t	p_t	$p(m/t)$	$p(n/t)$	$I(m/t)$	$I(n/t)$

With the new period n as a base,

$$p(n/k) = p(k)/p(n) \qquad\qquad \textbf{(10A.19)}$$

It follows, then, that the price relatives with the new base period may be obtained simply by dividing the former $p(m/k)$ by the price relative for the new base period, $p(m/n)$.

$$\frac{p(m/k)}{p(m/n)} = \frac{p(k)/p(m)}{p(n)/p(m)}$$

$$= p(k)/p(n)$$

$$= p(n/k) \qquad\qquad \textbf{(10A.20)}$$

This procedure is applicable to *index numbers* only if the modified cyclical test is satisfied.[2] That is, $I(n/k) = I(m/k)/I(m/n)$ if and only if

$$I(m/1) = I(m/n)I(n/1)$$
$$I(m/2) = I(m/n)I(n/2)$$
$$\vdots$$

This is an example of the modified cyclical property, discussed in Section 10A.1. The test is satisfied exactly by the weighted aggregate method using fixed period weights; it is not satisfied by the Laspeyres, Paasche, Fisher, or Marshall-Edgeworth methods, although they generally result in useful approximations.

---

2. This test is also known as the *circular test*.

A_j = pretax cash flow in period j in "now (present) dollars

A_j^* = pretax cash flow in period j in "then" (current) dollars

$\hat{A}_j$ = after-tax cash flow in period j, assuming uninflated pretax cash flow

$\hat{A}_j^*$ = after-tax cash flow in period j in "then" (current) dollars, assuming inflated pretax cash flow

$\hat{A}_{jd}^*$ = after-tax cash flow in period j in "now" (present) dollars, assuming inflated pretax cash flow

$\hat{A}_{ji}^*$ = after-tax cash flow in period j in "then" (current) dollars, assuming inflated pretax cash flow and depreciation expenses indexed to inflation rate

D_j = allowance for depreciation (depreciation expense) in period j

D_j^* = indexed allowance for depreciation (depreciation expense) in period j

d = differential rate of inflation—the incremental rate of price increase/decrease (h) over the general inflation rate (f)

f = general rate of inflation—the uniform periodic rate of increase/decrease of all prices in the economy

h = specific rate of inflation—the uniform periodic rate of increase/decrease in the price of a specific good or service

i = discount rate

i^* = inflation-adjusted discount rate (also written k^*)

I_j = index number for period j

k = minimum attractive rate of return in the absence of inflation

k^* = investment/reinvestment rate incorporating general inflation

N = depreciable and/or service life of capitalized investment

P = initial cost (basis) of capitalized investment

$p(a/b)$ = index number describing the change in relative prices between periods a and b (price relative)

$p(j)$ = price of commodity in period j

$p(j,0)$ = price of commodity j in base period

$p(j,n)$ = price of commodity j in given period n

$q(a/b)$ = quantity relative [See $p(a/b)$]

CHAPTER TEN APPENDIX B NOTATION USED IN CHAPTER 10 AND APPENDIX 10A

$q(j,0)$ = quantity of commodity j in base period

$q(j,n)$ = quantity of commodity j in given period n

$q(j,t)$ = quantity of commodity j in "typical" period t

R_j = net cash flow from operations *before* taxes in period j in "now" dollars

R_j^* = net cash flow from operations *before* taxes in period j in "then" dollars

$\hat{R}_j$ = net cash flow from operations *after* taxes in period j in "now" dollars

$\hat{R}_j^*$ = net cash flow from operations *after* taxes in period j in "then" dollars

S = salvage value at end of depreciable and/or service life in "now" dollars

S^* = salvage value at end of depreciable and/or service life in "then" dollars

$\hat{S}^*$ = after-tax salvage value in "then" (current) dollars

t_j = incremental income tax rate in year j

$v(a/b)$ = value relative [See $p(a/b)$]

$w(j)$ = arithmetic mean of base period and given period quantities

$\emptyset$ = do-nothing alternative

ρ = internal rate of return assuming no inflation ($f = 0$)
ρ_0 = before (income) taxes
ρ_1 = after (income) taxes

α = internal rate of return under inflation ($f > 0$)
α_0 = before (income) taxes
α_1 = after (income) taxes

β = internal rate of return under inflation ($f > 0$) but after deflating current ("then") dollars to present ("now") dollars
β_0 = before (income) taxes
β_1 = after (income) taxes

γ = internal rate of return under inflation ($f > 0$) but assuming that depreciation allowances are indexed to inflation
γ_0 = before (income) taxes
γ_1 = after (income) taxes

Equation 10.1, the modified Laspeyres index, is related to index numbers, which are discussed in Appendix 10A. The equation is omitted from this summary.

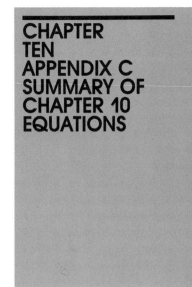

CHAPTER
TEN
APPENDIX C
SUMMARY OF
CHAPTER 10
EQUATIONS

10.2 $PW = \sum_{j=0}^{N} A_j (1 + i)^{-j}$

10.3 $AW = (PW)(A/P, i, N) = (PW)\left[\dfrac{i(1 + i)^N}{(1 + i)^N - 1}\right]$

10.4 $FW = \sum_{j=0}^{N} A_j (1 + i)^{N-j}$

10.5 $A_j^* = A_j (1 + f)^j$

10.6 $FW = \sum_{j=0}^{N} A_j^* (1 + i)^{N-j}$

10.7 $i^* = (1 + i)(1 + f) - 1 = i + f + if$

10.8 $FW = \sum_{j=0}^{N} A_j^* (1 + i^*)^{N-j}$

10.9 $PW = \sum_{j=0}^{N} A_j^* (1 + i)^{-j}$

where
$$A_j^* = A_j (1 + f)^j$$

10.10 $PW = \sum_{j=0}^{N} A_j^* (1 + i^*)^{-j}$

10.11 $FW = \sum_{j=0}^{N} A_j (1 + h)^j (1 + i^*)^{N-j}$

10.12 $PW = \sum_{j=0}^{N} A_j (1 + h)^j (1 + i^*)^{-j}$

10.13 $d = \left[\dfrac{1 + h}{1 + f}\right] - 1$

10.14 $PW = \sum_{j=0}^{N} A_j (1 + d)^j (1 + i)^{-j}$

10.15 $FW = (1 + f)^N \sum_{j=0}^{N} A_j (1 + d)^j (1 + i)^{N-j}$

10.16 $PW(\rho_0) \equiv 0 = \sum_{j=0}^{N} A_j (1 + \rho_0)^{-j}$

where
$$A_0 = -P$$
$$A_j = R_j \quad \text{for } j = 1, 2, \ldots, N - 1$$
$$A_N = R_N + S$$

10.17 $\hat{R}_j = R_j - t_j(R_j - D_j) = R_j(1 - t_j) + t_j D_j$
for $j = 1, 2, \ldots, N$

10.18 $PW(\rho_1) \equiv 0 = \displaystyle\sum_{j=0}^{N} \hat{A}_j(1 + \rho_1)^{-j}$

$\hat{A}_0 = -P$
$\hat{A}_j = \hat{R}_j$ for $j = 1, 2, \ldots, N - 1$
$\hat{A}_N = \hat{R}_N + S$

10.19 $PW(\rho_1) \equiv 0 = [R(1 - t) + tD](P/A, \rho_1, N)$
         $- P + S(P/F, \rho_1, N)$

10.20 $\hat{R}_j^* = R_j^* - t_j(R_j^* - D_j)$
     $= R_j(1 + f)^j(1 - t_j) + t_j D_j$
for $j = 1, 2, \ldots, N$

10.21 $\hat{S}^* = S^* - t_N(S^* - S) = S[(1 + f)^N(1 - t_N) + t_N]$

10.22 $PW(\alpha_0) \equiv \displaystyle\sum_{j=0}^{N} A_j^*(1 + \alpha_0)^{-j}$

where
$A_0^* = -P$
$A_j^* = R_j^*(1 + f)^j$ for $j = 1, 2, \ldots, N - 1$
$A_N^* = R_N(1 + f)^N + S$

10.23 $PW(\alpha_1) \equiv 0 = \displaystyle\sum_{j=0}^{N} \hat{A}_j^*(1 + \alpha_1)^{-j}$

where
$\hat{A}_0^* = -P$
$\hat{A}_j^* = R_j(1 + f)^j(1 - t_j) + t_j D_j$
     for $j = 1, 2, \ldots, N$
$\hat{A}_N^* = R_N(1 + f)^N(1 - t_N) + t_N D_N$
     $+ S[(1 + f)^N(1 - t_N) + t_N]$

10.24 $PW(\beta_1) \equiv 0 = \displaystyle\sum_{j=0}^{N} \hat{A}_{jd}^*(1 + \beta_1)^{-j}$

where
$\hat{A}_{0d}^* = -P$
$\hat{A}_{jd}^* = R_j(1 - t_j) + t_j D_j(1 + f)^{-j}$
     for $j = 1, 2, \ldots, N - 1$
$\hat{A}_{Nd}^* = R_N(1 - t_N) + t_N D_N(1 + f)^N$
     $+ S[(1 - t_N) + t_N(1 + f)^{-N}]$

10.25 $D_j^* = D_j(1 + f)^j$

10.26 $PW(\gamma_1) \equiv 0 = \displaystyle\sum_{j=0}^{N} \hat{A}_j^*(1 + \gamma_1)^{-j}$

where
$\hat{A}_{0i}^* = -P$
$\hat{A}_{ji}^* = (1 + f)^j[R_j(1 - t_j) + t_j D_j]$
     for $j = 1, 2, \ldots, N - 1$
$\hat{A}_{Ni}^* = (1 + f)^N[R_N(1 - t_N) + t_N D_N] + S(1 + f)^N$

10.27 $PW(\rho_1) \equiv 0 = -P + \sum_{j=1}^{N} [R_j(1 - t_j) + t_j D_j](1 + \rho_1)^{-j}$
$+ S(1 + \rho_1)^{-N}$

10.28 $PW(\gamma_1) \equiv 0 = -P + \sum_{j=1}^{N} [R_j(1 - t_j) + t_j D_j]\left(\dfrac{1 + f}{1 + \gamma_1}\right)^j$
$+ S\left(\dfrac{1 + f}{1 + \gamma_1}\right)^N$

10.29 $FW(\mathrm{I}:k) = (1 + k)^N[PW(\mathrm{I}:k)] = \sum_{j=0}^{N} A_j(1 + k)^{N-j}$

10.30 $FW(\mathrm{I}:k) = (1 + k)^N[PW(\mathrm{I}:k)] = \sum_{j=0}^{N} A_j^*(1 + k)^{N-j}$

10.31 $FW(\varnothing:k) = (1 + k)^N[PW(\varnothing:k)] = 0$

10.32 $\sum_{j=0}^{N} \left[\dfrac{1 + f}{1 + k}\right] > \sum_{j=0}^{N} A_j\left[\dfrac{1}{1 + \beta}\right]^j$

10.33 $\beta > p = \left[\dfrac{1 + k}{1 + f}\right] - 1$

10.34 $PW(\mathrm{I}:p) = \sum_{j=0}^{N} A_j(1 + p)^{-j}$

10.35 $FW(\mathrm{I}:k^*) = \sum_{j=0}^{N} A_j^*(1 + k^*)^{N-j} = (1 + k^*)^N[PW(\mathrm{I}:k^*)]$

10.1 The nation of Otherland has experienced a high inflation rate (25 percent per year). Recently, the exchange rate was 11 pesos per U.S. dollar, that is, one dollar will buy 11 pesos in the foreign exchange market.

Assume that Otherland will continue to experience an inflation rate of 25 percent per year and that the United States will experience an average rate of inflation of 10 percent per year over the next five years.

If the exchange rate varies with the rate of inflation, how many pesos will one U.S. dollar buy five years from now? (Note that this type of calculation enables the analyst to combine inflation rates for two or more countries.) (*Answer:* $1 = 20.844 pesos)

10.2 a. Mr. Johnson was hired by XYZ Company in July 1963 at a salary of $1,000 per month. At that time the CPI-W was 92.1. If his salary had "kept up with inflation," what must his monthly salary be in July 1983, when the CPI-W was 298.2?

b. Mr. Johnson's actual salary in July 1983 was $4,000 per month. What was the average annual rate of increase?

10.3 I. M. Smart, an engineer for the Hardrock Construction Company, is reviewing a previous investment in a certain heavy-duty compressor. The compressor was purchased in October 1982 for $5,000. At that time, the price index for compressors was 215.6; the base for this index was January 1979. Three months later, in January 1983, the price index was 220.2. The base for the index was revised in January 1983 to 100. In January 1984, twelve months later, the new index was at 114.6. In summary,

Time	Old index	New index
October 1982	215.6	——
January 1983	220.2	100.0
January 1984	——	114.6

a. What was the average inflation rate *per month,* stated as a percentage, over the fifteen-month period between October 1982 and January 1984?

b. Convert the answer from (a) to an average inflation rate *per year.*

c. If the heavy-duty compressor cost $5,000 in October 1982, how much would it cost in January 1984? Assume that the price increase for this specific compressor reflects the price index for compressors in general.

(*Answers:* (a) 1.05 percent; (b) 13.42 percent; (c) $5,852)

10.4 The Universal Steel Company (USC), located in Anaheim, California, purchased type 52-A lubricant in December 1967 for $18.20 per gallon. At that time the CPI-W was 102.0. In December 1982 the price for the same lubricant had risen to $58.00; the CPI-W in December 1982 was 292.0.

a. Determine the average *monthly* rate of increase in the CPI-W during this fifteen-year period.

b. Determine the average *annual* rate of increase in the price of type 52-A lubricant during this period.

c. Suppose that we want to correct the rate of price increase for the lubricant to account for the general level of price increase as reflected by the CPI-W. Determine the appropriate differential rate of increase *per year*.

d. If the price of the type 52-A lubricant is expected to continue to increase at the same rate as experienced over the 1967–1982 period, what would be the price per gallon in June 1990?

10.5* An alumnus has offered to make a donation to his alma mater. Three alternative plans for the donation have been offered: (1) donate $65,000 now, (2) donate $16,000 at the end of every year for twelve years, or (3) donate $50,000 three years from now with an additional $80,000 five years from now. Currently, the university can earn 12 percent per year on its "ready assets account." Inflation is expected to average 11 percent per year over the next twelve years.

a. Determine which plan yields maximum future worth at the end of twelve years, assuming that short-term investment of funds in the ready assets account will continue to earn at the rate of 12 percent per year.

b. Assume that the rate of return on funds in the ready assets account will increase with the rate of inflation; that is, today's rate, 12 percent per year, will increase at the rate of 11 percent per year. With this assumption, determine which plan yields maximum future worth at the end of twelve years.

(Answers: (a) *FW*(2) = $386,130; (b) *FW*(1) = $885,942)

10.6 Assume that the donation described in Problem 10.5(a) will be used to purchase microcomputers that currently cost $2,500 each. Assume that computers will be purchased as soon as the funds are available.

a. If this cost per unit remains constant over the next twelve years and if invested funds continue to earn at 12 percent per year, determine the number of microcomputers that can be purchased over the twelve-year period under the three plans. Round your answers to the nearest integer.

b. If the cost of microcomputers were to increase at the rate of 11 percent per year and if invested funds continue to earn at 12 percent per year, determine how many microcomputers could be purchased over the twelve-year period.

c. Solve (b) assuming that the rate of return on funds in the ready assets account will also increase with the rate of inflation.

*Adapted from L. T. Blank and A. J. Tarquin, *Engineering Economy*, 2nd ed. (New York: McGraw-Hill, 1983) p. 240.

10.7 Assume that $1,000 is invested in an account earning 8 percent per year over a five-year period. During this period, general inflation is expected to remain at 6 percent per year.
   a. Determine the amount in the fund at the end of five years. That is, what is the future worth after five years in "then" (current) dollars?
   b. Considering the eroding effect of inflation, what is the dollar value of the fund after five years in terms of today's buying power? In other words, what is the future worth after five years in "now" (present) dollars?
   c. What is the interest rate at which "now" dollars will expand, with their same buying power, into equivalent "then" dollars? (In the literature of finance, this is known as the *real interest rate*.)
   (*Answers:* (a) $1,469; (b) $1,098; (c) 1.89 percent)

10.8 A small manufacturing firm is considering a proposal to fully automate an assembly operation that is currently done entirely by manual operation at a cost of $150,000 annually. The automatic equipment will cost $500,000, will have an estimated economic life of ten years, and will have an expected net value of $20,000 on disposal. The cost of operation and maintenance for this equipment is estimated to be $50,000 per year. The firm's inflation-free pretax minimum attractive rate of return is 15 percent per year. The general rate of inflation is estimated to be 8 percent per year over the next ten years, and the *MARR* is expected to reflect this inflationary effect. The cost of labor (for current assembly operations as well as for proposed operations and maintenance of the automatic equipment) is expected to increase at the rate of 10 percent per year over the study period. For simplicity, assume that all cash flows, other than the initial investment, are end-of-year.
   After adjusting for inflation, determine the pretax equivalent uniform annual costs of the two alternatives.

10.9 Consider a proposed investment with the following anticipated cash flows in "now" (present) dollars:

End of year	Pretax cash flow
0	−$100,000
1	10,000
2	20,000
3	30,000
4	40,000
5	50,000

   a. Determine the cash flows in "then" (current) dollars if the inflation rate is 7 percent per year.
   b. Assuming that the firm's pretax *MARR* is 25 percent— this *includes* expectation of 7 percent inflation over the study period—use the rate of return method to determine

whether this proposal is preferable to the do-nothing alternative.

(*Answers:* (a) −$100,000, $10,700, $22,898, $36,751, $52,432, $70,128; (b) Proposal is *not* preferable to do-nothing)

10.10 The $100,000 negative cash flow in Problem 10.9 represents an investment in a five-year recovery property to be depreciated under ACRS. The firm's incremental tax rate is 55 percent. Disregard the investment tax credit and assume that the full depreciation allowance may be taken in the fifth year. Other assumptions are as given in Problem 10.9.

If the firm's after-tax *MARR* is 14 percent—again, this includes 7 percent inflation—use the rate of return method to determine whether this proposal is preferable to the do-nothing alternative.

Problems for Appendix 10A

10A.1 The average prices per kilogram for a certain commodity during a recent five-year period were

Year:	1979	1980	1981	1982	1983
Price per kg:	$10.48	$11.15	$11.65	$11.40	$12.70

Find the price relatives corresponding to each given year:
a. using 1979 as the base year.
b. using 1983 as the base year.
(*Answers:* (a) 100.00, 106.39, 111.16, 108.78, 121.18; (b) 82.52, 87.79, 91.73, 89.76, 100.00)

10A.2 The quantities of the commodity in metric tons (1,000 kg) in Problem 10A.1 sold during the period 1979 through 1983 are

Year:	1979	1980	1981	1982	1983
Quantity sold (metric tons):	10,000	10,800	11,000	13,500	14,000

a. The quantity relative for 1979, with 1975 as the base year, is 120.6. Find the quantity relatives for all other given years (1980–1983).
b. Find the quantity relatives for the given years if 1981 is the base year.

10A.3 The Combo Company produces a certain product from three raw materials. In an effort to establish an overall materials price index (MPI), consider the following data for a recent three-year period:

Commodity	Average annual prices (per kilogram)			Annual consumption (metric tons)		
	1981	1982	1983	1981	1982	1983
A	$0.38	$0.45	$0.50	500	550	600
B	2.40	2.20	2.00	20	24	26
C	3.15	3.40	4.00	18	20	25

a. Compute a simple aggregate price index for this three-commodity product using 1981 as the base year.

b. Explain why the simple aggregate price index is not an appropriate measure of the MPI in this instance.

(*Answer:* (a) 1.00, 1.02, 1.096)

10A.4 With the data from Problem 10A.3 and using 1981 as the base year, find the year-by-year MPI using
a. the simple average of relatives method.
b. the modified average of relatives method, with the median instead of the mean.

10A.5 With the data from Problem 10A.3 and using 1981 as the base year, find the year-by-year MPI using
a. the Laspeyres price index.
b. the Paasche price index.

(*Answers:* (a) 1.00, 1.12, 1.23; (b) 1.00, 1.12, 1.22)

10A.6 With the data from Problem 10A.3 and using 1981 as the base year, find the year-by-year MPI using
a. Fisher's ideal index.
b. the Marshall-Edgeworth Index.

10A.7 Determine the weighted average of price relatives for the data of Problem 10A.3 using
a. given year weights.
b. base year weights, using 1981 as the base year.
c. base year weights, using 1983 as the base year.

(*Answers:* (a) 1.00, 1.13, 1.25; (b) 1.00, 1.12, 1.23; (c) 0.82, 0.91, 1.00)

10A.8 Using the data from Problem 10A.3, and assuming 1981 as the base year.
a. find the weighted aggregate *volume* indexes with base year price weights. (This index is known as the Laspeyres volume index.)
b. find the weighted aggregate indexes with given year price weights. (This index is known as the Paasche volume index.)

10A.9 The following table shows the average hourly wages of the production workers at the Combo Company during the period 1973 through 1982. The CPI-W is also shown for the same period.

Year:	1973	1974	1975	1976	1977	1978	1979	1980	1981	1982
Average wage ($ per hour):	6.47	6.70	7.00	7.45	7.90	8.15	8.60	9.50	10.50	12.00
CPI-W (1967 = 100):	133.1	147.7	161.2	170.5	181.5	195.3	217.7	247.0	272.3	288.6

Determine the "real" wages of these workers compared to 1967 dollars. That is, "deflate" the actual wages assuming that the CPI-W is the appropriate deflator.

(*Answer:* $4.86, $4.54, . . . , $4.16)

10A.10 With 1967 as the base year, the average annual CPI-U for the years 1978 to 1982 were

Year:	1978	1979	1980	1981	1982
CPI-U:	195.4	217.4	246.8	272.4	289.1

a. Shifting the base year from 1967 to, say, 1980, the revised index numbers can be determined by dividing each of the old numbers by 246.8, the base period index number. Determine the new revised index numbers for the given years.

b. Prove that the above method is applicable only if the index numbers satisfy the circular test.

c. Prove that the weighted aggregate price index numbers with fixed quantity weights satisfy the circular test.

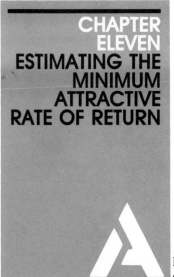

CHAPTER ELEVEN
ESTIMATING THE MINIMUM ATTRACTIVE RATE OF RETURN

ll of the principal methods used to assess the economic desirability of one or more investment proposals or used to determine the optimal allocation of a capital budget require an explicit or implicit determination of the minimum attractive rate of return. This value is used directly as the interest rate, i, in the present worth, annual worth, and future worth methods. For example;

$$PW = \sum_{j=0}^{N} A_j (1 + i)^{-j} \tag{11.1}$$

$$AW = PW(A/P, i, N) \tag{11.2}$$

$$FW = PW(F/P, i, N) \tag{11.3}$$

In the rate of return method, the internal rate of return (IRR) for a given investment proposal is compared to the minimum attractive rate of return ($MARR$). Generally, if $IRR > MARR$, the investment is economically justified; if $IRR < MARR$, it is not.[1]

The **minimum attractive rate of return** is a term widely used in the literature of engineering economy. There are other descriptive titles given to this value, among them *financial standard, target rate of return, hurdle rate,* and *required rate of return*. In any event, it should be understood that the *MARR* represents the rate of return available to the investor if he or she chooses to invest funds "elsewhere" rather than in the specific proposal under consideration; it is the rate of return provided by the do-nothing alternative. If the investor elects to invest funds in a certain project or program, he or she forgoes the opportunity to earn a return that would be available if those same funds had been

1. This decision rule does not hold for all cash flow sequences. (See Chapter 3.)

invested in some other way. The *MARR* represents a cost of opportunities forgone, expressed as a rate (percentage). Thus the concept of *opportunity cost* is central to understanding the minimum attractive rate of return.

The *MARR* is directly related to the costs of the individual elements of capital available for investment. Simply stated, if funds can be obtained at a cost of, say, 10 percent per year, these funds must be invested so as to return at least 10 percent; otherwise the cost of capital would not be warranted (at least from the perspective of economic efficiency). The cost of capital, then, establishes the minimum return required to justify investment. To estimate which *MARR* to use in capital allocation decisions, it is necessary to estimate the costs of the constituent elements of capital available to the investor. Procedures for estimating the costs of capital are explored in this chapter.

There is currently considerable controversy as to the manner in which the cost of capital should be determined. The following discussion parallels the so-called "traditional view," although some important alternative formulations are also discussed. For additional readings, see the list of selected references in Appendix A. Of particular interest are the presentations of J. C. T. Mao and David Quirin and John Wiginton.

11.1 SOURCES OF CAPITAL

Each dollar invested within a firm may be broadly classified as either debt or equity, according to the nature of its source. In the case of privately owned (investor-owned) enterprises, **equity capital** may be defined as funds that are provided by the owners of the firm in the form of profits realized from prior operations and retained for reinvestment or from the infusion of new capital raised from new owners through the sale of additional shares of *common stock*.[2]

The concept of equity capital requires a somewhat different interpretation for publicly owned enterprises, for which the "owners" are the citizens of the political entity having jurisdiction over the enterprise. The Los Angeles Department of Water and Power (LADWP), for example, is the creature of the citizens of the city of Los Angeles. The citizens are the "owners" of the LADWP. Thus investment decisions made by this publicly owned utility should reflect citizens' interests in the same sense that decisions of privately owned firms should reflect the aspirations of their stockholders.

Debt capital, in general, consists of funds provided to the enterprise by individuals and institutions in the form of loans that

---

2. The assumption of the corporate form of business enterprise for privately owned firms is used throughout this chapter. For partnerships or individual proprietorships, there are, of course, no ownership shares in the form of stock.

must be repaid according to some contractual arrangements. *Bonds* and *notes* are the principal kinds of debt capital.

For the purpose of our classification scheme, *preferred stock* may be considered another form of debt capital in the investor-owned firm.

Dividends on preferred stocks, even though they are not legal obligations to the same extent as debt-servicing charges, are a cost to the common stockholder. Few corporations would issue preferred stock without the intention of paying regular dividends, and the consequences of failing to do so can be extremely serious. . . . While the obligation to pay is moral, rather than legal, the cost is economically relevant.[3]

Leases are often viewed as another method for obtaining investment capital. In a sense, leasing can be considered a form of debt capital, because the terms of the lease generally impose some financial obligations on the part of the lessee (the "borrower"). However, care must be taken to ensure that two separable decisions are treated separately, namely, (1) whether to acquire the asset and (2) whether to use leasing or some other source of capital to finance the acquisition. Analysts frequently treat the questions as one, thereby overstating the acquisition's desirability.

There are a variety of other financial instruments that may be used to acquire investment funds, including debentures, trade credits, factoring, sale-and-leaseback arrangements, and installment financing. An exhaustive discussion of all sources of investment capital is beyond the scope of this text. Instead, the following sections prescribe a procedure for estimating the cost of any capital source, in general, and then demonstrate the application of this procedure to certain sources of equity and debt capital.

Determining the cost of a given increment of capital is analogous to determining the internal rate of return for a given investment proposal. Suppose that an increment of capital, Q, is "loaned" to a "borrower" with the understanding that the borrower will repay the loan over N periods. The amounts of repayment will be C_j at the end of periods $1, 2, \ldots, j, \ldots, N$. In this case, the lender's rate of return is that interest rate, k, such that the present worth of all future cash flows is exactly equal to the amount of the original loan. That is,

**11.2
MEASURING THE COST
OF CAPITAL: GENERAL
FORMULATION**

$$Q = \sum_{j=1}^{N} C_j (1 + k)^{-j} \qquad \textbf{(11.4)}$$

3. David G. Quirin and John C. Wiginton, *Analyzing Capital Expenditures* (Homewood, Ill.: Richard D. Irwin, 1981), p. 158.

The value of k in this equation is the return to the lender and the cost to the borrower. Put somewhat differently, from the viewpoint of the borrower, k represents the **cost of capital.** With respect to any specific capital of amount Q, it remains only to identify the amount and timing of associated dollar costs, C_j, thereby measuring the cost of that increment of capital. Clearly, if the borrower were to invest Q in a project, the project's internal rate of return should exceed k; otherwise the transaction would not be economically justifiable.

11.3
THE COST OF BONDS
OR NOTES

There are many types of bonds or notes: municipal bonds, industrial bonds, mortgage bonds, construction bonds, revenue bonds or notes, and so forth. There are certain differences among these debt instruments, but they need not concern us in this context. It is sufficient to observe at this point that bonds or notes are financial instruments in which the borrower agrees to repay the lender specified amounts of money at prescribed points in time. In general, the *pretax* cost of this capital is that value of k_b such that

$$Q = C_0 + \sum_{j=1}^{N} C_j (1 + k_b)^{-j} \tag{11.5}$$

where

Q = funds received by borrower

C_0 = costs associated with the borrowing (underwriting costs, for example)

C_j = repayment at end of period j ($j = 1, 2, \ldots, N$)

To illustrate, suppose that a firm obtains debt capital by issuing 5,000 bonds, each with a par value of $1,000, and promises to pay interest quarterly at the nominal rate of 12 percent per year over a twenty-year period. The initial underwriting costs are $200,000. The cost to the firm to process each quarterly interest payment is estimated to be $25,000. The price paid by bond purchasers is $900 each, thereby yielding 5,000($900) = $4,500,000 initially, before subtracting the underwriting costs. The future costs to the firm are 5,000(0.12/4)($5,000) + $25,000 = $175,000 quarterly and 5,000($1,000) = $5,000,000 at the end of twenty years. To determine the cost of this bond issue,

$$\$4{,}500{,}000 = \$200{,}000 + \$175{,}000(P/A, k_{bq}, 80)$$

$$+ \$5{,}000{,}000(P/F, k_{bq}, 80)$$

where k_{bq} = cost per quarter. Solving the equation, k_{bq} = 4.0977%. Converting this quarterly cost to the cost per year, generally a more useful base of reference,

$$k_b = (1.040977)^4 - 1 = 17.426\%$$

Note that the effective cost of this debt is *not* equal to the nominal or "coupon rate," 12 percent per year. That would have been the case only if the initial net receipts had been exactly $5,000,000 and each of the quarterly payments had been exactly $5,000,000 ($\sqrt[4]{1.12} - 1$) = $143,687. Frequently, the cost of bonds is assumed to be equal to the coupon rate of those bonds, but as this example illustrates, the true effective cost can differ substantially.

The pretax cost of bonds or notes, k_b, is equal to the nominal or coupon rate, r, only under special conditions. First, $k_b = r$ when the annual interest payments, I, are uniform and occur at the end of each year over N years and, further, the terminal payment after N years is equal to the amount of the initial borrowing. These assumptions are reflected in Figure 11.1. In this case,

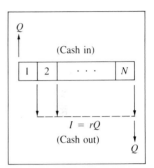

Figure 11.1

$$Q = I(P/A, k_b, N) + Q(1 + k_b)^{-N}$$

$$= rQ\left[\frac{(1 + k_b)^N - 1}{k_b(1 + k_b)^N}\right] + Q(1 + k_b)^{-N} \qquad \textbf{(11.6)}$$

Simplifying, $r = k_b$.

Secondly, $k_b = r$ if the annual interest payments are uniform and occur at the end of each year forever, as illustrated in Figure 11.2. (These are known as *perpetual bonds*.) As before, assume that the initial receipts to the borrower are exactly Q and that the annual end-of-year interest payments are rQ. In this case,

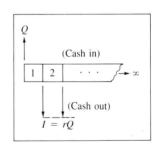

Figure 11.2

$$Q = I(P/A, k_b, \infty)$$

$$= rQ(1/k_b) \qquad \textbf{(11.7)}$$

Again simplifying, $r = k_b$.

The formulations in Equations 11.5, 11.6, and 11.7 are pretax formulations; that is, the effects of income taxes are ignored. Interest payments on debt are deductions from taxable income, so the after-tax cost of debt capital is somewhat less than the pretax cost. If the income tax rate (t) is constant over the period of the loan, then the *after-tax cost of bonds* is

$$k_b = (1 - t)r \qquad \textbf{(11.8)}$$

given the conditions underlying Equations 11.6 and 11.7.

The after-tax formulation is more complex, of course, given the general case in Equation 11.5. Interest payments each year are deductible in that year. Also deductible are expenses associated with processing interest payments as well as amortization of the difference between the par value of the bonds, V, and the funds received, Q. An approximate after-tax formulation is

$$k_b \simeq \frac{(1 - t)[r + (1/N)(V - Q)]}{(1/2)(V + Q)} \qquad \textbf{(11.9)}$$

11.4 THE COST OF PREFERRED STOCK

As indicated previously, preferred (or preference) stockholders have a prior right to stipulated dividends as well as assets in many cases. There are many forms of preferred stock: participating preferred, convertible preferred, and so on. Although the specific differences among these various classes are beyond the scope of this discussion, the approximate cost of preferred stock may be given as

$$k_p \simeq H_p/Q \qquad \textbf{(11.10)}$$

Where H_p is the annual dividend to preferred stockholders and Q is the initial sum received from the issue of the preferred stock. This formulation assumes that there is a perpetual obligation to pay these dividends and that dividends are paid at the end of each year. When all or parts of the issue are expected to be called at specified dates, and/or when there is a sinking fund involved, a more precise formulation may be appropriate. Note that dividends are not tax-deductible expenses, and hence the after-tax costs to the firm is the same as the pretax cost of capital raised in the form of preferred stock.

11.5 THE COST OF LEASES

The cost of a lease, stated as a percentage rate, is the cost of the leasing alternative compared to the cost of outright purchase. That is, the cost of the lease is the interest rate at which the investor would be indifferent between the lease alternative and the purchase alternative. A general formulation for this calculation is quite complex, because the cash flows due to the purchase alternative, as well as the lease arrangements, can be complicated. The following is an approximate formulation, given certain qualifying assumptions.

The equivalent present worth of costs of the *purchase* alternative, after income taxes, is approximately

$$PWOC \cong P - TC - [S - t(S - B_N)](1 + k_1)^{-N}$$

$$+ \sum_{j=1}^{N} [C_j - t(C_j + D_j)](1 + k_1)^{-j} \qquad \textbf{(11.11)}$$

where P is the initial cost of the asset if it is purchased, TC is the investment tax credit associated with the purchase, S is the terminal salvage value of the asset if it is purchased and then sold after N periods, B_N is the prospective book value (adjusted basis) at the time of disposition, t is the effective tax rate, C_j is the end-of-period costs that would be incurred each period (j = 1, 2, ..., N) if the asset were purchased rather than leased, D_j is the depreciation expense in period j, and k_1 is the after-tax cost of the lease. At this point, all values other than k_1 can be estimated.

There are a number of assumptions underlying this formulation. Among the most prominent: (a) cash flows and discounting are end-of-period; (b) if purchased, the asset will be retained in service for the same number of periods as the period of the lease; (c) pretax expenses, depreciation expenses, and cash flows for taxes occur at the same points in time; (d) any gain or loss on disposal will be treated as ordinary income; and (e) the timing of the investment tax credit coincides with the purchase.

Similarly, the present worth of *lease* costs, after taxes, is approximately

$$PWOC \cong \sum_{j=0}^{N-1} L_j(1 + k_1)^{-j} - t \sum_{j=1}^{N} L_{j-1}(1 + k_1)^{-j} \qquad \textbf{(11.12)}$$

where L_j is the lease payment at the end of period j and where t and k_1 are as defined earlier. Note that this formulation assumes that (a) lease payments occur at the *start* of each period, (b) tax consequences due to these leasing expenses occur at the *end* of each period, and (c) there are no other expenses incurred if the asset is leased rather than purchased.

The approximate cost of the lease is determined by setting Equation 11.11 equal to Equation 11.12 and solving for k_1. Writing out the equations and rearranging terms, the after-tax cost is that value of k_1 for which

$$P - TC - [S - t(S - B_N)](1 + k_1)^{-N}$$

$$- \sum_{j=0}^{N-1} L_j(1 + k_1)^{-j} + \sum_{j=1}^{N} [C_j(1 - t)$$

$$+ t(L_{j-1} - D_j)](1 + k_1)^{-j} = 0 \qquad \textbf{(11.13)}$$

Table 11.1
Costs Associated with Purchase and Lease Alternatives

	End of period (1)	Cash flow before taxes (2)	Depreciation (3)	Effect on taxable income (4) = (2) − (3)	Effect on income taxes^e (5) = t(4)	Cash flow after taxes (6) = (2) − (5)
Purchase	0	− $10,000	$ ———	$ ———	− $1,000^d	− $9,000
	1–10	− 500	760	− 1,260	− 630	130
	10	2,000^a	2,400^b	− 400^c	− 200	2,200
	10-year total =					− $5,500
Lease	0	− $ 1,500	$ ———	$ ———	$ ———	− $1,500
	1–9	− 1,500	$ ———	− 1,500	− 750	− 750
	10	$ ———	$ ———	− 1,500	− 750	750
	10-year total =					− $7,500

a. Expected salvage value is $2,000.
b. Book value is $2,400 after 10 years, or $10,000 − 10($760).
c. Loss on disposal is $400, treated as an ordinary loss.
d. Investment tax credit is 10 percent of original cost.
e. Tax rate = 0.50.

This calculation may be illustrated by referring to the numerical example shown in Table 11.1. If purchased, the initial cost of the asset, P, is $10,000; the investment tax credit, TC, is 10 percent of the purchase price; the expected salvage value, S, at the end of ten years is $2,000; annual depreciation expenses, D_j, are $760 each year; the book value of the asset after ten years, B_{10}, will be $10,000 − 10($760) = $2,400; the marginal income tax rate, t, is 50 percent; and the additional annual costs due to the purchase of the equipment, C_j, are $500 each year. This last figure reflects insurance and maintenance costs that must be borne by the firm if the asset is purchased but that would otherwise be included in the lease arrangement.

In the example, lease costs are $1,500, payable in ten equal payments at the start of each year over ten years. (The start of year j is the end of year $j − 1$.) The effect on taxable income of an expense at the start of each year is assumed to occur at the end of that year.

The total after-tax costs (negative cash flows) are $5,500 if the asset is purchased and $7,500 if it is leased. If the asset is leased rather than purchased, the differences in after-tax cash flows are

End of year	Cash flows if purchased	Cash flows if leased	Differential cash flows if leased rather than purchased
0	− $9,000	− $1,500	$7,500
1–9	130	− 750	− 880
10	2,330	750	− 1,580
Total	− $5,500	− $7,500	− $2,000

The investor, in effect, "saves" $7,500 immediately if the asset is leased rather than purchased. However, the funds so obtained must be "repaid" at the rate of \$880 at the end of every year for nine years, with an additional payment of \$1,580 at the end of the tenth year. The after-tax cost of capital, then, is the solution to the equation

$$\$7,500 - \$880(P/A, k_l, 9) - \$1,580(P/F, k_l, 10) = 0$$

from which $k_l = 4.3\%$. (Of course, this is also the solution to Equation 11.13.) By choosing to lease rather than purchase, the investor has effectively "borrowed" \$7,500 over a ten-year period at an after-tax cost of 4.3 percent per year.

In stating basic principles to guide investment decisions, it is first necessary to establish from whose point of view the relative merits of the decisions are to be judged. It would seem to be reasonable to assume that the view point of the *current* owners of the firm should be adopted. Moreover, let's assume that owners are profit maximizers and investment decisions are rational, and thus the principal investment criterion is maximization, over the long run, of the interests of the existing group of owners. This statement of the fundamental investment principle is not intended to preclude such management objectives as level employment, product safety, community service, and so on. Indeed, management decisions are affected in a variety of ways. It may be argued, however, that in order to continue as a viable enterprise in a competitive economy, there are at least *some* decisions that have an economic justification as their basis. Thus the discussion of the rationale for capital allocation belongs in this frame of reference.

11.6 THE COST OF EQUITY CAPITAL: INVESTOR-OWNED FIRMS

New Issues of Common Stock

Expected future earnings resulting from new issues of common stock, the shares of ownership of the corporation, should be sufficient to prevent the loss of value to current owners of the firm. The market value of the stock, as well as the future stream of dividend income, must not be diluted, otherwise there is no economic advantage in obtaining new equity funds. There are several approaches to determining the rate of return, k_e, that will prevent dilution. The following discussion examines only a few.

First, there is the *no-growth* case, wherein the firm's existing assets are expected to maintain a constant stream of annual net income. If the new issue is sold entirely to existing stockholders, then

$$k_{eo} = e/p \qquad\qquad (11.14)$$

where e is the expected after-tax earnings per share, exclusive of additional financing, and p is the current market price of a share of common stock. If the new issue is sold entirely to outside investors, then

$$k_{en} = e/p'$$
(11.15)

where p' is the price at which new shares are sold. If x_1 shares are sold to existing stockholders and x_2 shares are sold to new stockholders, and if x_0 is the number of shares before the new stock issue, then it may be shown that the overall cost of this new equity is

$$k_e = k_{en}\left[\frac{x_0 x_2}{(x_0 + x_1)(x_1 + x_2)}\right] + k_{eo}\left[\frac{x_1(x_0 + x_1 + x_2)}{(x_0 + x_1)(x_1 + k_2)}\right]$$
(11.16)

Second, there is the *growth* case, wherein the market price of common stock reflects investors' expectations of future growth in dividends as well as in current earnings and dividends. In this model, assume that a fraction, a, of total earnings is retained within the firm each period and that these retained earnings are reinvested to earn a rate of return, b, on book value. Let h_j be the dividends per share in period j; and let e_j be the earnings per share in period j. (Assume that all dividends are paid annually at the end of each year.) Then

$$h_j = (1 - a)e_j$$
(11.17)

and

$$e_j = e_{j-1} + b(ae_{j-1})$$
$$= e_{j-1}(1 + ab)$$
(11.18)

Clearly, earnings are growing at rate ab, and thus dividends also grow at the same rate. That is

$$h_j = h_0(1 + ab)^j$$
(11.19)

If we can assume that the current price per share of common stock, p, is solely a function of expected future dividends and if the dividends are expected to increase at rate ab perpetually, it follows that

$$p = \sum_{j=1}^{\infty} h_j (1 + k_e)^{-j}$$
(11.20)

where h_j is as given in Equation 11.19 and k_e represents the **equity**

capitalization rate. Combining Equations 11.19 and 11.20,

$$p = \sum_{j=1}^{\infty} h_0 \left[\frac{1 + ab}{1 + k_e} \right]^j \tag{11.21}$$

If $k_e > ab$, then $p = h_0/(k_e - ab)$, or

$$k_e = \frac{h_0}{p} + ab \tag{11.22}$$

The above model assumes that the growth rate remains constant perpetually. However, especially in regulated industries, it may be appropriate to consider a *finite* growth period, n, after which $b = k_e$. Since there is no growth, all earnings are distributed as dividends; that is $a = 0$, so $h_j = e_j$ for $j > n$. This situation is shown in Figure 11.3. With the assumptions specified previously, the current value per share of common stock, p, is the sum of the present value of the dividend stream during the growth period, p_1, as well as the present value of the stock price at the end of n periods, p_2. During the growth period, dividends are growing at rate ab. Thus

$$p_1 = \sum_{j=1}^{n} h_j (1 + k_e)^{-j}$$

$$= h_0 \sum_{j=1}^{n} \left[\frac{1 + ab}{1 + k_e} \right]^j \tag{11.23}$$

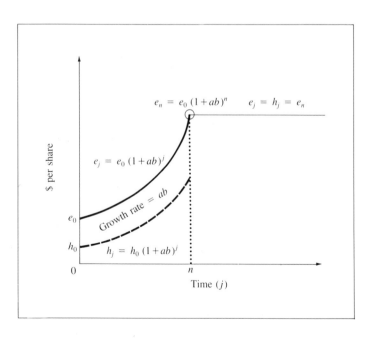

Figure 11.3. Growth of Earnings (e) and Dividends (h)

When growth stops, however, earnings remain constant at $e_j = e_n$ for $j > n$. This is the "no-growth" situation described in Equations 11.14 and 11.15. The value of the infinite stream of future earnings and dividends, measured at the end of period n, is e_n/k_e. The present worth is

$$p_2 = (e_n/k_e)(1 + k_e)^{-n}$$

$$= \frac{e_0}{k_e}\left[\frac{1 + ab}{1 + k_e}\right]^n \tag{11.24}$$

Combining Equations 11.23 and 11.24, the total present worth of one share of common stock, given a finite period of growth at rate ab, is

$$p = p_1 + p_2$$

$$= h_0 \sum_{j=1}^{n}\left[\frac{1 + ab}{1 + k_e}\right]^j + \frac{e_0}{k_e}\left[\frac{1 + ab}{1 + k_e}\right]^n \tag{11.25}$$

The cost of capital (equity capitalization rate) is the value of k_e that satisfies Equation 11.25. (This model assumes that earnings are growing exponentially over the interval $0 < j \le n$. Other models may be formulated for alternative assumptions, for example, for *constant* growth or *declining* growth.)[4]

Retained Earnings

Rather than paying dividends to current stockholders, the firm may retain earnings for reinvestment. To protect current stockholders, it is necessary that all funds employed within the firm provide a return at least as great as the return that could be earned by stockholders if these same funds were invested elsewhere.

What is the "value" of earnings to individual stockholders? Ignoring the question of personal income taxes for the moment, this figure may be inferred by examining an alternative that is always available, namely, sale of the ownership share in the stock market. If, as before, it is assumed that the current market value of equity shares is solely a function of perceived future dividends, then the opportunity cost of retained earnings, k_r, is similar to that given by Equation 11.22: $k_r = h_0/p + ab$. This formulation, of course, assumes that retained earnings are growing indefinitely at rate ab per period. But dividends are subject to personal income taxes, and thus the *after-tax cost of retained earnings* is more properly given by

---

4. Alternative models are developed in J. C. T. Mao, *Quantitative Analysis of Financial Decisions* (New York: Macmillan, 1969), Chapter 10.

$$k_r = (1 - t)\left[\frac{h_0}{p} + ab\right] \qquad\qquad (11.26)$$

where t is the effective personal income tax rate.

Note that any increase in the value of a share of common stock is a *capital gain* and is subject to a capital gains tax. (In certain cases, however, a capital gain may be taxed as though it were ordinary income.) Thus a refinement of Equation 11.26, recognizing that the capital gain is growing at rate ab with the growth assumptions underlying this model, is

$$k_r = (h_0/p)(1 - t_1) + ab(1 - t_2) \qquad\qquad (11.27)$$

where t_1 is the investor's income tax rate and t_2 is his or her appropriate tax rate on capital gains.

It is extremely difficult to determine a representative incremental personal income tax rate and corresponding capital gains rates for all stockholders in large, publicly held corporations. Nevertheless, the after-tax value of k_r is sensitive to these tax rates. To illustrate, consider the case in which $h_0 = \$5$, $p = \$40$, $a = 0.60$, and $b = 0.25$. Consider the cost of retained earnings, after personal income taxes, for three representative investors:

	Tax rates		k_r from
Investor	t_1	t_2	Equation 11.27
A	0%	0%	0.275
B	30	15	0.215
C	50	25	0.175

For the tax-exempt stockholder, Investor A, the cost of retained earnings is the same both before and after personal income taxes.

Consideration of individual income taxes, as reflected by Equation 11.27, is generally infeasible as a practical matter, but there is an alternative view that may prove helpful. Suppose that retained earnings are used for outside investment. That is, assume that earnings, rather than being returned to the stockholders in the form of dividends, are invested elsewhere in the capital market so as to provide a return equal to the equity capitalization rate of the tax-exempt stockholder. For example, retained earnings from Company X may be used to purchase shares in Company Y or Company Z. When the outside investment represents majority control—that is, when the interest is sufficient to permit a consolidated corporate income tax return—there is no immediate tax effect and the equity capitalization rate is consequently the rate given by Equation 11.22. If the firm invests so as to meet the requirements of tax-exempt stockholders, then it will exceed the requirements of all taxable stockholders. Some stockholders will be better off; none will be worse off.

11.7 COSTS OF CAPITAL IN THE PUBLIC SECTOR

Allocation of resources in the public sector is based on a somewhat different perspective from that of private enterprises. For the latter, it is generally agreed that the appropriate point of view to adopt with respect to investment decisions is that of the firm's current owners: The objective is to maximize their wealth. In the public sector, however, society itself is the "owner" of the enterprise. The public decision maker must choose from investment alternatives so as to maximize the social yield to the members of society, individually and collectively. There is extensive public-finance literature addressed to the question of which **social discount rate** is appropriate to use in public sector investment decisions.[5] Unfortunately, there is no single view held by all scholars on this question, although a consensus appears to be emerging. Historically, one group of economists has agreed that the appropriate discount rate should reflect only the prevailing government borrowing rate. Indeed, this is the position that has been adopted by the federal government with respect to water resource (and certain land resource) planning:

The interest rate to be used in plan formulation and evaluation for discounting future benefits and costs, or otherwise converting benefits and costs to a common time basis, shall be based upon the estimated average cost of federal borrowing as determined by the Secretary of the Treasury taking into consideration the average yield during the twelve months preceding his determination on interest-bearing marketable securities of the United States with remaining periods to maturity comparable to a 50-year period of investment: Provided, however, that the rate shall be raised or lowered by no more than or less than one-half percentage point for any year.

When the average cost of federal borrowing as determined by the Secretary of the Treasury exceeds the established discount rate by more than 0.25 percentage points, the rate shall be raised 0.5 percentage points. When the average cost is less than the established rate by more than 0.25 percentage points, the rate shall be lowered 0.5 percentage points.

The Water Resources Council shall determine, as of July 1, the discount rate to be used during the fiscal year. The Director of the Water Resources Council shall annually request the Secretary of the Treasury during the month of June to advise the Water

5. For an excellent summary of contemporary views, see Quirin and Wiginton, *Analyzing Capital Expenditures*, Chapters 8 and 9.

Resources Council of his determination of the average cost of federal borrowing during the preceding twelve months.[6]

Most observers, both within the government and outside it, criticize this view as inappropriate. It implies a diversion of higher return funds from the private sector to relatively lower returns in the public sector.

The view most widely held is that the social rate of discount should represent the costs to the private sector of opportunities forgone. Since individual members of society have investment opportunities in the private sector available to them, any funds diverted to the public sector should bear a cost that is equivalent to the opportunities forgone. Citizens can satisfy their private wants through private-sector investments. The value of a marginal dollar removed from the private sector should be equal to the value of a marginal dollar spent in the public sector to satisfy public wants. The appropriate interest rate to use in discounting future costs and benefits of public projects should be the opportunity cost of displaced private spending.

In the late 1960s, criticism of the "cost-of-debt-only" view became so intense that a Congressional committee held a series of hearings to further investigate this issue.[7] The committee found that the cost to the government of federal long-term borrowing (in 1969) was approximately $7\frac{1}{2}$ percent, after adjusting for income taxes on government interest payments. It also concluded that the average rate of return in the private sector, *before* income taxes, was in the 8–10 percent range. (This conclusion was based on a report by Jacob A. Stockfisch in which he found that the average rate of return in the private sector was about 15.4 percent for manufacturing companies and 9.3 percent for electric utilities in the years 1961–1965.[8] In this same period, rates ranged from 4.1 percent for railroads to 15.6 percent for oil pipelines.) In 1972, largely because of these findings by the Congress, the federal Office of Management and Budget (OMB) established the following discount rate policy:

The discount rates to be used for evaluations of programs and projects subject to the guidance of this Circular are as follows:

6. U.S. Congress, *Principles, Standards and Procedures for Water and Related Land Resource Planning,* S. Doc. 97, 87th Cong., 2nd sess., 1962.

7. U.S. Congress, Joint Economic Committee, *Interest Rate Guidelines for Federal Decision-making. Hearing before the subcommittee on economy in government,* 90th Cong., 2nd sess., 29 January 1968. See also U.S. Congress, Joint Economic Committee, *Economic Analysis of Public Investment Decisions: Interest Rate Policy and Discounting Analysis,* a report prepared by the subcommittee on economy in government, 90th Cong., 2nd sess., 23 September 1968.

8. Jacob A. Stockfisch, *Measuring the Opportunity Cost of Government Investment,* Research Paper P-490 (Washington, D.C.: Institute for Defense Analysis, March 1969).

a. *A rate of 10 percent; and, where relevant,*

b. *Any other rate prescribed by or pursuant to law, Executive order, or other relevant Circulars.*

The prescribed discount rate of 10 percent represents an estimate of the average rate of return on private investment, before taxes and after inflation.[9]

This policy was *suggested* for use in internal planning documents of agencies in the executive branch of government and *required* for use in program analyses submitted to OMB in support of legislative and budget programs. An exception to this requirement—an important exception—is water-resource projects. (As of this writing, the federal 10 percent rate has not been changed since it was first promulgated in 1972.)

Conceptually, the opportunity-cost approach to the social-discount rate question requires that rates of return in the private sector be weighted by proportion of funds currently invested in each investment class. Returns should be *before income taxes,* since taxes are merely transfers from individuals to society as a whole. To understand this principle, consider a proposed *private* project that could earn, say, 30 percent before taxes and 20 percent after taxes. If a *public* project were proposed to displace this private project, it would have to earn 30 percent, reflecting the after-tax return to the private investor(s) and the income taxes forgone.

Before leaving this section, two fundamental problems of this approach should be emphasized. (These are in addition to the obvious difficulty of measuring private-sector opportunity costs.) First, the principle of maximizing the excess of discounted benefits over discounted costs deals solely with the *efficiency* of capital investment. The *equity* issue is ignored entirely. It may be socially desirable to redistribute the wealth of society, rather than, or in addition to, maximizing the aggregate. Examples include special facilities for the physically handicapped, subsidized transit, and environmental controls. It is especially important in public-sector decision making to emphasize the implicit limitation of economic analysis: Only one aspect of the problem is examined. The social and environmental considerations, as well as the desirability of redistributing society's resources, are also relevant to the decision.

The second problem concerns the discounting process itself. Maximization of net present worth implies that the current "investor" is the beneficiary. Consider the following illustration:

9. U.S. Office of Management and Budget, *Discount Rates to Be Used in Evaluating Time-Distributed Costs and Benefits,* circular no. A-94 (rev.), 27 March 1972.

End of year	Cash flows for alternatives	
	A	B
0	− $1,000	− $1,000
5	3,000	0
50	0	75,000

Assuming a social discount rate of, say, 10 per cent, the equivalent present worths of A and B are $863 and − $361 respectively. Alternative A, as well as the do-nothing alternative, are economically superior to Alternative B. Thus Alternative A might make sense to the current investors, but what about future generations? Is it reasonable to reward those now living at the expense of progeny? Is it morally responsible? These disturbing questions are raised by, but not solved by, the discounting process. There is systematic discrimination against generations yet unborn, but there is no simple solution to this apparent conundrum. Perhaps all that may be said is that the problem exists and that socially responsible decision makers should consider the time profile of costs and benefits as an additional element in the decision.

Reducing the discount rate is *not* a reasonable way to protect the interests of future generations. To do so would result in the acceptance of lower-return projects at the expense of higher-return projects, thus lowering the rate of national economic growth and, in the long run, working to the disadvantage of future generations.

The cost of capital for *investor*-owned utilities need not be treated differently from that of other firms in the private sector. However, the cost of capital for *publicly* owned utilities creates a special problem with respect to equity capital. As mentioned previously, the "owners" are the citizens of the community having jurisdiction over the utility. The Los Angeles Department of Water and Power (LADWP), for example, is owned by the city of Los Angeles. It is operated by professional managers, but general policy is set by the citizens of Los Angeles through their elected city council.[10] Therefore investment decisions of the LADWP should reflect the private opportunities of the citizens of the community.

Conceptually, the cost of capital in this context is not substantially different from the social discount rate discussed above. There is one important difference, however. Whereas the preceding section addressed the broad question of capital allocation decisions in the public sector, the focus now is on a specific governmental activity. Inasmuch as the delivery of utilities (water, power, gas, and so on) to the public is a practical necessity in

11.8
PUBLICLY OWNED
REGULATED UTILITIES

10. Administration of the LADWP, the largest municipal utility in the United States, is under the direction of a five-member Board of Water and Power Commissioners, appointed by the mayor and confirmed by the city council. Rates are subject to the approval of the mayor and the city council; they are *not* regulated by the California Public Utilities Commission.

modern society, it may be argued that the delivery of these services by the private sector is a viable alternative. Indeed, in many communities, certain of these services are provided by investor-owned utilities. Thus the cost of equity capital for publicly owned utilities may be *estimated* by the rates of return available to investor-owned utilities. As a general principle, "projects which displace private investments should be evaluated by the rate of return prevailing in the sector from which the specific investment is displaced."[11]

11.9 WEIGHTED AVERAGE COST OF CAPITAL

Given a specific capital structure defined by proportions w_c of capital class c, the overall cost of capital, k, may be determined by computing the **weighted average cost of capital** of all funds:

$$k = \sum_c w_c k_c \qquad (11.28)$$

where k_c is the cost of capital class c. When there are only two classes of capital, debt and equity, Equation 11.28 reduces to

$$k = w_d k_d + w_e k_e \qquad (11.29)$$

where w_d and w_e are the proportions and k_d and k_e are the costs of debt and equity, respectively.

To illustrate this procedure, consider the capital structure for the XYZ Corporation, shown in Table 11.2. The cost of the debt (bonds), after adjusting for corporate income taxes, is determined from Equation 11.8 to be approximately 12.3 percent. This figure represents the anticipated *future* cost of debt, not the current average cost of outstanding debt (*embedded cost*).

The cost of preferred stock is determined from Equation 11.10 to be 9.6 percent. The cost of preferred stock is the same before and after income taxes, since preferred stock dividends are not tax-deductible expenses as is interest on debt.

Table 11.2
Determining the After-tax Cost of Capital for the XYZ Corporation

Capital class (c)	Current market value	Relative amount w_c	Cost of capital k_c	Weighted costs $w_c k_c$
1. Bonds	$30,000,000	0.375	0.123	0.046
2. Preferred stock	10,000,000	0.125	0.096	0.012
3. Common stock	40,000,000	0.500	0.240	0.120
Totals	$80,000,000	1.000		0.178

11. U.S. Congress, Joint Economic Committee, *Economic Analysis of Public Investment Decisions*, p. 15.

Equation 11.22 has been used to estimate the cost of common stock, k_e. The current ratio of dividends to market price per share is approximately 0.165. The firm believes that earnings will grow at the rate of about 10 percent per year over the indefinite future, assuming that three-fourths of the dividends will be reinvested in the firm. With these assumptions, $k_e = 0.165 + 0.75(0.10) = 0.24$ after corporate income taxes but before stockholders' personal income taxes.

The resulting overall after-tax weighted average cost of capital, $k = 0.178$, is the estimated "hurdle rate" as measured at that particular point in time. If used in economy studies as the minimum attractive rate of return, this rate is normally assumed to remain constant over the time interval in which the economic consequences of alternatives are assumed to occur. The cost of capital should be reestimated periodically to adjust for new information as it becomes available. Perceived equity capitalization rates, in particular, are likely to vary over time.

The preceding discussion assumes a specific mixture of debt and equity in the capital structure of the firm. The resulting cost of capital is the weighted average of the individual components. It should be recognized, however, that *changes* in the capital structure—for example, substituting debt for equity—may affect the cost of all the capital components. Specifically, increasing the proportion of debt in the capital structure may increase the equity capitalization rate, since investors will view the firm, and hence their investment, as being more risky. Thus the total cost of increased debt is determined from the direct cost of the increment of debt (as determined from the incremental interest charges) in addition to the indirect costs as measured by the increases in overall interest charges and in the equity capitalization rate. That is, the *marginal* cost of borrowing is based on the increment of interest required to repay the debt plus the increment of earnings required to maintain the equity value of stockholders.

11.10 THE MARGINAL COST OF CAPITAL

To illustrate this view, let's define *leverage, λ*, as the ratio of debt to the total value of the firm. (With this definition, leverage is equal to w_d as shown in Equation 11.29. An alternate definition for leverage is the ratio of debt to equity. The following comments are appropriate using either definition.) Table 11.3 summarizes the relevant data for a firm that is considering increasing its leverage from $\lambda_1 = 0.455$ to $\lambda_2 = 0.565$. Of course, the expected annual earnings after income taxes, but before interest payments to lenders, are equal in both cases.

The additional $1,000 of debt capital (replacing $1,000 of equity) results in an increase of $160 in interest payments as well as a slight increase in the equity capitalization rate (from 0.15 to 0.16). To preserve the equity value of existing stockholders, the

Table 11.3
Cost of Capital Analysis for a Firm at Two levels of Leverage

	Smaller leverage λ_1	Larger leverage λ_2
Outstanding debt (D)	$5,000	$6,000
Average interest rate paid on all debt (k_d)	0.10	0.11
Equity capitalization rate (k_e)	0.15	0.16
Expected annual earnings *after* income taxes but *before* return to lenders ($\hat{E}$)	$1,400	$1,400
Interest on debt ($I = k_d D$)	$500	$660
Annual earnings available to stockholders ($E = \hat{E} - I$)	$900	$740
Market value of equity ($M = E/k_e$)	$6,000	$4,625
Total market value of the firm ($V = D + M$)	$11,000	$10,625
Average cost of capital ($k = \hat{E}/V$)	0.127	0.132
Leverage ($\lambda = D/V$)	0.455	0.565

market value of equity should not fall below $5,000, since $1,000 of equity has been replaced by a like amount of debt. With the new capitalization rate, $k_e = 0.16$, expected annual earnings should be $0.16(\$5,000) = \$800$, or $60 more than the estimated $740. Thus the additional $1,000 of debt results in a total marginal cost of $160 + \$60 = \$220$. Stated as a percentage, the marginal cost of the increment of debt capital is 22 percent when leverage is increased from 0.455 to 0.565.

According to some scholars, the situation described above should, in fact, not occur. They argue that the overall cost of capital is independent of the capital structure and, further, that it is also independent of the amount of the total capital (debt and equity) in the firm. Let's examine the alternative views more closely.

The Traditional View

Figure 11.4a graphically illustrates the traditional view of the response of the market value of a firm to changes in leverage. Essentially, there are three general phases. During the first phase, market value is increased as leverage increases, since the cost of debt, k_d, rises rather slowly and the cost of equity capital, k_e, remains virtually constant. Recall from Equation 11.29 that

$$k_o = w_d k_d + w_e k_e$$
$$= \lambda k_d + (1 - \lambda)k_e$$

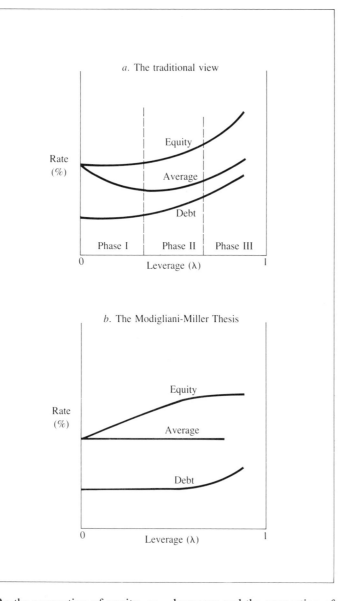

Figure 11.4. The Effects of Leverage on the Cost of Capital

a. The traditional view

Rate (%)

Equity

Average

Debt

Phase I | Phase II | Phase III

0

Leverage (λ)

1

b. The Modigliani-Miller Thesis

Rate (%)

Equity

Average

Debt

0

Leverage (λ)

1

As the proportion of equity, w_e, decreases and the proportion of debt, w_d, increases, the weighted average cost of capital, k_o, will decrease as long as k_d is less than k_e and k_e does not increase. In the third phase, k_o increases because of the reaction of the market to the large proportion of debt incurred by the firm; both stockholders and lenders become more apprehensive as the relative proportion of debt increases, and both groups require a higher price for their capital as the result of added risk. The market price of stock decreases as debt is increased beyond the point that investors consider acceptable. Lenders react in a similar manner, charging ever higher interest rates for their loans. During the second phase, total market value remains virtually constant as the opposite effects of the first and third phases tend to cancel each other. Thus the traditional view holds that the cost of capital as a function of leverage is U-shaped, and an optimum (minimum) point does indeed exist.[12]

The Modigliani-Miller Thesis

An important alternative to the traditional view has been articulately presented by Professors Franco Modigliani and Merton H. Miller. Their position is that the average cost of capital is independent of the capital structure of the firm and, as a corollary, the total market value of the firm is unaffected by the relative proportions of debt and equity. Thus there is no optimal capital structure defined by the minimum average cost of capital.

Although space does not permit a complete discussion of the **Modigliani-Miller thesis,** a simple example will serve to illustrate an essential feature of their position.[13] Consider two firms with the following financial data:

	Company A (lower leverage)	Company B (higher leverage)
Expected annual earnings after income taxes but before return to lenders ($\hat{E}$)	$1,000,000	$1,000,000
Market value of debt (D)	——	$5,000,000
Interest on debt ($I = k_d D = 0.06M$)	——	$300,000
Annual earnings available to stockholders ($E = \hat{E} - I$)	$1,000,000	$700,000
Total number of shares (s)	1,000,000	500,000
Earnings per share = dividends per share ($h = e = E/s$)	$1.00	$1.40
Market price per share (p)	$10.00	$12.00
Equity capitalization rate ($k_e = h/p$)	0.100	0.117
Market value of equity ($M = sp$)	$10,000,000	$6,000,000
Total value of the firm ($V = D + M$)	$10,000,000	$11,000,000
Leverage ($\lambda = D/V$)	0	0.455

12. Over the range of values wherein k_e remains constant and $k_d < k_e$, it may be shown that k continues to decrease until the point where the marginal rate of interest on debt equals the average cost of equity capital. The proof is left as an exercise.

13. This example is adapted from Quirin and Wiginton, *Analyzing Capital Expenditures*, pp. 195–196.

The situation described above is perfectly reasonable under the traditional view: Stockholders in Company B require a higher return on their equity because Company B is more highly leveraged than Company A, and hence the total value of B is greater than that of A. Under the Modigliani-Miller thesis, however, this situation is not possible given rational investors and perfect capital markets. An investor holding 1,000 shares of Company B could sell his or her shares for $1,000 \times \$12 = \$12,000$; borrow $8,000 at 6 percent, the same interest rate as that available to Company B, and then purchase 2,000 shares in Company A at $10 per share. (Note that the investor's percentage of ownership in Company B was $1,000/500,000 = 0.002$. The new percentage of ownership in Company A is $2,000/1,000,000 = 0.002$, so the relative ownership, or degree of risk participation, remains unchanged.)

Before, the investor earned $1,000 \times \$1.40 = \$1,400$ per year in dividends in B. When the investment is shifted to Company A,

Dividends: $2,000 \times \$1.00$ = $2,000

Less interest: $8,000 @ 6%   =   480

$1,520

The investor would be better off by $120 if he or she shifted from B to A: Income is increased and risk is unchanged. This exchange is available to all stockholders, of course, and thus rational investors in search of maximum return will switch their holdings from Company B to Company A, thereby driving down the price of stock in B and bidding up the price of the stock in A. (By a similar argument, it may be shown that, if Company B had a *lower* total value than Company A, investors in A would be better off shifting from A to B.)

These exchanges, known as **arbitrage,** will continue until the total values of the two companies are equal, at which point there will no longer be any advantage in switching ownership from one company to the other. At equilibrium, the values of the two companies must be equal, and so must their respective costs of capital. Therefore, according to the Modigliani-Miller thesis, the cost of capital should be independent of leverage. Because of the arbitrage process, the debt-equity mix in the capital structure has no effect on the weighted average cost of capital.

There are a number of theoretical arguments that both support and refute the Modigliani-Miller thesis. Attempts to support one position or another through empirical evidence are inconclusive. An argument can be made, however, that the cost of capital is independent of the total volume of funds employed and, further, that the average cost of capital in the long run is the same as the marginal cost.[14] Further examination of this question is beyond the scope of this book. For the purpose of capital allocation decisions, the analyst should incorporate the weighted average cost of capital approach as previously discussed. The question of *optimal* capital structure is a separate issue.

11.11
THE EFFECT OF INFLATION ON THE COST OF CAPITAL

The costs of debt and equity capital are influenced by general inflation in the economy. These effects are explored in this section.

All Equity Capital

Consider the case in which the firm's investment capital consists entirely of equity in the amount of $150,000. There is no debt ($w_d = \lambda = 0$). Let's assume that the firm is expected to earn $30,000 after taxes per year into the indefinite future and that all earnings are returned to stockholders in the form of dividends. Assume also that the investment environment is inflation-free. For the sake of simplifying calculations, assume that earnings and dividend payments occur at the end of the year. The implied equity capitalization rate, k_e, is $30,000/$150,000 = 0.20. Of course, since $\lambda = 0$, this is also the inflation-free overall cost of capital for the firm.

Now assume that there is a general inflation rate, f, of 5 percent per year. Because the general price level is increasing, investors expect dividends to increase at that same rate in order to preserve buying power. Therefore, if the real rate of return on equity for stockholders is to be maintained at 20 percent per year, earnings must be

$$1.05(\$30,000) = \$31,500 \text{ at the end of the first year}$$

$$1.05(\$31,500) = \$33,075 \text{ at the end of the second year}$$

and so on.

The value of equity must also increase at this same rate in order to preserve the value of the stockholder's equity in the firm. For example, the "inflated" value of equity must be $1.05(\$150,000) = \$157,500$ at the start of the second year. In the presence of inflation, the total earnings that must be effected in the first year are $39,000, consisting of $31,500 in dividends and $7,500 in retained earnings. Data for the first three years are shown in Table 11.4. In this example, the inflation-adjusted cost of capital, k^*, is the ratio of the end-of-year earnings required, E_j^*, to the start-of-year equity, M_{j-1}^*.

---

14. Quirin and Wiginton, *Analyzing Capital Expenditures*, pp. 201–211.

Table 11.4
Effect of Inflation on Cost of Capital for All-Equity Firm ($\lambda = 0$) with Inflation Rate (f) = 0.05

End of period (j)	Equity (M_j^*)	Increase in equity (R_j^*)	Dividends paid (H_j^*)	Earnings required (E_j^*)
0	$150,000		$30,000	$30,000
1	157,500	$7,500	31,500	39,000
2	165,375	7,875	33,075	40,950
3	.	8,269	34,729	42,998
$\vdots$	$\vdots$	$\vdots$	$\vdots$	$\vdots$

$M_j^* = 1.05 M_{j-1}^*$
$R_j^* = P_j^* - P_{j-1}^*$
$H_j^* = 1.05 H_{j-1}^*$
$E_j^* = R_j^* + H_j^*$

$$\frac{\$39,000}{\$150,000} = \frac{\$40,950}{\$157,500} = \frac{\$42,998}{\$165,375} = 0.26$$

To develop the generalized model for this case, note that the "inflated" equity value at the end of year j, M_j^*, is

$$M_j^* = M_0(1 + f)^j \tag{11.30}$$

where f is the inflation rate per period. The total "inflated" dividends at the end of period j, H_j^*, is

$$H_j^* = H_0(1 + f)^j \tag{11.31}$$

and the total earnings at the end of period j, E_j^*, necessary to provide for dividends and to preserve the value of equity is

$$E_j^* = M_j^* - M_{j-1}^* + H_j^* \tag{11.32}$$

At this point, note that $H_0 = M_0 k_e$, where k_e is the equity capitalization rate as well as (in this case) the overall cost of capital k_o. The inflation-adjusted cost of capital in year j, k_j^*, is given by

$$k_j = E_j^*/M_{j-1}^* \tag{11.33}$$

Combining Equations 11.30–11.33 and simplifying,

$$k_j^* = \frac{M_0(1 + f)^j - M_0(1 + f)^{j-1} + M_0 k_e(1 + f)^j}{M_0(1 + f)^{j-1}}$$

$$= (1 + k_e)(1 + f) - 1 \tag{11.34}$$

Since k_j^* is constant for all j and since $k_e = k_o$ in the absence of debt, it follows that

$$k^* = (1 + k_o)(1 + f) - 1 \tag{11.35}$$

which is, of course, the result obtained in Chapter 10 for defining the relationship between the inflation-free cost of capital, $k = k_o$, and the inflation-adjusted cost of capital, k^*. In this example, with $k_o = \$30,000/\$150,000 = 0.20$ and $f = 0.05$, $k^* = (1.20)(1.05) - 1 = 0.26$.

The Leveraged Firm

An example illustrating the cost of capital for a leveraged firm is summarized in Table 11.5. The value of the firm is $150,000, consisting of 40 percent debt $(D = \$60,000)$ and 60 percent equity $(M = \$90,000)$. The after-tax costs of debt and equity are 0.10 and 0.20, respectively. This is a "no-growth" case: All earnings are distributed in the form of dividends, and earnings are stable over time. The weighted average cost of capital is

$$k_o = w_d k_d + w_e k_e$$

$$= 0.4(0.10) + 0.6(0.20)$$

$$= 0.16$$

Note that the cost of capital can also be determined by

$$k_o = \frac{E_j + I_j}{V_{j-1}} \qquad \text{for } j = 1, 2, \ldots \tag{11.36}$$

The numerator is the return to stockholders and lenders at the end of year j: the denominator is the total value of the firm at the start of year j. In this example, $E_j = 0.20(\$90,000) = \$18,000$, and $I_j = 0.10(\$60,000) = \$6,000$. Thus

Table 11.5
Cost of Capital for the Leveraged Firm

End of period (j)	Debt (D_j)	Equity (M_j)	Value of the firm (V_j)	After-tax interest (I_j)	Dividends (H_j)	Increased equity (R_j)	Earnings (E_j)	Leverage (λ_j)	Overall cost of capital (k_j)
0	$60,000	$90,000	$150,000						
1	60,000	90,000	150,000	$6,000	$18,000	———	$18,000	0.4	0.16
2	60,000	90,000	150,000	6,000	18,000	———	18,000	0.4	0.16
3	:	:	:	6,000	18,000	———	18,000	0.4	0.16
:	:	:	:	:	:	:	:	:	:

$$k_o = \frac{\$18,000 + \$6,000}{\$150,000} = 0.16$$

(The interest received by lenders is *not* I_j, or $6,000. Rather, $6,000 is the after-tax cost to the firm. The actual interest paid to lenders is $I_j/(1 - t)$, where t is the effective income tax rate. If $t = 0.4$, for example, the pretax interest payment to lenders is $10,000.)

The effect of inflation on the cost of capital is summarized in Table 11.6. Here, the general inflation rate, f, is assumed to be 5 percent per period, and lenders and stockholders are affected equally. This is the same problem as in Table 11.5, except that a general rate of inflation has been introduced.

In order for interest payments to retain their purchasing power in the presence of general price inflation, the after-tax cost of *debt* becomes

$$\begin{aligned} k_d^* &= (1 + k_d)(1 + f) - 1 \\ &= (1.10)(1.05) - 1 = 0.155 \end{aligned} \qquad \textbf{(11.37)}$$

Whereas the after-tax interest expense is only $0.10(\$60,000) = \$6,000$ in the first year, assuming no inflation, it is $0.155(\$60,000) = \$9,300$ in the presence of 5 percent inflation. The difference, $\$9,300 - \$6,000 = \$3,300$, is a result of the 5 percent loss in purchasing power of the original $60,000 debt ($3,000) plus the 5 percent loss in purchasing power of the $6,000 after-tax cost of interest ($300). That is, for the first year,

Inflation-free interest on principal	$6,000
Loss in purchasing power of principal	3,000
Loss in purchasing power of interest	300
Total interest (after tax) at the end of year 1	$9,300

Table 11.6
Effect of Inflation on Cost of Capital for the Leveraged Firm
with Inflation Rate $(f) = 0.05$

End of period (j)	Debt (D_j^*)	Equity (M_j^*)	Value of the firm (V_j^*)	After-tax interest (I_j^*)	Dividends (H_j^*)	Increased equity (R_j^*)	Earnings (E_j^*)	Leverage (λ_j^*)	Cost of capital (k_j^*)
0	$60,000	$90,000	$150,000						
1	63,000	94,500	157,500	$9,300	$18,900	$4,500	$23,400	0.4	0.218
2	66,150	99,225	165,375	9,765	19,845	4,725	24,570	0.4	0.218
⋮	⋮	⋮	⋮	⋮	⋮	⋮	⋮	⋮	⋮

$D_j^* = D_0(1 + f)^j = \$60,000(1.05)^j$
$I_j^* = k_d^* D_{j-1}^* = 0.155 D_{j-1}^*$
$H_j^* = H_0(1 + f)^j = \$18,000(1.05)^j$
$P_j^* = P_0(1 + f)^j = \$90,000(1.05)^j$

$R_j^* = M_j^* - M_{j-1}^* = \$4,500(1.05)^j$
$E_j^* = H_j^* + R_j^* = \$23,400(1.05)^{j-1}$
$V_j^* = D_j^* + M_j^* = \$150,000(1.05)^j$
$\lambda_j^* = D_j^*/V_j^* = D_0/V_0$
$k_j^* = (E_j^* + I_j^*)/V_{j-1}^*$

In general, the inflation-adjusted interest expense, after taxes, at the end of year j is

$$I_j^* = k_d^* D_{j-1}^* \qquad (11.38)$$

where D_{j-1}^* is the inflation-adjusted debt at the start of year j. Moreover,

$$D_j^* = D_0(1 + f)^j \qquad (11.39)$$

By a similar argument, the cost of *equity* in the presence of inflation is

$$\begin{aligned} k_e^* &= (1 + k_e)(1 + f) - 1 \\ &= (1.20)(1.05) - 1 = 0.26 \end{aligned} \qquad (11.40)$$

The inflation-adjusted dividends at the end of year j are given by

$$H_j^* = k_e M_0(1 + f)^j \qquad (11.41)$$

The value of equity must also grow in order to compensate for the effects of inflation:

$$M_j^* = M_0(1 + f)^j \qquad (11.42)$$

and the year-to-year increase in equity is given by

$$\begin{aligned} R_j^* &= M_j^* - M_{j-1}^* \\ &= M_0 f(1 + f)^{j-1} \end{aligned} \qquad (11.43)$$

Earnings must provide for both dividends and increased equity. Thus earnings at the end of year j are

$$E_j^* = H_j^* + R_j^* \qquad (11.44)$$

By substituting Equations 11.41 and 11.43 and simplifying, an alternative formulation is

$$E_j^* = k_e^* M_0(1 + f)^{j-1} \qquad (11.45)$$

Of course, the value of the firm (V_j^*) is simply the sum of debt (D_j^*) and equity (M_j^*). Since both D_j^* and M_j^* are increasing at rate f per period, V_j^* also increases at that same rate.

The weighted average cost of capital in the presence of inflation is given by

$$k^* = \lambda\, k_d^* + (1 - \lambda)k_e^* \qquad (11.46)$$

Substituting the values from this example,

$$k* = 0.4(0.155) + 0.6(0.26) = 0.218$$

An alternative view of the inflation-adjusted cost of capital may be instructive here. From the viewpoint of the firm, the total after-tax return to stockholders and lenders at the end of period j is $E_j^* + I_j^*$, where E_j^* is the sum of dividends, H_j^*, and increased equity, R_j^*. The total value of the firm at the start of period j is V_{j-1}^*. Thus

$$k* = \frac{H_j^* + R_j^* + I_j^*}{V_{j-1}^*} \tag{11.47}$$

Using the data for $j = 1$ in our example,

$$k* = \frac{\$18,900 + \$4,500 + \$9,300}{\$150,000} = 0.218$$

For $j = 2$,

$$k* = \frac{\$19,845 + \$4,725 + \$9,765}{\$157,000} = 0.218$$

and so on, for all j.

Using the Inflation-Adjusted Cost of Capital to Evaluate Investment Proposals

Consider a proposed investment that requires an initial outlay of $150 and promises a return of $174 one period from now. Assume that inflation is negligible and that the firm's overall cost of capital is 0.16. The internal rate of return (IRR) for this project is exactly ($174/$150) $- 1 = 0.16$. Both the equivalent present worth and future worth are zero when discounted using 16 percent as the minimum attractive rate of return ($MARR$). From the viewpoint of economic efficiency, this proposed investment offers neither an advantage nor a disadvantage. The firm is as well off with the proposed investment as without it.

Now suppose that there is general inflation at the rate of 5 percent per period and that inflation affects the proposal's cash flows as well as the cost of capital. In other words, assume that, because of inflation, the cash flow to be expected one period hence (A_1^*) is $174(1.05) = 182.70. The inflation-adjusted rate of return is now ($182.70/$150) $- 1 = 0.218$. But the inflation-adjusted cost of capital is also 0.218:

$$k* = (1 + k)(1 + f) - 1 = (1.16)(1.05) - 1 = 0.218$$

Thus the firm should also be indifferent to the proposed investment.

This simple example illustrates the confusion that may arise from considering cash flows (the proposal) and the cost of capital (the do-nothing alternative) in the presence of inflation. Observe that

End of period	Cash flows	
(j)	A_j	$A_j^* = A_j(1.05)^j$
0	$-\$150.00$	$\$150.00$
1	174.00	182.70
IRR	0.16	0.218
PW @ 0.16 (k)	$\$\ 0$	$\$\ 7.50$
PW @ 0.218 (k*)	$-\$\ 7.14$	$\$\ 0$

If the firm is to use an inflation-adjusted cost of capital—if the project is to provide a 21.8 percent return on invested capital—then the *inflated* cash flows (A_j^*) must be used in the analysis. Otherwise, the "inflation-free" cash flows (A_j) must be used together with an inflation-free cost of capital. Used properly, both approaches lead to the correct conclusion.

This principle may be further illustrated by examining a two-period example. Consider a project with the expected cash flows $A_0 = -\$150$, $A_1 = \$100$, and $A_2 = \$85.84$. Assume that there is no inflation and that the firm's cost of capital (k) is 0.16. It may be shown that the internal rate of return for this project is also 0.16, and thus we are again indifferent between the project and the do-nothing alternative.

Now suppose that there is a general inflation rate of 5 percent per period, affecting cash flows as well as the cost of capital. Consider the equivalent value (future worth) of this project at the end of two periods. Note the results:

End of period	Cash flows	
(j)	A_j	$A_j^* = A_j(1.05)^j$
0	$-\$150.00$	$-\$150.00$
1	100.00	105.00
2	85.84	94.64
IRR	0.16	0.218
FW @ 0.16 (k)	$\$\ 0$	$\$\ 14.60$
FW @ 0.218 (k*)	$-\$\ 14.89$	$\$\ 0$

Assuming that opportunity costs during the first and second periods will be 0.218 each period, then the future worth of the project is exactly zero. As before, this is a break-even project. Otherwise, if the reinvestment rate is only 0.16, the future worth is positive ($FW = \$14.60$), indicating that the project is attractive under conditions of inflation. *The reinvestment assumption is critical.*

Given the conditions outlined above, however, it follows that $k*$ should be used with A_j*, since lenders and stockholders expect their returns to reflect inflation also. The cost of capital is higher—$k*$ rather than k.

The governing principle here is that the correct return is obtained using either the inflation-free cost of capital and cash flows in real ("now") dollars or, alternatively, using the inflation-adjusted cost of capital and cash flows in inflated ("then") dollars. But this is true only if the general inflation rate affects cash flows and opportunity costs equally. The appropriate use of differential inflation rates is discussed in greater detail in Chapter 10.

A fundamental issue in economic decision making is determining the appropriate discount rate to use in analyses. This rate, the minimum attractive rate of return, is explicit in the present worth, annual worth, and future worth methods, since it serves to discount or compound cash flows. It is implicit in the rate of return method because it serves as the standard against which the internal rate of return is compared.

11.12 SUMMARY

The *MARR* may be viewed as a cost of opportunities forgone, stated as a percentage or a decimal fraction. It is approximated by the costs of the capital components, debt and equity, which constitute the funds available for allocation among competing investment opportunities. The weighted average cost of capital is the estimate of the *MARR* to use in economy studies.

Several of the more common sources of investment capital are addressed in this chapter, including bonds or notes, preferred stock, leases, new issues of common stock, and retained earnings. This discussion is by no means exhaustive, however. If you are interested in a more extensive discussion, see the references included in Appendix A.

In closing, it should be emphasized that there is considerable disagreement as to the appropriate measurement, or estimation, of the cost of capital. (The traditional view concerning the cost of capital as a function of leverage, contrasted with the position of Modigliani and Miller, is but one example of the controversy that remains unresolved.) Although the level of disagreement seems to be diminishing in recent years, it would be misleading to suggest that all of the procedures outlined in this chapter are universally followed. Nevertheless, despite the continuing controversy concerning certain aspects of the cost of capital, the weighted average cost of capital approach described in this chapter has been adopted by most authors of engineering economy textbooks.

Analysts should bear in mind that the procedures presented here result only in an *estimate* of the minimum attractive rate of return. When using this estimate, a liberal dose of common sense and reasoned judgment should be applied. With this in mind, it would be prudent to examine the sensitivity of study results to reasonable variations in the estimate of the cost of capital.

11.1 Ms. Jones has an opportunity to purchase a bond from the Los Angeles Public Utility Company on February 1, 1985. The terms of the bond are (1) $1,000 face value due and payable January 31, 1988; (2) 8 percent coupon rate; and (3) interest payable quarterly each January 31, April 30, July 31, October 31, up to and including January 31, 1988 (12 future interest payments).

a. If Ms. Jones's effective opportunity cost (discount rate) is 3 percent per quarter, determine the equivalent present worth of this bond.

b. What is Ms. Jones's *effective* opportunity cost per year? (*Answers:* (a) $900.48; (b) 12.55 percent)

11.2 The Tri-X Corporation is planning to borrow $100,000 to pay for certain major construction equipment. The loan will be repaid in four equal beginning-of-year payments, each of which includes principal as well as interest computed at 10 percent on the unpaid balance. Note that the first payment occurs at the time of the loan; hence it will simply be subtracted from the amount loaned. The firm's incremental state and federal income tax rates are 9 percent and 46 percent, respectively.

a. What is the appropriate combined incremental income tax rate to use in economy studies for this firm?

b. What is the amount of the uniform annual loan payments?

c. Write the equation that will yield the after-tax cost of this loan, stated as a percentage.

11.3 A certain municipally owned utility is raising investment funds through a bond issue: $1,000, 25-year, 12 percent bonds, paying interest quarterly.

a. The utility believes that the average rate of return required by investors is 4 percent per quarter. What is the expected average price for each bond sold?

b. The utility will sell 25,000 bonds. The cost to the utility to underwrite the sale of these bonds is $500,000. In addition, the utility will pay $50,000 each quarter for administrative costs associated with paying interest. Determine the cost *per quarter* of funds raised through the sale of these bonds. (Note that "cost" will be in the form of an interest rate, effective, per quarter.)

c. What is the resulting effective rate *per year*? (*Answers:* (a) $754.95; (b) 4.43 percent; (c) 18.92 percent)

11.4 In order to raise funds for its plant expansion program, the XYZ Corporation will sell 3,000 bonds, each with a face value of $1,000. These bonds will pay 9 percent (nominal) interest annually, with quarterly dividend payments. The bonds will be due and payable in twenty years. The cost of selling the bonds—the fee charged by the brokerage firm handling the sale of the bonds—will be $200,000. It is expected that purchasers of the bonds will pay $950 at the time of the initial sale. Determine the *effective rate per year* paid by XYZ to borrow money in this way.

11.5 A firm sells a $1,000, 10-year bond for $950. The "coupon rate" is 6 percent and dividends are paid annually. (Ignore underwriting and administrative costs.) The firm's incremental income tax rate is 50 percent.
   a. Determine the *before-tax* cost of this debt to the firm.
   b. Determine the *after-tax* cost of this debt to the firm.
   (*Answers:* (a) 6.7 percent; (b) 3.6 percent)

11.6 Consider the following data for the XYZ Corporation:

Total cash revenues (earnings)	$5,000,000
Cash expenses to maintain earnings level	3,600,000
Net operating expenses	$1,400,000
Interest on debt	200,000
Net earnings before taxes	$1,200,000
Income taxes	480,000
	$ 720,000

The total market value of the firm's equity is $6,000,000; the total market value of debt is $4,000,000. With this information, determine
   a. the before-tax cost of equity capital.
   b. the before-tax cost of debt capital.
   c. the before-tax weighted average cost of capital.
   d. the after-tax cost of debt capital.

11.7 The Bigditch Construction Company has a contract to build a major hydroelectric project over a three-year period. Bigditch has most of the requisite equipment, but it will need a mobile van to serve as a field office during the three-year period. There are three alternative ways of acquiring the van, as outlined in this problem and in Problems 11.8 and 11.9. The firm's minimum attractive rate of return is 20 percent after income taxes. Its combined incremental corporate income tax rate is 50 percent.
   The van may be purchased for an initial cost of $50,000. It will be sold at the end of three years for an estimated $20,000. The van may be depreciated by the DDB method, but the IRS will probably insist that the depreciable life be at least five years. (Note that that allowable depreciable life may be greater than the number of years the taxpayer expects to actually own the property.) Thus Bigditch will depreciate the van by DDB over five years using a 20% salvage value at the end of five years. Property taxes, paid at the *end* of each year, will be 3 percent of the initial cost; insurance,

payable at the *start* of each year, will be $500 per year. Find the equivalent present value of the after-tax cost of all cash flows.

(*Answer:* $28,717)

11.8 The van in Problem 11.7 may also be leased from the CEL Company. Under the terms of the lease, Bigditch will make an initial deposit of $5,000, which will be recovered in full at the end of the three-year lease. Lease payments are $3,000 per month, payable at the start of each month. The leasing company, CEL, will provide all insurance coverage at no additional charge. Since Bigditch makes quarterly tax payments, the tax consequences of this lease will take place at the end of every three-month period, that is, at the end of months 3, 6, 9, . . . , 36. Find the equivalent present value, after taxes, of this lease.

11.9 The van in Problem 11.7 may also be purchased entirely with borrowed funds. Under the terms of the loan, Bigditch must repay the loan in six semiannual payments at the start of each six-month period. Each payment will be 1/6 of the principal plus 8 percent on the unpaid balance.* What is the cost to Bigditch of this loan, stated as a percentage (effective, per year)?

(*Answer:* 11.98 percent)

11.10 Which of the financing plans outlined in Problems 11.7, 11.8, and 11.9 is the most desirable economically? Explain your answer fully and show all work.

11.11 Consider the following prospective (assumed) capitalization rates for various leverage positions:

Market value of debt (D)	Effective interest rate paid on debt (k_d)	Expected annual net operating earnings (E)	Equity capitalization rate (k_e)
$ 0	———	$1,000	0.120
1,000	0.045	1,000	0.120
2,000	0.045	1,000	0.122
3,000	0.047	1,000	0.125
4,000	0.050	1,000	0.130
5,000	0.054	1,000	0.140
6,000	0.069	1,000	0.150
7,000	0.075	1,000	0.170
8,000	0.082	1,000	0.200
9,000	uncertain	1,000	uncertain

a. Using these data, determine (1) the annual interest charge, (2) the expected annual net earnings, (3) the market value of the stock, and (4) the total market value of the firm for each of the ten leverage positions shown above. Show the results in tabular form.

*Assume that interest is paid at the *start* of each period, based on the unpaid principal remaining at the *start* of the period. Thus, at the start of the first period, for example, interest is paid on the entire $50,000 loan.

b. Determine the overall cost of capital, k_o, and leverage, λ, for each of the indicated situations. Prepare a graph of k_o as a function of λ and discuss the results. (Define λ as the ratio of debt to total market value of the firm, that is, $\lambda = D/V$.)

c. Determine the overall average cost of capital, k_o, and the marginal cost of borrowing for each indicated leverage position. Plot these marginal costs and average costs as a function of leverage and discuss the results.

11.12 Construct a numerical example, similar to the one in the text, indicating the effect of arbitrage. Discuss the arbitrage process within the context of your example.

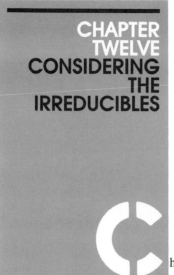

CHAPTER
TWELVE
CONSIDERING
THE
IRREDUCIBLES

hapter 1, as well as other sections throughout this book, emphasize that economic analysis in general, and engineering economy in particular, are decision-*assisting* activities; they focus only on the prospective economic consequences of alternative investment decisions. Consequences that are not measured in monetary terms, for whatever reasons, are omitted from the analysis. Since decisions should consider *all* differences between alternatives, economic analysis alone is insufficient for decision-*making*.

Consequences that are not measured in monetary terms or for which economic equivalence cannot readily be established are known as **irreducibles;** they cannot be reduced to a common monetary unit of measurement.[1]

Irreducibles are of particular interest in the public sector, since many social services are either free (education and fire protection, for example) or subsidized (fees at recreation facilities). Generally, no market mechanism exists to establish economic equivalence. The problem of irreducibles is also of considerable interest in the private sector, since many important consequences of investments are frequently difficult to measure directly in monetary terms. Examples include investments affecting employee morale, customer loyalty, or product appearance.

The irreducibles issue has received increasing attention in recent years, and there is considerable literature addressed to associated problems. (A select group of references is included in

---

Dr. Peter C. Gardiner, University of Southern California, is coauthor of this chapter.

---

1. Some authors refer to nonmonetary consequences as *intangibles*. This term is inaccurate, because many irreducibles, such as traffic fatalities, are in fact quite tangible.

Appendix A.) This chapter presents several analytical techniques designed to include *all* consequences, monetary as well as irreducible. But first we will explore the problem of irreducibles in more detail.

**12.1
CHARACTERISTICS
OF THE PROBLEM
OF IRREDUCIBLES**

There are two kinds of nonmonetary problems. The first problem occurs when the entire analysis is conducted in monetary units, and even then some nonmonetary considerations must be heeded. The second problem occurs when the analysis itself must take into account some inherently nonmonetary dimensions. As an example of the former, suppose that three programs, A, B, and C, have been underway over the past ten years, resulting in the sequences of cash flows indicated in Figure 12.1. The cash flows for the three programs are such that the equivalent worths now, at the end of the ten-year cycle, are equal for the three programs. Nevertheless, we should not be indifferent to the various performances of the three programs. Although the equivalent worths are equal, the patterns of cash flows reflect quite different levels of performance: Program A shows uniform results from year to year, Program B's performance has been erratic over time, and Program C has been steadily deteriorating. The *patterns* of these cash flows contain information that is beyond the information captured in the equivalent worth calculation.

An additional problem with analyses in which only monetary consequences are considered is related to the difficulty in completely describing all of these consequences. As a general principle, analysts should estimate, in monetary terms, the costs and benefits to whomever they may accrue. Consider the example of building a dam in a certain community. The costs of the dam might be straightforward, but how about the benefits? To whom do they accrue? The wheat farmer benefits because his land now has water and he can increase his wheat crop; the miller enjoys increased demand for services as a result of the farmer's increased productivity; a new bakery is constructed to take advantage of the increased flour availability; the delivery service gains from the increased volume of bakery goods; the grocery stores that receive the bakery goods gain additional profit; the owner of the grocery store enlarges the store and employs unemployed carpenters, who in turn purchase goods from the local hardware store; and so on. To whom, then, do the benefits of the dam accrue? And suppose the dam collapses and floods the towns and farms? Should this event, which has nonzero probability, be included? How? Should this cost be added to the dam's construction costs?

Some costs can be estimated only with great difficulty. Consider the original cost estimates used in the cost-benefit analyses of nuclear power plants. Were the costs of the Three Mile Island

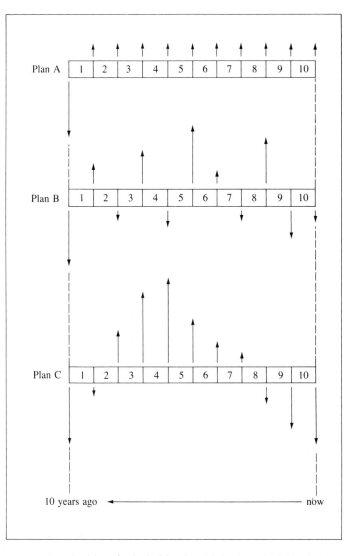

Figure 12.1. Cash Flow Diagrams for Three Alternatives

10 years ago ⟵⎯⎯⎯⎯⎯⎯⎯⎯⎯⎯⎯ now

power plant incident included in the original analysis? It is hard for analysts to imagine the eventual costs (and benefits) of alternatives involved in new technology. What would the analysis have looked like for Ford's Model T automobile? Could anyone have foreseen the costs of the contributions to smog caused by automobile emissions?

Economic analyses are concerned with benefits and costs in the aggregate; they do not address the issue of the distribution of costs and benefits to the citizens of society. Suppose that someone's personal analysis of trash disposal leads to a decision to dump it in a neighboring yard? His or her benefit is the neighbor's cost. And this is not such an extreme example. The City of Los

Angeles and other cities proposed through their utility companies that a huge power-generating plant be constructed at a place called Four Corners, so named because it would be built near the four-corner intersection of New Mexico, Colorado, Arizona, and Utah. The benefits would have accrued to those living several hundred miles away from Four Corners. The costs (especially the air pollution costs) would have been borne by the local population, Indians for the most part. To the population being served by the power-generating plant, the economic analysis made sense; from the viewpoint of the local population, however, the same analysis made no sense at all.

Economic efficiency has to do with selecting a course of action that maximizes wealth (PW, AW, and FW). Put somewhat differently, resources, measured in monetary terms, are conserved. The question of **equity**, on the other hand, has to do with the distribution of wealth among the participants. Consider for example, a prospective investment under consideration by Central City, located in Cow County, with the following estimated monetary consequences:

	Population affected	Present worth (benefits − costs)
Central City	10,000	$50,000
Remainder of Cow County	990,000	− 30,000
Total county	1,000,000	$20,000

Here, the proposed investment is *efficient* from the city's point of view and from that of the total county population, yet the *equity* question is significant. The residents of Central City are subsidized by the other residents of Cow County, so equity is an important irreducible effect in this case.

Let's turn now to problems that arise when certain consequences—significant differences between alternatives—are not readily reducible to a monetary dimension. For example, some analysts have established a monetary estimate for the cost of human life, based on the present worth of expected earnings forgone. But if someone were to offer you that amount (or more), would you end your life? Air is free, so what dollar proxy would you assign to the air we breathe? If someone cornered the market on air would you pay any price to have some? Clearly, there are some consequences for which it is extremely difficult, if not entirely infeasible, to establish monetary equivalence. For these consequences, it is necessary to adopt reasonable procedures for including the irreducibles along with the result of the economic analysis. These procedures, or approaches, may be described under two general categories, (1) cost effectiveness and (2) multi-attribute analyses, the topics of the following sections.

One of the first attempts to formally treat both monetary and nonmonetary values was referred to as **cost-effectiveness analysis.**[2] Consequences that naturally lend themselves to monetary dimensions are so measured. Typically, these are project *costs:* initial investment, operating expenses, support (maintenance) expenses, and phase-out costs. Consequences that do not readily lend themselves to monetary dimensions are evaluated in terms of *effectiveness,* a statistic that describes the (nonmonetary) quality of the project. Effectiveness models were designed to faithfully represent the effectiveness analog of all the elements of the system being analyzed, just as the cost model was designed to faithfully represent the (monetary) cost analog of all the elements of that system.

Establishing effectiveness involves three distinct but related steps: (1) establish one or more effectiveness scales, having each scale relate directly to a goal or contribute to meeting that goal; (2) establish, by measurement or estimation, a position on an effectiveness scale reached by some alternative system under consideration; and (3) in the case of more than one effectiveness scale, aggregate individual effectiveness scales into a single measure of effectiveness or worth that indicates overall effectiveness for the system being analyzed.

A key to cost-effectiveness analysis is to establish the *effectiveness scales* on which system effectiveness will be measured. Effectiveness scales, sometimes known as **measures of effectiveness** (MOEs), are the measure of what is received in return for the funds expended (costs). For example, if an automobile is purchased, certain qualities are expected in return for the purchase price: transportation, safety, style, room, comfort, reliability, maintainability, and so on. None of these qualities lends itself to direct monetary measurement. But maintainability can be described in terms of the mean time (months) or distance (miles) between tuneups, depending on the automobile. An automobile that requires 10,000 miles between tuneups is more effective (or scores higher on that effectiveness scale) than one that requires only 6,000 miles between tuneups. If miles between tuneups were the only dimension of interest, one automobile could be said to be more effective than another if it scored higher on the same effectiveness scale. Another common scale used in automobiles is "mileage," that is, miles per gallon (mpg) of fuel. An auto that

12.2 COST EFFECTIVENESS

2. Much of the early work on cost effectiveness emerged in the late 1950s and during the 1960s. See, for example, Roland McKean, *Efficiency in Government through Systems Analysis.* (New York: John Wiley and Sons, 1958); Charles J. Hitch and Roland McKean, *The Economics of Defense in the Nuclear Age* (Santa Monica, CA: The Rand Corporation, 1960); E. S. Quade and W. I. Boucher, eds., *Systems Analysis and Policy Planning* (New York: American Elsevier, 1968); and Karl Seiler, *Introduction to Systems Cost Effectiveness* (New York: John Wiley and Sons, 1969).

achieves 50 mpg is said to be more effective than one that achieves 12 mpg. Notice that both a scale and a measurement establishing performance on that scale are required before effectiveness can be presented.

The first task in developing a measure of effectiveness for a given system under analysis is to identify the scales on which the system's effectiveness is to be determined—not always an easy task. Often there are more than one scale, and depending on who you talk to, there may be conflicting ideas about which scales to use and how important, relative to each other, various scales are. As you will see, this problem led to the development of multi-criteria decision making (discussed in Section 12.3). For now, let's restrict ourselves to instances for which there is an agreed-upon single measure of effectiveness, other than monetary cost.

Once a scale has been established, the level of performance of each system under consideration must be measured, estimated, or otherwise established on that scale. In many instances, level of performance can simply be measured, provided that the scale of effectiveness is sufficiently well defined to permit unambiguous measurement. In other instances, measurement might be impossible, for a number of reasons. For example, the system may exist only on paper, making direct measurements impossible. In such cases analysts typically use estimates. There are a variety of techniques for establishing these estimates, including the Delphi method, computer simulation, and analytical models.

There may be very different scales of effectiveness involved in the analysis. For example, levels of vehicle performance can be measured in miles between maintenance as well as in miles per gallon. In such cases there are *objective* measurements on *objective* scales. In instances in which there are objective scales but the level of performance cannot be measured directly, estimated numbers that result in *subjective* estimates on *objective* scales are used. When developing effectiveness on esthetic or style scales, the scale and measurements must necessarily be *subjective*.

Regardless of how data are collected, analyzed, and displayed to indicate levels of performance by position on an effectiveness scale, the process is not yet an evaluation or a decision. It has simply established one dimension—effectiveness. Only after both cost and effectiveness have been established for the alternatives under consideration can a decision be made using cost-effectiveness procedures. The goal is to select the "most cost-effective alternative." As you will see, the simplicity of the cost-effectiveness statistic is deceptive and frequently leads to incorrect or insupportable decisions about an alternative's relative attractiveness.

Consider the four mutually exclusive highway safety projects

Table 12.1

Cost-Effectiveness Analysis for Four Traffic Safety Improvements

Improvement	Equivalent uniform annual cost	Effectiveness (accidents reduced per year)	Cost effectiveness (per accident reduced)
W	$27,150	32	$848
X	14,850	22	675
Y	9,150	10	915
Z	12,200	16	762

described in Table 12.1.[3] This set of alternatives is *exhaustive:* Other than the do-nothing alternative, the list of feasible options is complete. Here, cost effectiveness is defined as the equivalent uniform annual cost per accident reduced, the ratio of (monetary) cost to (nonmonetary) effectiveness. (The E/C formulation is sometimes known as "bang for the buck," especially with respect to military systems.) If we accept the premise that the objective is to maximize cost effectiveness—that is, if we want to select the alternative that minimizes the ratio C/E or maximizes E/C—it follows that the preference order for these four alternatives is X > Z > W > Y.

Now let's take a closer look at this problem. Observe the costs and effectiveness measures (accidents reduced) displayed in Figure 12.2. Note that the C/E ratio is the reciprocal of the effectiveness/cost (E/C) ratio and that minimum C/E is equivalent to maximum E/C. Note too that the slopes of the lines drawn from the origin to each point (W, X, Y, and Z) are the E/C ratios for the four alternatives. If we accept the principle that only differences among alternatives are relevant, the differences between alternatives W and Y should be examined as follows:

Alternative	Cost (EUAC)	Effectiveness (accidents reduced per year)
W	$27,150	32
Y	9,150	10
Difference	$18,000	22

Everything else being equal, the choice between W and Y depends entirely on the relationship between the additional $18,000 expenditure and the additional twenty-two accidents per year reduced. Specifically, if the *utility*, or worth, of an additional twenty-two accidents reduced is in excess of the utility of saving $18,000, then the more expensive alternative (W) is preferable.

3. This example is adapted from G. A. Fleischer, "Significance of Benefit/Cost and Cost/Effectiveness Ratios in Analyses of Traffic Safety Programs and Projects," in *Transportation Research Record 635, Price and Subsidy in Intercity Transportation and Issues of Benefits and Costs* (Washington, D.C.: Transportation Research Board, 1977), pp. 32–36.

Figure 12.2. Effectiveness versus Cost for Traffic Safety Alternatives

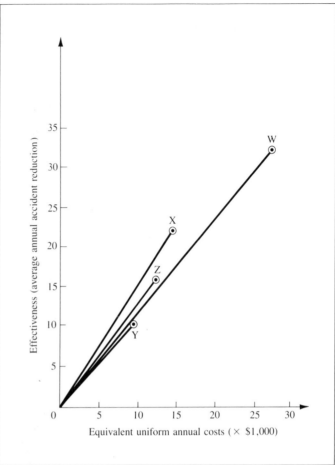

Equivalent uniform annual costs ($\times$ $1,000)

Otherwise, the less expensive alternative (Y) is preferable. In the absence of the relevant *utility function*, that is, without knowing anything about the tradeoff between costs and accidents reduced, the alternatives cannot be ordered and no preferences can be determined.

The preceding argument relates to only one comparison (W versus Y) in the example. The same argument can be applied to all six comparisons: W versus X, W versus Y, W versus Z, X versus Y, X versus Z, and Y versus Z.

Consider investment alternatives with costs and effectiveness given by $C(\cdot)$ and $E(\cdot)$ respectively. The index of cost effectiveness for each alternative is designated by $I(\cdot) = C(\cdot)/E(\cdot)$. There are only three conditions under which ranking cost-effectiveness ratios (minimizing C/E or maximizing E/C) is appropriate:

1. *Effectiveness for all alternatives is equal*, in which case rank-

ing by ratios is equivalent to ranking on the basis of decreasing costs. (See Figure 12.3.)

2. *Costs for all alternatives are equal*, in which case ranking by ratios is equivalent to ranking on the basis of increasing effectiveness. (See Figure 12.4.)

3. *Dominance obtains*. Consider any pair of alternatives, x and y. If $E(x) \geq E(y)$ and $C(x) \leq C(y)$, then alternative x is said to dominate alternative y. Under these conditions, of course, $E/C \ (x) \geq E/C(y)$ and likewise $C/E(x) \leq C/E(y)$. (See Figure 12.5.) Unfortunately, these conditions, known as **dominance,** rarely occur in the real world. Usually, an increase in effectiveness is brought about by an increase in project or program costs. When this occurs, it is not possible to establish a unique, unambiguous ordering of alternatives without knowing the utility function relating cost and effectiveness.

If the C/E ratios may not be used to directly rank-order investment alternatives, the cost-effectiveness approach can still be helpful to the decision maker. First, establish the range of feasibility for costs. Any alternative having costs that exceed the maximum allowable cost is infeasible and removed from further consideration. Next, establish the allowable (desired) range of effectiveness. Any alternative that fails to produce a level of effectiveness in the allowable range is infeasible and removed from further consideration. Remaining alternatives are considered feasible, since they meet all cost and effectiveness constraints.

Next, observe all alternatives in the domain of feasibility to search for dominance. Figure 12.6 shows some examples of dominance in the feasible domain. Notice that E is dominated by D, G is dominated by F, and J is dominated by H. (K and L are infeasible.) Dominance can be used to narrow the number of alternatives in the feasible domain to the undominated set of contenders. Of the eight original alternatives in Figure 12.6, two are infeasible and three are dominated. Now, how can the optimal remaining alternative be determined?

One possible approach is simply to disregard one dimension and rank on the other. For example, assuming that effectiveness is a constraint rather than a criterion, the least-cost alternative that meets the minimum level of effectiveness is D. Similarly, assuming that cost is a constraint rather than a criterion, then the cost-feasible alternative that maximizes effectiveness is H. This approach is known as **lexicographic ordering,** or the **pivot rule.**

For instances wherein neither dominance nor minimum levels of effectiveness or cost targets can be established, the analysis must eventually turn to asking the decision maker whether the additional (incremental) effectiveness of an alternative justifies the

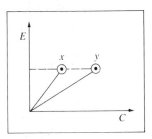

Figure 12.3

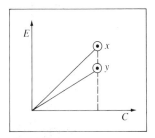

Figure 12.4

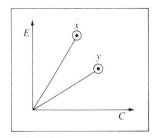

Figure 12.5

Figure 12.6. Identifying
Feasible, Undominated
Alternatives

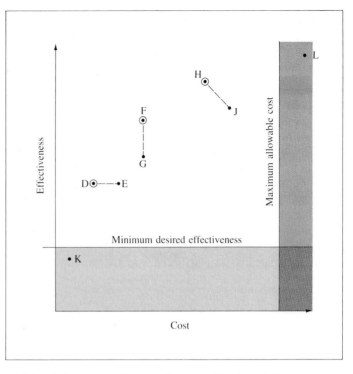

additional (incremental) cost. This step involves judgments as to the relative worths (utilities) of increments of costs and effectiveness, as noted previously.

A single measure of effectiveness can seldom adequately represent a system under consideration. In general, complex environments require decision makers to select from a set of alternatives that produce many nonmonetary consequences. Some theorists refer to these problems as **multicriteria,** or **multiattribute,** problems. Problems of this type can be described in general as follows:

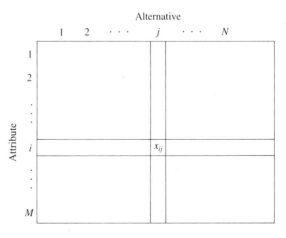

The attributes shown here represent multiple scales of effectiveness (each attribute is represented by one scale of effectiveness). Cells within the matrix contain quantitative or qualitative measures that report the performance (level of effectiveness) measured or estimated for the alternatives under consideration on each scale of effectiveness. These levels of performance are called **outcomes**, or **impacts**, and are shown as the x_{ij} in the above figure. The alternatives are plans, projects, programs, designs, and so on. One of the attributes represents the monetary consequences of the N alternatives; the remaining M-1 attributes represent nonmonetary attributes. The decision maker's task is to examine all the impacts displayed for each alternative and to select the "best" alternative for adoption and implementation.

Clearly, the multiple-criteria problem presents a substantial challenge to decision makers and, by extension, to analysts assisting in the decision-making process. The next section briefly reviews several approaches to such problems and, in the process, answers the question posed earlier concerning the appropriate course of action when there is more than one scale of effectiveness.

The issues associated with multicriteria decision making are essentially those that occur in any decision-making setting: (1) Reduce the number of alternatives to the set of feasible, nondominated contenders and (2) somehow rank order remaining contenders to identify the preferred alternative for adoption and implementation. It should be stated at the outset that the following discussion takes place in the context of *certainty*. For multicriteria decision making under risk and uncertainty, you are invited to review a more advanced text.[4]

Consider the problem of selecting from three alternative new product designs having the following attributes and impacts:

Attributes	Design I	Design II	Design III	Design IV
Cost per unit	$10.00	$10.50	$11.25	$9.75
Reliability: mean time between failures (hours)	580	580	540	480
Appearance	good	excellent	fair	fair

This problem presents multiple alternatives, multiple attributes, quantitative impacts (cost and reliability), and a qualitative impact (appearance).

4. See, for example, H. Raiffa and R. Keeney, *Decisions with Multiple Objectives* (New York: Wiley, 1976).

Narrowing the Scope of the Problem

The scope of the problem may be reduced by subjecting each alternative to feasibility and dominance tests.

First, ask if there are any constraints for the attributes and, if so, whether any of the prospective impacts render the alternatives infeasible. For the product design example, suppose that there are no constraints on cost or appearance but that the major customer for this product requires the mean time between failures (MTBF) to be no less than 500 hours. In that case, Design IV is infeasible, because its MTBF is only 480 hours.

Next, search for dominance among the feasible alternatives. For any pair of alternatives, x and y, x is said to dominate y if x is at least as good as y for every attribute. Examining the three feasible designs shows that Design III is dominated by Design I: Design III is more costly, it has poorer reliability, and its appearance is not as good as that of Design I. Viewing Designs I and II, neither dominates the other.

At this point, then, the number of alternatives has been reduced from four to two. This is substantial progress, but a unique solution has not yet been obtained; the problem of identifying the optimal, feasible, nondominated alternative still remains.

Lexicographic Ordering

In the two-dimensional case, cost versus effectiveness, it was suggested that the preferred alternative can be identified by adopting the principle of lexicographic ordering: Determine which attribute is the most important and compare all alternatives on the basis of this attribute. If one alternative has a higher level of performance for this attribute than all the other alternatives, it is optimal. In our example, if cost is the most important attribute, lexicographic ordering leads to choosing Design I, since it has the lower cost. If no single alternative predominates for the most important attribute, then determine the second most important attribute and proceed as before. This process should be continued until one alternative is selected or until all attributes are exhausted.

Lexicographic ordering is obviously simple to use, but it may introduce significant errors into the decision. This approach assumes that, when the most important attribute is identified, the entire decision can rest on an alternative's performance for that attribute alone. This is another way of saying that the most important attribute outweighs all the other attributes and impacts combined. This is rarely true. Some of the obvious weaknesses of lexicographic ordering have been addressed by somewhat more sophisticated techniques, one of which is presented below.

Simple Multiattribute Rating Techniques

To meet the requirements of multicriteria decision making, a major line of research has emerged in recent years, namely, **decision analysis,** or **behavioral decision theory.** The relevant research deals with evaluation, especially for contexts in which there are multiple decision makers, multiple alternatives, and multiple attributes. This approach has also been called *multiattribute utility measurement* (MAUM), *multiattribute utility technology* (MAUT), and **simple multiattribute rating technique** (SMART), depending on which version is presented. There are many versions of this utility-based evaluation technology in the literature, each producing more or less the same results. The following discussion presents the SMART version.[5]

In essence, the various utility theory approaches model the preferences of a decision maker and then use this preference model to evaluate the alternatives under review in terms of their levels of performance across the attributes (scales of effectiveness) in the analysis. But the SMART approach to evaluation takes the analysis one step further: *It involves the decision makers in determining the attributes.*

The basic idea is quite simple: Every alternative has value in terms of its outcomes (levels of performance). All versions of evaluation technology are designed to ''discover'' those values, one attribute at a time, and then aggregate them across attributes using a suitable aggregation rule and weighting procedure. The most widely used and easily communicated method is a simple weighted, linear average. This approach allows individuals or groups to shift their attention from ''holistic'' evaluations of a set of impacts to the values those impacts serve.[6] By using SMART, groups and individuals can concentrate on identifying the relevant **attributes,** their relative importance, the various levels of performance of each attribute, and the values for the various levels of performance. Then attribute value components can be aggregated into an overall measure of value for the alternative being considered. Thus this technique permits differences in points of view bearing on how to assign overall worth to an alternative to be discussed and resolved explicitly and numerically at the level of evaluation rules rather than at the level of the individual evaluations themselves.

---

5. For an additional discussion of SMART, see P. C. Gardiner and W. Edwards, ''Public Values: Multiattribute-Utility Measurement for Social Decision Making,'' *Human Judgment and Decision Processes* (New York: Academic Press, 1975).

6. The terms *impacts* and *attributes* are frequently confused. To avoid ambiguity, the term *attribute* is used here to describe a specific quality characteristic or dimension, for example, travel time between two points. An *impact* is an estimated or observed level of performance along that dimension, for example, a travel time of four hours.

SMART consists of ten steps:

Step 1 Identify the person or organization whose values are involved in the decision under consideration. If, as is often the case, there are several individuals or groups that have stakes in the decision (or "axes to grind"), they must be identified, and people who can and will speak for these interests should be induced to participate.

Step 2 Identify the issue for which the values are relevant. This step clarifies the nature of the decision to be made. For example, if a car is to be purchased, it is essential to specify whether the car is to be a family car, a second car, a car to race at the racetrack on weekends, and so on. Each use tends to bring different sets of values to mind.

Step 3 Identify which entities are to be evaluated. In this step, the alternatives are identified, although this is not necessary as a precondition to developing the evaluation model. In SMART, value models can be developed before any alternatives are developed. And after the evaluation model has been developed, additional alternatives can be proposed for consideration without necessitating a revision of the value model.

Step 4 Identify the relevant attributes of value. The first two steps are more or less philosophical, addressing the questions "whose values?" and "values for what purpose?" This step poses the first technical question: What dimensions of value (that is, what attributes or scales of effectiveness) are important in helping to choose among the alternatives to be considered? At this point, anyone can suggest dimensions for inclusion. If someone considers an attribute important, then it should be considered. As a general rule, fifteen attributes or so should be ample, but some analyses may have significantly more.

Step 5 Rank the attributes in order of importance. Ranking can be performed either individually, so that each representative of potentially conflicting viewpoints acts individually, or collectively, by representatives acting as a group. Here, value differences first begin to show up. Group rankings are usually not attempted; there is no attempt to force a consensus for its own sake. Judgments in this step (and in others) are expected to differ, reflecting the differing preferences and value systems of the individuals involved. The interesting question, from an evaluation perspective, is how these differences will affect an overall evaluation, if at all. The effect of including all the attributes considered important is resolved here. Attributes considered unimportant are simply ranked least important.

Step 6 Rate the attributes in importance, preserving ratios. There are many variations to this step, as well as important technical considerations. One way to establish weights that have a certain face validity is to have participants start by assigning an arbitrary value, or score, of 10 to the least important attribute. Then, the next most important attribute is compared to it, with participants assigning a number reflecting the relative importance. For example, if the next most important attribute is twice as important, it should be assigned a 20. If it is four times as important, it should be assigned a 40. Continuing up the list of attributes, repeat the process, checking for consistency. For example, suppose that participants assign the following weights to a set of attributes:

Attributes ranked	Importance weights
X	30
Y	20
Z	10

A consistency check should be made to ensure that attribute X is as important as Attributes Y and Z combined.

As you can see, attributes are ranked by their importance. Attribute Z is the least important; Y is twice as important as Z; and X is three times as important as Z, as important as Z and Y combined, and half again as important as Y. There is no upper limit on weights, and if the list of attributes is large, the most important attribute might have a weight many times larger than the weight of the least important attribute. Participants should feel free to change weights until they feel that the weights accurately reflect their preferences.

Step 7 Normalize the weights. This computational step converts the weights into numbers that sum to 1.00 or 100. The choice of a 0 to 100 scale is arbitrary but helpful to decision makers for readability and understandability. In fact, after seeing the normalized weights, it is not at all unusual for a decision maker to adjust the weights to more accurately reflect preferences.

Step 8 Construct value curves for each dimension. The best way to obtain these curves is simply to ask the participants to draw graphs. The horizontal axis of each graph represents the plausible range of levels of performance for the scale of effectiveness for the attribute under consideration. The vertical axis represents the range of values, or scores, associated with the corresponding levels of performance recorded on the horizontal axis. In a sense, the value curves are a device to translate an alternative's performance from its natural scale to a scale that presents the value of that performance. When value curves are available for all attributes, the level of performance of an alternative for any attribute can be converted to the value of that level of performance. Once converted, all attributes are measured in the same value-oriented scale.

Step 9 Calculate the values, or weighted scores, for each alternative. That is, find

$$V_j = \sum_{i=1}^{M} w_i s_{ij} \qquad\qquad (12.1)$$

where

V_j = total weighted score for alternative j

w_i = importance weight of the ith attribute, or dimension

s_{ij} = unweighted score for the ith attribute and the jth alternative

Here, the unweighted scores, s_{ij}, result from impacts x_{ij} using the value curves for the ith attribute developed in Step 8. The attribute weights are those developed in Step 7.

Step 10 Decide. The decision is made by rank-ordering the alternatives evaluated by the overall values determined in Step 9. If there are multiple decision makers or multiple participants, then there will be more than one set of rank orderings open for inspection. Often, but not always, the participants will agree on which alternative or which set of alternatives seems to be desirable for adoption and implementation. When disagreement occurs about which alternative is best, an explicit numerical method is available to find the sources of disagreement.

In summary, SMART offers a formal evaluation process that can be used when there are multiple scales of effectiveness (attributes), when there are multiple decision makers, and when not all impacts can be reduced to monetary measurements. Participants with strong points of view tend to concentrate on aspects of the alternatives that most strongly engage their biases. Using SMART avoids this problem, because in developing the importance weights, participants make tradeoffs among different types of impacts. The values associated with each impact are discovered in the next step, when the value curves are drawn. The overall worths of proposed alternatives are then calculated using the weighted average process. SMART does not permit one or two impacts to become so salient that they aggravate existing sources of conflict and disagreement and overshadow what might otherwise produce overall agreement.

Of course, SMART cannot and should not be expected to eliminate all differences of opinion. Some value conflicts are indeed genuine, and any formal value-measurement procedure should respect and reflect them. When conflicts do occur, SMART offers a clear set of procedures to promote orderly debate. By inspecting the calculations, a decision maker can select the best areas for compromise and improvement.

EXAMPLE To illustrate the SMART process of evaluation, consider the following example, inspired by an attempt to purchase a new car in the early 1970s. The example is a study actually conducted by a group of students using SMART on a term project to evaluate, after the fact, the soundness of an automobile purchase by one member of the group. Although the data are not current, the experience of this student group in implementing SMART is instructive.

Step 1 *Determine whose values*. Who has a stake in the decision? (In this instance, one of the team members was identified as the decision maker, since he had just purchased a new car. The identity of his selection was withheld from the group.)

Step 2 *Determine the purposes, or objectives*. In this example, the purpose is to select a car. But why is the car wanted? Will it be a family car, a car to commute to work, a car to enter the drag races on Saturdays? Each purpose should suggest very different kinds of impacts that are important to a decision maker. In this example, a number of different candidate automobiles are being considered for use as a family car.

Step 3 *Identify the alternatives*. In this example, eight automobiles are candidates for a family car: a Ford Pinto, a Chevrolet Nova, an AMC Matador, a Plymouth Fury, a Cadillac El Dorado, a Toyota Corona, a Dodge Dart, and a Lincoln Continental.

Step 4 *Identify the dimension, or attributes, of value*. Sixteen attributes are considered important, as shown in Table 12.2. (This figure shows the form used to establish the list of initial attributes, and subsequently, to establish their rankings and importance weights.)

Table 12.2
Attributes and Their Relative Ranking

Attribute	Initial attribute list	Rank
1	List price	1
2	Operating cost (yearly)	7
3	Resale value after 5 years	13
4	Maintainability	9
5	City mileage	2
6	Highway mileage	3
7	Acceleration (from 0–60 mph)	8
8	Braking (from 60 mph to stop)	14
9	Ride	6
10	Handling	11
11	Maneuverability	10
12	Trunk space	12
13	Front seat room	5
14	Front seat comfort	4
15	Rear seat room	15
16	Rear seat comfort	16

Step 5 *Rank-order the attributes*. As shown in Table 12.2, the list price of the cars is the most important attribute, whereas rear seat comfort is the least important. (The decision maker is married but has no children. And obviously he rarely, if ever, has to sit in the rear seat.)

Step 6 *Weight the attributes*. This process is summarized in Table 12.3. The initial importance weight assigned to the least important attribute is 10 (an arbitrary starting point), which leads, in this example, to an unnormalized weight of 4,000 for the most important attribute.

Step 7 *Normalize the weights*. Refer to the final column of Table 12.3. Note that, when the importance weights are normalized, 85 percent of the "weight" rests on the six most important attributes and less than 2 percent is accounted for by the five least important attributes. This example shows the doubtful benefit of including a large number of attributes. When the weighting process is completed, the decision tends, almost exclusively, to be a function of the top five or six attributes. Even dramatic changes in the level of performance for the low-weighted attributes tends to have little effect on the overall worth computed. Indeed, low-ranked attributes may be included so that decision makers or other participants can see their own attribute(s) included or to ensure that an alternative's performance at least meets minimum acceptable levels of performance for each attribute. Of course, when there are multiple decision makers, a low-ranking attribute for one interest group may be a highly important one for another group. In that case the logic of including numerous attributes is sound.

Table 12.3
Weighting the Attributes

Rank number	Attribute identification	Attribute name (ranked order)	Weighted ranking	Percent total
1	1	List price	4,000	21.6%
2	5	City mileage	3,000	16.2
3	6	Highway mileage	3,000	16.2
4	14	Front seat comfort	2,000	10.8
5	13	Front seat room	2,000	10.8
6	9	Ride	1,700	9.2
7	2	Operating cost (yearly)	700	3.8
8	7	Acceleration (0–60)	600	3.2
9	4	Maintainability	500	2.7
10	11	Maneuverability	400	2.2
11	10	Handling	320	1.7
12	12	Trunk space	160	0.9
13	3	Resale value after 5 years	80	0.4
14	8	Braking (60–0)	40	0.2
15	15	Rear seat room	20	0.1
16	16	Rear seat comfort	10	0.1
		Totals	18,530	100.0%

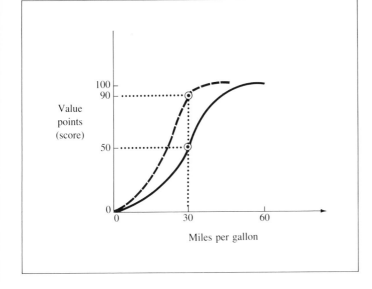

Figure 12.7. Sample Value Curves: Miles per Gallon

Step 8 *Develop the decision maker's value curves for each attribute.* Figure 12.7 shows a sample value curve. In this curve, values (*scores*) for various levels of performance on a "miles per gallon" scale of effectiveness are shown. There are two value curves, to illustrate the effect of difference preferences by two different decision makers. For one decision maker, 30 mpg is worth 50 value points; for another, the same level of performance is worth 95 value points, almost the maximum. These value differences should be carried through the analysis when multiple decision makers are involved to see what effect, if any, they have on the ultimate alternative rankings.

Note that the miles per gallon scale in Figure 12.7 ranges from 6 mpg to 60 mpg. In terms of the candidate cars, 60 mpg may seem excessive. However, it is useful to include a plausible range (maximum and minimum) to develop the value curves. If a smaller range is used, a candidate may emerge in the late stages of the analysis with impacts that lie outside the scale, thereby necessitating a complete reworking of the scale and related computations. If the scale represents the plausible range, late candidates can easily be accommodated; the means to evaluate them have already been established and no reworking of the evaluation model is required.

Figure 12.8 shows the value curves used in this example for eight of the sixteen attributes. Measurements on the "natural" scales, the abscissas, are converted to values, or scores, on the ordinates. All impacts can now be stated in commensurable units.

Step 9 *Find the total value, or score, for each alternative.*

Figure 12.8. Value Curves for
First Eight Attributes

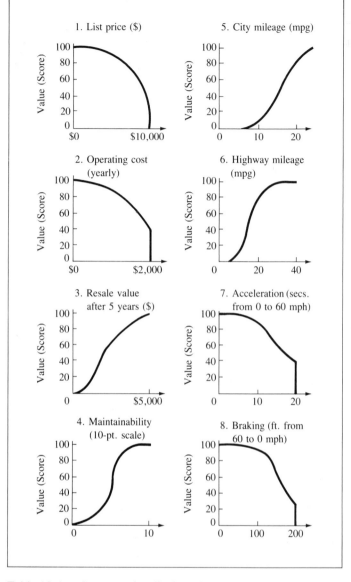

Table 12.4, a **data matrix,** displays the levels of performance for
each candidate car (alternative) for each attribute. For example,
the city mileage in miles per gallon for a Plymouth Fury was 9
mpg. A number of attributes do not lend themselves to direct
numerical measurement; they call for subjective estimates. All
subjective estimates are done on a ten-point scale (zero being
worst and ten being best). Attributes 4, 9, 10, 11, 13, 14, 15, and
16 are subjective scales, and performances on these scales are
subjectively established.

Two comments can be made at this point, both relevant to

Table 12.4

Data Matrix

Attribute identification	Attribute	Ford Pinto	Chevrolet Nova	AMC Matador	Plymouth Fury	Cadillac El Dorado	Toyota Corona	Dodge Dart	Lincoln Continental
1	List price ($)	$2,769	$3,222	$3,526	$4,565	$9,948	$3,679	$3,269	$9,214
2	Operating Cost ($/yr)	$1,170	$1,540	$1,790	$1,980	$2,000	$1,540	$1,540	$2,000
3	Resale value after 5 yrs ($)	$1,000	$1,200	$1,200	$2,000	$4,000	$1,200	$1,200	$4,000
4	Maintainabilty (*)	5.0	5.0	7.0	2.5	6.0	7.0	5.0	6.0
5	City mileage (mpg)	18.0	12.0	10.0	9.0	11.0	19.0	12.0	10.0
6	Highway mileage (mpg)	26.0	19.0	16.0	18.0	16.0	28.0	19.0	15.0
7	Acceleration (0–60; sec)	14.7	19.1	10.8	13.0	13.7	16.2	15.3	13.5
8	Braking (60–0; ft)	180.0	165.0	155.6	141.6	178.0	170.0	165.0	190.6
9	Ride (*)	4.3	8.6	8.6	5.7	10.0	5.7	8.6	5.7
10	Handling (*)	7.0	7.0	7.0	7.0	4.0	4.0	7.0	1.0
11	Maneuverability (*)	7.0	7.0	8.0	6.0	5.0	6.0	7.0	3.0
12	Trunk space (cu. ft)	6.3	13.0	14.3	20.4	13.6	10.9	16.0	19.3
13	Front seat room (*)	7.5	7.5	10.0	7.5	10.0	7.5	7.5	10.0
14	Front seat comfort (*)	7.1	8.6	7.5	7.1	7.1	8.6	7.1	7.1
15	Rear seat room (*)	7.5	7.5	10.0	7.5	10.0	10.0	7.5	10.0
16	Rear seat comfort (*)	5.0	7.5	7.5	7.5	7.5	10.0	7.5	7.5

*Units from 10-point scale, from 0 (worst) to 10 (best).

evaluations using SMART. First, some attributes tend to be measured subjectively when in fact they could be measured quantitatively on a natural, objective scale. For example, maintainability can be described subjectively or it can be measured in monetary cost per unit time or per mile. Clearly, objective measurements on objective scales are preferred wherever possible. Second, the numbers in the data matrix, which were current in the early 1970s, seem odd to us today. The list price and the miles per gallon seem very low. This emphasizes the point that SMART evaluation models need to be continually updated to reflect current values and real-world conditions. If an ''old'' SMART model is used without revision or recalibration to evaluate candidates in a ''new'' situation, analysts and decision makers are asking for trouble.

The unweighted scores for the example are summarized in Table 12.5. The individual scores for each alternative-attribute pair are determined by entering the value curves of Figure 12.8 with the attribute values in Table 12.4. Table 12.6 summarizes the scores after they have been multiplied by their respective importance weights. That is, the weight contribution of value for each alternative across each attribute is shown in the cells of the matrix. The grand totals represent the overall values, or scores, of the candidate cars.

Note that the economy cars, such as the Ford Pinto and the

Table 12.5
Summary of Unweighted Scores

Attribute (ranked order)	Ford Pinto	Chevrolet Nova	AMC Matador	Plymouth Fury	Cadillac El Dorado	Toyota Corona	Dodge Dart	Lincoln Continental
1. List price	99.0	98.2	97.4	94.1	21.2	97.0	98.1	37.4
2. City mileage	68.2	19.2	8.0	4.3	12.9	74.8	19.2	8.0
3. Highway mileage	93.2	74.8	52.2	68.2	52.2	94.5	74.8	43.5
4. Front seat comfort	92.1	98.5	93.9	92.1	92.1	98.5	92.1	92.1
5. Front seat room	93.9	93.9	100.0	93.9	100.0	93.9	93.9	100.0
6. Ride	27.0	98.5	98.5	73.0	100.0	73.0	98.5	73.0
7. Operating cost (yearly)	85.2	67.3	53.1	41.3	40.0	67.3	67.3	40.0
8. Acceleration (0–60)	66.6	44.5	85.5	75.3	71.8	59.0	63.6	72.8
9. Maintainability	50.0	50.0	91.5	6.1	80.0	91.5	50.0	80.0
10. Maneuverability	91.5	91.5	96.0	80.0	50.0	80.0	91.5	8.5
11. Handling	91.5	91.5	91.5	91.5	20.0	20.0	91.5	0.5
12. Trunk space	24.8	95.5	98.5	99.5	97.4	78.6	100.0	100.7
13. Resale value	12.6	20.0	20.0	50.0	87.4	20.0	20.0	87.4
14. Braking (60–0)	41.8	55.8	63.6	73.8	43.8	51.3	55.8	30.7
15. Rear seat room	93.9	93.9	100.0	93.9	100.0	100.0	93.9	100.0
16. Rear seat comfort	50.0	93.9	93.9	93.9	93.9	100.0	93.9	93.9

Table 12.6
Summary of Weighted Scores

Attribute (ranked order)	Weight factor	Ford Pinto	Chevrolet Nova	AMC Matador	Plymouth Fury	Cadillac El Dorado	Toyota Corona	Dodge Dart	Lincoln Continental
1. List price	0.216	21.4	21.2	21.0	20.3	4.6	20.9	21.2	8.1
2. City mileage	0.162	11.0	3.1	1.3	0.7	2.1	12.1	3.1	1.3
3. Highway mileage	0.162	15.1	12.1	8.4	11.0	8.4	15.3	12.1	7.0
4. Front seat comfort	0.108	9.9	10.6	10.1	10.0	9.9	10.6	9.9	9.9
5. Front seat room	0.108	10.1	10.1	10.8	10.1	10.8	10.1	10.1	10.8
6. Ride	0.092	2.5	9.0	9.0	6.7	9.2	6.7	9.0	6.7
7. Operating cost (yearly)	0.038	3.2	2.5	2.0	1.6	1.5	2.5	2.5	1.5
8. Acceleration (0–60)	0.032	2.2	1.4	2.8	2.4	2.3	1.9	2.1	2.4
9. Maintainability	0.027	1.3	1.3	2.5	0.2	2.2	2.5	1.3	2.2
10. Maneuverability	0.022	2.0	2.0	2.1	1.7	1.1	1.7	2.0	0.2
11. Handling	0.017	1.6	1.6	1.6	1.6	0.3	0.3	1.6	0.0
12. Trunk space	0.009	0.2	0.8	0.9	0.9	0.8	0.7	0.9	0.9
13. Resale value	0.004	0.1	0.1	0.1	0.2	0.4	0.1	0.1	0.4
14. Braking (60–0)	0.002	0.1	0.1	0.1	0.2	0.1	0.1	0.1	0.1
15. Rear seat room	0.001	0.1	0.1	0.1	0.1	0.1	0.1	0.1	0.1
16. Rear seat comfort	0.001	0.0	0.1	0.1	0.1	0.1	0.1	0.1	0.1
Grand totals:		80.8	76.3	72.9	67.7	53.9	85.8	76.2	51.5

Toyota Corona, all score approximately the same, and the luxury cars, the Cadillac El Dorado and the Lincoln Continental, both scored poorly. This tends to lend face validity to the process. Note too that, if the decision maker wishes to find out what contributes to the differences among candidates, the data are available for inspection. For example, the difference in overall value between the Ford Pinto and the Toyota Corona shows up primarily in the "Ride" attribute (2.5 weighted score versus 6.7).

Step 10 *Rank-order the alternatives on the basis of their respective scores.* (In this example, the decision maker had actually purchased a Ford Pinto, the second-ranked car on the basis of aggregate weighted score.) The decision maker may have been unable to consider all sixteen attributes simultaneously when making the actual decision and hence did not give proper consideration to the "Ride" attribute.

As this example shows, SMART does offer one way to handle multiple attributes, most of which are monetary, in a systematic, explicit, and numerical way. Using this basic evaluation process, any set of alternatives under discussion by any individual or set of decision makers can be formally evaluated. In this example, cost was included as one dimension and weighted along with the others. If the decision maker prefers, cost can be handled separately—then the analysis becomes a cost-effectiveness analysis.

It should be emphasized that this presentation is only introductory. There are a number of technical issues that analysts should be aware of before using such an evaluation technique in a real-world setting. For most applications, the technical issues have little or no effect. But in instances in which they do influence the evaluation results, the conclusions could be an artifact of the methodology and may not reflect actual preferences accurately.

A marked departure from the weighted utility aproach is the presentation of attribute-alternative impacts (x_{ij}) in their original dimensions. No attempt is made to seek a single figure of merit for each alternative. Rather, the profile of attribute values for each alternative is presented and the decision maker is free to provide his or her personal utility function, subjectively, at the time of the decision. This is a "holistic" process in the sense that all attributes and impacts are considered concurrently. Several such techniques are described in this section, along with a discussion of their principal advantages and disadvantages.

12.4 SYMBOLIC SCORECARDS

Colored Scorecards

The basic format for the **colored scorecard** approach is illustrated in Figure 12.9. Each cell of the table (x_{ij}) contains, in addition to the original quantitative or qualitative data, a color-coded transparent overlay that indicates the relative desirability of the alternative (j) with respect to the attribute (i) under consideration. (Because of printing costs, it is not possible to reproduce the overlays in their original colors. Accordingly, we have simply indicated the appropriate colors in Figures 12.9 and 12.11.)

The color code may vary from one analyst to another, of course, but the scheme that is most often used is

Green = best

Yellow = comparable

Red = worst

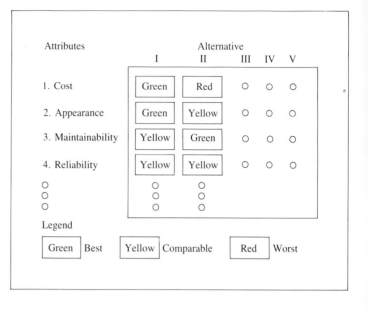

Figure 12.9. General Form for the Colored Scorecard

Attributes · Alternative

	I	II	III	IV	V
1. Cost	Green	Red	O	O	O
2. Appearance	Green	Yellow	O	O	O
3. Maintainability	Yellow	Green	O	O	O
4. Reliability	Yellow	Yellow	O	O	O
O		O	O		
O		O	O		
O		O	O		

Legend

Green	Best	Yellow	Comparable	Red	Worst

Thus by examining alternatives across the attribute row, the analyst assigns colors that are consistent with the alternatives' positions on this three-point ordinal scale. Note that more than one alternative can take on a specific color. "Ties" will occur when, in the opinion of the analyst, the impacts for two or more alternatives are approximately equivalent for any given attribute.

The colored scorecard technique has several distinct advantages. *First,* and most obviously, the technique reduces the clutter and complexity of numbers and/or words into a simple color-coded scheme. *Second,* the use and choice of these colors have a psychological advantage in that, to most people, red means "stop," yellow suggests "caution," and green signifies "go." *Third,* instead of the usual dull black and white media, colors are extensively used in data representation to act as a psychological stimulus on the decision maker. The color-coded scheme serves to stimulate more participation and interest in the decision-making process. *Fourth,* the color-coded scheme makes it easier *visually* for the decision maker to differentiate, compare, evaluate, and trade off among and within the attributes and alternatives. If more definition is required, numerical values are included within each colored box, wherever applicable. *Fifth,* the color code serves as a common base or common language for decision makers, who might represent diverse fields of interest; the color code serves as a substitute for technical language. *Sixth,* colors are readily compatible with nonquantitative rankings. And *seventh,* this technique enhances and encourages the tradeoff process. It defers to the decision maker the task of making tradeoffs, that is, recognizing and seeking the complementary or functional relationships between attributes, if they exist. The technique does not make these relationships visually apparent to the decision maker, however.

Several disadvantages, or weaknesses, of the colored scorecard technique should also be mentioned. *First,* color codes may be confusing to the color-blind viewer, although this problem could be overcome by using light, medium, and dark tones of one hue. *Second,* the decision maker must exercise extreme caution in comparing attributes. For example, when comparing a red (or yellow or green) valuation for one attribute under Alternative I with a green (or red or yellow) valuation for another attribute under Alternative I or any other alternative, the decision maker can easily be misled if he or she allows the color coding to overcome rational judgment. No equality of utility is implied. The lowest-cost alternative, for example, may not appear as desirable as the alternative that results in the most attractive appearance. *Third,* color coding is inherently dangerous in that a decision maker might tend to sum the number of cells of each color under each alternative and make comparisons on this basis. Doing so implies that the various attributes are weighted equally, an assumption that may not be warranted. *Fourth,* within the attribute, there is the problem of differentiating between attribute values so as to assign the "appropriate" color. The determination of the color—the position on the ordinal scale for the attribute—is selected by the analyst, and thus the decision maker's assessment of the overall utility of the alternative is directly influenced at this point by the judgment of the analyst. (Note that this "imposition" by the analyst in the decision-making process is quite apart from his or her role in selecting alternatives, choosing attributes, identifying data, and the like.) *Fifth,* the assignment of colors to each score assumes that the attributes are mutually independent. That is, there is no accounting for complementary or functional relationships among attributes.

Geometric Scorecards

A slight variation of the colored scorecard tachnique is the **geometric scorecard.** Here, instead of using color overlays, simple geometric figures are used to characterize the relative desirability of alternatives across an attribute. For example,

$\bigcirc$ = best

$\triangle$ = comparable

$\square$ = worst

As with colored scorecards, the viewer's attention is directed to relative desirability within attributes through a visual device.

The principal advantages and disadvantages of the geometric

scorecard are similar to those of the colored scorecard. With this technique, however, the geometric figures do not have the same visual impact that colors have and the figures do not connote "stop" or "go" as colors do. On the other hand, the geometric figures are easier to reproduce, since only black-on-white printing is required. This can be an important advantage if large numbers of scorecards must be reproduced for distribution to a wide audience. Another advantage of the geometric scorecard versus the colored scorecard is that geometric figures can be generated as part of a computer printout, thus enabling the analyst to develop scorecards directly while conducting tests for sensitivity to inputs.

Polar Graphs

The **polar graph** is another device for presenting multiple attributes to the decision maker. As illustrated in Figure 12.10, a circle is drawn with M radii representing M relevant attributes. Normally, but not necessarily, the radii (attributes) are equally spaced around the circle so that the angle between radii is equal to $360°/M$. Each radius is then appropriately scaled so that either the maximum or minimum possible attribute value is located on the

Figure 12.10. The Polar Graph

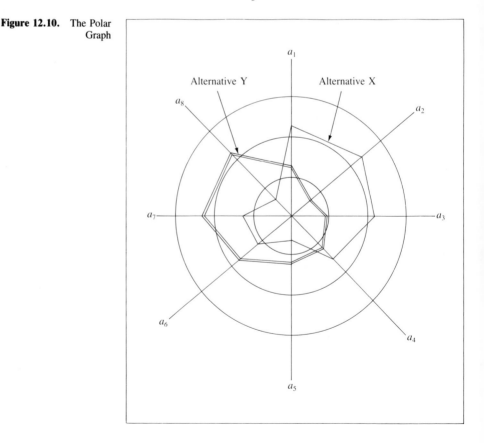

outermost ring. The convention selected for scaling must be consistent, of course, for all attributes. Next, for each alternative, the attribute value for that alternative are plotted on the radii and the points are then connected. The resulting figure represents the profile, or polar graph, for that alternative.

Constraints (minimum allowable values for each attribute) can be graphically portrayed as shown in Figure 12.11. The constraint profile is shown here as a dashed line, but if multicolor presentations are possible, it is useful to represent the constraint profile in a vivid color, say, red.

Dominant or dominated alternatives are also readily identified by referring to the graph. To illustrate, in Figure 12.11, it is assumed that the minimum, or least desirable, attribute value is represented by the outer ring; the most desirable position for all attributes is the center. Thus Alternative Y dominates Alternative X in the sense that Y is preferable to X across all attributes. Note, in this case, that Alternative X is infeasible inasmuch as the constraint for Attribute a_1 is violated by X.

An interesting variation of the polar graph technique is the use of color-coded (or shaded) "bands of acceptability." For exam-

Figure 12.11. Polar Graph with Constraints

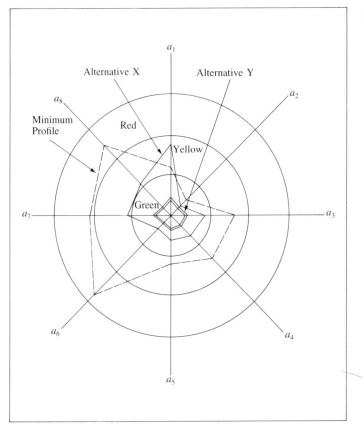

ple, a red band may be generated from the locus of all points on the radii that represent "poor" values. Similarly, a green band may represent "good" values and a yellow band may represent "acceptable" values. To illustrate, Figure 12.11 should be interpreted as follows:

Attribute	Alternative X	Alternative Y
a_1	poor	acceptable
a_2	poor	good
a_3	poor	good
a_4	acceptable	acceptable
a_5	good	acceptable
a_6	acceptable	acceptable
a_7	acceptable	poor
a_8	good	poor

The polar graph enjoys many of the advantages and suffers many of the disadvantages of the tabular scorecard techniques discussed previously. In addition, the form of the polar graph is such that its proper use may require substantial explanation. Decision makers who are unfamiliar with (or unaccustomed to) data presented in the polar graph format may find the procedure unintelligible. And, unlike tabular techniques, it is not possible to display large numbers of alternatives with a polar graph. Therefore, it is doubtful that a polar graph can readily be used to evaluate the profiles of more than four or five alternatives.

12.5 SUMMARY

Evaluation of the *economic* consequences of alternative plans, programs, and projects is only one part of the overall decision-making process. The central thrust of this book, of course, is the maximization of "worth" based on consequences (impacts) that can be evaluated in monetary terms. But this is only one dimension of the decision maker's problem. There will almost invariably be other consequences that cannot be reduced to monetary equivalence. These nonmonetary attributes, known as irreducibles, are of direct interest in analyses.

In a broad sense, decision problems in which two or more attributes are considered are categorized as cost-effectiveness problems. A narrower definition, one that is adopted in this chapter, restricts consideration to only two dimensions: (1) costs, the monetary consequences, and (2) effectiveness, a single measure of the nonmonetary quality of alternatives. (In some cases, two or more nonmonetary attributes can be combined into a single measure of effectiveness.) Although cost-effectiveness ratios are frequently used to demonstrate "bang for the buck," it is generally incorrect to rank-order alternatives on the basis of either decreasing E/C (or increasing C/E). Increments of costs and effectiveness must be compared.

Given multiple alternatives and multiple attributes, the decision problem may be simplified by examining the alternatives to see whether they meet feasibility constraints for each attribute. The list of candidate alternatives may be reduced further by eliminating the dominated alternatives, if there are any.

This chapter presents several techniques for identifying the optimal alternative from the set of feasible, nondominated alternatives. The first technique is simple lexicographic ordering, that is, ranking on the basis of the most important attribute. The second provides a model that combines the weighted scores, or utilities, of the impacts. There are a variety of scoring models: The simple multi-attribute rating technique (SMART) is presented here. The third approach discussed in this chapter is the holistic consideration of symbolic scorecards, in which all attributes and impacts are considered concurrently. Three types of scorecards are presented: colored, symbolic, and geometric.

The discussion of the problem of multiple attributes, monetary as well as irreducible, is not intended to be an exhaustive survey of this complex issue, and thorough treatment is well beyond the scope of this chapter. Moreover, experts do not agree on the efficacy of the various approaches. If you are interested in additional material about this topic, see the references given in the Bibliography.

12.1 The *equity* issue was raised early in this chapter. The example cited was an investment under consideration by Central City, located in Cow County, with the following prospective consequences:

	Population affected	Present worth $(B - C)$
Community subunit	1,000	$50,000
Rest of community	99,000	− 30,000

Is this project justifiable?

12.2 Suppose that you are instructed by your employer to find the maximum amount of coal you can purchase at the minimum total cost; that is, you must find the most cost-effective coal. Can this mission be accomplished? How would you restate this mission assignment? What was wrong with the original mission statement?

12.3 Suppose that you are the associate director of planning for the Regional Environmental Management Authority. You have awarded a contract to a consulting firm to synthesize and evaluate a number of alternative plans for meeting air-quality standards for your region. The consultant has determined that the appropriate measure of effectiveness for each plan is the percent reduction in the number of days per year in which the air-quality standard will be exceeded. This figure is determined for each potential plan by comparing the number of "excess days" for the plan with the number of "excess days" that can be expected if no control plan is implemented. The consultant has determined the initial cost for each plan as well as the present worth of all costs associated with each plan. These latter figures include initial costs:

PW of all costs = (initial costs)
+ (sum of all other costs appropriately discounted)

Finally, the consultant has defined the figure of merit for each plan to be the ratio of effectiveness to total (discounted) system costs. All data are summarized below for the existing system as well as the twelve alternative plans. All costs are in millions of dollars.

Plan	Effectiveness	Initial costs	Total cost	E:C ratio
A (existing)	0	0	0	——
B	0.12	4	5	2.40
C	0.20	4	15	1.33
D	0.30	4	20	1.50
E	0.40	5	15	2.67
F	0.40	5	30	1.33
G	0.50	6	25	2.00
H	0.50	6	40	1.25
J	0.60	7	35	1.71
K	0.70	8	35	2.00
L	0.80	8	40	2.00
M	0.85	9	45	1.89
N	0.90	10	50	1.80

Any acceptable plan must provide an improvement of at least a 33 percent reduction in the number of days in which the air-quality standard is currently exceeded. Moreover, the initial cost of any acceptable plan must not exceed $8,000,000.

a. What are the units, or dimensions, of the figure of merit in this example?
b. Which, if any, alternatives are infeasible because they do not meet the minimum effectiveness requirements?
c. Which, if any, alternatives are infeasible because they do not meet the maximum cost requirements?
d. Consider the feasible alternatives in view of the above two constraints. Which of these, if any, are dominated with reference to the measures of effectiveness and total cost? Include all cases of dominance.
e. Which, if any, alternatives would you recommend for selection? Why? (Omit consideration of any factors or information other than what is given.)

12.4 As an engineer for a medium-sized manufacturing firm, you are asked to evaluate the introduction of a robotic system into an assembly line. What scales of effectiveness, or attributes, would you establish to evaluate the current system (the do-nothing alternative) with the proposed robotics alternative? Describe what kinds of scales they are and whether objective measures can be obtained for each. Are there any subjective measurements involved? How would you collect the data? Are there any subjective scales involved? How would you collect the data on them? How many decision makers would be involved in this proposal and how can they be identified?

12.5 Invent an example for which the use of C/E or E/C ratios *are* appropriate for ranking alternatives. Illustrate with a graph. Why is the use of the ratio appropriate in this instance?

12.6 Using the automobile example presented in the text, convert the analysis to a color-coded scorecard analysis using three colors, red, yellow, and green. (See Table 12.4.) Compare and contrast the two approaches in terms of their strengths and weaknesses from the point of view of (a) an analyst and (b) a decision maker.

12.7 Show, by changing data in the automobile example in the text, that you understand how the results of an evaluation can be altered by shifts in the data used in the evaluation process. Specifically, determine what changes in the original data are necessary to shift the decision from the Ford Pinto to the Cadillac El Dorado.

12.8 Conduct a SMART evaluation for the purchase of a personal computer from the point of view of the decision maker. Show all attributes, candidate computers, importance weights, and value curves used in the analysis. Show the final results of the analysis by ranking at least five personal computers currently on the market.

12.9 The state highway department is considering five mutually exclusive projects for improving motor vehicle safety. Effectiveness is defined as the percent reduction in accident rate per 100 million vehicle-miles. Costs and effectiveness for the alternative projects are

Plan	Cost	Effectiveness
A	$2,000,000	4
B	3,000,000	15
C	5,000,000	10
D	6,000,000	18
E	8,000,000	20

Because of budget constraints, the maximum allowable cost is $7,000,000. The minimum acceptable level of effectiveness is an 8 percent reduction in the accident rate. Determine which plans are infeasible and/or dominated. Of the remaining plans, which is optimum? Explain your reasoning.

12.10 A manufacturing firm wants to locate a satellite distribution center somewhere on the West Coast. Four alternative sites are under consideration. Preliminary analysis indicates the following impacts, with importance ratings (weights) for each of the four attributes:

Attribute	Weight	Site I	Site II	Site III	Site IV
1. *PW* of costs of construction and operation	0.50	$900,000	$800,000	$950,000	$700,000
2. Availability of skilled labor	0.25	good	fair	excellent	poor
3. Distance from central warehouse	0.15	800 mi.	1,200 mi.	600 mi.	1,400 mi.
4. Community services and attitude (based on 10-point scale)	0.10	9	8	9	7

Using the SMART method described in the chapter, complete the evaluation. Make all necessary assumptions, but state them explicitly.

SUMMARY OF NOTATION

Included here are symbols and key abbreviations used throughout the book. Not included, however, are the notation used in Chapter 9 (Revenue Requirement Method) and Chapter 10 (Inflation). The notation specific to those chapters may be found in Appendix 9A and Appendix 10B.

A	Cash flow, or equivalent cash flow, occurring uniformly at the *end* of every period for a specified number of periods.
A_j	Cash flow at *end* of period j.
$\overline{A}$	Amount of money (or equivalent value) flowing continuously and uniformly *during* each period for a specified number of periods.
$\overline{A}_j$	Cash flow occurring continuously and uniformly *during* the jth period.
a	Depreciation rate used with the declining balance method. Fraction of total earnings retained within the firm each period (Chapter 11).
a_i	Alternative i.
AC	(Equivalent uniform) annual cost.
$AC(C, N)$	(Equivalent uniform) annual cost for challenger if kept in service N periods.
$AC(D, N)$	(Equivalent uniform) annual cost for defender if kept in service N periods.
$ACRS$	Accelerated Cost Recovery System.
AIC	Allowable investment credit.
$\overline{AP}$	Average annual accounting profit.
AW	(Equivalent uniform) annual worth.
B	(Incremental) benefits. Unadjusted basis (when computing ACRS depreciation).
B_j	Book value after j years of depreciation. (Includes the depreciation expense for the jth year.)
B_0	Original book value of depreciable asset.
B_N	Book value of depreciable asset at end of period N.
b	Rate of return on book value for retained earnings.
$B:C$	Benefit-cost ratio.

C	(Incremental) costs.
C_j	Loan repayment at end of period j. (C_0 is cost associated with underwriting.)
C_{ij}	Cost associated with the ith alternative and the jth state of nature.
CAM	Challenger's adverse minimum.
CC	Capitalized cost: the equivalent present value of an infinite series of cash flows.
CDF	Cumulative distribution function.
C/E	Ratio of system cost to system effectiveness.
CN^*	Challenger's economic life.
CR	Capital recovery.
$CR(C, j)$	Cost of capital recovery for challenger if kept j periods.
D	Annual depreciation expense. Outstanding debt (Chapter 11).
d	Imaginary uniform deposit into an imaginary sinking fund.
D_c	Cost depletion.
D_j	Depreciation expense for year j. (No subscript if constant for all j).
D_j^*	Debt at start of year j, reflecting inflation.
D_p	Percentage depletion.
D_s	Uniform annual ''deposit'' into imaginary sinking fund.
DDB	Double declining balance.
DN	Defender's current age.
E	(Accounting) expenses. Annual earnings available to stockholders (Chapter 11).
$\hat{E}$	Expected annual earnings after income taxes but before return to lenders.
E_j^*	Inflated earnings at end of year j.
e	Base of the Naperian logarithm system, the ''exponential,'' approximately equal to 2.71828.
e_j	Earnings per share in period j.

E/C	Ratio of system effectiveness to system cost. (Sometimes known as "bang for the buck.")
$EUAC$	Equivalent uniform annual cost.
ERR	External rate of return.
ERTA	Economic Recovery Tax Act of 1981.
$E[X]$	Expected value of random variable X.
F	Amount of cash flow at end of Nth period. Equivalent future value (measured at end of Nth period) of prior cash flows.
$\overline{F}$	Amount of money (or equivalent value) flowing continuously and uniformly during the Nth period.
f	General rate of inflation: uniform rate of increase/decrease of prices.
FW	Future worth.
$f(x)$	Probability density function for continuous random variable.
G	Arithmetic gradient: amount of cash flow increase/decrease from period to period. Inferiority gradient (Appendix 6A).
g	Geometric gradient: rate of cash flow increase/decrease from period to period.
H_j^*	Inflated dividends paid at end of year j.
h_j	Dividends per share in period j.
H_p	Annual dividends on preferred stock.
$H(a_i)$	Hurwicz value for alternative a_i.
I	Interest paid on debt. Annual interest payment (uniform over each year).
I_j	Loan interest accumulated in period j.
I_j^*	Inflated after-tax interest expense in year j.
$\overline{I}$	Average interest paid over N payments.
$I(\cdot)$	Index of cost-effectiveness, C/E.
i	Effective interest rate per interest period.
i_a	Effective interest rate per year (per *annum*).
i_m	Effective interest rate per subperiod.
i_s	Rate of interest "earned" by imaginary sinking fund.

i^*	Internal rate of return.
i_e^*	External rate of return.
i	After-tax (internal) rate of return.
IRR	(Internal) rate of return. Sometimes written as *RoR* for "rate of return."
j	Index, generally used to denote interest period, for example, A_j.
k	Auxiliary interest rate used when computing the external rate of return. The minimum attractive rate of return. Index for replacement (Chapter 6). Index for ACRS property class (Chapter 7).
k^*	Cost of capital, reflecting inflation.
k_b	Prefax cost of bonds or notes.
k_c	Cost of capital of class c.
k_d	Cost of capital—debt.
k_d^*	After-tax cost of debt capital, reflecting effects of inflation.
k_e	Cost of capital—equity. The equity capitalization rate.
k_{en}	Cost of capital—equity (new stockholders).
k_{eo}	Cost of capital—equity (existing stockholders).
k_l	Cost of capital—lease.
k_o	Weighted average cost of capital. (Usually written without subscript.)
k_p	Cost of capital—preferred stock.
k_r	Cost of capital—retained earnings.
L_j	Lease payment at end of period j.
M	Number of compounding subperiods per period (each of which is assumed to be of equal length). Market value of the firm's equity.
M_j^*	Inflated market value of equity at end of year j.
m_j	Number of mutually exclusive proposals within the jth set of prospective investments. (See Chapter 4.)
MARR	Minimum attractive rate of return.
N	Number of compounding periods (each of which is assumed to be of equal length): the length of the planning horizon (study period). Life of investment.

N_{max}	Maximum physical life.
N^*	Number of periods for "payback" of original investment (Chapter 5). Economic life (Chapter 6).
n	Number of mutually exclusive budget packages that can be obtained from a set of independent investment proposals (Chapter 4). Length of finite growth period.
$n(k)$	(ACRS) recovery period for property class k.
$NB:C$	Net benefit-cost ratio.
$P.$	Initial investment. Equivalent present value of future cash flow(s). Loan principal.
$\overline{P}$	Amount of money (or equivalent value) flowing continuously and uniformly during the first period of the planning horizon.
P_j	Amount of loan principal unpaid at the start of period j.
p	Percent deviation from estimate (Chapter 8). Current price per share of common stock (Chapter 11).
p_1	Present value of dividend stream during the growth period.
p_2	Present value of stock price at the end of n periods.
$p_j(k)$	Applicable (ACRS) percentage in year j for property class k.
$p(x)$	Probability mass function for discrete random variable.
$P(E)$	Discounted present value of earnings over the life of the project.
$P(I)$	Present value of initial investment.
$P(S)$	Present value of terminal salvage value.
$P[E]$	Probability of the event E. Sometimes written $Pr[E]$ or Prob $[E]$.
$P[E \cup F]$	Probability of the union of events E and F.
$P[EF]$	Probability of the intersection of events E and F.
$P[S]$	Probability of the sum of all possible outcomes in the sample space.
PI	Profitability index.
PW	(Equivalent) present worth, or (net) present value, of present and future cash flows.

PWOC	Present worth of costs.
PWP	Premium worth percentage.
Q	Amount of loan.
R	(Accounting) revenue.
r	Nominal interest rate per period; usually, the nominal interest rate per year. Rate of return under the "truth in lending formula" (Chapter 5); widely known as the *annual percentage rate*.
R_j^*	Increase in equity in year j reflecting effects of inflation.
R_{ij}	Revenue or profit associated with the ith alternative and the jth state of nature.
RND	Random normal deviate.
RoR	Rate of return.
S	Net salvage value of capital investment.
s	Number of independent investment proposals under consideration (Chapter 4).
s_j	Outcome, or state of nature, j.
s_{ij}	Unweighted score for the ith attribute and the jth alternative.
SIR	Savings-investment ratio.
SYD	Sum of the years digits.
T	Taxable income (Chapter 7).
t	Effective income tax rate.
t_f	Effective federal income tax rate.
t_s	Effective state income tax rate.
t_1	Investor's income tax rate.
t_2	Investor's tax rate on capital gains.
TC	(Investment) tax credit.
U_e	Number of units of output expected over the depreciable life of the asset.
U_j	Number of units of actual output during period j.
V	Total market value of the firm. Par value of bonds.
V_j	Total weighted score for alternative j.

$\text{Var}[x]$	Variance of x.
w_c	Proportion of capital of class c.
w_d	Proportion of total available capital that is debt.
w_e	Proportion of total available capital that is equity.
w_i	Importance weight of the ith attribute, or dimension.
X	Random variable.
x	$(1 + i)^{-1}$ in Chapter 3. In Chapter 8, a number assumed by random variable X.
x_0	Number of shares before new stock issue.
x_1	Number of new shares sold to existing stockholders.
x_2	Number of new shares sold to new stockholders.
x_{ij}	Impact of the jth alternative as measured on the ith attribute.
$\emptyset$	The do-nothing alternative.
α	Index of optimism.
Δ	Incremental.
λ	Leverage, the ratio of debt to the total value of the firm. (Alternatively, the ratio of debt to equity.)
μ	Mean, or expected value.
σ	Standard deviation. ($\sigma^2 = $ variance.)

Compound Interest Factors $i = 0.01$

	SINGLE PAYMENT			UNIFORM SERIES			
	COMPOUND AMOUNT	PRESENT WORTH		COMPOUND AMOUNT		PRESENT WORTH	
N	F/P	P/F	P/$\overline{F}$	F/A	F/$\overline{A}$	P/A	P/$\overline{A}$
1	1.010	0.9901	0.9950	1.000	1.005	0.990	0.995
2	1.020	.9803	.9852	2.010	2.020	1.970	1.980
3	1.030	.9706	.9754	3.030	3.045	2.941	2.956
4	1.041	.9610	.9658	4.060	4.081	3.902	3.921
5	1.051	.9515	.9562	5.101	5.126	4.853	4.878
6	1.062	.9420	.9467	6.152	6.183	5.795	5.824
7	1.072	.9327	.9374	7.214	7.250	6.728	6.762
8	1.083	.9235	.9281	8.286	8.327	7.652	7.690
9	1.094	.9143	.9189	9.369	9.415	8.566	8.609
10	1.105	.9053	.9098	10.462	10.514	9.471	9.519
11	1.116	.8963	.9008	11.567	11.625	10.368	10.419
12	1.127	.8874	.8919	12.682	12.746	11.255	11.311
13	1.138	.8787	.8830	13.809	13.878	12.134	12.194
14	1.149	.8700	.8743	14.947	15.022	13.004	13.069
15	1.161	.8613	.8656	16.097	16.177	13.865	13.934
16	1.173	.8528	.8571	17.258	17.344	14.718	14.791
17	1.184	.8444	.8486	18.430	18.522	15.562	15.640
18	1.196	.8360	.8402	19.615	19.713	16.398	16.480
19	1.208	.8277	.8319	20.811	20.915	17.226	17.312
20	1.220	.8195	.8236	22.019	22.129	18.046	18.136
21	1.232	.8114	.8155	23.239	23.355	18.857	18.951
22	1.245	.8034	.8074	24.472	24.594	19.660	19.759
23	1.257	.7954	.7994	25.716	25.845	20.456	20.558
24	1.270	.7876	.7915	26.973	27.108	21.243	21.349
25	1.282	.7798	.7837	28.243	28.384	22.023	22.133
26	1.295	.7720	.7759	29.526	29.673	22.795	22.909
27	1.308	.7644	.7682	30.821	30.975	23.560	23.677
28	1.321	.7568	.7606	32.129	32.289	24.316	24.438
29	1.335	.7493	.7531	33.450	33.617	25.066	25.191
30	1.348	.7419	.7456	34.785	34.959	25.808	25.937
31	1.361	.7346	.7382	36.133	36.313	26.542	26.675
32	1.375	.7273	.7309	37.494	37.681	27.270	27.406
33	1.389	.7201	.7237	38.869	39.063	27.990	28.129
34	1.403	.7130	.7165	40.258	40.459	28.703	28.846
35	1.417	.7059	.7094	41.660	41.868	29.409	29.555
40	1.489	.6717	.6750	48.886	49.130	32.835	32.999
45	1.565	.6391	.6422	56.481	56.763	36.095	36.275
50	1.645	.6080	.6111	64.463	64.785	39.196	39.392
55	1.729	.5785	.5814	72.852	73.216	42.147	42.358
60	1.817	.5504	.5532	81.670	82.077	44.955	45.179
65	1.909	.5237	.5263	90.937	91.391	47.627	47.864
70	2.007	.4983	.5008	100.676	101.179	50.168	50.419
80	2.217	.4511	.4534	121.671	122.279	54.888	55.162
90	2.449	.4084	.4104	144.863	145.586	59.161	59.456
100	2.705	.3697	.3716	170.481	171.332	63.029	63.344

UNIFORM SERIES		GRADIENT SERIES		
SINKING FUND	**CAPITAL RECOVERY**	**UNIFORM SERIES**	**PRESENT WORTH**	
A/F	*A/P*	*A/G*	*P/G*	*N*
1.0000	1.0100	0.000	0.000	1
0.4975	0.5075	0.497	0.980	2
.3300	.3400	0.993	2.921	3
.2463	.2563	1.488	5.804	4
.1960	.2060	1.980	9.610	5
.1625	.1725	2.471	14.320	6
.1386	.1486	2.960	19.917	7
.1207	.1307	3.448	26.381	8
.1067	.1167	3.934	33.696	9
.0956	.1056	4.418	41.843	10
.0865	.0965	4.900	50.806	11
.0788	.0888	5.381	60.568	12
.0724	.0824	5.861	71.112	13
.0669	.0769	6.338	82.422	14
.0621	.0721	6.814	94.481	15
.0579	.0679	7.289	107.273	16
.0543	.0643	7.761	120.783	17
.0510	.0610	8.232	134.995	18
.0481	.0581	8.702	149.895	19
.0454	.0554	9.169	165.465	20
.0430	.0530	9.635	181.694	21
.0409	.0509	10.100	198.565	22
.0389	.0489	10.563	216.065	23
.0371	.0471	11.024	234.179	24
.0354	.0454	11.483	252.894	25
.0339	.0439	11.941	272.195	26
.0324	.0424	12.397	292.069	27
.0311	.0411	12.852	312.504	28
.0299	.0399	13.304	333.485	29
.0287	.0387	13.756	355.001	30
.0277	.0377	14.205	377.039	31
.0267	.0367	14.653	399.585	32
.0257	.0357	15.099	422.628	33
.0248	.0348	15.544	446.156	34
.0240	.0340	15.987	470.157	35
.0205	.0305	18.178	596.854	40
.0177	.0277	20.327	733.702	45
.0155	.0255	22.436	879.417	50
.0137	.0237	24.505	1032.813	55
.0122	.0222	26.533	1192.804	60
.0110	.0210	28.522	1358.388	65
.0099	.0199	30.470	1528.645	70
.0082	.0182	34.249	1879.875	80
.0069	.0169	37.872	2240.565	90
.0059	.0159	41.343	2605.774	100

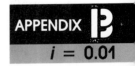

APPENDIX **B**

i = **0.01**

Compound Interest Factors $i = 0.02$

	SINGLE PAYMENT			UNIFORM SERIES			
	COMPOUND AMOUNT	PRESENT WORTH		COMPOUND AMOUNT		PRESENT WORTH	
N	F/P	P/F	$P/\overline{F}$	F/A	$F/\overline{A}$	P/A	$P/\overline{A}$
1	1.020	0.9804	0.9902	1.000	1.010	0.980	0.990
2	1.040	.9612	.9707	2.020	2.040	1.942	1.961
3	1.061	.9423	.9517	3.060	3.091	2.884	2.913
4	1.082	.9238	.9331	4.122	4.163	3.808	3.846
5	1.104	.9057	.9148	5.204	5.256	4.713	4.760
6	1.126	.8880	.8968	6.308	6.371	5.601	5.657
7	1.149	.8706	.8792	7.434	7.508	6.472	6.536
8	1.172	.8535	.8620	8.583	8.668	7.325	7.398
9	1.195	.8368	.8451	9.755	9.852	8.162	8.244
10	1.219	.8203	.8285	10.950	11.059	8.983	9.072
11	1.243	.8043	.8123	12.169	12.290	9.787	9.884
12	1.268	.7885	.7964	13.412	13.546	10.575	10.681
13	1.294	.7730	.7807	14.680	14.827	11.348	11.461
14	1.319	.7579	.7654	15.974	16.133	12.106	12.227
15	1.346	.7430	.7504	17.293	17.466	12.849	12.977
16	1.373	.7284	.7357	18.639	18.825	13.578	13.713
17	1.400	.7142	.7213	20.012	20.211	14.292	14.434
18	1.428	.7002	.7071	21.412	21.626	14.992	15.141
19	1.457	.6864	.6933	22.840	23.068	15.678	15.835
20	1.486	.6730	.6797	24.297	24.539	16.351	16.514
21	1.516	.6598	.6664	25.783	26.040	17.011	17.181
22	1.546	.6468	.6533	27.299	27.571	17.658	17.834
23	1.577	.6342	.6405	28.845	29.132	18.292	18.475
24	1.608	.6217	.6279	30.422	30.725	18.914	19.102
25	1.641	.6095	.6156	32.030	32.349	19.523	19.718
26	1.673	.5976	.6035	33.671	34.006	20.121	20.322
27	1.707	.5859	.5917	35.344	35.697	20.707	20.913
28	1.741	.5744	.5801	37.051	37.420	21.281	21.493
29	1.776	.5631	.5687	38.792	39.179	21.844	22.062
30	1.811	.5521	.5576	40.568	40.972	22.396	22.620
31	1.848	.5412	.5466	42.379	42.802	22.938	23.166
32	1.885	.5306	.5359	44.227	44.668	23.468	23.702
33	1.922	.5202	.5254	46.111	46.571	23.989	24.228
34	1.961	.5100	.5151	48.034	48.512	24.499	24.743
35	2.000	.5000	.5050	49.994	50.493	24.999	25.248
40	2.208	.4529	.4574	60.402	61.004	27.355	27.628
45	2.438	.4102	.4143	71.892	72.609	29.490	29.784
50	2.692	.3715	.3752	84.579	85.422	31.424	31.737
55	2.972	.3365	.3399	98.586	99.569	33.175	33.505
60	3.281	.3048	.3078	114.051	115.188	34.761	35.107
65	3.623	.2761	.2788	131.126	132.433	36.197	36.558
70	4.000	.2500	.2525	149.977	151.472	37.499	37.872
80	4.875	.2051	.2072	193.771	195.702	39.744	40.141
90	5.943	.1683	.1699	247.155	249.619	41.587	42.001
100	7.245	.1380	.1394	312.230	315.343	43.098	43.528

UNIFORM SERIES		GRADIENT SERIES		
SINKING FUND	CAPITAL RECOVERY	UNIFORM SERIES	PRESENT WORTH	
A/F	A/P	A/G	P/G	N
1.0000	1.0200	0.000	0.000	1
0.4951	0.5151	0.495	0.961	2
.3268	.3468	0.987	2.846	3
.2426	.2626	1.475	5.617	4
.1922	.2122	1.960	9.240	5
.1585	.1785	2.442	13.679	6
.1345	.1545	2.921	18.903	7
.1165	.1365	3.396	24.877	8
.1025	.1225	3.868	31.571	9
.0913	.1113	4.337	38.954	10
.0822	.1022	4.802	46.996	11
.0746	.0946	5.264	55.670	12
.0681	.0881	5.723	64.946	13
.0626	.0826	6.178	74.798	14
.0578	.0778	6.631	85.200	15
.0537	.0737	7.080	96.127	16
.0500	.0700	7.526	107.553	17
.0467	.0667	7.968	119.456	18
.0438	.0638	8.407	131.812	19
.0412	.0612	8.843	144.598	20
.0388	.0588	9.276	157.793	21
.0366	.0566	9.705	171.377	22
.0347	.0547	10.132	185.328	23
.0329	.0529	10.555	199.628	24
.0312	.0512	10.974	214.256	25
.0297	.0497	11.391	229.196	26
.0283	.0483	11.804	244.428	27
.0270	.0470	12.214	259.936	28
.0258	.0458	12.621	275.703	29
.0247	.0447	13.025	291.713	30
.0236	.0436	13.426	307.950	31
.0226	.0426	13.823	324.400	32
.0217	.0417	14.217	341.047	33
.0208	.0408	14.608	357.878	34
.0200	.0400	14.996	374.879	35
.0166	.0366	16.888	461.989	40
.0139	.0339	18.703	551.561	45
.0118	.0318	20.442	642.355	50
.0101	.0301	22.106	733.348	55
.0088	.0288	23.696	823.692	60
.0076	.0276	25.215	912.703	65
.0067	.0267	26.663	999.829	70
.0052	.0252	29.357	1166.781	80
.0040	.0240	31.793	1322.164	90
.0032	.0232	33.986	1464.747	100

i = 0.02

Compound Interest Factors $i = 0.03$

	SINGLE PAYMENT			UNIFORM SERIES			
	COMPOUND AMOUNT	PRESENT WORTH		COMPOUND AMOUNT		PRESENT WORTH	
N	F/P	P/F	P/$\overline{F}$	F/A	F/$\overline{A}$	P/A	P/$\overline{A}$
1	1.030	0.9709	0.9854	1.000	1.015	0.971	0.985
2	1.061	.9426	.9567	2.030	2.060	1.913	1.942
3	1.093	.9151	.9288	3.091	3.137	2.829	2.871
4	1.126	.8885	.9017	4.184	4.246	3.717	3.773
5	1.159	.8626	.8755	5.309	5.388	4.580	4.648
6	1.194	.8375	.8500	6.468	6.565	5.417	5.498
7	1.230	.8131	.8252	7.662	7.777	6.230	6.323
8	1.267	.7894	.8012	8.892	9.025	7.020	7.124
9	1.305	.7664	.7779	10.159	10.311	7.786	7.902
10	1.344	.7441	.7552	11.464	11.635	8.530	8.658
11	1.384	.7224	.7332	12.808	12.999	9.253	9.391
12	1.426	.7014	.7118	14.192	14.404	9.954	10.103
13	1.469	.6810	.6911	15.618	15.851	10.635	10.794
14	1.513	.6611	.6710	17.086	17.341	11.296	11.465
15	1.558	.6419	.6514	18.599	18.877	11.938	12.116
16	1.605	.6232	.6325	20.157	20.458	12.561	12.749
17	1.653	.6050	.6140	21.762	22.086	13.166	13.363
18	1.702	.5874	.5962	23.414	23.764	13.754	13.959
19	1.754	.5703	.5788	25.117	25.492	14.324	14.538
20	1.806	.5537	.5619	26.870	27.271	14.877	15.100
21	1.860	.5375	.5456	28.676	29.105	15.415	15.645
22	1.916	.5219	.5297	30.537	30.993	15.937	16.175
23	1.974	.5067	.5143	32.453	32.937	16.444	16.689
24	2.033	.4919	.4993	34.426	34.940	16.936	17.188
25	2.094	.4776	.4847	36.459	37.003	17.413	17.673
26	2.157	.4637	.4706	38.553	39.129	17.877	18.144
27	2.221	.4502	.4569	40.710	41.317	18.327	18.601
28	2.288	.4371	.4436	42.931	43.572	18.764	19.044
29	2.357	.4243	.4307	45.219	45.894	19.188	19.475
30	2.427	.4120	.4181	47.575	48.286	19.600	19.893
31	2.500	.4000	.4060	50.003	50.749	20.000	20.299
32	2.575	.3883	.3941	52.503	53.286	20.389	20.693
33	2.652	.3770	.3827	55.078	55.900	20.766	21.076
34	2.732	.3660	.3715	57.730	58.592	21.132	21.447
35	2.814	.3554	.3607	60.462	61.365	21.487	21.808
40	3.262	.3066	.3111	75.401	76.527	23.115	23.460
45	3.782	.2644	.2684	92.720	94.104	24.519	24.885
50	4.384	.2281	.2315	112.797	114.481	25.730	26.114
55	5.082	.1968	.1997	136.072	138.103	26.774	27.174
60	5.892	.1697	.1723	163.053	165.487	27.676	28.089
65	6.830	.1464	.1486	194.333	197.233	28.453	28.878
70	7.918	.1263	.1282	230.594	234.036	29.123	29.558
80	10.641	.0940	.0954	321.363	326.160	30.201	30.652
90	14.300	.0699	.0710	443.349	449.967	31.002	31.465
100	19.219	.0520	.0528	607.287	616.352	31.599	32.071

UNIFORM SERIES		GRADIENT SERIES		
SINKING FUND	CAPITAL RECOVERY	UNIFORM SERIES	PRESENT WORTH	
A/F	A/P	A/G	P/G	N
1.0000	1.0300	0.000	0.000	1
0.4926	0.5226	0.493	0.943	2
.3235	.3535	0.980	2.773	3
.2390	.2690	1.463	5.438	4
.1884	.2184	1.941	8.889	5
.1546	.1846	2.414	13.076	6
.1305	.1605	2.882	17.955	7
.1125	.1425	3.345	23.481	8
.0984	.1284	3.803	29.612	9
.0872	.1172	4.256	36.309	10
.0781	.1081	4.705	43.533	11
.0705	.1005	5.148	51.248	12
.0640	.0940	5.587	59.419	13
.0585	.0885	6.021	68.014	14
.0538	.0838	6.450	77.000	15
.0496	.0796	6.874	86.348	16
.0460	.0760	7.294	96.028	17
.0427	.0727	7.708	106.014	18
.0398	.0698	8.118	116.279	19
.0372	.0672	8.523	126.799	20
.0349	.0649	8.923	137.549	21
.0327	.0627	9.319	148.509	22
.0308	.0608	9.709	159.657	23
.0290	.0590	10.095	170.971	24
.0274	.0574	10.477	182.434	25
.0259	.0559	10.853	194.026	26
.0246	.0546	11.226	205.731	27
.0233	.0533	11.593	217.532	28
.0221	.0521	11.956	229.414	29
.0210	.0510	12.314	241.361	30
.0200	.0500	12.668	253.361	31
.0190	.0490	13.017	265.399	32
.0182	.0482	13.362	277.464	33
.0173	.0473	13.702	289.544	34
.0165	.0465	14.037	301.627	35
.0133	.0433	15.650	361.750	40
.0108	.0408	17.156	420.632	45
.0089	.0389	18.558	477.480	50
.0073	.0373	19.860	531.741	55
.0061	.0361	21.067	583.053	60
.0051	.0351	22.184	631.201	65
.0043	.0343	23.215	676.087	70
.0031	.0331	25.035	756.086	80
.0023	.0323	26.567	823.630	90
.0016	.0316	27.844	879.854	100

i = 0.03

Compound Interest Factors $i = 0.04$

	SINGLE PAYMENT			UNIFORM SERIES			
	COMPOUND AMOUNT	PRESENT WORTH		COMPOUND AMOUNT		PRESENT WORTH	
N	F/P	P/F	P/$\overline{F}$	F/A	F/$\overline{A}$	P/A	P/$\overline{A}$
1	1.040	0.9615	0.9806	1.000	1.020	0.962	0.981
2	1.082	.9246	.9429	2.040	2.081	1.886	1.924
3	1.125	.8890	.9067	3.122	3.184	2.775	2.830
4	1.170	.8548	.8718	4.246	4.331	3.630	3.702
5	1.217	.8219	.8383	5.416	5.524	4.452	4.540
6	1.265	.7903	.8060	6.633	6.765	5.242	5.346
7	1.316	.7599	.7750	7.898	8.055	6.002	6.121
8	1.369	.7307	.7452	9.214	9.397	6.733	6.867
9	1.423	.7026	.7165	10.583	10.793	7.435	7.583
10	1.480	.6756	.6890	12.006	12.245	8.111	8.272
11	1.539	.6496	.6625	13.486	13.754	8.760	8.935
12	1.601	.6246	.6370	15.026	15.324	9.385	9.572
13	1.665	.6006	.6125	16.627	16.957	9.986	10.184
14	1.732	.5775	.5889	18.292	18.655	10.563	10.773
15	1.801	.5553	.5663	20.024	20.421	11.118	11.339
16	1.873	.5339	.5445	21.825	22.258	11.652	11.884
17	1.948	.5134	.5236	23.697	24.168	12.166	12.407
18	2.026	.4936	.5034	25.645	26.155	12.659	12.911
19	2.107	.4746	.4841	27.671	28.221	13.134	13.395
20	2.191	.4564	.4655	29.778	30.370	13.590	13.860
21	2.279	.4388	.4476	31.969	32.604	14.029	14.308
22	2.370	.4220	.4303	34.248	34.928	14.451	14.738
23	2.465	.4057	.4138	36.618	37.345	14.857	15.152
24	2.563	.3901	.3979	39.083	39.859	15.247	15.550
25	2.666	.3751	.3826	41.646	42.473	15.622	15.932
26	2.772	.3607	.3679	44.312	45.192	15.983	16.300
27	2.883	.3468	.3537	47.084	48.020	16.330	16.654
28	2.999	.3335	.3401	49.968	50.960	16.663	16.994
29	3.119	.3207	.3270	52.966	54.019	16.984	17.321
30	3.243	.3083	.3144	56.085	57.199	17.292	17.636
31	3.373	.2965	.3024	59.328	60.507	17.588	17.938
32	3.508	.2851	.2907	62.701	63.947	17.874	18.229
33	3.648	.2741	.2795	66.209	67.525	18.148	18.508
34	3.794	.2636	.2688	69.858	71.246	18.411	18.777
35	3.946	.2534	.2585	73.652	75.116	18.665	19.035
40	4.801	.2083	.2124	95.025	96.914	19.793	20.186
45	5.841	.1712	.1746	121.029	123.434	20.720	21.132
50	7.107	.1407	.1435	152.667	155.700	21.482	21.909
55	8.646	.1157	.1180	191.159	194.957	22.109	22.548
60	10.520	.0951	.0969	237.990	242.719	22.623	23.073
65	12.799	.0781	.0797	294.968	300.829	23.047	23.505
70	15.572	.0642	.0655	364.290	371.528	23.395	23.859
80	23.050	.0434	.0442	551.244	562.197	23.915	24.391
90	34.119	.0293	.0299	827.981	844.434	24.267	24.749
100	50.505	.0198	.0202	1237.621	1262.212	24.505	24.992

UNIFORM SERIES		GRADIENT SERIES		
SINKING FUND	CAPITAL RECOVERY	UNIFORM SERIES	PRESENT WORTH	
A/F	A/P	A/G	P/G	N
1.0000	1.0400	0.000	0.000	1
0.4902	0.5302	0.490	0.925	2
.3203	.3603	0.974	2.702	3
.2355	.2755	1.451	5.267	4
.1846	.2246	1.922	8.555	5
.1508	.1908	2.386	12.506	6
.1266	.1666	2.843	17.066	7
.1085	.1485	3.294	22.180	8
.0945	.1345	3.739	27.801	9
.0833	.1233	4.177	33.881	10
.0741	.1141	4.609	40.377	11
.0666	.1066	5.034	47.248	12
.0601	.1001	5.453	54.454	13
.0547	.0947	5.866	61.962	14
.0499	.0899	6.272	69.735	15
.0458	.0858	6.672	77.744	16
.0422	.0822	7.066	85.958	17
.0390	.0790	7.453	94.349	18
.0361	.0761	7.834	102.893	19
.0336	.0736	8.209	111.564	20
.0313	.0713	8.578	120.341	21
.0292	.0692	8.941	129.202	22
.0273	.0673	9.297	138.128	23
.0256	.0656	9.648	147.101	24
.0240	.0640	9.993	156.104	25
.0226	.0626	10.331	165.121	26
.0212	.0612	10.664	174.138	27
.0200	.0600	10.991	183.142	28
.0189	.0589	11.312	192.120	29
.0178	.0578	11.627	201.062	30
.0169	.0569	11.937	209.955	31
.0159	.0559	12.241	218.792	32
.0151	.0551	12.540	227.563	33
.0143	.0543	12.832	236.260	34
.0136	.0536	13.120	244.876	35
.0105	.0505	14.476	286.530	40
.0083	.0483	15.705	325.402	45
.0066	.0466	16.812	361.164	50
.0052	.0452	17.807	393.689	55
.0042	.0442	18.697	422.996	60
.0034	.0434	19.491	449.201	65
.0027	.0427	20.196	472.479	70
.0018	.0418	21.372	511.116	80
.0012	.0412	22.283	540.737	90
.0008	.0408	22.980	563.125	100

i = 0.04

Compound Interest Factors $i = 0.05$

	SINGLE PAYMENT			UNIFORM SERIES			
	COMPOUND AMOUNT	PRESENT WORTH		COMPOUND AMOUNT		PRESENT WORTH	
N	F/P	P/F	$P/\bar{F}$	F/A	$F/\bar{A}$	P/A	$P/\bar{A}$
1	1.050	0.9524	0.9760	1.000	1.025	0.952	0.976
2	1.102	.9070	.9295	2.050	2.101	1.859	1.906
3	1.158	.8638	.8853	3.152	3.231	2.723	2.791
4	1.216	.8227	.8431	4.310	4.417	3.546	3.634
5	1.276	.7835	.8030	5.526	5.663	4.329	4.437
6	1.340	.7462	.7647	6.802	6.971	5.076	5.202
7	1.407	.7107	.7283	8.142	8.344	5.786	5.930
8	1.477	.6768	.6936	9.549	9.786	6.463	6.623
9	1.551	.6446	.6606	11.027	11.300	7.108	7.284
10	1.629	.6139	.6291	12.578	12.890	7.722	7.913
11	1.710	.5847	.5992	14.207	14.559	8.306	8.512
12	1.796	.5568	.5706	15.917	16.312	8.863	9.083
13	1.886	.5303	.5435	17.713	18.152	9.394	9.627
14	1.980	.5051	.5176	19.599	20.085	9.899	10.144
15	2.079	.4810	.4929	21.579	22.114	10.380	10.637
16	2.183	.4581	.4695	23.657	24.244	10.838	11.107
17	2.292	.4363	.4471	25.840	26.481	11.274	11.554
18	2.407	.4155	.4258	28.132	28.830	11.690	11.979
19	2.527	.3957	.4055	30.539	31.296	12.085	12.385
20	2.653	.3769	.3862	33.066	33.886	12.462	12.771
21	2.786	.3589	.3678	35.719	36.605	12.821	13.139
22	2.925	.3419	.3503	38.505	39.460	13.163	13.489
23	3.072	.3256	.3336	41.430	42.458	13.489	13.823
24	3.225	.3101	.3178	44.502	45.605	13.799	14.141
25	3.386	.2953	.3026	47.727	48.911	14.094	14.443
26	3.556	.2812	.2882	51.113	52.381	14.375	14.732
27	3.733	.2678	.2745	54.669	56.025	14.643	15.006
28	3.920	.2551	.2614	58.402	59.851	14.898	15.268
29	4.116	.2429	.2490	62.323	63.868	15.141	15.517
30	4.322	.2314	.2371	66.439	68.086	15.372	15.754
31	4.538	.2204	.2258	70.761	72.515	15.593	15.979
32	4.765	.2099	.2151	75.299	77.166	15.803	16.195
33	5.003	.1999	.2048	80.064	82.049	16.003	16.399
34	5.253	.1904	.1951	85.067	87.176	16.193	16.594
35	5.516	.1813	.1858	90.320	92.560	16.374	16.780
40	7.040	.1420	.1456	120.799	123.795	17.159	17.585
45	8.985	.1113	.1141	159.700	163.660	17.774	18.215
50	11.467	.0872	.0894	209.347	214.539	18.256	18.709
55	14.636	.0683	.0700	272.712	279.474	18.633	19.096
60	18.679	.0535	.0549	353.582	362.350	18.929	19.399
65	23.840	.0419	.0430	456.796	468.124	19.161	19.636
70	30.426	.0329	.0337	588.526	603.120	19.343	19.822
80	49.561	.0202	.0207	971.224	995.308	19.596	20.082
90	80.730	.0124	.0127	1594.599	1634.141	19.752	20.242
100	131.500	.0076	.0078	2610.010	2674.732	19.848	20.340

UNIFORM SERIES		GRADIENT SERIES		
SINKING FUND	CAPITAL RECOVERY	UNIFORM SERIES	PRESENT WORTH	
A/F	A/P	A/G	P/G	N
1.0000	1.0500	0.000	0.000	1
0.4878	0.5378	0.488	0.907	2
.3172	.3672	0.967	2.635	3
.2320	.2820	1.439	5.103	4
.1810	.2310	1.902	8.237	5
.1470	.1970	2.358	11.968	6
.1228	.1728	2.805	16.232	7
.1047	.1547	3.244	20.970	8
.0907	.1407	3.676	26.127	9
.0795	.1295	4.099	31.652	10
.0704	.1204	4.514	37.499	11
.0628	.1128	4.922	43.624	12
.0565	.1065	5.321	49.988	13
.0510	.1010	5.713	56.554	14
.0463	.0963	6.097	63.288	15
.0423	.0923	6.474	70.159	16
.0387	.0887	6.842	77.140	17
.0355	.0855	7.203	84.204	18
.0327	.0827	7.557	91.327	19
.0302	.0802	7.903	98.488	20
.0280	.0780	8.242	105.667	21
.0260	.0760	8.573	112.846	22
.0241	.0741	8.897	120.008	23
.0225	.0725	9.214	127.140	24
.0210	.0710	9.524	134.227	25
.0196	.0696	9.827	141.258	26
.0183	.0683	10.122	148.222	27
.0171	.0671	10.411	155.110	28
.0160	.0660	10.694	161.912	29
.0151	.0651	10.969	168.622	30
.0141	.0641	11.238	175.233	31
.0133	.0633	11.501	181.739	32
.0125	.0625	11.757	188.135	33
.0118	.0618	12.006	194.416	34
.0111	.0611	12.250	200.580	35
.0083	.0583	13.377	229.545	40
.0063	.0563	14.364	255.314	45
.0048	.0548	15.223	277.914	50
.0037	.0537	15.966	297.510	55
.0028	.0528	16.606	314.343	60
.0022	.0522	17.154	328.691	65
.0017	.0517	17.621	340.841	70
.0010	.0510	18.353	359.646	80
.0006	.0506	18.871	372.749	90
.0004	.0504	19.234	381.749	100

$i = 0.05$

Compound Interest Factors $i = 0.06$

	SINGLE PAYMENT			UNIFORM SERIES			
	COMPOUND AMOUNT	PRESENT WORTH		COMPOUND AMOUNT		PRESENT WORTH	
N	F/P	P/F	$P/\overline{F}$	F/A	$F/\overline{A}$	P/A	$P/\overline{A}$
1	1.060	0.9434	0.9714	1.000	1.030	0.943	0.971
2	1.124	.8900	.9164	2.060	2.121	1.833	1.888
3	1.191	.8396	.8646	3.184	3.278	2.673	2.752
4	1.262	.7921	.8156	4.375	4.505	3.465	3.568
5	1.338	.7473	.7695	5.637	5.805	4.212	4.338
6	1.419	.7050	.7259	6.975	7.183	4.917	5.063
7	1.504	.6651	.6848	8.394	8.643	5.582	5.748
8	1.594	.6274	.6461	9.897	10.192	6.210	6.394
9	1.689	.5919	.6095	11.491	11.833	6.802	7.004
10	1.791	.5584	.5750	13.181	13.572	7.360	7.579
11	1.898	.5268	.5424	14.972	15.416	7.887	8.121
12	2.012	.4970	.5117	16.870	17.371	8.384	8.633
13	2.133	.4688	.4828	18.882	19.443	8.853	9.116
14	2.261	.4423	.4554	21.015	21.639	9.295	9.571
15	2.397	.4173	.4297	23.276	23.967	9.712	10.001
16	2.540	.3936	.4053	25.672	26.435	10.106	10.406
17	2.693	.3714	.3824	28.213	29.051	10.477	10.789
18	2.854	.3503	.3608	30.906	31.824	10.828	11.149
19	3.026	.3305	.3403	33.760	34.763	11.158	11.490
20	3.207	.3118	.3211	36.786	37.878	11.470	11.811
21	3.400	.2942	.3029	39.993	41.181	11.764	12.114
22	3.604	.2775	.2858	43.392	44.681	12.042	12.399
23	3.820	.2618	.2696	46.996	48.392	12.303	12.669
24	4.049	.2470	.2543	50.815	52.325	12.550	12.923
25	4.292	.2330	.2399	54.864	56.494	12.783	13.163
26	4.549	.2198	.2263	59.156	60.914	13.003	13.389
27	4.822	.2074	.2135	63.706	65.598	13.211	13.603
28	5.112	.1956	.2014	68.528	70.564	13.406	13.804
29	5.418	.1846	.1900	73.640	75.827	13.591	13.994
30	5.743	.1741	.1793	79.058	81.407	13.765	14.174
31	6.088	.1643	.1691	84.801	87.321	13.929	14.343
32	6.453	.1550	.1596	90.890	93.590	14.084	14.502
33	6.841	.1462	.1505	97.343	100.235	14.230	14.653
34	7.251	.1379	.1420	104.184	107.279	14.368	14.795
35	7.686	.1301	.1340	111.435	114.745	14.498	14.929
40	10.286	.0972	.1001	154.762	159.359	15.046	15.493
45	13.765	.0727	.0748	212.743	219.063	15.456	15.915
50	18.420	.0543	.0559	290.335	298.961	15.762	16.230
55	24.650	.0406	.0418	394.171	405.882	15.991	16.466
60	32.988	.0303	.0312	533.126	548.965	16.161	16.642
65	44.145	.0227	.0233	719.080	740.444	16.289	16.773
70	59.076	.0169	.0174	967.928	996.685	16.385	16.871
80	105.796	.0095	.0097	1746.592	1798.483	16.509	17.000
90	189.464	.0053	.0054	3141.060	3234.379	16.579	17.071
100	339.300	.0029	.0030	5638.337	5805.851	16.618	17.111

UNIFORM SERIES		GRADIENT SERIES		
SINKING FUND	**CAPITAL RECOVERY**	**UNIFORM SERIES**	**PRESENT WORTH**	
A/F	*A/P*	*A/G*	*P/G*	*N*
1.0000	1.0600	0.000	0.000	1
0.4854	0.5454	0.485	0.890	2
.3141	.3741	0.961	2.569	3
.2286	.2886	1.427	4.945	4
.1774	.2374	1.884	7.934	5
.1434	.2034	2.330	11.459	6
.1191	.1791	2.768	15.450	7
.1010	.1610	3.195	19.841	8
.0870	.1470	3.613	24.577	9
.0759	.1359	4.022	29.602	10
.0668	.1268	4.421	34.870	11
.0593	.1193	4.811	40.337	12
.0530	.1130	5.192	45.963	13
.0476	.1076	5.564	51.713	14
.0430	.1030	5.926	57.554	15
.0390	.0990	6.279	63.459	16
.0354	.0954	6.624	69.401	17
.0324	.0924	6.960	75.357	18
.0296	.0896	7.287	81.306	19
.0272	.0872	7.605	87.230	20
.0250	.0850	7.915	93.113	21
.0230	.0830	8.217	98.941	22
.0213	.0813	8.510	104.700	23
.0197	.0797	8.795	110.381	24
.0182	.0782	9.072	115.973	25
.0169	.0769	9.341	121.468	26
.0157	.0757	9.603	126.860	27
.0146	.0746	9.857	132.142	28
.0136	.0736	10.103	137.309	29
.0126	.0726	10.342	142.359	30
.0118	.0718	10.574	147.286	31
.0110	.0710	10.799	152.090	32
.0103	.0703	11.017	156.768	33
.0096	.0696	11.228	161.319	34
.0090	.0690	11.432	165.743	35
.0065	.0665	12.359	185.957	40
.0047	.0647	13.141	203.109	45
.0034	.0634	13.796	217.457	50
.0025	.0625	14.341	229.322	55
.0019	.0619	14.791	239.043	60
.0014	.0614	15.160	246.945	65
.0010	.0610	15.461	253.327	70
.0006	.0606	15.903	262.549	80
.0003	.0603	16.189	268.395	90
.0002	.0602	16.371	272.047	100

i = 0.06

Compound Interest Factors $i = 0.07$

	SINGLE PAYMENT			UNIFORM SERIES			
	COMPOUND AMOUNT	PRESENT WORTH		COMPOUND AMOUNT		PRESENT WORTH	
N	F/P	P/F	P/$\overline{F}$	F/A	F/$\overline{A}$	P/A	P/$\overline{A}$
1	1.070	0.9346	0.9669	1.000	1.035	0.935	0.967
2	1.145	.8734	.9037	2.070	2.142	1.808	1.871
3	1.225	.8163	.8445	3.215	3.326	2.624	2.715
4	1.311	.7629	.7893	4.440	4.594	3.387	3.504
5	1.403	.7130	.7377	5.751	5.950	4.100	4.242
6	1.501	.6663	.6894	7.153	7.401	4.767	4:931
7	1.606	.6227	.6443	8.654	8.953	5.389	5.576
8	1.718	.5820	.6021	10.260	10.615	5.971	6.178
9	1.838	.5439	.5628	11.978	12.392	6.515	6.741
10	1.967	.5083	.5259	13.816	14.295	7.024	7.267
11	2.105	.4751	.4915	15.784	16.330	7.499	7.758
12	2.252	.4440	.4594	17.888	18.507	7.943	8.218
13	2.410	.4150	.4293	20.141	20.838	8.358	8.647
14	2.579	.3878	.4012	22.551	23.331	8.745	9.048
15	2.759	.3624	.3750	25.129	25.999	9.108	9.423
16	2.952	.3387	.3505	27.888	28.853	9.447	9.774
17	3.159	.3166	.3275	30.840	31.907	9.763	10.101
18	3.380	.2959	.3061	33.999	35.176	10.059	10.407
19	3.617	.2765	.2861	37.379	38.672	10.336	10.693
20	3.870	.2584	.2674	40.996	42.414	10.594	10.961
21	4.141	.2415	.2499	44.865	46.418	10.836	11.210
22	4.430	.2257	.2335	49.006	50.702	11.061	11.444
23	4.741	.2109	.2182	53.436	55.285	11.272	11.662
24	5.072	.1971	.2040	58.177	60.190	11.469	11.866
25	5.427	.1842	.1906	63.249	65.438	11.654	12.057
26	5.807	.1722	.1782	68.677	71.053	11.826	12.235
27	6.214	.1609	.1665	74.484	77.061	11.987	12.402
28	6.649	.1504	.1556	80.698	83.490	12.137	12.557
29	7.114	.1406	.1454	87.347	90.369	12.278	12.703
30	7.612	.1314	.1359	94.461	97.730	12.409	12.838
31	8.145	.1228	.1270	102.073	105.605	12.532	12.965
32	8.715	.1147	.1187	110.218	114.032	12.647	13.084
33	9.325	.1072	.1109	118.934	123.049	12.754	13.195
34	9.978	.1002	.1037	128.259	132.697	12.854	13.299
35	10.677	.0937	.0969	138.237	143.021	12.948	13.396
40	14.974	.0668	.0691	199.636	206.544	13.332	13.793
45	21.002	.0476	.0493	285.750	295.638	13.606	14.076
50	29.457	.0339	.0351	406.530	420.598	13.801	14.278
55	41.315	.0242	.0250	575.930	595.860	13.940	14.422
60	57.947	.0173	.0179	813.523	841.674	14.039	14.525
65	81.273	.0123	.0127	1146.759	1186.442	14.110	14.598
70	113.990	.0088	.0091	1614.140	1669.996	14.160	14.650
80	224.235	.0045	.0046	3189.075	3299.432	14.222	14.714
90	441.105	.0023	.0023	6287.213	6504.779	14.253	14.747
100	867.721	.0012	.0012	12381.723	12810.187	14.269	14.763

UNIFORM SERIES		GRADIENT SERIES		
SINKING FUND	CAPITAL RECOVERY	UNIFORM SERIES	PRESENT WORTH	
A/F	A/P	A/G	P/G	N
1.0000	1.0700	0.000	0.000	1
0.4831	0.5531	0.483	0.873	2
.3111	.3811	0.955	2.506	3
.2252	.2952	1.416	4.795	4
.1739	.2439	1.865	7.647	5
.1398	.2098	2.303	10.978	6
.1156	.1856	2.730	14.715	7
.0975	.1675	3.147	18.789	8
.0835	.1535	3.552	23.140	9
.0724	.1424	3.946	27.716	10
.0634	.1334	4.330	32.467	11
.0559	.1259	4.703	37.351	12
.0497	.1197	5.065	42.330	13
.0443	.1143	5.417	47.372	14
.0398	.1098	5.758	52.446	15
.0359	.1059	6.090	57.527	16
.0324	.1024	6.411	62.592	17
.0294	.0994	6.722	67.622	18
.0268	.0968	7.024	72.599	19
.0244	.0944	7.316	77.509	20
.0223	.0923	7.599	82.339	21
.0204	.0904	7.872	87.079	22
.0187	.0887	8.137	91.720	23
.0172	.0872	8.392	96.255	24
.0158	.0858	8.639	100.677	25
.0146	.0846	8.877	104.981	26
.0134	.0834	9.107	109.166	27
.0124	.0824	9.329	113.227	28
.0114	.0814	9.543	117.162	29
.0106	.0806	9.749	120.972	30
.0098	.0798	9.947	124.655	31
.0091	.0791	10.138	128.212	32
.0084	.0784	10.322	131.644	33
.0078	.0778	10.499	134.951	34
.0072	.0772	10.669	138.135	35
.0050	.0750	11.423	152.293	40
.0035	.0735	12.036	163.756	45
.0025	.0725	12.529	172.905	50
.0017	.0717	12.921	180.124	55
.0012	.0712	13.232	185.768	60
.0009	.0709	13.476	190.145	65
.0006	.0706	13.666	193.519	70
.0003	.0703	13.927	198.075	80
.0002	.0702	14.081	200.704	90
.0000	.0701	14.170	202.200	100

$i = 0.07$

Compound Interest Factors $i = 0.08$

	SINGLE PAYMENT			UNIFORM SERIES			
	COMPOUND AMOUNT	PRESENT WORTH		COMPOUND AMOUNT		PRESENT WORTH	
N	F/P	P/F	P/$\overline{F}$	F/A	F/$\overline{A}$	P/A	P/$\overline{A}$
1	1.080	0.9259	0.9625	1.000	1.039	0.926	0.962
2	1.166	.8573	.8912	2.080	2.162	1.783	1.854
3	1.260	.7938	.8252	3.246	3.375	2.577	2.679
4	1.360	.7350	.7641	4.506	4.684	3.312	3.443
5	1.469	.6806	.7075	5.867	6.098	3.993	4.150
6	1.587	.6302	.6551	7.336	7.626	4.623	4.805
7	1.714	.5835	.6065	8.923	9.275	5.206	5.412
8	1.851	.5403	.5616	10.637	11.057	5.747	5.974
9	1.999	.5002	.5200	12.488	12.981	6.247	6.494
10	2.159	.4632	.4815	14.487	15.059	6.710	6.975
11	2.332	.4289	.4458	16.645	17.303	7.139	7.421
12	2.518	.3971	.4128	18.977	19.726	7.536	7.834
13	2.720	.3677	.3822	21.495	22.344	7.904	8.216
14	2.937	.3405	.3539	24.215	25.171	8.244	8.570
15	3.172	.3152	.3277	27.152	28.224	8.559	8.897
16	3.426	.2919	.3034	30.324	31.522	8.851	9.201
17	3.700	.2703	.2809	33.750	35.083	9.122	9.482
18	3.996	.2502	.2601	37.450	38.929	9.372	9.742
19	4.316	.2317	.2409	41.446	43.083	9.604	9.983
20	4.661	.2145	.2230	45.762	47.569	9.818	10.206
21	5.034	.1987	.2065	50.423	52.414	10.017	10.412
22	5.437	.1839	.1912	55.457	57.647	10.201	10.604
23	5.871	.1703	.1770	60.893	63.298	10.371	10.781
24	6.341	.1577	.1639	66.765	69.401	10.529	10.945
25	6.848	.1460	.1518	73.106	75.993	10.675	11.096
26	7.396	.1352	.1405	79.954	83.112	10.810	11.237
27	7.988	.1252	.1301	87.351	90.800	10.935	11.367
28	8.627	.1159	.1205	95.339	99.104	11.051	11.487
29	9.317	.1073	.1116	103.966	108.071	11.158	11.599
30	10.063	.0994	.1033	113.283	117.756	11.258	11.702
31	10.868	.0920	.0956	123.346	128.217	11.350	11.798
32	11.737	.0852	.0886	134.214	139.513	11.435	11.887
33	12.676	.0789	.0820	145.951	151.714	11.514	11.969
34	13.690	.0730	.0759	158.627	164.890	11.587	12.044
35	14.785	.0676	.0703	172.317	179.121	11.655	12.115
40	21.725	.0460	.0478	259.057	269.286	11.925	12.395
45	31.920	.0313	.0326	386.506	401.768	12.108	12.587
50	46.902	.0213	.0222	573.771	596.427	12.233	12.717
55	68.914	.0145	.0151	848.925	882.446	12.319	12.805
60	101.257	.0099	.0103	1253.216	1302.701	12.377	12.865
65	148.780	.0067	.0070	1847.252	1920.193	12.416	12.906
70	218.607	.0046	.0048	2720.086	2827.493	12.443	12.934
80	471.956	.0021	.0022	5886.950	6119.405	12.474	12.966
90	1018.918	.0010	.0010	12723.976	13226.401	12.488	12.981
100	2199.768	.0005	.0005	27484.605	28569.877	12.494	12.988

UNIFORM SERIES		GRADIENT SERIES		
SINKING FUND	CAPITAL RECOVERY	UNIFORM SERIES	PRESENT WORTH	
A/F	A/P	A/G	P/G	N
1.0000	1.0800	0.000	0.000	1
0.4808	0.5608	0.481	0.857	2
.3080	.3880	0.949	2.445	3
.2219	.3019	1.404	4.650	4
.1705	.2505	1.846	7.372	5
.1363	.2163	2.276	10.523	6
.1121	.1921	2.694	14.024	7
.0940	.1740	3.099	17.806	8
.0801	.1601	3.491	21.808	9
.0690	.1490	3.871	25.977	10
.0601	.1401	4.240	30.266	11
.0527	.1327	4.596	34.634	12
.0465	.1265	4.940	39.046	13
.0413	.1213	5.273	43.472	14
.0368	.1168	5.594	47.886	15
.0330	.1130	5.905	52.264	16
.0296	.1096	6.204	56.588	17
.0267	.1067	6.492	60.843	18
.0241	.1041	6.770	65.013	19
.0219	.1019	7.037	69.090	20
.0198	.0998	7.294	73.063	21
.0180	.0980	7.541	76.926	22
.0164	.0964	7.779	80.673	23
.0150	.0950	8.007	84.300	24
.0137	.0937	8.225	87.804	25
.0125	.0925	8.435	91.184	26
.0114	.0914	8.636	94.439	27
.0105	.0905	8.829	97.569	28
.0096	.0896	9.013	100.574	29
.0088	.0888	9.190	103.456	30
.0081	.0881	9.358	106.216	31
.0075	.0875	9.520	108.858	32
.0069	.0869	9.674	111.382	33
.0063	.0863	9.821	113.792	34
.0058	.0858	9.961	116.092	35
.0039	.0839	10.570	126.042	40
.0026	.0826	11.045	133.733	45
.0017	.0817	11.411	139.593	50
.0012	.0812	11.690	144.006	55
.0008	.0808	11.902	147.300	60
.0005	.0805	12.060	149.739	65
.0004	.0804	12.178	151.533	70
.0002	.0802	12.330	153.800	80
.0000	.0801	12.412	154.993	90
.0000	.0800	12.455	155.611	100

$i = 0.08$

Compound Interest Factors $i = 0.09$

	SINGLE PAYMENT			UNIFORM SERIES			
	COMPOUND AMOUNT	PRESENT WORTH		COMPOUND AMOUNT		PRESENT WORTH	
N	F/P	P/F	P/$\overline{F}$	F/A	F/$\overline{A}$	P/A	P/$\overline{A}$
1	1.090	0.9174	0.9581	1.000	1.044	0.917	0.958
2	1.188	.8417	.8790	2.090	2.183	1.759	1.837
3	1.295	.7722	.8064	3.278	3.423	2.531	2.644
4	1.412	.7084	.7398	4.573	4.776	3.240	3.383
5	1.539	.6499	.6788	5.985	6.250	3.890	4.062
6	1.677	.5963	.6227	7.523	7.857	4.486	4.685
7	1.828	.5470	.5713	9.200	9.609	5.033	5.256
8	1.993	.5019	.5241	11.028	11.518	5.535	5.780
9	2.172	.4604	.4808	13.021	13.599	5.995	6.261
10	2.367	.4224	.4411	15.193	15.867	6.418	6.702
11	2.580	.3875	.4047	17.560	18.339	6.805	7.107
12	2.813	.3555	.3713	20.141	21.034	7.161	7.478
13	3.066	.3262	.3406	22.953	23.971	7.487	7.819
14	3.342	.2992	.3125	26.019	27.173	7.786	8.131
15	3.642	.2745	.2867	29.361	30.663	8.061	8.418
16	3.970	.2519	.2630	33.003	34.467	8.313	8.681
17	4.328	.2311	.2413	36.974	38.614	8.544	8.923
18	4.717	.2120	.2214	41.301	43.133	8.756	9.144
19	5.142	.1945	.2031	46.019	48.060	8.950	9.347
20	5.604	.1784	.1863	51.160	53.429	9.129	9.533
21	6.109	.1637	.1710	56.765	59.282	9.292	9.704
22	6.659	.1502	.1568	62.873	65.662	9.442	9.861
23	7.258	.1378	.1439	69.532	72.616	9.580	10.005
24	7.911	.1264	.1320	76.790	80.196	9.707	10.137
25	8.623	.1160	.1211	84.701	88.458	9.823	10.258
26	9.399	.1064	.1111	93.324	97.463	9.929	10.369
27	10.245	.0976	.1019	102.723	107.279	10.027	10.471
28	11.167	.0895	.0935	112.968	117.979	10.116	10.565
29	12.172	.0822	.0858	124.136	129.641	10.198	10.651
30	13.268	.0754	.0787	136.308	142.353	10.274	10.729
31	14.462	.0691	.0722	149.575	156.210	10.343	10.802
32	15.763	.0634	.0663	164.037	171.313	10.406	10.868
33	17.182	.0582	.0608	179.801	187.775	10.464	10.929
34	18.728	.0534	.0558	196.983	205.720	10.518	10.984
35	20.414	.0490	.0512	215.711	225.279	10.567	11.035
40	31.409	.0318	.0332	337.883	352.869	10.757	11.234
45	48.327	.0207	.0216	525.860	549.184	10.881	11.364
50	74.358	.0134	.0140	815.085	851.237	10.962	11.448
55	114.409	.0087	.0091	1260.095	1315.984	11.014	11.503
60	176.032	.0057	.0059	1944.797	2031.056	11.048	11.538
65	270.847	.0037	.0039	2998.297	3131.282	11.070	11.561
70	416.731	.0024	.0025	4619.237	4824.117	11.084	11.576
80	986.555	.0010	.0011	10950.613	11436.312	11.100	11.592
90	2335.536	.0004	.0004	25939.289	27089.787	11.106	11.599
100	5529.065	.0002	.0002	61422.945	64147.266	11.109	11.602

UNIFORM SERIES		GRADIENT SERIES		
SINKING FUND	CAPITAL RECOVERY	UNIFORM SERIES	PRESENT WORTH	
A/F	A/P	A/G	P/G	N
1.0000	1.0900	0.000	0.000	1
0.4785	0.5685	0.478	0.842	2
.3051	.3951	0.943	2.386	3
.2187	.3087	1.393	4.511	4
.1671	.2571	1.828	7.111	5
.1329	.2229	2.250	10.092	6
.1087	.1987	2.657	13.375	7
.0907	.1807	3.051	16.888	8
.0768	.1668	3.431	20.571	9
.0658	.1558	3.798	24.373	10
.0569	.1469	4.151	28.248	11
.0497	.1397	4.491	32.159	12
.0436	.1336	4.818	36.073	13
.0384	.1284	5.133	39.963	14
.0341	.1241	5.435	43.807	15
.0303	.1203	5.724	47.585	16
.0270	.1170	6.002	51.282	17
.0242	.1142	6.269	54.886	18
.0217	.1117	6.524	58.387	19
.0195	.1095	6.767	61.777	20
.0176	.1076	7.001	65.051	21
.0159	.1059	7.223	68.205	22
.0144	.1044	7.436	71.236	23
.0130	.1030	7.638	74.143	24
.0118	.1018	7.832	76.927	25
.0107	.1007	8.016	79.586	26
.0097	.0997	8.191	82.124	27
.0089	.0989	8.357	84.542	28
.0081	.0981	8.515	86.842	29
.0073	.0973	8.666	89.028	30
.0067	.0967	8.808	91.102	31
.0061	.0961	8.944	93.069	32
.0056	.0956	9.072	94.931	33
.0051	.0951	9.193	96.693	34
.0046	.0946	9.308	98.359	35
.0030	.0930	9.796	105.376	40
.0019	.0919	10.160	110.556	45
.0012	.0912	10.430	114.325	50
.0008	.0908	10.626	117.036	55
.0005	.0905	10.768	118.968	60
.0003	.0903	10.870	120.334	65
.0002	.0902	10.943	121.294	70
.0000	.0901	11.030	122.431	80
.0000	.0900	11.073	122.976	90
.0000	.0900	11.093	123.233	100

$i = 0.09$

Compound Interest Factors $i = 0.10$

	SINGLE PAYMENT			UNIFORM SERIES			
	COMPOUND AMOUNT	PRESENT WORTH		COMPOUND AMOUNT		PRESENT WORTH	
N	F/P	P/F	P/$\overline{F}$	F/A	F/$\overline{A}$	P/A	P/$\overline{A}$
1	1.100	0.9091	0.9538	1.000	1.049	0.909	0.954
2	1.210	.8264	.8671	2.100	2.203	1.736	1.821
3	1.331	.7513	.7883	3.310	3.473	2.487	2.609
4	1.464	.6830	.7166	4.641	4.869	3.170	3.326
5	1.611	.6209	.6515	6.105	6.406	3.791	3.977
6	1.772	.5645	.5922	7.716	8.095	4.355	4.570
7	1.949	.5132	.5384	9.487	9.954	4.868	5.108
8	2.144	.4665	.4895	11.436	11.999	5.335	5.597
9	2.358	.4241	.4450	13.579	14.248	5.759	6.042
10	2.594	.3855	.4045	15.937	16.722	6.145	6.447
11	2.853	.3505	.3677	18.531	19.443	6.495	6.815
12	3.138	.3186	.3343	21.384	22.437	6.814	7.149
13	3.452	.2897	.3039	24.523	25.729	7.103	7.453
14	3.797	.2633	.2763	27.975	29.352	7.367	7.729
15	4.177	.2394	.2512	31.772	33.336	7.606	7.980
16	4.595	.2176	.2283	35.950	37.719	7.824	8.209
17	5.054	.1978	.2076	40.545	42.540	8.022	8.416
18	5.560	.1799	.1887	45.599	47.843	8.201	8.605
19	6.116	.1635	.1716	51.159	53.676	8.365	8.777
20	6.728	.1486	.1560	57.275	60.093	8.514	8.932
21	7.400	.1351	.1418	64.003	67.152	8.649	9.074
22	8.140	.1228	.1289	71.403	74.916	8.772	9.203
23	8.954	.1117	.1172	79.543	83.457	8.883	9.320
24	9.850	.1015	.1065	88.497	92.852	8.985	9.427
25	10.835	.0923	.0968	98.347	103.186	9.077	9.524
26	11.918	.0839	.0880	109.182	114.554	9.161	9.612
27	13.110	.0763	.0800	121.100	127.059	9.237	9.692
28	14.421	.0693	.0728	134.210	140.814	9.307	9.765
29	15.863	.0630	.0661	148.631	155.945	9.370	9.831
30	17.449	.0573	.0601	164.494	172.588	9.427	9.891
31	19.194	.0521	.0547	181.944	190.896	9.479	9.945
32	21.114	.0474	.0497	201.138	211.035	9.526	9.995
33	23.225	.0431	.0452	222.252	233.188	9.569	10.040
34	25.548	.0391	.0411	245.477	257.556	9.609	10.081
35	28.102	.0356	.0373	271.025	284.361	9.644	10.119
40	45.259	.0221	.0232	442.593	464.371	9.779	10.260
45	72.891	.0137	.0144	718.906	754.280	9.863	10.348
50	117.391	.0085	.0089	1163.910	1221.181	9.915	10.403
55	189.059	.0053	.0055	1880.594	1973.130	9.947	10.437
60	304.482	.0033	.0034	3034.821	3184.151	9.967	10.458
65	490.372	.0020	.0021	4893.715	5134.514	9.980	10.471
70	789.748	.0013	.0013	7887.483	8275.592	9.987	10.479
80	2048.405	.0005	.0005	20474.045	21481.484	9.995	10.487
90	5313.035	.0002	.0002	53120.348	55734.168	9.998	10.490
∞	∞	∞	0	∞	∞	∞	$\frac{1}{i}$

UNIFORM SERIES		GRADIENT SERIES			
SINKING FUND	CAPITAL RECOVERY	UNIFORM SERIES	PRESENT WORTH		
A/F	A/P	A/G	P/G		N
1.0000	1.1000	0.000	0.000		1
0.4762	0.5762	0.476	0.826		2
.3021	.4021	0.937	2.329		3
.2155	.3155	1.381	4.378		4
.1638	.2638	1.810	6.862		5
.1296	.2296	2.224	9.684		6
.1054	.2054	2.622	12.763		7
.0874	.1874	3.004	16.029		8
.0736	.1736	3.372	19.421		9
.0627	.1627	3.725	22.891		10
.0540	.1540	4.064	26.396		11
.0468	.1468	4.388	29.901		12
.0408	.1408	4.699	33.377		13
.0357	.1357	4.996	36.801		14
.0315	.1315	5.279	40.152		15
.0278	.1278	5.549	43.416		16
.0247	.1247	5.807	46.582		17
.0219	.1219	6.053	49.640		18
.0195	.1195	6.286	52.583		19
.0175	.1175	6.508	55.407		20
.0156	.1156	6.719	58.110		21
.0140	.1140	6.919	60.689		22
.0126	.1126	7.108	63.146		23
.0113	.1113	7.288	65.481		24
.0102	.1102	7.458	67.696		25
.0092	.1092	7.619	69.794		26
.0083	.1083	7.770	71.777		27
.0075	.1075	7.914	73.650		28
.0067	.1067	8.049	75.415		29
.0061	.1061	8.176	77.077		30
.0055	.1055	8.296	78.640		31
.0050	.1050	8.409	80.108		32
.0045	.1045	8.515	81.486		33
.0041	.1041	8.615	82.777		34
.0037	.1037	8.709	83.987		35
.0023	.1023	9.096	88.953		40
.0014	.1014	9.374	92.454		45
.0009	.1009	9.570	94.889		50
.0005	.1005	9.708	96.562		55
.0003	.1003	9.802	97.701		60
.0002	.1002	9.867	98.471		65
.0001	.1001	9.911	98.987		70
.0000	.1000	9.961	99.561		80
.0000	.1000	9.983	99.812		90

$i = 0.10$

Compound Interest Factors $i = 0.12$

	SINGLE PAYMENT			UNIFORM SERIES			
	COMPOUND AMOUNT	PRESENT WORTH		COMPOUND AMOUNT		PRESENT WORTH	
N	F/P	P/F	$P/\overline{F}$	F/A	$F/\overline{A}$	P/A	$P/\overline{A}$
1	1.120	0.8929	0.9454	1.000	1.059	0.893	0.945
2	1.254	.7972	.8441	2.120	2.245	1.690	1.790
3	1.405	.7118	.7537	3.374	3.573	2.402	2.543
4	1.574	.6355	.6729	4.779	5.061	3.037	3.216
5	1.762	.5674	.6008	6.353	6.727	3.605	3.817
6	1.974	.5066	.5365	8.115	8.593	4.111	4.353
7	2.211	.4523	.4790	10.089	10.683	4.564	4.832
8	2.476	.4039	.4277	12.300	13.024	4.968	5.260
9	2.773	.3606	.3818	14.776	15.645	5.328	5.642
10	3.106	.3220	.3409	17.549	18.582	5.650	5.983
11	3.479	.2875	.3044	20.655	21.870	5.938	6.287
12	3.896	.2567	.2718	24.133	25.554	6.194	6.559
13	4.363	.2292	.2427	28.029	29.679	6.424	6.802
14	4.887	.2046	.2167	32.393	34.299	6.628	7.018
15	5.474	.1827	.1935	37.280	39.474	6.811	7.212
16	6.130	.1631	.1727	42.753	45.270	6.974	7.385
17	6.866	.1456	.1542	48.884	51.761	7.120	7.539
18	7.690	.1300	.1377	55.750	59.032	7.250	7.676
19	8.613	.1161	.1229	63.440	67.174	7.366	7.799
20	9.646	.1037	.1098	72.052	76.294	7.469	7.909
21	10.804	.0926	.0980	81.699	86.508	7.562	8.007
22	12.100	.0826	.0875	92.503	97.948	7.645	8.095
23	13.552	.0738	.0781	104.603	110.761	7.718	8.173
24	15.179	.0659	.0698	118.155	125.111	7.784	8.243
25	17.000	.0588	.0623	133.334	141.183	7.843	8.305
26	19.040	.0525	.0556	150.334	159.184	7.896	8.360
27	21.325	.0469	.0497	169.374	179.345	7.943	8.410
28	23.884	.0419	.0443	190.699	201.925	7.984	8.454
29	26.750	.0374	.0396	214.583	227.215	8.022	8.494
30	29.960	.0334	.0353	241.333	255.539	8.055	8.529
31	33.555	.0298	.0316	271.293	287.263	8.085	8.561
32	37.582	.0266	.0282	304.848	322.793	8.112	8.589
33	42.092	.0238	.0252	342.429	362.587	8.135	8.614
34	47.143	.0212	.0225	384.521	407.157	8.157	8.637
35	52.800	.0189	.0201	431.663	457.074	8.176	8.657
40	93.051	.0107	.0114	767.091	812.248	8.244	8.729
45	163.988	.0061	.0065	1358.230	1438.185	8.283	8.770
50	289.002	.0035	.0037	2400.018	2541.301	8.304	8.793
55	509.321	.0020	.0021	4236.005	4485.366	8.317	8.807
60	897.597	.0011	.0012	7471.641	7911.475	8.324	8.814
65	1581.872	.0006	.0007	13173.938	13949.449	8.328	8.818
70	2787.800	.0004	.0004	23223.332	24590.422	8.330	8.821
80	8658.482	.0001	.0001	72145.688	76392.695	8.332	8.823

UNIFORM SERIES		GRADIENT SERIES			
SINKING FUND	CAPITAL RECOVERY	UNIFORM SERIES	PRESENT WORTH		
A/F	A/P	A/G	P/G		N
1.0000	· 1.1200	0.000	0.000		1
0.4717	0.5917	0.472	0.797		2
.2963	.4163	0.925	2.221		3
.2092	.3292	1.359	4.127		4
.1574	.2774	1.775	6.397		5
.1232	.2432	2.172	8.930		6
.0991	.2191	2.551	11.644		7
.0813	.2013	2.913	14.471		8
.0677	.1877	3.257	17.356		9
.0570	.1770	3.585	20.254		10
.0484	.1684	3.895	23.129		11
.0414	.1614	4.190	25.952		12
.0357	.1557	4.468	28.702		13
.0309	.1509	4.732	31.362		14
.0268	.1468	4.980	33.920		15
.0234	.1434	5.215	36.367		16
.0205	.1405	5.435	38.697		17
.0179	.1379	5.643	40.908		18
.0158	.1358	5.838	42.998		19
.0139	.1339	6.020	44.968		20
.0122	.1322	6.191	46.819		21
.0108	.1308	6.351	48.554		22
.0096	.1296	6.501	50.178		23
.0085	.1285	6.641	51.693		24
.0075	.1275	6.771	53.105		25
.0067	.1267	6.892	54.418		26
.0059	.1259	7.005	55.637		27
.0052	.1252	7.110	56.767		28
.0047	.1247	7.207	57.814		29
.0041	.1241	7.297	58.782		30
.0037	.1237	7.381	59.676		31
.0033	.1233	7.459	60.501		32
.0029	.1229	7.530	61.261		33
.0026	.1226	7.596	61.961		34
.0023	.1223	7.658	62.605		35
.0013	.1213	7.899	65.116		40
.0007	.1207	8.057	66.734		45
.0004	.1204	8.160	67.762		50
.0002	.1202	8.225	68.408		55
.0001	.1201	8.266	68.810		60
.0000	.1201	8.292	69.058		65
.0000	.1200	8.308	69.210		70
.0000	.1200	8.324	69.359		80

i = 0.12

Compound Interest Factors $i = 0.15$

	SINGLE PAYMENT			UNIFORM SERIES			
	COMPOUND AMOUNT	PRESENT WORTH		COMPOUND AMOUNT		PRESENT WORTH	
N	F/P	P/F	$P/\overline{F}$	F/A	$F/\overline{A}$	P/A	$P/\overline{A}$
1	1.150	0.8696	0.9333	1.000	1.073	0.870	0.933
2	1.322	.7561	.8115	2.150	2.307	1.626	1.745
3	1.521	.6575	.7057	3.472	3.727	2.283	2.450
4	1.749	.5718	.6136	4.993	5.359	2.855	3.064
5	2.011	.4972	.5336	6.742	7.236	3.352	3.598
6	2.313	.4323	.4640	8.754	9.395	3.784	4.062
7	2.660	.3759	.4035	11.067	11.877	4.160	4.465
8	3.059	.3269	.3508	13.727	14.732	4.487	4.816
9	3.518	.2843	.3051	16.786	18.015	4.772	5.121
10	4.046	.2472	.2653	20.304	21.791	5.019	5.386
11	4.652	.2149	.2307	24.349	26.133	5.234	5.617
12	5.350	.1869	.2006	29.002	31.126	5.421	5.818
13	6.153	.1625	.1744	34.352	36.868	5.583	5.992
14	7.076	.1413	.1517	40.505	43.472	5.724	6.144
15	8.137	.1229	.1319	47.580	51.066	5.847	6.276
16	9.358	.1069	.1147	55.717	59.799	5.954	6.390
17	10.761	.0929	.0997	65.075	69.842	6.047	6.490
18	12.375	.0808	.0867	75.836	81.392	6.128	6.577
19	14.232	.0703	.0754	88.212	94.674	6.198	6.652
20	16.367	.0611	.0656	102.444	109.948	6.259	6.718
21	18.822	.0531	.0570	118.810	127.513	6.312	6.775
22	21.645	.0462	.0496	137.632	147.714	6.359	6.824
23	24.891	.0402	.0431	159.276	170.944	6.399	6.868
24	28.625	.0349	.0375	184.168	197.659	6.434	6.905
25	32.919	.0304	.0326	212.793	228.381	6.464	6.938
26	37.857	.0264	.0284	245.712	263.711	6.491	6.966
27	43.535	.0230	.0247	283.569	304.341	6.514	6.991
28	50.066	.0200	.0214	327.104	351.066	6.534	7.012
29	57.575	.0174	.0186	377.170	404.799	6.551	7.031
30	66.212	.0151	.0162	434.745	466.592	6.566	7.047
31	76.144	.0131	.0141	500.957	537.654	6.579	7.061
32	87.565	.0114	.0123	577.100	619.375	6.591	7.073
33	100.700	.0099	.0107	664.666	713.355	6.600	7.084
34	115.805	.0086	.0093	765.365	821.431	6.609	7.093
35	133.176	.0075	.0081	881.170	945.719	6.617	7.101
40	267.864	.0037	.0040	1779.090	1909.416	6.642	7.128
45	538.769	.0019	.0020	3585.128	3847.752	6.654	7.142
50	1083.657	.0009	.0010	7217.716	7746.442	6.661	7.148
55	2179.622	.0005	.0005	14524.148	15588.099	6.664	7.152
60	4384.000	.0002	.0002	29219.996	31360.475	6.665	7.153
65	8817.788	.0001	.0001	58778.586	63084.344	6.666	7.154

UNIFORM SERIES		GRADIENT SERIES		
SINKING FUND	CAPITAL RECOVERY	UNIFORM SERIES	PRESENT WORTH	
A/F	A/P	A/G	P/G	N
1.0000	1.1500	0.000	0.000	1
0.4651	0.6151	0.465	0.756	2
.2880	.4380	0.907	2.071	3
.2003	.3503	1.326	3.786	4
.1483	.2983	1.723	5.775	5
.1142	.2642	2.097	7.937	6
.0904	.2404	2.450	10.192	7
.0729	.2229	2.781	12.481	8
.0596	.2096	3.092	14.755	9
.0493	.1993	3.383	16.979	10
.0411	.1911	3.655	19.129	11
.0345	.1845	3.908	21.185	12
.0291	.1791	4.144	23.135	13
.0247	.1747	4.362	24.972	14
.0210	.1710	4.565	26.693	15
.0179	.1679	4.752	28.296	16
.0154	.1654	4.925	29.783	17
.0132	.1632	5.084	31.156	18
.0113	.1613	5.231	32.421	19
.0098	.1598	5.365	33.582	20
.0084	.1584	5.488	34.645	21
.0073	.1573	5.601	35.615	22
.0063	.1563	5.704	36.499	23
.0054	.1554	5.798	37.302	24
.0047	.1547	5.883	38.031	25
.0041	.1541	5.961	38.692	26
.0035	.1535	6.032	39.289	27
.0031	.1531	6.096	39.828	28
.0027	.1527	6.154	40.315	29
.0023	.1523	6.207	40.753	30
.0020	.1520	6.254	41.147	31
.0017	.1517	6.297	41.501	32
.0015	.1515	6.336	41.818	33
.0013	.1513	6.371	42.103	34
.0011	.1511	6.402	42.359	35
.0006	.1506	6.517	43.283	40
.0003	.1503	6.583	43.805	45
.0001	.1501	6.620	44.096	50
.0000	.1501	6.641	44.256	55
.0000	.1500	6.653	44.343	60
.0000	.1500	6.659	44.390	65

$i = 0.15$

Compound Interest Factors $i = 0.20$

| | SINGLE PAYMENT | | | UNIFORM SERIES | | | |
| | COMPOUND AMOUNT | PRESENT WORTH | | COMPOUND AMOUNT | | PRESENT WORTH | |
N	F/P	P/F	$P/\overline{F}$	F/A	$F/\overline{A}$	P/A	$P/\overline{A}$
1	1.200	0.8333	0.9141	1.000	1.097	0.833	0.914
2	1.440	.6944	.7618	2.200	2.413	1.528	1.676
3	1.728	.5787	.6348	3.640	3.993	2.106	2.311
4	2.074	.4823	.5290	5.368	5.888	2.589	2.840
5	2.488	.4019	.4408	7.442	8.163	2.991	3.281
6	2.986	.3349	.3674	9.930	10.893	3.326	3.648
7	3.583	.2791	.3061	12.916	14.168	3.605	3.954
8	4.300	.2326	.2551	16.499	18.099	3.837	4.209
9	5.160	.1938	.2126	20.799	22.816	4.031	4.422
10	6.192	.1615	.1772	25.959	28.476	4.192	4.599
11	7.430	.1346	.1476	32.150	35.268	4.327	4.747
12	8.916	.1122	.1230	39.581	43.418	4.439	4.870
13	10.699	.0935	.1025	48.497	53.199	4.533	4.972
14	12.839	.0779	.0854	59.196	64.936	4.611	5.058
15	15.407	.0649	.0712	72.035	79.020	4.675	5.129
16	18.488	.0541	.0593	87.442	95.921	4.730	5.188
17	22.186	.0451	.0494	105.931	116.202	4.775	5.238
18	26.623	.0376	.0412	128.117	140.539	4.812	5.279
19	31.948	.0313	.0343	154.740	169.744	4.843	5.313
20	38.338	.0261	.0286	186.688	204.790	4.870	5.342
21	46.005	.0217	.0238	225.026	246.845	4.891	5.366
22	55.206	.0181	.0199	271.031	297.311	4.909	5.385
23	66.247	.0151	.0166	326.237	357.870	4.925	5.402
24	79.497	.0126	.0138	392.484	430.541	4.937	5.416
25	95.396	.0105	.0115	471.981	517.746	4.948	5.427
26	114.475	.0087	.0096	567.377	622.392	4.956	5.437
27	137.371	.0073	.0080	681.853	747.967	4.964	5.445
28	164.845	.0061	.0067	819.223	898.658	4.970	5.452
29	197.814	.0051	.0055	984.068	1079.486	4.975	5.457
30	237.376	.0042	.0046	1181.882	1296.480	4.979	5.462
31	284.852	.0035	.0039	1419.258	1556.874	4.982	5.466
32	341.822	.0029	.0032	1704.109	1869.345	4.985	5.469
33	410.186	.0024	.0027	2045.932	2244.311	4.988	5.471
34	492.224	.0020	.0022	2456.118	2694.270	4.990	5.474
35	590.668	.0017	.0019	2948.341	3234.221	4.992	5.476
40	1469.772	.0007	.0007	7343.858	8055.940	4.997	5.481
45	3657.262	.0003	.0003	18281.313	20053.922	4.999	5.483
50	9100.438	.0001	.0001	45497.191	49908.730	4.999	5.484

UNIFORM SERIES		GRADIENT SERIES		
SINKING FUND	CAPITAL RECOVERY	UNIFORM SERIES	PRESENT WORTH	
A/F	A/P	A/G	P/G	N
1.0000	1.2000	0.000	0.000	1
0.4545	0.6545	0.455	0.694	2
.2747	.4747	0.879	1.852	3
.1863	.3863	1.274	3.299	4
.1344	.3344	1.641	4.906	5
.1007	.3007	1.979	6.581	6
.0774	.2774	2.290	8.255	7
.0606	.2606	2.576	9.883	8
.0481	.2481	2.836	11.434	9
.0385	.2385	3.074	12.887	10
.0311	.2311	3.289	14.233	11
.0253	.2253	3.484	15.467	12
.0206	.2206	3.660	16.588	13
.0169	.2169	3.817	17.601	14
.0139	.2139	3.959	18.509	15
.0114	.2114	4.085	19.321	16
.0094	.2094	4.198	20.042	17
.0078	.2078	4.298	20.680	18
.0065	.2065	4.386	21.244	19
.0054	.2054	4.464	21.739	20
.0044	.2044	4.533	22.174	21
.0037	.2037	4.594	22.555	22
.0031	.2031	4.647	22.887	23
.0025	.2025	4.694	23.176	24
.0021	.2021	4.735	23.428	25
.0018	.2018	4.771	23.646	26
.0015	.2015	4.802	23.835	27
.0012	.2012	4.829	23.999	28
.0010	.2010	4.853	24.141	29
.0008	.2008	4.873	24.263	30
.0007	.2007	4.891	24.368	31
.0006	.2006	4.906	24.459	32
.0005	.2005	4.919	24.537	33
.0004	.2004	4.931	24.604	34
.0003	.2003	4.941	24.661	35
.0001	.2001	4.973	24.847	40
.0000	.2001	4.988	24.932	45
.0000	.2000	4.995	24.970	50

$i = 0.20$

Compound Interest Factors $i = 0.25$

	SINGLE PAYMENT			UNIFORM SERIES			
	COMPOUND AMOUNT	PRESENT WORTH		COMPOUND AMOUNT		PRESENT WORTH	
N	F/P	P/F	$P/\overline{F}$	F/A	$F/\overline{A}$	P/A	$P/\overline{A}$
1	1.250	0.8000	0.8963	1.000	1.120	0.800	0.896
2	1.563	.6400	.7170	2.250	2.521	1.440	1.613
3	1.953	.5120	.5736	3.813	4.271	1.952	2.187
4	2.441	.4096	.4589	5.766	6.460	2.362	2.646
5	3.052	.3277	.3671	8.207	9.195	2.689	3.013
6	3.815	.2621	.2937	11.259	12.614	2.951	3.307
7	4.768	.2097	.2350	15.073	16.888	3.161	3.542
8	5.960	.1678	.1880	19.842	22.230	3.329	3.730
9	7.451	.1342	.1504	25.802	28.908	3.463	3.880
10	9.313	.1074	.1203	33.253	37.255	3.571	4.000
11	11.642	.0859	.0962	42.566	47.689	3.656	4.096
12	14.552	.0687	.0770	54.208	60.732	3.725	4.173
13	18.190	.0550	.0616	68.760	77.035	3.780	4.235
14	22.737	.0440	.0493	86.949	97.414	3.824	4.284
15	28.422	.0352	.0394	109.687	122.888	3.859	4.324
16	35.527	.0281	.0315	138.109	154.731	3.887	4.355
17	44.409	.0225	.0252	173.636	194.534	3.910	4.381
18	55.511	.0180	.0202	218.045	244.287	3.928	4.401
19	69.389	.0144	.0161	273.556	306.480	3.942	4.417
20	86.736	.0115	.0129	342.945	384.220	3.954	4.430
21	108.420	.0092	.0103	429.681	481.395	3.963	4.440
22	135.525	.0074	.0083	538.101	602.864	3.970	4.448
23	169.407	.0059	.0066	673.626	754.701	3.976	4.455
24	211.758	.0047	.0053	843.033	944.496	3.981	4.460
25	264.698	.0038	.0042	1054.791	1181.741	3.985	4.464
26	330.872	.0030	.0034	1319.489	1478.296	3.988	4.468
27	413.590	.0024	.0027	1650.361	1848.990	3.990	4.471
28	516.988	.0019	.0022	2063.952	2312.358	3.992	4.473
29	646.235	.0015	.0017	2580.939	2891.568	3.994	4.474
30	807.794	.0012	.0014	3227.174	3615.581	3.995	4.476
31	1009.742	.0010	.0011	4034.968	4520.596	3.996	4.477
32	1262.177	.0008	.0009	5044.710	5651.865	3.997	4.478
33	1577.722	.0006	.0007	6306.888	7065.953	3.997	4.479
34	1972.152	.0005	.0006	7884.609	8833.561	3.998	4.479
35	2465.190	.0004	.0005	9856.762	11043.071	3.998	4.480
40	7523.164	.0001	.0001	30088.656	33709.973	3.999	4.481

UNIFORM SERIES		GRADIENT SERIES		
SINKING FUND	CAPITAL RECOVERY	UNIFORM SERIES	PRESENT WORTH	
A/F	*A/P*	*A/G*	*P/G*	*N*
1.0000	1.2500	0.000	0.000	1
0.4444	0.6944	0.444	0.640	2
.2623	.5123	0.852	1.664	3
.1734	.4234	1.225	2.893	4
.1218	.3718	1.563	4.204	5
.0888	.3388	1.868	5.514	6
.0663	.3163	2.142	6.773	7
.0504	.3004	2.387	7.947	8
.0388	.2888	2.605	9.021	9
.0301	.2801	2.797	9.987	10
.0235	.2735	2.966	10.846	11
.0184	.2684	3.115	11.602	12
.0145	.2645	3.244	12.262	13
.0115	.2615	3.356	12.833	14
.0091	.2591	3.453	13.326	15
.0072	.2572	3.537	13.748	16
.0058	.2558	3.608	14.108	17
.0046	.2546	3.670	14.415	18
.0037	.2537	3.722	14.674	19
.0029	.2529	3.767	14.893	20
.0023	.2523	3.805	15.078	21
.0019	.2519	3.836	15.233	22
.0015	.2515	3.863	15.362	23
.0012	.2512	3.886	15.471	24
.0009	.2509	3.905	15.562	25
.0008	.2508	3.921	15.637	26
.0006	.2506	3.935	15.700	27
.0005	.2505	3.946	15.752	28
.0004	.2504	3.955	15.796	29
.0003	.2503	3.963	15.832	30
.0002	.2502	3.969	15.861	31
.0002	.2502	3.975	15.886	32
.0002	.2502	3.979	15.906	33
.0001	.2501	3.983	15.923	34
.0001	.2501	3.986	15.937	35
.0000	.2500	3.995	15.977	40

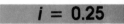

$i = 0.25$

Compound Interest Factors $i = 0.30$

	SINGLE PAYMENT			UNIFORM SERIES			
	COMPOUND AMOUNT	PRESENT WORTH		COMPOUND AMOUNT		PRESENT WORTH	
N	F/P	P/F	P/$\overline{F}$	F/A	F/$\overline{A}$	P/A	P/$\overline{A}$
1	1.300	0.7692	0.8796	1.000	1.143	0.769	0.880
2	1.690	.5917	.6766	2.300	2.630	1.361	1.556
3	2.197	.4552	.5205	3.990	4.562	1.816	2.077
4	2.856	.3501	.4004	6.187	7.075	2.166	2.477
5	3.713	.2693	.3080	9.043	10.340	2.436	2.785
6	4.827	.2072	.2369	12.756	14.586	2.643	3.022
7	6.275	.1594	.1822	17.583	20.105	2.802	3.204
8	8.157	.1226	.1402	23.858	27.280	2.925	3.344
9	10.605	.0943	.1078	32.015	36.608	3.019	3.452
10	13.786	.0725	.0829	42.620	48.733	3.092	3.535
11	17.922	.0558	.0638	56.405	64.497	3.147	3.599
12	23.298	.0429	.0491	74.327	84.989	3.190	3.648
13	30.288	.0330	.0378	97.625	111.629	3.223	3.686
14	39.374	.0254	.0290	127.913	146.261	3.249	3.715
15	51.186	.0195	.0223	167.286	191.283	3.268	3.737
16	66.542	.0150	.0172	218.472	249.812	3.283	3.754
17	86.504	.0116	.0132	285.014	325.899	3.295	3.767
18	112.456	.0089	.0102	371.518	424.812	3.304	3.778
19	146.192	.0068	.0078	483.974	553.399	3.311	3.785
20	190.050	.0053	.0060	630.166	720.562	3.316	3.791
21	247.065	.0040	.0046	820.216	937.875	3.320	3.796
22	321.184	.0031	.0036	1067.281	1220.380	3.323	3.800
23	417.540	.0024	.0027	1388.465	1587.638	3.325	3.802
24	542.801	.0018	.0021	1806.005	2065.073	3.327	3.804
25	705.642	.0014	.0016	2348.806	2685.739	3.329	3.806
26	917.335	.0011	.0012	3054.448	3492.604	3.330	3.807
27	1192.535	.0008	.0010	3971.783	4541.528	3.331	3.808
28	1550.296	.0006	.0007	5164.318	5905.131	3.331	3.809
29	2015.384	.0005	.0006	6714.614	7677.813	3.332	3.810
30	2620.000	.0004	.0004	8729.999	9982.302	3.332	3.810
31	3406.000	.0003	.0003	11349.998	12978.135	3.332	3.810
32	4427.800	.0002	.0003	14756.000	16872.723	3.333	3.811
33	5756.141	.0002	.0002	19183.801	21935.684	3.333	3.811
34	7482.983	.0001	.0002	24939.943	28517.535	3.333	3.811
35	9727.878	.0001	.0001	32422.926	37073.938	3.333	3.811

UNIFORM SERIES		GRADIENT SERIES		
SINKING FUND	CAPITAL RECOVERY	UNIFORM SERIES	PRESENT WORTH	
A/F	A/P	A/G	P/G	N
1.0000	1.3000	0.000	0.000	1
0.4348	0.7348	0.435	0.592	2
.2506	.5506	0.827	1.502	3
.1616	.4616	1.178	2.552	4
.1106	.4106	1.490	3.630	5
.0784	.3784	1.765	4.666	6
.0569	.3569	2.006	5.622	7
.0419	.3419	2.216	6.480	8
.0312	.3312	2.396	7.234	9
.0235	.3235	2.551	7.887	10
.0177	.3177	2.683	8.445	11
.0135	.3135	2.795	8.917	12
.0102	.3102	2.889	9.314	13
.0078	.3078	2.969	9.644	14
.0060	.3060	3.034	9.917	15
.0046	.3046	3.089	10.143	16
.0035	.3035	3.135	10.328	17
.0027	.3027	3.172	10.479	18
.0021	.3021	3.202	10.602	19
.0016	.3016	3.228	10.702	20
.0012	.3012	3.248	10.783	21
.0009	.3009	3.265	10.848	22
.0007	.3007	3.278	10.901	23
.0006	.3006	3.289	10.943	24
.0004	.3004	3.298	10.977	25
.0003	.3003	3.305	11.005	26
.0003	.3003	3.311	11.026	27
.0002	.3002	3.315	11.044	28
.0001	.3001	3.319	11.058	29
.0001	.3001	3.322	11.069	30
.0000	.3001	3.324	11.078	31
.0000	.3001	3.326	11.085	32
.0000	.3001	3.328	11.090	33
.0000	.3000	3.329	11.094	34
.0000	.3000	3.330	11.098	35

$i = 0.30$

Compound Interest Factors $i = 0.35$

	SINGLE PAYMENT			UNIFORM SERIES			
	COMPOUND AMOUNT	PRESENT WORTH		COMPOUND AMOUNT		PRESENT WORTH	
N	F/P	P/F	$P/\overline{F}$	F/A	$F/\overline{A}$	P/A	$P/\overline{A}$
1	1.350	0.7407	0.8639	1.000	1.166	0.741	0.864
2	1.823	.5487	.6399	2.350	2.741	1.289	1.504
3	2.460	.4064	.4740	4.173	4.866	1.699	1.978
4	3.322	.3011	.3511	6.633	7.736	1.997	2.329
5	4.484	.2230	.2601	9.954	11.609	2.220	2.589
6	6.053	.1652	.1927	14.438	16.839	2.385	2.782
7	8.172	.1224	.1427	20.492	23.899	2.508	2.924
8	11.032	.0906	.1057	28.664	33.430	2.598	3.030
9	14.894	.0671	.0783	39.696	46.296	2.665	3.108
10	20.107	.0497	.0580	54.590	63.666	2.715	3.166
11	27.144	.0368	.0430	74.697	87.116	2.752	3.209
12	36.644	.0273	.0318	101.841	118.773	2.779	3.241
13	49.470	.0202	.0236	138.485	161.509	2.799	3.265
14	66.784	.0150	.0175	187.955	219.204	2.814	3.282
15	90.159	.0111	.0129	254.739	297.091	2.825	3.295
16	121.714	.0082	.0096	344.897	402.240	2.834	3.305
17	164.314	.0061	.0071	466.611	544.190	2.840	3.312
18	221.824	.0045	.0053	630.925	735.823	2.844	3.317
19	299.462	.0033	.0039	852.749	994.527	2.848	3.321
20	404.274	.0025	.0029	1152.211	1343.778	2.850	3.324
21	545.770	.0018	.0021	1556.485	1815.266	2.852	3.326
22	736.789	.0014	.0016	2102.255	2451.775	2.853	3.328
23	994.665	.0010	.0012	2839.044	3311.063	2.854	3.329
24	1342.798	.0007	.0009	3833.710	4471.102	2.855	3.330
25	1812.778	.0006	.0006	5176.508	6037.153	2.856	3.330
26	2447.250	.0004	.0005	6989.287	8151.325	2.856	3.331
27	3303.788	.0003	.0004	9436.537	11005.455	2.856	3.331
28	4460.114	.0002	.0003	12740.325	14858.530	2.857	3.331
29	6021.153	.0002	.0002	17200.438	20060.180	2.857	3.332
30	8128.557	.0001	.0001	23221.592	27082.410	2.857	3.332

UNIFORM SERIES		GRADIENT SERIES		
SINKING FUND	CAPITAL RECOVERY	UNIFORM SERIES	PRESENT WORTH	
A/F	A/P	A/G	P/G	N
1.0000	1.3500	0.000	0.000	1
0.4255	0.7755	0.426	0.549	2
.2397	.5897	0.803	1.362	3
.1508	.5008	1.134	2.265	4
.1005	.4505	1.422	3.157	5
.0693	.4193	1.670	3.983	6
.0488	.3988	1.881	4.717	7
.0349	.3849	2.060	5.352	8
.0252	.3752	2.209	5.889	9
.0183	.3683	2.334	6.336	10
.0134	.3634	2.436	6.705	11
.0098	.3598	2.520	7.005	12
.0072	.3572	2.589	7.247	13
.0053	.3553	2.644	7.442	14
.0039	.3539	2.689	7.597	15
.0029	.3529	2.725	7.721	16
.0021	.3521	2.753	7.818	17
.0016	.3516	2.776	7.895	18
.0012	.3512	2.793	7.955	19
.0009	.3509	2.808	8.002	20
.0006	.3506	2.819	8.038	21
.0005	.3505	2.827	8.067	22
.0004	.3504	2.834	8.089	23
.0003	.3503	2.839	8.106	24
.0002	.3502	2.843	8.119	25
.0001	.3501	2.847	8.130	26
.0001	.3501	2.849	8.137	27
.0000	.3501	2.851	8.143	28
.0000	.3501	2.852	8.148	29
.0000	.3500	2.853	8.152	30

$i = 0.35$

Compound Interest Factors $i = 0.40$

	SINGLE PAYMENT			UNIFORM SERIES			
	COMPOUND AMOUNT	PRESENT WORTH		COMPOUND AMOUNT		PRESENT WORTH	
N	F/P	P/F	$P/\overline{F}$	F/A	$F/\overline{A}$	P/A	$P/\overline{A}$
1	1.400	0.7143	0.8491	1.000	1.189	0.714	0.849
2	1.960	.5102	.6065	2.400	2.853	1.224	1.456
3	2.744	.3644	.4332	4.360	5.183	1.589	1.889
4	3.842	.2603	.3095	7.104	8.445	1.849	2.198
5	5.378	.1859	.2210	10.946	13.012	2.035	2.419
6	7.530	.1328	.1579	16.324	19.406	2.168	2.577
7	10.541	.0949	.1128	23.853	28.357	2.263	2.690
8	14.758	.0678	.0806	34.395	40.889	2.331	2.771
9	20.661	.0484	.0575	49.153	58.433	2.379	2.828
10	28.925	.0346	.0411	69.814	82.995	2.414	2.869
11	40.496	.0247	.0294	98.739	117.382	2.438	2.899
12	56.694	.0176	.0210	139.235	165.523	2.456	2.920
13	79.371	.0126	.0150	195.929	232.921	2.469	2.935
14	111.120	.0090	.0107	275.300	327.278	2.478	2.945
15	155.568	.0064	.0076	386.420	459.378	2.484	2.953
16	217.795	.0046	.0055	541.988	644.318	2.489	2.958
17	304.913	.0033	.0039	759.783	903.235	2.492	2.962
18	426.879	.0023	.0028	1064.697	1265.717	2.494	2.965
19	597.630	.0017	.0020	1491.575	1773.193	2.496	2.967
20	836.682	.0012	.0014	2089.205	2483.658	2.497	2.968
21	1171.355	.0009	.0010	2925.888	3478.311	2.498	2.969
22	1639.897	.0006	.0007	4097.243	4870.824	2.498	2.970
23	2295.856	.0004	.0005	5737.140	6820.342	2.499	2.971
24	3214.198	.0003	.0004	8032.995	9549.668	2.499	2.971
25	4499.877	.0002	.0003	11247.193	13370.724	2.499	2.971
26	6299.828	.0002	.0002	15747.069	18720.199	2.500	2.972
27	8819.759	.0001	.0001	22046.896	26209.469	2.500	2.972

UNIFORM SERIES		GRADIENT SERIES			
SINKING FUND	CAPITAL RECOVERY	UNIFORM SERIES	PRESENT WORTH		
A/F	A/P	A/G	P/G		N
1.0000	1.4000	0.000	0.000		1
0.4167	0.8167	0.417	0.510		2
.2294	.6294	0.780	1.239		3
.1408	.5408	1.092	2.020		4
.0914	.4914	1.358	2.764		5
.0613	.4613	1.581	3.428		6
.0419	.4419	1.766	3.997		7
.0291	.4291	1.919	4.471		8
.0203	.4203	2.042	4.858		9
.0143	.4143	2.142	5.170		10
.0101	.4101	2.221	5.417		11
.0072	.4072	2.285	5.611		12
.0051	.4051	2.334	5.762		13
.0036	.4036	2.373	5.879		14
.0026	.4026	2.403	5.969		15
.0018	.4018	2.426	6.038		16
.0013	.4013	2.444	6.090		17
.0009	.4009	2.458	6.130		18
.0007	.4007	2.468	6.160		19
.0005	.4005	2.476	6.183		20
.0003	.4003	2.482	6.200		21
.0002	.4002	2.487	6.213		22
.0002	.4002	2.490	6.222		23
.0001	.4001	2.493	6.229		24
.0000	.4001	2.494	6.235		25
.0000	.4001	2.496	6.239		26
.0000	.4000	2.497	6.242		27

$i = 0.40$

Compound Interest Factors $i = 0.45$

	SINGLE PAYMENT			UNIFORM SERIES			
	COMPOUND AMOUNT	PRESENT WORTH		COMPOUND AMOUNT		PRESENT WORTH	
N	F/P	P/F	$P/\overline{F}$	F/A	$F/\overline{A}$	P/A	$P/\overline{A}$
1	1.450	0.6897	0.8352	1.000	1.211	0.690	0.835
2	2.103	.4756	.5760	2.450	2.967	1.165	1.411
3	3.049	.3280	.3973	4.553	5.514	1.493	1.809
4	4.421	.2262	.2740	7.601	9.206	1.720	2.083
5	6.410	.1560	.1889	12.022	14.559	1.876	2.271
6	9.294	.1076	.1303	18.431	22.322	1.983	2.402
7	13.476	.0742	.0899	27.725	33.578	2.057	2.492
8	19.541	.0512	.0620	41.202	49.900	2.109	2.554
9	28.334	.0353	.0427	60.743	73.566	2.144	2.596
10	41.085	.0243	.0295	89.077	107.881	2.168	2.626
11	59.573	.0168	.0203	130.162	157.639	2.185	2.646
12	86.381	.0116	.0140	189.735	229.787	2.196	2.660
13	125.252	.0080	.0097	276.115	334.403	2.204	2.670
14	181.615	.0055	.0067	401.367	486.095	2.210	2.677
15	263.342	.0038	.0046	582.982	706.049	2.214	2.681
16	381.846	.0026	.0032	846.325	1024.982	2.216	2.684
17	553.677	.0018	.0022	1228.171	1487.435	2.218	2.686
18	802.831	.0012	.0015	1781.848	2157.992	2.219	2.688
19	1164.106	.0009	.0010	2584.679	3130.300	2.220	2.689
20	1687.953	.0006	.0007	3748.785	4540.146	2.221	2.690
21	2447.532	.0004	.0005	5436.738	6584.423	2.221	2.690
22	3548.922	.0003	.0003	7884.271	9548.626	2.222	2.691
23	5145.937	.0002	.0002	11433.193	13846.719	2.222	2.691
24	7461.609	.0001	.0002	16579.131	20078.953	2.222	2.691

UNIFORM SERIES		GRADIENT SERIES		
SINKING FUND	CAPITAL RECOVERY	UNIFORM SERIES	PRESENT WORTH	
A/F	A/P	A/G	P/G	N
1.0000	1.4500	0.000	0.000	1
0.4082	0.8582	0.408	0.476	2
.2197	.6697	0.758	1.132	3
.1316	.5816	1.053	1.810	4
.0832	.5332	1.298	2.434	5
.0543	.5043	1.499	2.972	6
.0361	.4861	1.661	3.418	7
.0243	.4743	1.791	3.776	8
.0165	.4665	1.893	4.058	9
.0112	.4612	1.973	4.277	10
.0077	.4577	2.034	4.445	11
.0053	.4553	2.082	4.572	12
.0036	.4536	2.118	4.668	13
.0025	.4525	2.145	4.740	14
.0017	.4517	2.165	4.793	15
.0012	.4512	2.180	4.832	16
.0008	.4508	2.191	4.861	17
.0006	.4506	2.200	4.882	18
.0004	.4504	2.206	4.898	19
.0003	.4503	2.210	4.909	20
.0002	.4502	2.214	4.917	21
.0001	.4501	2.216	4.923	22
.0000	.4501	2.218	4.927	23
.0000	.4501	2.219	4.930	24

$i = 0.45$

Compound Interest Factors $i = 0.50$

	SINGLE PAYMENT			UNIFORM SERIES			
	COMPOUND AMOUNT	PRESENT WORTH		COMPOUND AMOUNT		PRESENT WORTH	
N	F/P	P/F	$P/\overline{F}$	F/A	$F/\overline{A}$	P/A	$P/\overline{A}$
1	1.500	0.6667	0.8221	1.000	1.233	0.667	0.822
2	2.250	.4444	.5481	2.500	3.083	1.111	1.370
3	3.375	.2963	.3654	4.750	5.857	1.407	1.736
4	5.063	.1975	.2436	8.125	10.019	1.605	1.979
5	7.594	.1317	.1624	13.188	16.262	1.737	2.142
6	11.391	.0878	.1083	20.781	25.626	1.824	2.250
7	17.086	.0585	.0722	32.172	39.673	1.883	2.322
8	25.629	.0390	.0481	49.258	60.742	1.922	2.370
9	38.443	.0260	.0321	74.887	92.347	1.948	2.402
10	57.665	.0173	.0214	113.330	139.753	1.965	2.424
11	86.498	.0116	.0143	170.995	210.863	1.977	2.438
12	129.746	.0077	.0095	257.493	317.528	1.985	2.447
13	194.620	.0051	.0063	387.239	477.524	1.990	2.454
14	291.929	.0034	.0042	581.859	717.520	1.993	2.458
15	437.894	.0023	.0028	873.788	1077.513	1.995	2.461
16	656.841	.0015	.0019	1311.682	1617.502	1.997	2.463
17	985.261	.0010	.0013	1968.522	2427.487	1.998	2.464
18	1477.892	.0007	.0008	2953.784	3642.463	1.999	2.465
19	2216.838	.0005	.0006	4431.676	5464.928	1.999	2.465
20	3325.257	.0003	.0004	6648.513	8198.625	1.999	2.466
21	4987.885	.0002	.0002	9973.770	12299.171	2.000	2.466
22	7481.828	.0001	.0002	14961.655	18449.990	2.000	2.466

UNIFORM SERIES		GRADIENT SERIES		
SINKING FUND	CAPITAL RECOVERY	UNIFORM SERIES	PRESENT WORTH	
A/F	A/P	A/G	P/G	N
1.0000	1.5000	0.000	0.000	1
0.4000	0.9000	0.400	0.444	2
.2105	.7105	0.737	1.037	3
.1231	.6231	1.015	1.630	4
.0758	.5758	1.242	2.156	5
.0481	.5481	1.423	2.595	6
.0311	.5311	1.565	2.947	7
.0203	.5203	1.675	3.220	8
.0134	.5134	1.760	3.428	9
.0088	.5088	1.824	3.584	10
.0058	.5058	1.871	3.699	11
.0039	.5039	1.907	3.784	12
.0026	.5026	1.933	3.846	13
.0017	.5017	1.952	3.890	14
.0011	.5011	1.966	3.922	15
.0008	.5008	1.976	3.945	16
.0005	.5005	1.983	3.961	17
.0003	.5003	1.988	3.973	18
.0002	.5002	1.991	3.981	19
.0002	.5002	1.994	3.987	20
.0001	.5001	1.996	3.991	21
.0000	.5001	1.997	3.994	22

$i = 0.50$

TABLE OF RANDOM NUMBERS*

139	407	027	030	530	687	694	017	943	787
073	886	255	332	037	264	341	948	462	774
075	259	224	042	332	890	196	693	988	467
254	352	917	614	273	643	994	956	128	193
096	119	694	625	095	727	846	565	868	405
459	637	289	778	407	468	234	472	567	681
577	111	813	903	194	321	019	757	959	726
062	868	748	951	815	863	435	621	154	365
895	362	955	001	004	798	091	394	637	554
438	170	667	256	871	953	972	528	265	370
424	995	495	044	900	283	436	601	275	016
963	666	423	819	951	864	219	317	274	820
539	136	809	158	257	900	430	504	249	235
011	483	389	765	429	720	553	115	557	840
615	910	272	467	450	776	447	227	934	337
958	745	941	218	680	646	347	045	488	555
026	442	257	096	854	034	862	896	705	447
178	578	454	305	080	768	977	233	443	091
149	856	142	171	844	800	051	635	937	689
047	106	304	149	003	210	819	804	796	572
357	279	299	816	794	199	389	569	005	190
939	454	864	876	825	097	246	882	922	123
027	834	106	157	081	356	250	823	284	073
230	747	510	611	920	554	634	594	197	869
532	647	935	317	078	396	009	523	148	464
294	111	617	479	664	707	358	063	996	936
248	843	163	423	162	443	042	793	974	488
506	670	559	604	431	680	793	415	692	449
551	546	165	599	706	623	723	758	136	270
242	550	713	112	597	599	314	775	663	531
814	883	315	971	087	061	427	544	008	935
876	874	453	128	536	588	296	268	281	309
413	977	988	663	678	882	530	275	967	607
784	769	154	777	623	772	114	018	923	907
723	954	560	800	855	210	407	076	386	412
340	360	190	184	234	276	143	151	964	450
119	939	405	508	993	172	432	073	641	475
920	770	938	474	743	226	758	792	778	064
976	057	899	910	468	891	980	389	108	921
898	126	771	771	526	746	333	066	740	873
669	432	416	134	653	493	427	152	160	875
649	553	066	201	957	961	245	098	226	003
573	190	331	302	924	103	147	484	173	461
549	174	196	889	412	997	868	013	610	577
062	457	020	541	656	846	516	512	522	805

*This table was prepared by Mr. Ken Molay using a PDP 11 computer with a TOPS 10 operating system. To produce these random numbers, a congruential multiplicative generator was used, based on a seed of system time in milliseconds.

TABLE OF RANDOM NORMAL DEVIATES*

0.199	−0.066	−0.205	0.455	−2.023	−0.131	−0.032	1.050
1.344	0.421	−0.599	−0.575	0.231	−0.455	1.977	2.029
−0.362	1.112	−0.200	0.072	1.044	1.399	0.910	−1.630
−0.451	−0.413	−0.159	1.421	0.286	0.499	1.402	0.750
−1.477	−0.149	−1.234	−0.644	−1.753	−0.895	1.393	0.853
−0.392	0.977	0.603	0.851	−1.161	0.206	0.294	−0.270
1.341	0.009	−1.489	0.499	0.695	−1.284	−0.542	0.682
−0.993	1.078	0.194	0.231	0.615	−1.436	−0.19	0.928
−0.708	−0.134	−0.308	1.797	−0.354	−0.445	0.019	1.355
−0.336	2.044	0.199	−0.401	−0.929	−1.964	−0.746	−0.229
0.307	−0.998	−1.083	0.104	−1.385	−1.224	0.428	0.607
−1.361	−0.203	0.675	−0.761	−0.092	−1.309	−0.966	−0.335
0.467	−0.256	0.788	0.72	−0.349	−1.401	0.205	1.043
0.373	−1.472	0.334	−0.361	−2.519	−0.658	−0.249	−1.017
1.517	0.615	−1.414	−0.665	−0.701	−0.105	−0.78	−0.266
−1.659	−0.902	−0.883	−1.679	−0.197	−1.329	0.596	−0.419
1.078	0.274	0	0.926	−1.557	−0.610	1.554	−0.139
−0.388	−1.048	−1.135	−0.878	−1.705	0.275	0.535	−0.488
0.008	0.184	−0.208	0.236	−0.134	−0.705	0.202	0.354
−2.998	−0.165	−0.295	−0.282	−0.709	1.024	0.029	1.179
0.051	−1.229	−1.265	0.440	0.593	0.276	1.053	−0.125
0.536	−0.367	2.430	0.312	0.431	0.987	0.335	0.505
1.761	0.349	−1.039	−0.814	0.299	−0.057	0.970	1.705
0.365	0.250	1.426	−1.042	−0.822	−1.065	0.708	−0.144
0.921	0.190	0.385	1.674	0.483	−0.863	−0.743	2.513
−1.308	−0.892	1.333	−0.127	−0.590	−1.590	−0.470	0.159
0.647	0.879	0.094	−0.464	0.093	0	0.614	0.393
−0.603	−0.333	−0.373	−0.523	−0.058	−1.294	0.321	−1.855
−0.214	−0.699	−0.292	0.928	0.363	0.035	0.645	−1.243
1.223	−0.868	−0.397	−0.047	0.870	−0.613	0.174	1.602
−0.649	−0.244	0.008	−0.611	0.958	−0.940	2.080	0.964
−2.215	1.712	0.941	0.537	−1.221	0.263	0.893	1.171
0.630	−0.602	−0.401	0.922	−0.734	1.992	−0.310	−1.030
−0.516	0.539	1.148	−0.373	−0.805	1.855	−0.115	−0.773
0.764	−1.190	−0.150	0.396	1.620	0.575	−0.049	−0.279
−0.519	0.772	0.817	1.003	0.306	−1.761	−0.841	−1.099
−0.144	1.254	−0.661	0.890	0.645	1.618	−1.800	−0.297
0.469	0.514	−0.304	−0.166	1.145	1.018	−0.080	0.030
1.871	0.048	−0.075	0.105	−0.617	−1.945	1.378	0.782
−2.306	−1.901	1.636	−0.725	0.264	0.169	−0.337	−0.208

*This table was prepared by Mr. Shay Bao Lai using an Apple II computer. The algorithm for producing these random normal deviates is from A. M. Law and W. D. Kelton, *Simulation Modeling and Analysis* (McGraw-Hill, 1982), pp. 258–259.

Selected Bibliography

APPENDIX E

This appendix includes selected references that are addressed generally to engineering economy, economic analysis, cost-benefit analysis, capital budgeting, and capital allocation theory. The general references include many of the topics discussed in this book, especially in Chapters 1–6. In addition, references of special relevance to particular topics are identified separately. The bibliography is by no means complete. It is simply intended to serve as a starting point for students interested in additional study in these areas.

General References

The American Society for Engineering Education, and Institute of Industrial Engineers. *The Engineering Economist.* Norcross, Ga.: Institute of Industrial Engineers. Published quarterly by the Engineering Economy Division of the American Society for Engineering Education and the Institute of Industrial Engineers.

American Telephone and Telegraph, Engineering Department. *Engineering Economy.* 3d ed. New York: American Telephone and Telegraph, 1980.

Au, Tung, and Thomas P. Au. *Engineering Economics for Capital Investment Analysis.* Boston, Mass.: Allyn and Bacon, 1983.

Barish, Norman N., and Seymour Kaplan. *Economic Analysis: For Engineering and Managerial Decision Making.* 2d ed. New York: McGraw Hill, 1978.

Bierman, H., and S. Smidt. *The Capital Budgeting Decision.* 4th ed. New York: Macmillan, 1975.

Bussey, Lynn E. *The Economic Analysis of Industrial Projects.* Englewood Cliffs, N.J.: Prentice-Hall, 1978.

Canada, John R., and John A. White, Jr. *Capital Investment Decision Analysis for Management and Engineering.* Englewood Cliffs, N.J.: Prentice Hall, 1980.

Dasgupta, Agit K., and D. W. Pearce. *Cost-Benefit Analysis: Theory & Practice.* New York: Harper & Row, 1972.

DeGarmo, E. Paul, John R. Canada, and William G. Sullivan. *Engineering Economy.* 6th ed. New York: Macmillan, 1979.

Emerson, Robert, and William Taylor. *An Introduction to Engineering Economy.* 2d ed. Bozeman, Mont.: Cardinal, 1979.

Fabrycky, Wolter, and Gerald J. Thuesen. *Economic Decision Analysis.* 2d ed. Englewood Cliffs, N.J.: Prentice-Hall, 1980.

Fish, J. C. L. *Engineering Economics.* 2d ed. New York: McGraw-Hill, 1923. This book is of historical importance; it was the first general engineering economy textbook.

Fleischer; G. A. *Capital Allocation Theory.* New York: Appleton-Century-Crofts, 1969.

Frost, Michael J. *How To Use Benefit-Cost Analysis in Project Appraisal.* 2d ed. New York: Macmillan 1979.

Grant, Eugene L., W. Grant Ireson, and Richard S. Leavenworth. *Principles of Engineering Economy.* 7th ed. New York: John Wiley & Sons, 1982.

Kasner, Erick. *Essentials of Engineering Economics.* New York: McGraw-Hill, 1979.

Mallik, Arup K. *Engineering Economy with Computer Applications.* Mahomet, Ill. Engineering Technology, 1979.

Morris, William T. *Engineering Economic Analysis.* Englewood Cliffs, N.J.: Reston, 1976.

Newnan, Donald G. *Engineering Economic Analysis.* 2d ed. San Jose, Calif.: Engineering Press, 1983.

Oakford, R. V. *Capital Budgeting: A Quantitative Evaluation of Investment Alternatives.* New York: John Wiley & Sons, 1970.

Park, William R. *Cost Engineering Analysis.* New York: John Wiley & Sons, 1973.

Reisman, Arnold. *Managerial and Engineering Economics.* Boston, Mass.: Allyn and Bacon, 1971.

Riggs, James L. *Engineering Economics.* 2d ed. New York: McGraw-Hill, 1982.

Rose, L. M. *Engineering Investment Decisions.* Netherlands: Elsevier Scientific, 1976.

Sassone, Peter G., and William A. Schaffer. *Cost-Benefit Analysis: A Handbook.* New York: Academic Press, 1978.

Smith, Gerald W. *Engineering Economy: The Analysis of Capital Expenditures.* 3d ed. Ames, Iowa: Iowa State University Press, 1979.

Steiner, H. M. *Public and Private Investments—Socioeconomic Analysis.* New York: John Wiley & Sons, 1980.

Stevens, G. T. *Economic and Financial Analysis of Capital Investments.* New York: John Wiley & Sons, 1979.

Tarquin, Anthony J., and Leland T. Blank. *Engineering Economy: A Behavioral Approach.* 2d ed. New York: McGraw-Hill, 1983.

Taylor, George. *Managerial and Engineering Economy.* 3d ed. New York: D. Van Nostrand, 1980.

Thuesen, H. G., W. J. Fabrycky, and G. J. Thuesen. *Engineering Economy.* 5th ed. Englewood Cliffs, N.J.: Prentice-Hall, 1977.

Wellington, Arthur M. *The Economic Theory of Railway Location.* 2d ed. New York: John Wiley & Sons, 1887. This book is of historical importance; it was the first to address the issue of economic evaluation of capital investments due to engineering design decisions. Wellington is widely considered to be the "father of engineering economy."

White, John A., Marvin H. Agee, and Kenneth E. Case. *Principles of Engineering Economy Analysis.* New York: John Wiley & Sons, 1977.

Equivalent Methods for Selection Among Alternatives (Chapter 3)

American Association of Highway and Transportation Officials. *Manual on Road User Benefit Analysis of Highway and Bus-Transit Improvements.* Washington, D.C.: American Association of Highway and Transportation Officials, 1978.

American Association of State Highway Officials. *Road User Benefit Analysis for Highway Improvements.* Washington, D.C.: American Association of State Highway Officials, 1960. (The "Red Book")

Bernhard, Richard H. "'Modified' Rates of Return for Investment Project Evaluation—A Comparison and Critique." *The Engineering Economist,* vol. 24, no. 3, Spring 1979, pp. 161–167.

Hitch, Charles J., and Roland N. McKean. *The Economics of Defense in the Nuclear Age,* Report R-346. Santa Monica, Calif.: The Rand Corporation, March 1960.

Kendall, M. G., ed. *Cost-Benefit Analysis.* New York: American Elsevier 1971.

Landau, H. J. "On Comparison of Cash Flow Streams." *Management Science,* vol. 26, no. 12, December 1980, pp. 1218–1226.

Maas, Arthur. "Benefit-Cost Analysis: Its Relevance to Public Investment Decisions." In A. V. Kneese and S. C. Smith, *Water Research.* Baltimore, Md.: The Johns Hopkins Press, 1966, pp. 311–328.

Prest, A. R., and R. Turvey. "Cost-Benefit Analysis: A Survey." In *Surveys of Economic Theory: Vol. 3, Resource Allocation.* Prepared for the American Economic Association and the Royal Economic Society, New York: St. Martin's Press, 1966.

Solomon, Ezra. "The Arithmetic of Capital Budgeting Decisions." *Journal of Business,* vol. 29, 1956, p. 124.

United Nations. *Guidelines for Project Evaluation.* Sales No. E.72.II.B.11. New York: United Nations, 1972.

U.S. Congress. Joint Economic Committee, Subcommittee on Priorities in Government. *Benefit-Cost Analyses of Federal Programs, A Compendium of Papers.* 92nd Cong., 2d sess., 1973.

U.S. Government. *U.S. Code.* Washington, D.C.: Government Printing Office, 1940, p. 2964.

———. *The Analysis and Evaluation of Public Expenditures: The PPB System.* Washington, D.C.: Government Printing Office, 1969.

U.S. Government, Interagency Committee on Water Resources. *Proposed Practices for Economic Analysis of River Basin Projects.* Washington, D.C.: Government Printing Office, 1958. (The "Green Book")

Winfrey, Robley. *Economic Analysis for Highways*. Scranton, Pa.: International Textbook, 1969.

Multiple Alternatives (Chapter 4)

Charnes, A., W. W. Cooper, and M. H. Miller. "Application of Linear Programming to Financial Budgeting and the Costing of Funds." *Journal of Business,* vol. 32, no. 1, January 1959, pp. 20–46.

Everett, Hugh. "Generalized LaGrange Multiplier Method for Solving Problems of Optimum Allocation." *Journal of the Operations Research Society of America,* vol. 11, no. 3, May 1963, pp. 399–417.

Fleischer, Gerald A. "Two Major Issues Associated with the Rate of Return Method for Capital Allocation: The 'Ranking Error' and 'Preliminary Selection.'" *The Journal of Industrial Engineering,* vol. 17, no. 4, April 1966, pp. 202–208.

Freeland, J., and M. J. Rosenblatt. "An Analysis of Linear Programming Formulation for the Capital Rationing Problem." *The Engineering Economist,* vol. 24, no. 4, Fall 1978, pp. 49–61.

Kim, Suk, and Henry Guithnes. *Capital Expenditure Analysis*. Washington, D.C.: University Press, 1981.

Lorie, J. H., and L. J. Savage. "Three Problems in Rationing Capital." *Journal of Business,* vol. 28, no. 4, October 1955, pp. 229–239.

Radhakrishnan, S. R., and V. E. Unger. "Capital Budgeting and Mixed Zero-One Integer Programming: The Quadratic Case." *AIIE Transactions,* vol. 5, no. 4, December 1973, pp. 366–369.

Unger, Jr., V. E. "Capital Budgeting and Mixed Zero-One Integer Programming." *AIIE Transactions,* vol. 2, no. 1, March 1970, pp. 28–36.

Weingartner, H. Martin. *Mathematical Programming and the Analysis of Capital Budgeting Problems*. Englewood Cliffs, N.J.: Prentice-Hall, 1963. (Also, Chicago: Markham Publishing, 1967.)

——. "Capital Budgeting of Interrelated Projects: Survey and Synthesis." *Management Science,* vol. 12, no. 7, March 1966, pp. 485–516.

Some Incorrect and/or Approximate Methods (Chapter 5)

Baldwin, R. H. "How to Assess Investment Proposals." *Harvard Business Review,* vol. 37, no. 3, May/June 1959, pp. 98–104.

Barish, Norman N. *Economic Analysis*. New York: McGraw-Hill, 1962, pp. 141–143.

Brown, Richard S., et al. *Economic Analysis Handbook*. Alexandria, Va.: Naval Facilities Engineering Command, 1975, pp. 23–28.

Dryden, Miles. "The MAPI Urgency Rating as an Investment Ranking Criterion." *The Journal of Business,* vol. 33, no. 4, October 1960, pp. 327–341.

Hoskold, Henry. *Engineer's Valuing Assistant*. 2d ed. New York: Longman's Green, 1905.

Ijiri, Yuji. "Approximations to Interest Formulas." *Journal of Business,* vol. 45, no. 3, July 1972, pp. 398–402.

Weingartner, H. Martin. "The Excess Present Value Index—A Theoretical Basis and Critique." *The Journal of Accounting Research,* vol. 1, no. 2, 1963, pp. 213–224.

Wortham, A. W., and R. J. McNichols. "Return Analysis on Equipment Payout." *International Journal of Production Research,* vol. 7, no. 3, 1969, pp. 183–187.

Retirement and Replacement (Chapter 6)

Alchian, A. A. *Economic Replacement Policy*. Santa Monica, Calif.: The Rand Corporation, RM-2153, 1958.

Dean, Burton V. "Replacement Theory." *Progress in Operations Research,* vol. 1, ORSA Publication no. 5, (edited by R. L. Ackoff), 1963, chapter 9.

Dreyfus, Stuart E. "A Generalized Equipment Replacement Study." *Journal of the Society for Industrial and Applied Mathematics,* vol. 8, no. 3, September 1960, pp. 425–435.

Grinyer, Peter H. "The Effects of Technological Change on the Economic Life of Capital Equipment." *AIIE Transactions,* vol. 5, no. 3, September 1973, pp. 203–213.

Morris, W. T. *Engineering Economic Analysis.* Reston, Va.: Reston Publishing, 1976.

Naik, M. D., and K. P. Nair. "Multistage Replacement Strategies." *Journal of the Operations Research Society of America,* vol. 13, no. 2, March/April 1965, pp. 279–290.

Terborgh, George. *Dynamic Equipment Policy.* Washington, D.C.: Machinery and Allied Products Institute, 1949.

———. *Realistic Depreciation Policy.* Washington, D.C.: Machinery and Allied Products Institute, 1954.

———. *Business Investment Policy.* Washington, D.C.: Machinery and Allied Products Institute, 1958.

———. *Business Investment Management.* Washington, D.C.: Machinery and Allied Products Institute, 1967.

Depreciation, Taxation, and After-Tax Economy Studies (Chapter 7)

Blank, L. T., and D. R. Smith. "A Comparative Evaluation of the Accelerated Cost Recovery System as Enacted by the 1981 Economic Recovery Tax Act." *The Engineering Economist,* vol. 28, no. 1, Fall 1982, pp. 1–30.

Buck, James R., and T. W. Hill. "Generalized Depreciation Methods and After-Tax Project Evaluation." *The Engineering Economist,* vol. 22, no. 2, Winter 1977, pp. 79–96.

Commerce Clearing House. *Tax Equity and Fiscal Responsibility Tax Act of 1982.* Chicago, Ill.: Commerce Clearing House, 1982.

Frair, L. C., and M. D. Devine. "The Consideration of Depletion in After-Tax Optimization Models." *The Engineering Economist,* vol. 24, no. 1, Fall 1978, pp. 13–27.

Gehrlein, W. V., and T. K. Tiemann. "Optimal Depreciation Policies for Regulated Firms." *The Engineering Economist,* vol. 25, no. 2, Winter 1980, pp. 79–85.

Grant, E. L., and Paul T. Norton. *Depreciation.* New York: The Ronald Press, 1955. Revised printing.

Kopits, G. *International Comparison of Tax Depreciation Practices.* Paris, France: Organization for Economic Cooperation and Development, 1975.

Lasser, J. K. *Your Income Tax.* New York: Simon and Schuster. (See latest edition)

Lohmann, J. R., E. W. Foster, and D. J. Layman. "A Comparative Analysis of the Effect of ACRS on Replacement Economy Decisions." *The Engineering Economist,* vol. 27, no. 4, Summer 1982, pp. 247–260.

Prentice-Hall. *Complete Internal Revenue Code of 1954.* Englewood Cliffs, N.J.: Prentice-Hall. (See latest edition.)

Schoomer, B. Alva. "Optimal Depreciation Strategies for Income Tax Purposes." *Management Science,* vol. 12, no. 12, August 1966, pp. 552–580.

U.S. Department of the Treasury. *Your Federal Income Tax,* IRS Publication 17. Washington, D.C.: Government Printing Office. (Revised annually)

———. *Tax Guide for Small Business,* IRS Publication 334. Washington, D.C.: Government Printing Office. (Revised annually)

———. *Depreciation,* IRS Publication 534. Washington, D.C.: Government Printing Office. (See latest edition)

———. *Investment Credit,* IRS Publication 572. Washington, D.C.: Government Printing Office. (See latest edition)

Risk and Uncertainty (Chapter 8)

Bennion, Edward G. "Capital Budgeting and Game Theory." *Harvard Business Review,* vol. 34, no. 6, November/December 1956, pp. 115–123.

Farrar, Donald. *The Investment Decision under Uncertainty.* Englewood Cliffs, N.J.: Prentice-Hall, 1962.

Ferguson, Earl J., and James E. Shamblin. "Break-Even Analysis." *The Journal of Industrial Engineering,* vol. 18, no. 8, August 1967, pp. xvii–xx.

Fishburn, Peter C. *Decision and Value Theory.* New York: John Wiley & Sons, 1964.

Grayson, Jr., C. Jackson. *Decisions under Uncertainty.* Boston: Harvard Business School Press, 1960.

Hertz, David B. "Risk Analysis in Capital Investment." *Harvard Business Review,* vol. 42, no. 1, January/February 1964, pp. 89–106.

———. "Investment Policies That Pay Off." *Harvard Business Review,* vol. 46, no. 1. January/February 1968, pp. 96–108.

Hillier, F. S. "The Derivation of Probabilistic Information for Evaluation of Risky Investments." *Management Science,* vol. 9, no. 3, April 1963, pp. 443–457.

———. Supplement to "The Derivation of Probabilistic Information for the Evaluation of Risky Investments." *Management Science,* vol. 11, no. 3, January 1965, pp. 485–487.

———. *The Evaluation of Risky Interrelated Investments.* Amsterdam: North Holland Publishing, 1969.

Lipschutz, Seymour. *Theory and Problems of Probability.* Schaum's Outline Series. New York: McGraw-Hill, 1968.

Luce, R. Duncan, and Howard Raiffa. *Games and Decisions: Introduction and Critical Survey.* New York, John Wiley & Sons, 1957.

Markowitz, Harry M. *Portfolio Selection: Efficient Diversification of Investments.* Cowles Commision Monograph 16, New York: John Wiley & Sons, 1959.

Mendenhall, Richard, and R. L. Scheaffer. *Mathematical Statistics with Applications.* North Scituate, Mass.: Duxbury Press, 1973.

Oakford, R. V., A. Salazar, and H. A. DiGuilio. "The Long-Term Effectiveness of Expected Net Present Value Maximization in an Environment of Incomplete and Uncertain Information." *AIIE Transactions,* vol. 13, no. 3, September 1981, pp. 265–276.

Schlaifer, Robert. *Analysis of Decisions under Uncertainty.* vol. 1. New York: McGraw-Hill, 1967.

Shannon, Robert E. "Simulation: A Survey with Research Suggestions." *AIIE Transactions,* vol. 7, no. 3, September 1975, pp. 289–299.

Smith, L. D. "Quantifying Risk for Establishing Rates of Return in Regulated Industries." *The Engineering Economist,* vol. 28, no. 4. Summer 1983.

Sullivan, W. G., and R. G Orr. "Monte Carlo Simulation Analyzes Alternatives in Uncertain Economy." *Industrial Engineering,* vol. 14, no. 11, November 1982, pp. 43–49.

Teichroew, Robichek, and Montalbano. "Mathematical Analysis of Rates of Return under Uncertainty." *Management Science,* vol. 11, no. 3, January 1965, pp. 395–404.

Thrall, R. M., C. H. Coombs, and R. L. Davis. *Decision Processes.* New York: John Wiley & Sons, 1954.

Van Horne, J. "Capital Budgeting Decisions Involving Combinations of Risky Investments." *Management Science,* vol. 13, no. 10, October 1966, pp. 84–92.

von Neumann, J., and O. Morgenstern. *Theory of Games and Economic Behavior.* Princeton, N.J.: Princeton University Press, 1947.

The Revenue Requirement Method (Chapter 9)

Commonwealth Edison Company. *Engineering Economics.* Chicago, Ill.: Commonwealth Edison Company, 1975.

Jeynes, P. H. *Profitability and Economic Choice.* Ames, Iowa: The Iowa State University Press, 1968.

Mayer, R. R. "Finding Your Minimum Revenue Requirements." *Industrial Engineering,* vol. 9, no. 4, April 1977, pp. 16–22.

Sullivan, W. G. "Engineering Economy Studies in Investor-Owned Utilities." Proceedings of the American Society for Engineering Education, 87th Annual Conference. Louisiana State University, June 25–28, 1979.

Ward, T. L., and W. G. Sullivan. "Equivalence of the Present Worth and Revenue Requirement Methods of Capital Investment Analysis." *AIIE Transactions,* vol. 13, no. 1, March 1981, pp. 29–40.

Inflation and Index Numbers (Chapter 10)

Aaron, Henry. *Inflation and the Income Tax.* Washington, D.C.: The Brookings Institution, 1976.

Allen, R. G. D. *Index Numbers in Theory and Practice.* Chicago: Aldine, 1975.

Davidson, Sidney, Clyde Stickney, and Roman Weil. *Inflation Accounting.* New York: McGraw-Hill, 1976.

Duvall, R. M., and James Bulloch. "Adjusting Rate of Return and Present Value for Price Level Changes." *The Accounting Review,* vol. 40, no. 3, July 1965, pp. 569–573.

Elliott, J. W., and J. R. Baier. "Econometric Models and Current Interest Rates: How Well Do They Predict Future Rates?" *The Journal of Finance,* vol. 34, no. 4, September 1979, pp. 975–986.

Engineering News Record. *Construction Costs.* New York: McGraw-Hill. (Published weekly.) For subscriptions, write to Fulfillment Manager, *Engineering News Record,* P.O. Box 430, Hightstown, N.J. 08520.

Foster, Earl M. "Impact of Inflation on Capital Budgeting Decisions." *Quarterly Review of Economics and Business,* vol. 10, no. 3, Fall 1970, pp. 19–24.

Freidenfelds, John, and Michael Kennedy. "Price Inflation and Long-Term Present Worth Studies." *The Engineering Economist,* vol. 24, no. 3, Spring 1979, pp. 143–160.

Gainsburgh, Martin R., and Jules Backman. *Inflation and the Price Indexes.* New York: National Industrial Conference Board, 1966.

Industrial Engineering, vol. 12, no. 3, March 1980. The entire issue is devoted to "The Industrial Engineer and Inflation." Of particular interest are
   a. Estes, C. B., W. C. Turner, and K. E. Case. "Inflation—Its Role in Engineering-Economic Analysis," pp. 18–22.
   b. Bontadelli, J. A., and W. G. Sullivan. "How an IE Can Account for Inflation in Decision-Making," pp. 24–33.
   c. Ward, Tom L. "Leasing During Inflation: A Two-Edged Sword," pp. 34–37.

Jones, Byron W. *Inflation In Engineering Economic Analysis.* USA and Canada: Wiley-Interscience, 1982.

Jordan, Joe D. "Aspects of Engineering Economy Studies under Inflation." Ph.D. diss., Department of Industrial Engineering, Mississippi State University, May 1981.

Miller, Elwood L. "What's Wrong with Price Level Accounting?" *Harvard Business Review,* November/December 1978, pp. 111–118.

Oakford, R. V., and A. Salazar. "The Arithmetic of Inflation Corrections in Evaluating 'Real' Present Worths." *The Engineering Economist,* vol. 27, no. 2, Winter 1982, pp. 127–146.

Oskounejad, M. Mehdi. *"An Investigation of the Effects of Depreciation Methods on Pricing Policy during Inflation."* Ph.D. diss., Department of Industrial Engineering, Mississippi State University, May 1979.

Remer, D. S., and S. A. Ganiy. "The Role of Interest and Inflation Rates in Present-Worth Analysis in the United States." *The Engineering Economist,* vol. 28, no. 3, Spring 1983, pp. 173–190.

Revsine, Lawrence. *Replacement Cost Accounting.* Englewood Cliffs, N.J.: Prentice-Hall, 1973.

Spiegel, M. R. *Statistics.* Schaum's Outline Series. New York: Schaum Publishing, 1961. (See Chapter 17, "Index Numbers.")

Stockfisch, J. A. *Measuring the Opportunity Cost of Government Investments.* Research Paper P-490. Arlington, Va.: Institute for Defense Analysis, 1969.

Stuart, David O., and Robert D. Meckna. "Accounting for Inflation in Present Worth Studies." *Power Engineering,* February 1975, pp. 45–47.

Terborgh, George. "Effect of Anticipated Inflation on Investment Analysis." Reprinted in Gerald W. Smith, *Engineering Economy: Analysis of Capital Expenditures.* 2d ed. Ames, Iowa: Iowa State University Press, 1973, pp. 545–552.

U.S. Department of Commerce, Bureau of Economic Analysis. *Business Conditions Digest.* Washington, D.C.: U.S. Department of Commerce, Bureau of Economic Analysis. (Published monthly)

U.S. Department of Labor, Bureau of Labor Statistics. *Handbook of Labor Statistics*. Washington, D.C.: Government Printing Office. (Published periodically; obtain latest issue.)

U.S. Department of Labor, Bureau of Labor Statistics. *The Consumer Price Index: Concepts and Contents over the Years*. Report 517. Washington, D.C.: Government Printing Office, May 1978.

Watson, F. A., and F. A. Holland. "Profitability Assessment of Projects under Inflation." *Engineering and Process Economics,* vol. 2, no. 3, 1976, pp. 207–221.

Estimating the Minimum Attractive Rate of Return (Chapter 11)

Barges, Alexander. *The Effect of Capital Structure on the Cost of Capital.* Englewood Cliffs, N.J.: Prentice-Hall, 1963.

Baumol, William J. "On the Appropriate Discount Rate for Evaluation of Public Projects." In *Program Budgeting and Benefit-Cost Analysis,* edited by Hinrichs and Taylor. Pacific Palisades, Cal.: Goodyear Publishing, 1969, pp. 202–212.

Bierman, H., and C. P. Alderfer. "Estimating the Cost of Capital: A Different Approach." *Decision Sciences,* vol. 1, no. 1, January 1970, pp. 40–53.

Brigham, E. F., and R. H. Pettway. "Capital Budgeting by Public Utilities." *Financial Management,* vol. 2, Fall 1973, pp. 11–22.

Fisher, Lawrence, and James H. Lorie. "Rates of Return on Investments in Common Stock, the Year by Year Record 1926–65." *Journal of Business,* vol. 41, no. 3, July 1968, pp. 291–316.

Gordon, M. J. "A General Solution to the Buy or Lease Decision." *Journal of Finance,* March 1974, pp. 245–250.

Levy, Haim, and Yorem Landskroner. "Lease Financing: Cost versus Liquidity." *The Engineering Economist,* vol. 27, no. 1, Fall 1982, pp. 59–69.

Lintner, J. "The Cost of Capital and Optimal Financing of Corporate Growth." *Journal of Finance,* vol. 18, no. 2, May 1963, pp. 292–310.

Long, M. S. "Using a Before or After Tax Discount Rate in the Lease-Buy Decision." *The Engineering Economist,* vol. 26, no. 4, Summer 1981, pp. 263–274.

Mao, J. C. T. *Quantitative Analysis of Financial Decisions.* New York: Macmillan, 1969, pp. 374–499.

Marglin, Stephen A. "The Social Rate of Discount and the Optimal Rate of Investment." *Quarterly Journal of Economics,* vol. 77, no. 1, February 1963, pp. 95–111.

Miller, M. H., and F. Modigliani. "Dividend Policy, Growth and the Valuation of Shares." *Journal of Business,* vol. 34, no. 4, October 1961, pp. 411–433.

———. "The Cost of Capital, Corporation Finance and the Theory of Investment." *American Economic Review,* vol. 48, no. 3, June 1958, pp. 261–297.

Ofer, A. H. "The Valuation of the Lease versus Purchase Alternatives." *Financial Management,* Summer 1976, pp. 67–74.

Porterfield, James T. *Investment Decisions and Capital Costs.* Englewood Cliffs, N.J.: Prentice-Hall, 1965.

Quirin, G. D., and J. G. Wiginton. *Analyzing Capital Expenditures: Private and Public Perspectives.* Homewood, Ill.: Richard D. Irwin, 1981, chapters 5, 6, and 8.

Solomon, Ezra. *The Management of Corporate Capital.* Glencoe, Ill.: The Free Press, 1959.

———. *The Theory of Financial Management.* New York: Columbia University Press, 1963.

Thompson, H. E. "Estimating the Cost of Equity Capital for Electric Utilities: 1958–1976." *The Bell Journal of Economics,* vol. 10., no. 2, Fall 1979.

U.S. Congress. *Hearings before the Subcommittee on Economy of the Joint Economic Committee. Interest Rate Guidelines for Federal Decision-making.* 90th Cong., 2d sess., January 29, 1968.

U.S. Congress. *Report of the Subcommittee on Economy in Government of the Joint Economic Committee. Economic Analysis of Public Investment Deci-*

sions: Interest Rate Policy and Discounting Analysis. 90th Cong., 2d sess., September 23, 1968, pp. 10–16.

U.S. Government, Office of Management and Budget. Circular No. A-94, revised. Washington, D.C.: Government Printing Office, March 27, 1972.

Ward, Thomas L. "Engineering Economy Studies in Investor Owned Utilities: The Treatment of Inflation." Proceedings of the American Society for Engineering Education, 87th Annual Conference, Louisiana State University, June 25–28, 1979.

Considering the Irreducibles (Chapter 12)

Chankong, V., and Y. Haimes. *Multiobjective Decision Making: Theory and Methodology.* New York: North-Holland, 1983.

Edwards, W., and B. Newman. *Multi-Attribute Evaluation.* Beverly Hills, Calif.: Sage Publications, 1982.

Fleischer, G. A. *Symbolic Scorecard Methodology: Graphic Techniques for the Partial Resolution of the Multi-Attribute Decision Problem.* Industrial and Systems Engineering, Technical Report 76-4. Los Angeles, Calif.: Univ. of Southern California, August 1976.

Gardiner, P. C., and W. Edwards. "Public Values: Multiattribute-Utility Measurement for Social Decision Making." In *Human Judgment and Decision Processes.* New York: Academic Press, 1975.

Hwang, C. L., and K. Yoon. *Multiple Attribute Decision Making: Methods and Applications.* New York: Springer-Verlag, 1981.

Keeler, E. B., and S. Cretin. "Discounting Life-Saving and Other Nonmonetary Effects." *Management Science,* vol. 29, no. 3. March 1983, pp. 300–306.

Keeney, R. L., and H. Raiffa, eds. *Decisions with Multiple Objectives: Preferences and Value Trade-Offs.* New York: John Wiley & Sons, 1976.

Rand Corporation. *Systems Analysis: Methods, Techniques and Theory.* A bibliography of selected Rand Publications. SB-1055. Santa Monica, Calif.: The Rand Corporation, June 1982.

Starr, M. K., and M. Zeleny, eds. *Multiple Criteria Decision Making—TIMS Studies in the Management Sciences.* Amsterdam: North-Holland, 1977.

INDEX

V

Value curves, 443, 444
Value relative, 368

W

Weighted Aggregate Price Index
  (WAPI), 371–372
Weighted average cost of capital,
  315, 406–407

Weighted Average Relative Price
  Index (WARPI), 373–374
Wholesale Price Index, 343
Worth, defined, 51

Y

Year-by-year book method, 148

DISCRETE CASH FLOWS

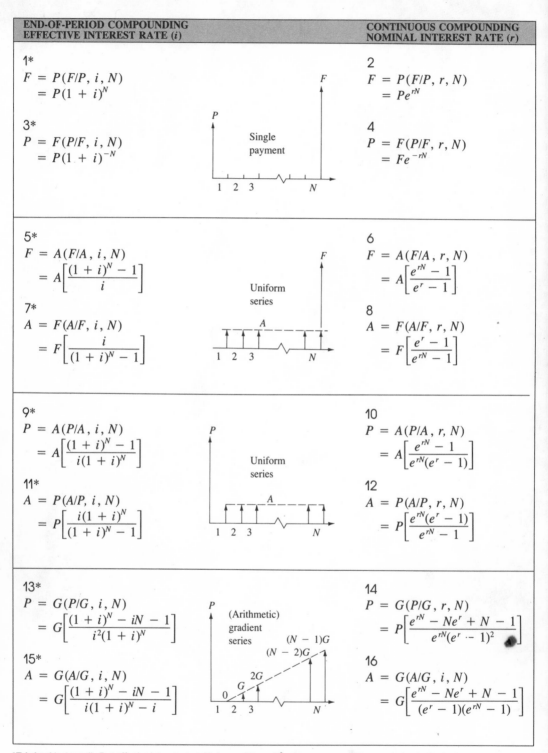

END-OF-PERIOD COMPOUNDING EFFECTIVE INTEREST RATE (i)		CONTINUOUS COMPOUNDING NOMINAL INTEREST RATE (r)
1* $$F = P(F/P, i, N)$$ $$= P(1 + i)^N$$	Single payment	2 $$F = P(F/P, r, N)$$ $$= Pe^{rN}$$
3* $$P = F(P/F, i, N)$$ $$= P(1 + i)^{-N}$$		4 $$P = F(P/F, r, N)$$ $$= Fe^{-rN}$$
5* $$F = A(F/A, i, N)$$ $$= A\left[\frac{(1 + i)^N - 1}{i}\right]$$	Uniform series	6 $$F = A(F/A, r, N)$$ $$= A\left[\frac{e^{rN} - 1}{e^r - 1}\right]$$
7* $$A = F(A/F, i, N)$$ $$= F\left[\frac{i}{(1 + i)^N - 1}\right]$$		8 $$A = F(A/F, r, N)$$ $$= F\left[\frac{e^r - 1}{e^{rN} - 1}\right]$$
9* $$P = A(P/A, i, N)$$ $$= A\left[\frac{(1 + i)^N - 1}{i(1 + i)^N}\right]$$	Uniform series	10 $$P = A(P/A, r, N)$$ $$= A\left[\frac{e^{rN} - 1}{e^{rN}(e^r - 1)}\right]$$
11* $$A = P(A/P, i, N)$$ $$= P\left[\frac{i(1 + i)^N}{(1 + i)^N - 1}\right]$$		12 $$A = P(A/P, r, N)$$ $$= P\left[\frac{e^{rN}(e^r - 1)}{e^{rN} - 1}\right]$$
13* $$P = G(P/G, i, N)$$ $$= G\left[\frac{(1 + i)^N - iN - 1}{i^2(1 + i)^N}\right]$$	(Arithmetic) gradient series	14 $$P = G(P/G, r, N)$$ $$= P\left[\frac{e^{rN} - Ne^r + N - 1}{e^{rN}(e^r - 1)^2}\right]$$
15* $$A = G(A/G, i, N)$$ $$= G\left[\frac{(1 + i)^N - iN - 1}{i(1 + i)^N - i}\right]$$		16 $$A = G(A/G, i, N)$$ $$= G\left[\frac{e^{rN} - Ne^r + N - 1}{(e^r - 1)(e^{rN} - 1)}\right]$$

*Tabulated in Appendix B at effective interest rate i. Under conditions of continuous compounding, nominal and effective interest rates are related by $i = e^r - 1$, or $r = \ln(1 + i)$.